Regional Geology Reviews

More information about this series at http://www.springer.com/series/8643

Akif A. Alizadeh · Ibrahim S. Guliyev
Fakhraddin A. Kadirov · Lev V. Eppelbaum

Geosciences of Azerbaijan

Volume I: Geology

Springer

Akif A. Alizadeh
Institute of Geology and Geophysics Sciences
Azerbaijan National Academy of Sciences
Baku
Azerbaijan

Ibrahim S. Guliyev
Institute of Geology and Geophysics Sciences
Azerbaijan National Academy of Sciences
Baku
Azerbaijan

Fakhraddin A. Kadirov
Institute of Geology and Geophysics Sciences
Azerbaijan National Academy of Sciences
Baku
Azerbaijan

Lev V. Eppelbaum
Department of Earth Sciences, Faculty of Exact Sciences
Tel Aviv University
Tel Aviv
Israel

ISSN 2364-6438 ISSN 2364-6446 (electronic)
Regional Geology Reviews
ISBN 978-3-319-27393-8 ISBN 978-3-319-27395-2 (eBook)
DOI 10.1007/978-3-319-27395-2

Library of Congress Control Number: 2016936973

Printed on acid-free paper

This Springer imprint is published by Springer Nature
The registered company is Springer International Publishing AG Switzerland

This book is dedicated to the 70th Anniversary of the Azerbaijan National Academy of Sciences foundation

Preface

Azerbaijan occupies an area of about 87,000 km^2 in the unique tectonic environment (the Greater and Lesser Caucasus, Talysh Mts., Kur and Caspian Basins, etc.) that finds a corresponding reflection in the studied geophysical fields. Geological–geophysical data on the geological structure of Azerbaijan can shed light on the basic principles of evolution of the Earth, the distribution of economic minerals and seismic activity. Geodynamically, this region can be considered as a result of the interaction of several microplates—fragments of the Afro-Arabian and Eurasian lithospheric plates. The deep understanding of both near-surface and deep geological structure of Azerbaijan may be realized only on the basis of comprehensive combined analysis of tectonic, stratigraphic, morphological, lithological features and various geophysical fields (ground, shipborne, airborne and satellite observed) examination. A number of important mineral resources are concentrated there: besides well-known hydrocarbon reserves (Azerbaijan is the oldest hydrocarbon province in the world), here was found a number of large polymetallic, copper, gold-bearing, iron ore, and other deposits.

Separate attention was paid to stratigraphy and lithology, geomorphology, neotectonics, hydrogeology, petrophysical parameters, seismological monitoring, qualitative and quantitative geophysical field examination and modelling. Environmental and near-surface investigations reflect examination of mud volcanoes, solving different ecological, technical and hydrogeological problems. Many tens of developed models clearly illustrate application of advanced geophysical–geological methodologies.

This book not only generalizes all previous publications in Russian and Azerbaijani languages (and partially—in English), but reflects many unpublished materials and principally new geoscientific data obtained during the recent years. Our view is that only skillful aggregation of various geological and geophysical methods, supported by the latest advances in the analysis of satellite data and extensive use of unmanned vehicles will make a significant leap in all the areas of the geosciences in Azerbaijan.

The authors of this book have been working in Azerbaijan and adjacent areas for many years and developed a lot of new methodologies, approaches and different kinds of models (geological, paleontological, geochemical, geophysical, etc.) for this region.

In the book preparing was used the results of long-term work of the Institute of Geology and Geophysics of the Azerbaijan National Academy of Sciences, including eight volumes of "Geology of Azerbaijan" and numerous domestic and International publications. Many chapters and sections were significantly reduced since the book volume is limited. Nevertheless, we hope that the presented materials allow to obtain a bright impression about the main directions of the geological–geophysical studies in Azerbaijan.

Chapter VI in second Volume "Geodynamics and Seismology" is prepared in cooperation with scientists from the Massachusetts Institute of Technology. Authors of this chapter are F. Kadirov, S. Mammadov, M. Floyd, R. Reilinger, A. Alizadeh and L. Eppelbaum.

A great assistance in the book preparing we received from our colleagues: E.K. Alizadeh, Ch.S. Aliyev, A.A. Feyzullayev, Yu.H. Israfilov, A.D. Ismailzadeh, T.N. Kangarli, A.Sh. Mukhtarov and G.J. Yetirmishli. We wish also to express our gratitude to all researchers from

Azerbaijan and other countries (first of all, from the USA, France, Italy, England and Germany) whose scientific results have been analysed in this book.

This book may be useful for geoscientists and engineers working in Azerbaijan and similar regions, students of different specialties studying geology, geophysics and environment of complex tectonic regions, and for a wide range of readers who want to get acquainted with the unusual geological–geophysical phenomena of this beautiful country.

Akif A. Alizadeh
Ibrahim S. Guliyev
Fakhraddin A. Kadirov
Lev V. Eppelbaum

Contents

About the Authors

Akif A. Alizadeh received M.Sc. (1956) from the Azerbaijan Oil Academy, Ph.D. (1961) and D.Sc. (1969) from the Geological Inst. of the Azerbaijan Academy Sciences. More than 40 years he occupied a position of Director of Geology and Geophysics Institute of the National Academy of Sciences (ANAS). In 2013 he was elected as President of ANAS. Professor Akif A. Alizadeh is the author of more than 370 publications including 8 books. Akif A. Alizadeh's main scientific filed is a regional stratigraphy of the Cretaceous deposits in Azerbaijan. He together with followers developed stratigraphic scheme of Cretaceous deposits, analyzed numerous paleontological–stratigraphic materials and generalized the modern ideas of geological processes. His researches designed the basis of sedimentary and volcanogenic–sedimentary formations, distribution and development of the Cretaceous fauna. Professor Akif A. Alizadeh's many years investigations enabled to formulate the principal concepts of study of the Cretaceous Mollusk fauna (belemnites) in Azerbaijan and adjacent regions. His researches cover classification, phylogeny and morphofunctional analysis, the principles of paleobiogeographic zoning, examination of ecological and paleontological problems. Professor Alizadeh himself identified the stages of spatial–temporal distribution for the Cretaceous Invertebrates in the Mediterranean and revealed here appropriate paleobiogeographic taxons. Study of morphogenesis in historical aspect, definition of tendency in population's nature alteration, identification of heritable morphological changes allowed him to restore a general evolution process of various belemnoide groups. Professor Alizadeh carried out fundamental research for the biogeochemical zonation of the Cretaceous basins in the Caucasus and for examination of geochemical conditions of Paleobasins as evolution factor. Analysing the microstructural, chemical and crystallochemical peculiarities of matters in skeleton formations of fossils, he revealed the patterns of migration and concentration for various chemical elements, paleophysiological peculiarities for some taxons. Professor Alizadeh created a description of geochemical ecology for the Cretaceous Invertebrates. Under his advising for the first time absolute values of paleotemperatures of Azerbaijan Cretaceous basins were determined by isotopic paleothermometry method.

Ibrahim S. Guliyev received M.Sc. from the Azerbaijan Oil Academy in 1970, and Ph.D. from the Moscow (VNIGNI) in 1978 and D.Sc. in 1989 from the Moscow State University. Since 1970s he has been working at the Geology and Geophysics Institute of the Azerbaijan National Academy of Sciences (ANAS) where he now occupies position of a Director of Oil and Gas Institute. Since 2013 he is a Vise President of ANAS. Professor Ibrahim S. Guliyev is the author of more than 100 publications including 10 books. Scientific interests of Ibrahim S. Guliyev cover geology and geochemistry of oil and gas, basin modelling and mud volcanism. Under his supervision and participation have been conducted original investigations of a gas regime in folded areas with the assessment of amount of discharge of carbonaceous gases and their distribution in the surficial atmosphere. His studies of the hydrocarbonaceous systems of rapidly subsiding basins (on example of the South Caspian basin) are of a special practical importance. Other scientific interests of Prof. Ibrahim S. Guliyev include results of basin's simulation, investigations of rates and range of generation, migration and accumulation of hydrocarbons in the terms of rapid subsidence as well as studies of isotopic–geochemical properties of the organic matter, oil, gas and water enabled to determine the depth of generation and preservation of hydrocarbons in the South Caspian basin.

Fakhraddin A. Kadirov received M.Sc. from the Azerbaijan State University in 1972, Ph.D. from the Geophysical Institute of the Georgian Academy of Sciences in 1984 and D.Sc. from the Geological and Geophysical Institute of the Azerbaijan National Academy of Sciences (ANAS) in 2000. Since 1974 up to now he has been working at the Geology and Geophysics Institute of ANAS where occupied positions of Researcher and Senior Researcher; now he is the Head of the Earth Physics Department. In 2013 he was elected as Academician-Secretary of the Earth's Science Division of the ANAS Presidium. Professor Fakhraddin A. Kadirov is the author of more than 100 publications including 6 books. Field of scientific interests of Fakhraddin A. Kadirov covers different aspects of the Earth's gravity field studies, application of high precise GPS measurements for examination of crustal deformation and tectonic reconstructions. His subsequent research has principally involved the Arabia–Africa–Eurasia plate system. Professor Fakhraddin A. Kadirov developed numerous gravity, seismological and integrated geophysical parameter maps of Azerbaijan, Caucasus and adjacent regions.

Lev V. Eppelbaum received M.Sc. from the Azerbaijan Oil Academy in 1982, and Ph.D. from the Inst. of Geophysics (Azerbaijan), Geophysical Inst. of Georgia and Mining Academy (Moscow) in 1989. In 1982–1990 he worked as geophysicist, Researcher and Senior Researcher at the Institute of Geophysics in Baku (Azerbaijan). In 1991–1993 Eppelbaum completed postdoctoral studies in the Dept. of Geophysics and Planetary Sciences at Tel Aviv University; at present he occupies a position of Assoc. Professor at the Dept. of Geosciences. He is the author of more than 330 publications including 7 books and about 130 articles. His scientific interests cover potential (magnetic, gravity, thermal) and quasi-potential geophysical field analysis in complex geological-geophysical environments, integrated interpretation of geophysical and geological data, tectonic, geodynamic and paleomagnetic reconstructions. Eppelbaum's research interests include also analysis of nonlinear geophysical processes, archaeological and environmental investigations as well as searching economic minerals.

History of Geological Research in Azerbaijan 1

Oil and gas shows in Azerbaijan have been known since the tenth century and are mentioned in the works of such famous scientists, historians, travelers like Al-Mas'udi, El Istakhri, Marco Polo, etc. The works by Marco Polo (the thirteenth century) contain the data on oil production (sources) at the Svyatoi Island (during the Soviet period, the island of Artiom; at present it is the Pirallakhy Island) and the morphological description of oil wells and natural oil shows with indication of their production, etc. is given for the Western Azerbaijan (at the border with Georgia).

In 1683, E. Kempfer visited Baku. He gave comparatively more complete description of oil well whereabouts, of nature and thickness of nearly all oil-containing beds that were then being developed, geomorphological features of oil and gas shows, occurrence nature of oil-containing rocks, quality of oils, yields of individual oil wells, etc. In 1654, in Balakhany, an Azerbaijani Al akhyar Mamed Nuri ogly dug the first 35-m depth oil well. Already by 1735, 52 wells could be counted in Balakhany, and oil was bailed from them by leather buckets. In 1816, 116 wells were already found in Baku with maximal depth not exceeding 35 m. More than that, already in the early nineteenth century, a Bakuvian Kasymbeq was the first in the world to produce oil from the Caspian bottom, from the wells constructed 20–30 m offshore. The first comparatively detailed data on oil wells at Balakhany area were given by N. Lerche (c. 1735). Descriptions of "eternal flames" near the village of Surakhany, of mud volcano eruptions that are widely developed at the Absheron peninsula, were first given by J. Hanvey (late eighteenth century), E.I. Eichwald, Gamba, and Deichman (the first quarter of the nineteenth century) and by F.O. Lenz (mid-nineteenth century).

Already in the eighteenth century, rich ore and nonmetallic deposits of mineral resources of Azerbaijan were investigated by Russian natural scientists (A.A. Musin-Pushkin, E.I. Eichwald, etc.). Vyatkin and E.I. Eichwald, who have studied the iron ore Dashkesan deposit and the alunite Zaglik deposit, started investigation of Azerbaijan ore deposits in the early nineteenth century.

The pioneering data on the Western Absheron geological structure are met in the work by A. Humboldt (early nineteenth century) dedicated to eruption description of Bozdagh (Kobiyski) mud volcano, as well as to the Azerbaijan oil fields. The 1820s should be thought as the commencement of scientific geological investigation of Azerbaijan. In 1827, mining engineer N.I. Voskoboinikov was the first to give the in-depth description of oil wells of the Baku region, thus starting the research of the Azerbaijan peninsula oil deposits.

E.I. Eichwald (first quarter of the nineteenth century) described a number of fossils from the Azerbaijan Pliocene and Miocene sediments. His 1834 research was dedicated to the geological description of the Caspian Sea and to the paleontology of the Azerbaijan Mesozoic sediments. E.I. Eichwald also gave detailed description of the Absheron peninsula oil wells and described the Zaglik alunite field. It is in his works that the stratigraphic level integrity of most Absheron peninsula oil wells was first mentioned.

A systematic study was started by the investigations of outstanding German (Russian) scientist G.W. Abich that began in the Caucasus in 1844 and continued until 1877. They were devoted to various issues of Caucasian geology, and, specifically, to Azerbaijan oil-bearing regions. Having started from the study of individual natural phenomena such as gas shows, changes in Caspian Sea level, earthquakes, eruptions of mud volcanoes, G.W. Abich posed a number of valuable propositions concerning the regularities of mud volcano distribution related to the region tectonics, confinement of oil accumulations to anticlinal elevation domes, formation of oil pools by distant radial migration, etc. G.W. Abich compiled and published the 1:168,000 scale geological map of the Absheron peninsula; he offered deposit schematic, stratigraphic, and lithological cross-sections, and their paleontological characteristics. His well-known book "Oil Industry in Pennsylvania and Caucasus" covers the

A.A. Alizadeh et al., *Geosciences of Azerbaijan*, Regional Geology Reviews,
DOI: 10.1007/978-3-319-27395-2_1

results of geological research conducted by the author for more than 30 years and dealing with the study of structure and potentials of the Absheron Peninsula oil fields and of adjacent Caucasian areas. The work paid special attention to the relationship between mud volcano manifestations and oil-bearing capacity.

G.W. Abich was the first to discover and give paleontological characteristic to the Devonian, Carboniferous, and Permian deposits of Nakhchivan. It was he who found the Jurassic, Cretaceous, and Neogene deposits in the Caucasus. The works by G.V. Abich laid the foundation for the paleontological and stratigraphic studies of the Caucasus, including Azerbaijan.

In addition to E.I. Eichwald and G.W. Abich, paleontological and stratigraphic studies of Azerbaijan territory were conducted by F. Frech and G. Artgaber. In 1872, G.G. Tzulukidze, R. Kraft, and V.I. Arkhipov published the geological description of the Absheron peninsula. Commercial development of Azerbaijan oil began in 1871. The issues of geology and oil-bearing capacity of individual Azerbaijan fields, essentially of the Absheron peninsula, are discussed in the papers by L.F. Batzevich, A. Konshin, N.I. Andrusov, A.I. Sorokin, N.I. Lebedev (they all were published in late nineteenth century) and others.

The following stage of geological investigation history starts from the establishment of Caucasian Mining Authority in 1880, Geological Committee in 1882, and various societies of natural scientists. As a result of conducted regional studies, L.F. Batzevich has published "*Geological Description of the Absheron peninsula*". N.I. Andrusov, K.I. Bogdanovich and D.V. Golubyatnikov showed stratigraphic diagrams of the Mesozoic and Tertiary complexes (late nineteenth–early twentieth century). N.I. Andrusov, having studied paleontology of the Miocene–Pliocene deposits, proposes the pioneering scheme of Azerbaijan Cenozoic stratigraphy. K.I. Bogdanovich gives first stratigraphic scheme of the Greater Caucasus Meso-Cenozoic period, etc. The publications of I.M. Gubkin (1914–1916), D.V. Nalivkin et al. contribute greatly to the stratigraphy of Azerbaijan Meso-Cenozoic deposits, inserting important clarifications and additions.

Geological investigations in oil-bearing regions were held under the guidance of D.V. Golubyatnikov and I.M. Gubkin, the Geological Committee geologists. Detailed descriptions of the Bibi-Heyibat field made by D.V. Golubyatnikov and of the Balakhany-Sabunchi-Ramany and Surakhany fields made by N.I. Usheikin, based on structural maps and detailed geological profiles, contributed greatly to the study of oil field morphologies. D.V. Golubyatnikov was the first in Russia who began investigating the Absheron peninsula oil-field waters and he pioneered in geothermal research conduct. The works by I.M. Gubkin dealt with the issues of regional studies of the Azerbaijan oil fields.

It is important to give prominence to Abdul Kadyr Efendiyev among these researchers, the first Azerbaijani mining engineer, who dealt with the geological structure study of the Fatmai oil-bearing region at the Absheron peninsula (1910–1913). It is also necessary to mention Gasanbeq Zardabi, an Azerbaijan enlightener and natural scientist, and Farrukhbeq Javanshir, a mining engineer.

During that period, the Geological Committee geologists, M.V. Abramovich, N.I. Andrusov, S.M. Apresov, K.I. Bogdanovich, P.G. Volarovich, K.P. Kalitsky, Ya.I. Lednev, K.V. Kharichkov, S.I. Charnotsky et al. made geological survey of oil-bearing areas both on the Absheron peninsula and in alternative Azerbaijan regions such as the near-Caspian region, the near-Kur Depression, and the Shamakhy-Gobustan Region.

The publications by G.W. Abich, A. Konshin, V. Meller et al. present pioneering data on the Azerbaijan underground and mineral water. And in 1902–1904, the irrigation construction attempts were held under P.P. Kozintsev's guidance in the Mughan Steppe. Along with the study of oil fields, certain attention was paid to exploration activities for other minerals. Thus, starting from 1830, increasingly more attention was paid to geological investigations oriented to exploration activities for ore minerals (K.I. Bogdanovich, N.N. Chervinsky, E.S. Fedorov, L.K. Konyushevsky, V.V. Nikitin, N.S. Uspenski, and others).

In 1864, the Gedabey copper deposits began to be developed, and continued up to 1914. In the 1890s, mining men made selective working of rich cobalt ores in Dashkesan region. In 1880–1890, geological prospecting was conducted at the Gyumyushlug lead-zinc deposit. Later, in 1908, mining engineer A. Ustimovich established a joint-stock society that held primitive ore extraction and zinc smelting at the deposit. In 1866, an alum plant, designed by mining engineer V.V. Bogachov, was constructed at the Zaglik alunite deposit and existed till the 1930s. In 1909, A.M. Margolius studied rock salt deposits in Caucasian Krai (Region), including the Nakhchivan deposits.

In May 1920, the Oil Industry Authority (Azneft) was established and A.P. Serebrovsky was its first manager. The Azneft Geological Survey Bureau was organized there and headed by M.V. Abramovich. The same year, the Azerbaijan Polytechnic Institute was founded that was later renamed to M. Azizbekov Azerbaijan Industrial Institute and played a unique role in the training of engineering and technical personnel, including geologists, for the Republican oil industry.

I.M. Gubkin made substantial contribution to workforce training and organization of geological prospecting in Azerbaijan. In 1920, editorial office of the "*Neftyanoye khozyaistvo*" (Oil Industry) was created under Azneft guidance that at present bears the title of "*Azerbaijanskoye neftyanoye khozyaistvo*" (ANK) (Azerbaijan Oil Industry). The

Azneft Geological Survey Bureau was in 1930 renamed as the Azneft Geological Prospecting Authority (GPA), and from 1933 until 1947, it was named the "*Aznefterazvedka*". A number of research institutions have been established such as AzNII (Azerbaijan Research Institute) of Oil Production (1929), I.M. Gubkin Institute of Geology (1933) and others. After nationalization of Azerbaijan oil industry (1920), a new era in survey of oil fields as well as in application of geological knowledge in solution of efficiency tasks of oil field development and prospecting of new pools came on.

Within this time interval, the Azneft geologists made detailed geological survey of vast Azerbaijan regions. At the Absheron peninsula, in Gobustan, at the near-Kur Depression, the Azneft geologists (M.V. Abramovich, S.M. Apresov, M.F. Mirchink, V.A. Sulin, N.M. Mullayev, I.A. Malinov, A.L. Putkaradze, B.P. Yasenev, M.P. Voronin, and others) carried out these works in collaboration with the Moscow Mining Academy and the Geological Committee (I.M. Gubkin, S.F. Fedorov, V.E. Ruzhentsov, N.S. Shatsky, V.V. Weber, and others). Structural core drilling is widely implemented.

In parallel, other deposit complexes are being investigated and different facies isolated, which would have permitted to compare the cross-sections of their various occurrence regions. Systematization of data is held for producing areas as well as cross-section parallelization with tracing of oil-bearing horizons. In 1924, basing on operation data, D.V. Golubyatnikov made an attempt to calculate oil reserves for the Absheron peninsula fields. In 1927, he published a detailed geological map of Atashki district, where he examined the most essential geological structure issues of the entire Absheron peninsula.

The research of A.Ya. Krems dealing with the study and comparison of the productive series cross-sections should be noted as well as works by V.V. Bilibin dedicated to the development and prospecting of the sub-Kirmakin suite at the Absheron peninsula, to calculations of underground oil pool (1933–1935).

Quite considerable are the services of M.V. Abramovich (1919–1928) dealing with detailed study of Azerbaijan oil fields, with pool prospecting and exploration, with development of oil geology theoretical problems. He was the first to examine the issues of improved oil field development. M. V. Abramovich was the first in the world literature who indicated the rhythmical structure of oil-bearing thickness, conducted works for the development of oil reserves calculation techniques, etc.

During 1924–1928, V.V. Weber carried out in Gobustan and Absheron peninsula works on compiling detailed geological maps of potential exploration targets. S.A. Kovalevsky, along with detailed study of some Absheron peninsular oil fields (Bibi-Heybat and some others), takes intensive measures to investigate the Eastern Transcaucasian mud volcanoes.

In 1924–1926, A.A. Melikov is engaged in the study of the Bibi-Heybat and alternative oil fields in the Absheron Peninsula. In 1930–1936, K.A. Mashkovich conducts reconnaissance works in the Lenkaran area, at the Absheron peninsula, in Gobustan and the near-Kur Depression, investigates the Sulutepe and the Khyrdalan districts, using shallow drilling. At the same period, I.I. Mullayev carries out detailed studies in Gobustan (geological survey of the Utalgi-Akhtarma, Klych-Kulkhut, Cheildag, Kalendartapa, Jamaleddin, and some other areas).

In 1926, first in the USSR, they started employment of geophysical methods of oil pool search at the Absheron peninsula. The methods played an important role in the solution of various tectonic problems and occurrence conditions of the Tertiary system abyssal layers.

In 1927, in Baku, the Research Laboratory of the Department of Petrography at the Azerbaijan Industrial Institute first started the investigation of rock lithology in the Azerbaijan oil fields. In 1929–1937, this research was held on a broader scale by P.P. Avdusin, B.V. Baturin, G.Yu. Fuks-Romanova et al. at the Azerbaijan Research Institute of Oil Production. Investigation of sedimentary rock lithology in the Tertiary formations was continued by A.D. Sultanov, G.A. Aliyev et al.

In 1930–1931, oil-field geology, a new oil geology branch emerges. In 1932, M.F. Mirchink wrote "A Course in Oil-Field Geology". Since 1930, structural core drilling is being widely developed at vast republican area, geophysical investigations are being conducted by means of electrometric, gravimetric, and seismic techniques. It is worth mentioning that seismics made invaluable contribution in search for oil fields at marine areas.

In 1931, electrical logging technique is introduced into practical use of the Azerbaijan oil industry (L.M. Alpin). In 1933–1935, tectonics of the Caspian coastal strip is being investigated by marine electrical prospecting method that was first developed in Azerbaijan. In 1929–1932, gravimetric survey in some Azerbaijan districts was conducted, and in 1932, the paper by A.D. Arkhangelsky and V.V. Fedynsky was published, where the results of gravimetric works in the Eastern Azerbaijan were presented. In 1937, seismic survey was implemented for the first time at the Absheron peninsula, which later had widespread development as one of the most essential geophysical methods for studying the abyssal tectonics of oil-and-gas bearing regions. Search for oil and gas in the near-Kur region was crowned in 1937 by the discovery of Pirsaat field. In 1938, the pioneering Maykop series oil was found in Siazan (the near-Caspian Region).

Discovery of stratigraphic oil pools, confined to thinning out zone of separate suites and horizons in the productive series lower division, is of primary importance. It resulted in the new stage of prospecting and exploration and expanded the boundaries of potential oil-and-gas bearing areas. Stratigraphic oil pools were discovered in the sub-Kirmakin suite (SK suite) and in the Kirmakin suite (KS) bottoms, at the Binagady fold southern margin, in Chakhnaglyuar, Sulutepe, Yasamaly Valley. In 1936 and later, stratigraphic pools were discovered in the Kalin suite (KaS), at Staraya Kala area (V. A. Gorin, B.K. Babazade). Still later, stratigraphic pools of KaS were discovered at the Surakhany-Karachukhur-Zykh anticline eastern margin, in Bina-Govsany depression, on Peschany (now Gum) and Zire islands, at anticlinal zone margins of Zhiloy Island (now Chilov)—Mud Salse (Palchyg Pilpilesi)—Neft Dashlary. Discovery, foremost owing to seismic survey, of a rich oil field, confined to the buried and low-rise structures of Buzovny, Zire type was of essential importance during that period.

V.V. Bogachyov, K.A. Alizadeh, Sh. Mamedzade, M.M. Aliyev et al. studied the fauna of Azerbaijan Meso-Cenozoic deposits. In 1927, V.E. Liventhal began micro fauna research in Azerbaijan Industrial Institute under the guidance of V.V. Bogachyov. Further on, the method was developed in the project led by D.A. Jafarov, D.M. Khalilov et al. In 1925, V.V. Bogachyov has published "The Outline of Azerbaijan Geology" where first for the republic he gave general stratigraphic characteristic and assessment based on still scarce paleontological material. The publications by I. M. Gubkin, D.V. Golubyatnikov, V.V. Weber, M.F. Mirchink elucidated the stratigraphy of the Cenozoic and Mesozoic deposits developed based on more extensive paleontological documentation.

In 1924–1928, V.V. Weber conducts geological survey of some Gobustan areas and compiles detailed geological map of the Absheron peninsula (separate map-boards). In 1930, V.P. Rengarten first publishes his scheme of Caucasian tectonics. The results of I.M. Gubkin's longstanding research found their reflection in his comprehensive treatise devoted to the tectonics of the South-Eastern Caucasus and to oil-bearing capacity of the region. I.M. Gubkin was the first to cover geological structure and geotectonic zonation of the Caucasus, oil deposit formation conditions and alternative issues. M.F. Mirchink publishes his "*Tectonic Problems of the Southeastern Caucasus*" book.

In 1923, the Geological Committee begins investigations of some Lesser Caucasus large deposits that were conducted by K.N. Paffenholtz. Based on the 1923–1926 research results, K. N. Paffenholtz publishes in 1928 his "*Dashkesan and Zaglik*" work, where he describes the geological structure of both deposits and the considerations concerning their genesis, reserves, and potentials of commercial use.

In 1928–1931, exploration works were started that were conducted by M.A. Kashkay and M.I. Bushuev from the Institute of Applied Mineralogy by order of Azgornokhimtrest. Search for nonferrous and rare metals is steadily growing: since 1930, it is made by Azgeolbaza (in 1931 it was renamed to Azgeolbyuro) attached to Zakgeoltrest (Tbilisi, 1929), and reorganized in 1938 to the Azerbaijan Geological Authority.

In 1932, in order to explore Azerbaijan natural riches, the Azerbaijan affiliation of the Transcaucasian branch of the USSR Academy of Sciences (AzOZFAN) was established on the basis of the Society for Azerbaijan Survey and Study (Research in geology and Geophysics). In 1935, AzOZFAN was reorganized into Azerbaijan branch of the USSR Academy of Sciences (AzFAN USSR) including a number of research institutes, and Geological Sector among them.

At that stage, certain achievements were made in regional geological and tectonic research. The works were carried out by M.G. Agabekov, M.M. Aliyev, Sh.A. Azizbekov, I.M. Gubkin, V.E. Khain, M.F. Mirchink, V.P. Rengarten, E.Sh. Shikhalibeyli, N.B. Vassoyevich, V.V. Weber, and others.

Starting from the 1930s of the twentieth century, geological survey and oil-and-gas exploration, and prospecting gain breadth, and it results in the discovery of novel oil and gas fields not only within the Absheron peninsula but also in the other oil-and-gas bearing regions of Azerbaijan. A lot of the credit must go the large team of Azerbaijan geologists, and first of all, to the Republic oil industry geological service. Geophysical survey methods such as gravity, magnetic, electrical surveys, and seismic prospecting of oil fields are widely used. Due to the engineering achievements and drilling success, prospecting is aimed at the exposure of novel, more abyssal horizons. In 1946, the first abyssal well has been drilled at 3904 m which opened the way to Azerbaijan superdeep drilling.

Since 1948, large-scale development of marine oil fields was started. It should be noted that pioneering marine wells have been drilled already in 1935, near Artiom Island. In 1948, the Gyurgyany-sea, Zhiloy Island (Chilov Island) deposits came online. In 1949, the Neft Dashlary multi-horizon, world famous deposit has been discovered. A little later, oil-bearing capacity was found for the Darwin Bank and Mud Salse (Palchyg Pilpilesi) areas.

In 1953, the Peschany Island field was discovered (the Serebrovsky Oil-field Assoc.). Discovery of marine fields greatly enhanced oil production gain for the entire Republic. Lucrative oil-and-gas bearing areas were revealed in the Baku Archipelago such as Duvanny, Sangachal, and Bulla-sea. A promising gas condensate field of Yuzhnaya was discovered. A large gas blowout at the Karadagh area in 1955 laid foundation to the new stage of prospecting and exploration for high-productive gas pools in Azerbaijan. Development of superdeep drilling in Azerbaijan resulted in the discovery of deep-seated gas fields at the Zire area (1956). The Duvanny (1949), the Kyanizadagh (1963), and other gas condensate fields were not less promising.

In 1938, by the initiative of such leading Azerbaijan scientists like Sh.A. Azizbekov, M.A. Kashkay, A.A.

Yakubov (and with the support from the USSR Academy of Sciences Presidium) Sector of Geology was reorganized to become the independent research institution, the Geological Institute, that gave an impulse to successful and rapid development of Azerbaijan geology.

Such outstanding Russian scientists like V.V. Bogachev, I.M. Gubkin, D.V. Golubyanikov, V.E. Khain, S.A. Kovalevsky, D.V. Nalivkin, Yu.V. Riznichenko, M.A. Sadovsky, E.F. Savarensky, B.S. Sokolov, V.V. Tikhomirov, N.B. Vasseyevich, A.V. Vinogradov, V.V. Weber et al. contributed greatly to the Institute establishment and development.

Outstanding scientists of Azerbaijan: academicians G.A. Akhmedov, M.M. Aliyev, A.A. Alizadeh, Ak.A. Alizadeh, K.A. Alizadeh, Sh.A. Azizbekov, G.Kh. Efendiyev, M.A. Kashkai, A.V. Mamedov, Sh.N. Mamedov, Sh.F. Mekhtiev, S.G. Salaev, E.Sh. Shikhalibeyli, A.D. Sultanov, K.M. Sultanov, doctors of sciences R.M. Gadjiev, T.A. Ismailzadeh et al. played a great role in the development of Azerbaijan geological science and in training of highly qualified specialists.

To date, Institute of Geology and Geophysics is the largest international research center, implementing numerous fundamental and applied researches. Here, research "schools" have been created, dealing with oil geology, mud volcanism, paleontology and stratigraphy, lithology, geophysics, geotectonics, geodynamics search and development of oil and gas fields. As a result of careful transfer of research achievements across the generations, the schools that conducted fundamental and applied research in prospecting and development of oil fields, in geology and geophysics have been thoroughly preserved. The Institute is justifiably proud of its scientific schools and fundamental researches, recognized by the scientific community.

Research in paleontology and stratigraphy that began in 1927 was further successfully developed at a fast race after the Department of Paleontology and Stratigraphy headed by Academician M.M. Aliyev was launched in 1950. In continuation, the said department hosted the Laboratories of Mesozoic, Cenozoic, and Paleozoic periods as well as that of microfauna and biochemistry and they were headed by such outstanding scientists like K.A. Aliyev, Kh. Aliyulla, M.M. Aliyev, K.A. Alizadeh, Sh.A. Babaev, M.A. Bagmanov, T.A. Gasanov, A.Yu. Khalilov, J.M. Khalilov, A.B. Mamedov and others.

Of particular interest are the researches in paleontology and stratigraphy of the Cretaceous deposits made by academician Ak.A. Alizadeh, one of the representatives of said school. Stratigraphic diagrams of the Azerbaijan Cretaceous deposits were developed under his guidance and with direct participation, abundant paleontological and stratigraphic material, reflecting modern concepts on geological processes, on generation of sedimentary and volcanogenic sedimentary formations, on fauna distribution and development during the Cretaceous period, etc., has been investigated, analyzed, and generalized. Based on the results of his longstanding detailed studies, Ak.A. Alizadeh has formulated the basic concepts of his study of the Azerbaijan Cretaceous molluscan fauna systems (belemnites), gave their classification and phylogeny, developed the principles of their paleobiogeographical zoning, elucidate the issues of ecology and paleopathology. And it is not accidental that the scientific community, taking into account the significant contribution into the study of Cretaceous belemnites, made by the Azerbaijan scientist, named after him a new family of belemnoids, i.e., the "Akifibelides".

The Institute foundation and research activity development coincided with rapid development of the Azerbaijan oil and gas industry. Particular development features of this leading national economy branch determined priorities in the Institute research activities. A school of Azerbaijan oil geologists was crystallized (M.T. Abasov, A.A. Alizadeh, Ak.A. Alizadeh, B.K. Babazade, F.G. Dadashev, I.S. Guliyev, G.I. Ismailov, G.I. Jalilov, Sh.F. Mekhtiev, A.Kh. Mirzajanzade, S.G. Salayev, Kh.B. Yusifzade, and others). The Institute of Geology and Geophysics researchers contributed greatly to the discovery and development of such oil-and-gas fields of world significance as Neft Dashlary, Azeri, Chyrag, Gyuneshli, Shakhdeniz, Bakhar, Khara-Zire, etc.

Results of investigation of natural hydrocarbon systems in rapidly subsiding basins conducted at the Institute on the example of the South Caspian Basin (SCB) have been recognized. Oil geologists have studied the processes of hydrocarbon migration and accumulation under rapid subsidence conditions, extreme rates and generation scope; basin simulation was made; geochemical and isotope studies were conducted; extensive deep scope of hydrocarbon generation and preservation and depth limits (down to 16–18 km) were established; generation models for formation and pore pressures and temperatures under extreme subsidence were created; regional zones with decreased compression (or decompression) and subvertical geological bodies were established.

S.M. Akselrod, L.A. Putkaradze, R.R. Rakhmanov, D.M. Sadykhov and others made substantial contribution to the development of oilfield geophysics techniques in Azerbaijan.

Development of seismic research in Azerbaijan is associated with the names of such leading specialists as R.A. Abdullayev, G.A. Gamburtsev, P.Z. Mamedov, M.M. Radjabov, N.I. Shapirovsky, and Yu.A. Shikhaliyev.

M.T. Abasov headed the well-known school, dealing with theoretical development foundations for various oil and gas deposits, in the Azerbaijan Academy of Sciences.

Institute of Geology and Geophysics also holds leading position in preparation and development of essentials and methodology of geochemical investigations by study of the Caspian Sea gas conditions, by prospecting of underwater mud volcanoes and gas hydrates.

Profound interest in the study of mud volcanoes, widely distributed throughout Azerbaijan, motivated the creation of a new "school" in this trend. Such scientists as Ad.A. Aliyev, Ya.A. Gadjiev, I.S. Guliyev, R.R. Rakhmanov, A.A. Yakubov et al. contributed greatly to the study of mud volcanism, and to corresponding scientific school creation which received the worldwide recognition. First, classical research of mud volcanism was carried out, including detailed description of violent eruptions, volume of mud volcano ejection products (breccia, gas, water, and oil) was assessed, catalogs of volcano eruptions were compiled, volcano morphogenetic classification was made, and detailed isotope and geochemical investigations of eruption products were held. Besides, some original studies were carried out in order to examine gas conditions in folded zones. Qualitative assessment of carbon-containing gas discharge was given, its distribution in lower atmospheric layers was inspected, and its role in the mechanism of the so-called "greenhouse effect" was estimated.

Since the date of the Geological institute establishment to the present-day quite considerable researches in regional geology, geotectonics, and neotectonics (F.S. Akhmedbeyli, A.J. Ismailzadeh, T.N. Kangarli, V.E. Khain, E.Sh. Shikhalibeyli, and others) were made. Geological, geotectonic, geodynamical, paleotectonic, geomorphological, neotectonic, and metallogenic maps plotted for Azerbaijan territory at various years hold a specific place among these works as well as international tectonic maps of the Caucasus, the Caspian Sea, and Eurasia plotted within the framework of integrated studies.

Study of ore and non-ore mineral fields is traditionally thought as the principal research activity trend of the Geological Institute. Complex geological and geochemical investigations of potential areas lent a weighty contribution to the solution of development challenges at large ore and non-ore mineral fields (Belakan-Zagatala, Dashkesan-Zeilik, Dash Salakhly, Ayidag, etc.).

Despite the fact that separate ore and non-ore mineral fields have been studied by Russian scientists in the first half of the twentieth century (I.V. Barkanov, Yu.S. Fedorov, V. N. Kotlyar, G.A. Krutov, K.N. Paffenholz, N.S. Uspensky et al.), detailed investigation of mineral resources is directly related to the works of Azerbaijan scientists. These researches have intensified after the AzFAN branch was established in the former USSR, and particularly, when in 1945, the Republican Academy of Sciences was founded in the Republic. The Geological Institute (at present, Geological and Geophysical Institute) researchers played an important role in the conduct of metallogenic and petrological investigations. They worked in collaboration with the scientists from the Baku State University (BSU) and the Geological Service Authority (R.N. Abdullaev, F.A. Akhundov, Sh.I. Allakhverdiev, M.M. Aliyev, J.A. Azadaliyev, Sh.A. Azizbekov, V.M. Babazade, A.T. Bairamalibeyli, S.A. Bektashi, G.G. Efendiyev, T.G. Gadjiev, A.D. Ismailzade, M.A. Kashkay, A.D. Kerimov, G.I. Kerimov, D.M. Khalifazade, G.I. Makhmudov, S.A. Makhmudov, A.I. Mamedov, M.N. Mamedov, Sh.N. Mamedov, G.V. Mustafaev, V.N. Nagiyev, V.G. Ramazanov, M.I. Rustamov, A.G. Seidov, I.N. I.A. Shirvanzade, Sitkovsky, E.S. Suleimanov, S.M. Suleimanov, A.D. Sultanov, and others).

Investigations of the Dashkesan ore region, believed to be the Azerbaijan Urals, and of pyrite deposits, are associated with the name of academician M.A. Kashkay. Another outstanding researcher, academician Sh.A. Azizbekov, who studied geology and petrography of the Lesser Caucasus northeastern slope, showed prospectivity for metal (ore) minerals in this area. Sh.A. Azizbekov, who for many years, together with his research team, had conducted scientific studies at Nakhchivan area, published a fundamental treatise dedicated to the region geology and substantiated prospectivity of search for copper-molybdenum, polymetallic, gold-bearing and alternative minerals in the SE Lesser Caucasus.

Geology of Daghliq Garabagh (Mountainous Garabagh) was always in the highlight of researchers. Discovery of polymetallic, sulfur, and copper-pyrite deposits, and, within the recent decades, of gold ore and copper-porphyry deposits, particularly, at the area embracing a portion of the Somkhit-Garabagh (Lok-Garabagh) and the Goycha-Kharakan folded zones in the Lesser Caucasus can be attributed to such scientists as R.N. Abdullayev, A.A. Bairamov, A.D. Kerimov, T.M. Mamedov, R.Kh. Rafibeyli, E.Sh. Shikhalibeyli, L.V. Zaitseva, et al.

Discovery of the unique Filizchay pyrite-polymetallic deposit and several smaller deposits of the same kind in the vicinity (Jihikh, Katekh, Katzdag, Katzmala, Mazumchay, etc.) at the Greater Caucasus southern slope in 1959, later turned this area into the Belakan-Sheki pyrite province, with high potentials for development and production of lead, copper, zinc, gold and some other valuable metals. From this point of view, the investigations by T.G. Gadjiev, I.M. Ibragimov, D.D. Mazanov and others are of high significance. It is worth noting that the Filizchay deposit occupies the fifth place in the world for its reserves of proven complex ores.

The contribution of Azerbaijan scientists into prospecting and investigation of the Lesser Caucasus ophiolite deposits, clarification of their role in the region geodynamical development, study of geological structure and development history of the Kur Intermountain Depression on the basis of the region geological, geophysical, and geochemical investigations, and development of combined model for the Saatly superdeep geological and prospecting borehole SD-1 are particularly noteworthy.

Abyssal structure of Azerbaijan territory studied by means of gravimagnetic, thermal and other geophysical techniques was illuminated in the papers by V.V. Alexeyev, T.S. Amiraslanov, L.V. Eppelbaum, V.V. Fedynsky, R.M. Gadjiev, F.A. Kadirov, B.E. Khesin, I.O. Tzimelzon, and some other researchers.

On January 23, 2008 the Azerbaijan Republic Council of Ministers has signed the Resolution (No. 22) on adoption of the plan of actions "Investigation, development and application of mineral production technology in the Azerbaijan ore deposits without environmental damage." Within this document, for the first time in Azerbaijan, basic concepts of geotechnological science were tested at three gold ore deposits. Conducted actions resulted in the development of new underground gold production techniques. One of these production techniques is metal washing by sand. This gold production technique is low cost and does not damage the environment (it was earlier emphasized by the well-known Azerbaijan geologist M.A. Kashkay).

Mining geophysics in Azerbaijan as in one of the world reference regions was developed in the researches by V.V. Alexeyev, L.V. Eppelbaum, D.A. Gasanov, B.E. Khesin, A. Sh. Mamedzadeh, S.A. Muradkhanov, and many other scientists.

Geophysical school is one of the prominent "schools" established at the Institute of Geology and Geophysics. The school establishment is associated with the names of Azerbaijan nationals who had acquired higher education at Moscow universities and institutions, such scientists like Sh. S. Ragimov (seismology), R.M. Gadjiev (gravity field of the Earth), T.A. Ismailzadeh (Earth's magnetism), F.T. Kuliyev (seismology), Sh.A. Balakishibeyli (high pressure and temperature), T.A. Zolotovitsky (radiometry) and others. Establishment of the "Earth Physics" section in the Geological Institute created favorable conditions for conducting fundamental and applied researches in various fields of Azerbaijan geophysics. The results of prime importance were obtained in such trends like seismology and geodynamics, gravity field analysis, radiometry, paleomagnetic investigations, geothermics and study of rock petrophysical properties.

Such researches as E.I. Byus, N.V. Malinovsky, Yu.V. Riznichenko and V.N. Weber played a significant role in the development of Azerbaijan seismology.

Azerbaijan scientists made great contributions to the study of Azerbaijan seismicity and they were: S.T. Agayeva, E.B. Agalarova, R.A. Agamirzoev, F.S. Akhmedbeyli, G.R. Babaev, O.B. Babazade, A.G. Gasanov, R.I. Gasanov, K.Sh. Islamov, D.Kh. Jafarov, I.G. Kerimov, F.T. Kuliyev, V. P. Kuznetsov, T.Ya. Mammedli, B.M. Panakhi, Sh.S. Ragimov, Z.Z. Sultanova and G.J. Yetirmishli. Their works were dedicated to the research in up-to-date tendencies in seismology: development of new techniques for the Earth internal structure investigation, basing on seismic data, interpretation of seismic waves; study of the Earth crust stress condition in Azerbaijan, investigations of seismicity from macroseismic and instrumental data; qualitative assessment of seismicity; seismic zonation of Azerbaijan territory; study of the Earth weak high-frequency noise; development of software packages and geological and geophysical data bases; seismic micro-zoning of individual cities and towns (Baku, Sumgayit, Shirvan, Shamakhy, Sabirabad, Salyan, etc.) and the most important industrial facilities (the Baku TV tower, potential areas for atomic power plant, cascade of water developments on the Kur River (Yenikend, Shamkir)). On the Tauzchay, Shamkirchay, Aljiganchay, Kudialchay and Bazarchay rivers, for the Novo-Baku heat and power plant, for the Azerbaijan hydro-electric station for separate Baku districts, etc.

Nowadays, principal seismological trends that are being essentially developed in Azerbaijan are: study of seismicity and risks, structure of the Earth interiors and earthquake source zones using integrated approaches, the issues of seismic zonation and microzonation, study of earthquake fractal features, etc. The investigations on elucidation of temporal fluctuations in earthquake emergence are being conducted for obtaining information on dynamical properties of seismic process (L. Telesca and F.A. Kadirov). Dynamics of seismic event sequence has been studied for the Absheron-near-Balkhan sill, non-Poisson nature of earthquake distribution in the area was determined; mathematical models integrating all factors into the cumulative seismic risk value for the nearby, distant, and local earthquake scenarios were computed; stress state of the Shamakhy-Gobustan zone mud volcanoes after violent earthquakes was found; microseismic zonation maps, maps of enhancement distribution for oscillation amplitude and for soil resonant frequencies for Baku from the results of microtremor measurements were compiled (F.A. Kadirov et al.).

Sh.S. Ragimov had made substantial contribution to the development of seismology: he has conducted pioneering generalizations with regard to the Earth crust abyssal structure, basing on seismological data, mainly associated with the study of surface waves; he has developed the method for finding direction to the earthquake focus from surface Rayleigh and Love waves. The techniques developed by Sh. S. Ragimov served as the basis for Azerbaijan seismic zonation. The *Seismology and Engineering Seismology* handbook published in 2002 in the USA included a short biographical sketch of Sh.S. Ragimov in the section dedicated to the major results obtained by seismologists in the twentieth century.

Afterward, the Republican Seismological Service Center (RSSC) was founded on the grounds of Azerbaijan seismological school. At present, 35 telemetering stations with satellite system of data transfer are operating at the Republic area (A. Gasanov and G. Yetirmishli).

Recently, server memory has been enlarged and the latest version of "Antelope 5.2-64" software was uploaded; a novel system is being used by processing of earthquakes and in alternative programs, by creation of databases and their archiving. Presence of "real time" interactive communication permits to control technical condition of seismic station facilities, deployed in various regions of the Republic, temperature conditions, some alternative problems. All data on the earthquakes collected by seismic stations arrive via satellite communication at the RSSC in Baku, at its "Bureau for Current Information on Earthquakes" (BCIE). The BCIE investigates the parameters of local, nearby, and distant earthquakes recorded via the network consisting of 35 telemetering stations and deployed at the Republic territory. Cumulative catalog of earthquakes is compiled on the basis of data from these telemetering stations. In addition, the RSSC headed by G. Yetirmishli carries out analysis of various geophysical fields, geochemical and geodynamical monitoring.

For the first time ever in Azerbaijan, the RSSC has prepared the map of significant earthquake foci for the time interval from 427 to 2013. At present, the RSSC main activity trends at the Republic territory and at neighboring territories are regular seismic monitoring and fundamental and applied researches in seismology.

Research foundations for the study of spatial and temporal distribution regularities in the Earth gravitational field, its internal structure, and dynamic processes were laid in the 1960–1980s by the outstanding Azerbaijan geophysicist R. M. Gadjiev. Conduct of long-term repeated high-precision gravimetric measurements and high-resolution leveling permitted to reveal the regularities of gravitational field nontidal variations at the Azerbaijan geodynamical polygons (R.M. Gadjiev, F.A. Kadirov and A.T. Nabiyev).

F.A. Kadirov with collaborators have prepared the numerical analysis technique for gravitational anomalies, using Hartley transformation, for the Azerbaijan territory, permitted to launch the 2D and 3D gravitational models, to make geological interpretation of regional gravitational anomalies. The Bouguer gravity map for the Caspian Sea region compiled in collaboration with the Russian scientists is of high practical significance (M.I. Lodzhevsky and F.A. Kadirov). Another activity, also of practical interest, is theoretical investigation of hydrocarbon migration mechanisms and gravitational instability processes in sedimentary layers of the Azerbaijan oil-and-gas bearing regions (A.G. Gadirov, I.S. Guliyev and F.A. Kadirov).

Numerous researches of geothermal conditions have been carried out (S.A. Aliyev, M.A. Kashkay, K.M. Kerimov, Sh. F. Mekhtiev, A.Sh. Mukhtarov, A.N. Pilchin, A.A. Yakubov, and others), a number of geothermal maps and sections that have been used for the solution of various scientific and production tasks has been constructed.

Recent A.Sh. Mukhtarov research was dedicated to the analysis of the SCB thermal field structure, to integration of numerical model of its thermal evolution as a rapidly subsiding Earth crust block, to elucidation of the liquid- and gaseous-phase hydrocarbon generation zone, to improvement of technique and device for geothermal gradient and thermal flow density measurement in the bottom sediments.

Starting from 1998, the *Physics of the Earth* section of the Institute of Geology and Geophysics of the Azerbaijan National Academy of Sciences (ANAS), carries out, in collaboration with the Massachusetts Institute of Technology (USA), works on launching the GPS monitoring network and investigation of the Earth crust horizontal movement (F. A. Kadirov and R.R. Reilinger). So far, a geodynamical network of 46 GPS stations has been established. Basing on the interpretation of novel data obtained from GPS monitoring of Earth crust movements, from the investigations of contemporary kinematics, abyssal structure models, mechanisms of earthquake foci, and studies of the Caucasian Caspian region fractures, the presence of the Caspian Plate was confirmed, its rotational motion (counterclockwise) in contemporary period was determined, a new geodynamical model of the Caucasus Azerbaijan sector, based on plate tectonics concept, was developed.

The *Physics of the Earth* section has constructed the reference magnetostratigraphic Phanerozoic scale for the Azerbaijan area, basing on the analysis and systematization of paleomagnetic data obtained for various stratigraphic intervals and on the study of geological sequences. Kinematic parameters of dynamics in various geological zones of the Greater and the Lesser Caucasus during the Cretaceous period basing on the comprehensive paleomagnetic data analysis were determined (M.I. Isayeva and A.A. Khalafly). Paleomagnetic investigations of the contemporary Caspian Sea bottom sediments, studies of rock physical properties under high pressures and temperatures, construction of physical models of medium, research on spatial and temporal distribution regularities of the Earth physical fields and seismicity are also noteworthy.

One of the leading places in the Geological Institute research activity is occupied by the investigations of interaction between radioactive fields and geological structure, radiometry of oil-and-gas areas and direct prospecting of oil-and-gas fields by radiometric techniques, activity assessment in abyssal fractures, and the problems of geological ecology (T.A. Zolotovitsky and Ch.S. Aliyev). They also study natural radioactivity of the Azerbaijan depression zones; construct normal gamma radiation background scale for Meso-Cenozoic system rocks and define geological boundaries; compile the natural radioactivity map for various Azerbaijan areas; define the regularities of radioactive field distributions in depression zones and determine geological formation conditions, define and clarify their relation to the Earth crust abyssal structure; develop methodological foundations for radioactive monitoring in seismically dangerous zones of the Greater Caucasus Azerbaijan section; reveal hydrocarbon fields; develop criteria for selecting the complex of efficient radiometric and geophysical techniques; identify the sites of radionuclide contamination and present their generation mechanism; compile the first in Azerbaijan "Map of distribution of radon volume activity at the Azerbaijan area" in order to investigate medical geology issues in Azerbaijan; establish the relationship between radon and methane flows of Azerbaijan mud volcanoes and seismicity.

At present, the Institute of Geology and Geophysics collaborates with many leading world research centers: the Universities of South Carolina, Utah, Wyoming, Maryland, Michigan, the Massachusetts Institute of Technology (USA), the Universities of Birmingham and Aberdeen (Great Britain), Toronto University (Canada), Pierre-and-Marie Curie University and IFPEN (France), Delft Technical University (the Netherlands), Karlsruhe and Hannover Universities (Germany), Scuola Universitaria professionale della Svizzera italiana (SUPSI) (The University of Applied Sciences and Arts of Southern Switzerland), the University of Sofia (Bulgaria), Universities of Istanbul and Ankara, the Moscow State University (Russia), Tel Aviv University, Chang Kong University (Taiwan), South Korean Institutes (KIQAM), University of Pavia, Bicocca University of Milano, Bologna University, Institute of Environmental Analysis Methodology (Italy), Institutes of Geology and Earth Physics of the Russian Academy of Sciences, etc.

After Azerbaijan has conquered its independence in 1991, international scientific relations have been widely expanded, and a new stage of fundamental research, with novel modalities of engineering implementations, is under way. In a short span of time, the Institute was successfully integrated into the world geological science. The process became even more active after signing the 1994 Oil Contract of the Century. Joint investigations with the largest world oil companies, visits to the research centers of these companies have radically altered the nature of international collaboration and the development pace.

Longstanding partnership with leading international oil companies such as "British Petroleum-Statoil", "Exxon", "Shell", "Amoco", "Mobil", "Yunocal", "Texaco", "Philips", "Total", "Elf", "Aqip", and some others (that started in the late twentieth century and is successfully continuing at present) can be called a breakthrough in the international scientific space. During that period, more than 50 research projects in geology, paleontology and stratigraphy, oil and gas geochemistry, and geophysics have been prepared. Joint field research as well as analysis and interpretation of research results, which were later conducted at the foreign research centers, permitted to adapt to the novel techniques applied in researches of Western scientists, accelerate the mastering of new software, on the one side, and speed up the training of new national scientific manpower, on the other side.

The Geological Institute scientists were consistently honored with the highest Soviet and Azerbaijan Republic awards, with state prizes of the USSR and Azerbaijan for their outstanding scientific achievements (M.T. Abasov, E. N. Alikhanov, A.A. Aliyev, G.A. Aliyev, E.G. Aliyeva, A. A. Alizadeh, Ak.A. Alizadeh, M.M. Alizadeh, F.G. Dadashev, A.A. Feizullayev, I.S. Guliyev, Yu.M. Kondryushkin, Sh.F. Mekhtiyev, A.A. Yakubov, etc.).

2 Geomorphology

2.1 Authors' Contribution: A Brief Review

Peculiar features of Azerbaijan geological and geomorphological development, its orographic plan and distribution of heights, and contemporary geomorphological processes have predetermined nearly all relief types at the Republic territory.

Intensive geomorphological researches started in Azerbaijan in mid-fifties of the twentieth and beginning of twenty-first century. Among the main publications can be noted Abasov (1955, 1970), Akhundov (1978), Alizadeh (1982, 1984, 1987a, b, 2007, 2011), Alizadeh and Tarikhazer (2014), Alizadeh et al. (2014), Antonov (1972, 1993), Budagov (1959, 1960, 1961, 1962, 1964, 1965, 1969, 1973, 1976, 1993a, b), Budagov and Alizadeh (1995, 1998), Budagov et al. (1984, 1988, 1993), Dumitrashko (1962), Dumitrashko et al. (1968), Gadjiev (1980), Guluzadeh (1982), Kerimov (1982), Khalilov (1975, 1980, 1993), Kuliyev (1977), Lilienberg (1953, 1955, 1957, 1958, 1959, 1985), Mardanov (1980, 1981), Milanovsky (1960), Museibov (1973), Shirinov (1965, 1973, 1975, 1979, 1980), Vekilov (1956), and others.

2.2 Orography

One of the most specific peculiarities of the Azerbaijan topography is the direct reflection of its large geotectonic structural features. The following basic orographic elements that simultaneously serve as its large morphostructural units are isolated in the Azerbaijan relief (Fig. 2.1).

Main morphostructural units of Azerbaijan are the Gusar sloping plane and the Samur-Devechi lowland, the Greater Caucasus, the Lesser Caucasus, the Kur intermountain Depression (trough), the Garabagh volcanic upland, the near-Araz Ridges; the Middle Araz Depression, and the Talysh Mts.

The Gusar sloping plane is a monocline plateau with its highest peak, Large Suval Mt. (1910 m), and step-like structure with erosional, erosional-denudational, and gravitational relief. The Samur, Gusarchay, Gudialchay, Garachay, Agchay, Velvelechay, and some other rivers that cross the plain, have well developed terraced valleys with landslide-complicated slopes. At the absolute height of 200 m the plain northeastern edge is limited by a bench, and northeastward of this bench, up to the Caspian Sea shore, the Samur-Devechi (now Shabran) lowland is situated, composed of the most recent alluvial-proluvial deposits (Budagov 1993a, b). The Samur-Devechi lowland is poorly broken, composed of clay loams, lying on the boulder-gravel deposits of the Samur, Gusarchay, Gudialchay, Velvelechay, etc., river fans. Along the seashore, in the lowland seaside part, bottom and hammock-ridgy sands are developed that prevent river water runoff to the Caspian Sea.

Within Azerbaijan borders, the Greater Caucasus is represented by the Main Caucasian and Lateral Ridges that gradually descend southeastward and are transformed into the Gobustan-Absheron peninsula with low undulated foothills.

In the southwest, along the Ganykh-Airichay valley northern edge and along the Ajichay-Alyat fracture zone, the Greater Caucasus borders the Kur intermountain Depression. The depression is composed by sedimentary rocks of the Mesozoic (Jurassic, Cretaceous), Paleogene, Neogene ages, and of the Quaternary deposits developed in river valleys, in water parting flat surfaces, and in intermountain depressions.

Main Caucasian Ridge with such peaks as Bazardyuzyu (4466 m), Tufan (Tfan) (4191 m), Guton (3648 m), Babadagh (3629 m), Akhvai (3481 m), Tinov-Rosso (3374 m), etc., has narrow dividing crest with Alpine highland, intensively broken nival-glacial relief. In plan, it corresponds to the Tufan horst-anticline composed of intensively metamorphically altered Jurassic and Cretaceous deposits, represented by clay slates, sandstone, and arenaceous limestone. From the south, the Tufan anticline is bound by the Malkamud thrust. At the Main Caucasian Ridge southern slope, the following morphological structures are outlined: the Govdagh (with inverted relief) and the Nialdagh (with direct relief) (Budagov 1993a).

Lateral Ridge serves orographically as the Greater Caucasus northern part. Its northwestern section with Shakhdagh

A.A. Alizadeh et al., *Geosciences of Azerbaijan*, Regional Geology Reviews,
DOI: 10.1007/978-3-319-27395-2_2

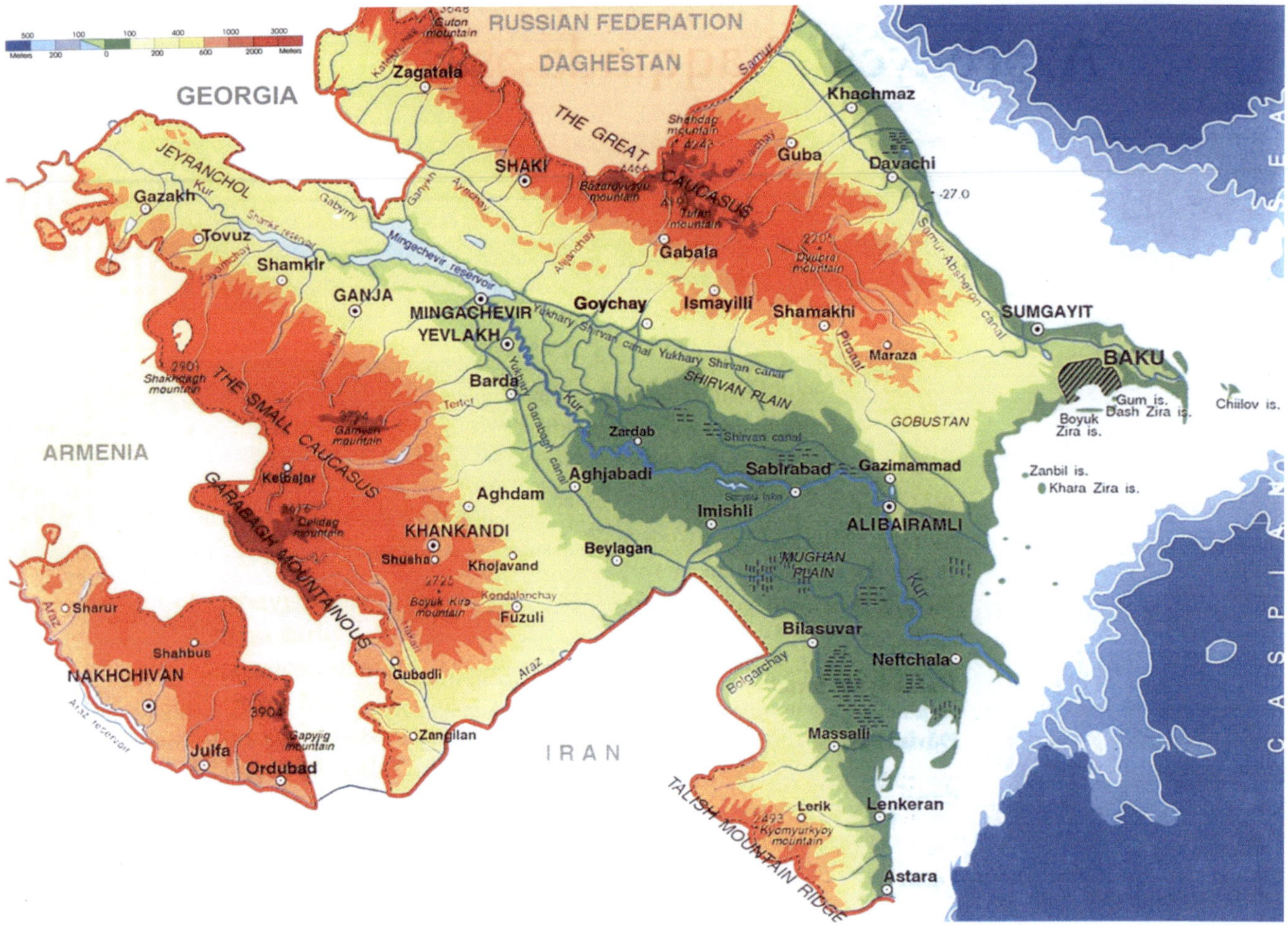

Fig. 2.1 Orographic map of Azerbaijan

Mt. (4243 m) corresponds in plan to the Shakhdagh-Khizi syncline (Fig. 2.2), composed of thick Upper Jurassic and Lower Cretaceous dolomite limestone that create a series of highland plateau with delimiting covered and steep slopes. The Lateral Ridge southeastern part—the Tengi-Beshbarmagh Ridge—looks like medium- and low-altitude mountain erosion–denudation relief, composed of the Jurassic and Cretaceous limestone and clays. The ridge extreme eastern peak is Beshbarmagh Mt. (440 m).

A series of mountain ridges (Gaitar-Goja, Shakhandagh, Aladash, Ilkhydagh, etc.) and intermountain troughs (Khynalyg, Gilgilchay, Khaltan, Lagich, Myudri, etc.) are prominent within the Greater Caucasus southeastern plunge characterized by the middle-altitude mountain denudation relief. Linearly elongated folded ridges and dome-shaped elevations alternating with large synclinal valleys (for instance, the Jeirankechmez depression corresponds in plan to the Shamakhy-Gobustan syncline eastern part) are typical for Gobustan (Fig. 2.3). The Jeirankechmez depression northwestern part is expressed in topography by the Baku age alluvial-proluvial plain, and its southeastern part—by a younger abrasion–accretion marine plain, its relief being complicated by a series of large brachyanticline elevations, broken by longitudinal ruptures with located mud volcanoes. Along with direct topographic shapes, Gobustan features the inverted ones as well—the Gyuzdek, Kyaftarandagh synclinal plateau, narrow intermountain valleys, etc. The Jeirankechmez depression is limited from the south by the Alyat range and the Lengebiz low-altitude mountain range, composed of the Paleogene, Miocene, Pliocene rocks. The Lengebiz Ridge faces the Kur-Araz lowland by its steep slope caused by the Ajichay-Alyat fracture. At the Alyat range, where the fault is broken by transversal ruptures, mud volcanoes are confined to their intersection point (Budagov 1993a, b).

The South-Eastern Caucasus ends in the Absheron Peninsula. Within the Peninsula, the northwestern site is prominent, westward bordering the Dyubrar system with preserved mountainous nature. Here, the Absheron peninsula maximal heights of 300–350 m are observed (at the border with Gobustan) such as Yunusdagh, Ylkhydagh, and Koun. The peninsula northern part topography is hummock-and-hollow, with mud volcano cones doming here and there. Salt

Fig. 2.2 The Shakhdagh-Khizi syncline

lake and flat salt marshes are met occasionally, as well. At the Peninsula southwest the synclinal plateaus are prominent, the Baku and Gyuzdek plateaus, etc., that drop en echelon to the flanking valleys such as the Gobi, Yasamal valleys, etc. Mud volcanoes are developed here as well—Akhtarma, Lokbatan, Bozdagh, etc.

The Absheron peninsula eastern part is flat, poorly hilled in places, with maximal height of 38 m. Some sites are situated below the sea level: Bina, Govsan, Tyrkyan, and Zirya.

The Lower Pliocene deposits, i.e., the productive strata, have extensive occurrence within the Absheron peninsula oil- and gas-bearing regions and serve as the major oil- and gas play. At the Absheron peninsula, except some local elevations, the entire productive strata section is oil- and gas bearing, at nearly 20 anticlinal structures. The most potential here is the following productive strata: supra-Kirmak sandy, sub-Kirmak. Kalin, associated with sandy horizon thinning zones in the plunged parts of syn-depositionally growing anticlinal zones and with individual local structures (Zire, Galin, Surakhany, Garachukhur, etc.) Horizon thinning in the suite productive strata is observed in the Absheron western part (Bibi-Heybat, Garadagh, Lokbatan areas, etc.) as well as in the bottom horizons of the eastern and central parts (Gala, Turkan, Zire, Surakhany, Garachukhur areas, etc.).

The Lesser Caucasus mountain ridges, dissimilarly to the Greater Caucasus Ridge, are less high, do not form a single axial elevation and have more pronounced folded-boulder structure, caused by the presence of abyssal fractures of transversal and longitudinal striking.

In Azerbaijan, within the Lesser Caucasus, the Murguz, Shakhdagh, Murovdagh (Fig. 2.4), Eastern Goycha, Garabagh, and Mykhtekyan Ridges are prominent.

The Shakhdagh basin range has inverted relief and corresponds to the same-name syncline composed of the Upper Cretaceous, Eocene carbonate, and volcanogenic rocks. From Shakhdagh Mt. (2901 m) to Large Ginaldagh Mt. (3367 m) the range has northwestern—southeastern stripping, and then joins en echelon the Murovdagh Ridge, having latitudinal strike. Its highest point is Gyamysh Mt. (3724 m). Kyapaz Mt. (3066 m) rises north of the Murovdagh Ridge.

Fig. 2.3 Gobustan

A number of spurs of northeastern strike and erosional origin such as Pant, Chyngyldagh, Kalteke, etc., are found at the Shakhdagh and Murovdagh Ridge northeastern and northern slopes.

The East Goycha Ridge, serving as water parting between the Lake Goycha and the Terterchay River left tributaries, extends meridionally south of Ginaldagh Mt. Structurally, it corresponds to the transversal elevation that dissects the Goycha-Hakeri syncline into two parts: the Goycha (northwestern), and the Hakeri (southeastern). The ridge is composed by the Eocene and volcanogenic-sedimentary rocks and of Miopliocene volcanogenic thickness. The ridge's highest point is the Ketidagh Mt. (3399 m). The Mykhtekyan Ridge with its Dalidagh Mt. (3616 m) is situated at the Terterchay and Akera river water parting. It is a large anticline elevation composed of granitoids of thick Upper Eocene intrusion.

The Garabagh anticline ridge with its Beyuk Kirs (2725 m), Ziyarat (2478 m), Gyrkhgyz (2823 m), etc., peaks extends in the general Caucasian direction in the central part of the Lesser Caucasus southeastern edge.

The Lesser Caucasus mountain slopes are characterized by intermontane troughs such as Bashkend-Dastafyur, Khankendi, Umudly, Sirik, etc. The Kelbajar superimposed depression, a young graben-syncline trough filled in its central part by the Garabagh volcanic upland Pleistocene lavas, is located in the Terterchay River upstream (Budagov 1993a, b; Antonov 1993; Alizadeh 1998; Khalilov 1999).

In the Shakhdagh, Murovdagh, and East Goycha Ridge highlands kars and troughs (the traces of the Upper Pleistocene glaciation) have developed. The river water parting have well preserved peneplanation surfaces of the Early-Middle Oligocene, Middle Miocene, and Late Pliocene (Akchagyl, Early Absheron) ages.

The Garabagh volcanic highland is situated southwestward of the Lesser Caucasus ridges and its relief is created by the Late Pliocene and Quaternary and andesite basaltic lava outflows. Major eruption centers are confined to abyssal

Fig. 2.4 The Murovdagh Ridge

fractures whereupon sit Upper Pliocene volcanoes and Quaternary cinder cones.

The Garabagh volcanic highland is the Lesser Caucasus most highly elevated mountain structure as well as a large water parting for the Kur and Araz basin rivers. Its principal topographic shapes are the Beyuk Ishygly (3550 m), Gyzylbogaz (3381 m) volcanic massifs, lava plateaus, and lava flows. Lava plateau relief is characterized by step structure, "*chyngyly*" boulder placers, having hummock surface. River valleys that deeply cut into lava plateaus have stepwise slopes and remind canyons. Trough valleys, kars, and morainic deposits (the traces of ancient glaciations) are developed at the highland high volcanic massifs and are preserved on Beyuk Ishygly and Gyzylbogaz mountain slopes.

The Daralagez and Zangazur folded-block and block ridges are located southwest of the Garabagh volcanic highland (Fig. 2.5).

The Daralagez Ridge, located in the Arpachay and Nakhchivanchay interfluve, has nearly latitudinal course in the northeastern part, and meridional course in the southwestern part. The ridge is composed of the Paleozoic carbonate and Jurassic volcanogenic-sedimentary rocks. The ridge highest section is at its abutment joint with the Zangazur Ridge (Kyukyudagh Mt., 3120 m). The Zangazur horst ridge (its highest point is Gapyjig Mt., 3904 m) is composed of the Megri-Ordubad fracture granitoids, and in the east, it is cut by the East Gapyjig fracture. The Zangazur Ridge water divide part is characterized by powerful and deep differentiation, by intensive physical weathering processes resulting in accumulation of coarse material of rock falls and talus at foot slopes. At the Ridge northeastern slope, the Late Pleistocene kars and troughs are developed, with preserved bottom terminal moraines.

The Daralagez and Zangazur Ridge middle-altitude and low-hill terrains, due to sharp continental climate and poor

Fig. 2.5 The Zangazur Ridge

vegetation, are characterized by vigorous development of arid-denudation processes. Monoclinic ranges characterize low-hill terrains with slopes covered by thick talus of coarse gravel and rock debris (Abasov 1970; Budagov 1993a, b; Antonov 1993).

Specific peculiarity of the Daralagez and Zangazur Ridge low-hill terrains is the presence of extrusive domes, laccolites, of diversified sizes and shapes, e.g., the Ilandagh Mt. (2410 m), Khanagya Mt. (1910 m), Alinja Mt. (1815 m), etc.

The Middle Araz kettle filled with thick Miocene and Pleistocene deposits is a large superimposed structure divided by transversal elevations into a series of intermountain pits and accumulation–denudation plateaus and the Negram High. At the Wolfish Entry transversal elevation (Dalidagh Mt., 1240 m; Degne Mt., 1150 m) the Middle and Upper Paleozoic limestone outcrops, while in the Negram High the Permian, Triassic, Middle Jurassic, Upper Cretaceous, Eocene, Oligocene sedimentary, and volcanogenic-sedimentary deposits outcrop.

The Gyvrag and Duzdagh plateau beds are composed of the Miocene salt-bearing thickness rocks and overlapped by thick Late Pliocene—Early Pleistocene bounder-gravel deposits. Their slopes are differentiated by dense network of badland.

The Sadarak Plain is located between Degne spur and the Araz River. Average altitude is 930 m. The Sharur plain stretches from the Degne spur to Gyvrag plateau, the Beyukdyuz plain stretches southeastward of Gyvrag plateau to Degne spur and is covered by low-rise hilly uplands. The Nakhchivan plain with its average altitude of 890 m is located between the Nakhchivanchay and the Jagrychay River lower courses. The plain is occasionally crossed by hilly uplands and low-rise ranges. The Gyulyustan plain is located in the Alinjachay River lower course (southeastward of the Darydagh Ridge). The Ordubad plain is found at the Nakhchivan AR extreme southeast.

The Kur-Araz lowland is situated between the Greater and Lesser Caucasus and occupies a large intermontane lowering with a number of sites differing in their geological structure and topographic peculiarities.

The Ganykh-Airichay synclinal valley stretches in the North, along the Main Caucasian Ridge bottoms, as a sloping plain where basic topographic shapes are expansive debris conduits of the rivers descending from the Main Caucasian Ridge southern slope.

The Jeiranchel-Ajinour low-hill terrain with direct tectonic relief is situated south of the Ganykh-Airichay valley, with low-rise linearly elongated monoclinic ridges composed of the Miocene and Upper Pliocene argillo-arenaceous rocks. The rivers that cross the Jeiranchel-Ajinour low-hill terrain eastern part create the antecedent valleys.

The Ganja sloping alluvial-proluvial plain is situated south of the Kur River valley, between the Khrami River entry and the Ganjachay River valley.

The Kur-Araz lowland itself is located southeast of the town of Mingachevir, up to the Caspian Sea shores and looks like a deep intermountain trough filled with thick mass of the Neogene and Quaternary argillo-arenaceous rocks. The Kur and Araz river valleys dissect the lowland into the Shirvan, Garabagh, Mil, Mughan, and Salyan depositional plains. A zone of young brachyanticlinal elevations such as the Large and the Small Kharami, Mishovdagh stretches submeridionally in the Kur-Araz lowland eastern part. The southeastern Shirvan, flat alluvial marine plain, is located between this zone and the Caspian Sea shore. In the south, the Mughan and Salyan plains merge with the Lenkaran lowland surface. The Kur-Araz lowland is characterized by numerous topographic micro shapes such as "*chala*", and elevations such as "*gobu*", winding dells, and "akhmaz", river-formed lakes. Eolian and salt marsh topographic shares are developed at the Mughan Plain and in the southeastern Shirvan. Beach barriers and dunes stretch as narrow strip along the Caspian Sea shore.

Talysh Mts. (Fig. 2.6) are the continuation of southeastern plunging of the Lesser Caucasus folded structures.

They are composed of thick Paleogene and Neogene volcanogenic deposits and by the Late Cretaceous deposits in the southeastern part. Unlike the Greater and Lesser Caucasus Mts., former glaciation relief shapes are absent in the Talysh Mts. part, since its highest peaks slightly exceed 2400 m (Kemyurkei, 2493 m). High mountains are situated in the development belt of poorly disseminated terrain, with flat water divides and smooth hillsides.

Representation of the Talysh hill features displays four ridges of northwestern—southeastern strike: the Talysh Ridge (with its highest peak of Kemyurkei Mt., 2493 m), the Peshtasar Ridge (Pashagol Mt., 2240 m), the Dizgoni Ridge (Dizgoni Mt., 1056 m), and the Burovar Ridge (peaks of up to 900–1000 m). The Dyman and Gosmalyan kettles are located between the Talysh and Peshtasar Ridges, and the Yardymly kettle, between the Peshtasar and Burovar Ridges. The kettles are composed of the thick Maykop argillo-arenaceous deposits that are superimposed by the Middle and Upper Miocene deposits in the northwest (Antonov 1993; Budagov 1993a).

The Lenkaran lowland (depression) is positioned between Talysh Mts. and the Caspian Sea, stretching as a narrow flat belt. The lowland seaside part lies below the sea level. The lowland is generated by sediments of the rivers descending from the Talysh Mts. Various lakes are extensively developed such as estuaries and bogs. The lowland seashore is generally flat, but steep and scarp in some places. Sand dunes are developed parallel to the sea coast.

Fig. 2.6 The Talysh Ridge

2.3 Morphostructural Configuration

Contemporary terrain of Azerbaijan Republic took shape within a prolonged geological period, under collision conditions of the Transcaucasian, Anatolian-Iranian, and Scythian geotectonic plates. Convergence of these plates resulted in horizontal compression of the Transcaucasian massif by the Anatolian-Iranian block from the south and by the Scythian Plate from the north. Crustal folding, compartmentalization, and warping are increased in these suture zones that also resulted in vertical differentiation. Intensive seismic and volcanic processes accompanied generation of large mountain systems and intermountain troughs. Enhancement of horizontal compression and increased vertical differentiation amplitude in relatively weakened Earth crust sites resulted in the formation of tectonic fissures and fractures. They created conditions for volcano linear eruptions, played the role of migration channels for outcropping of magmatic-origin mineral deposits, and delineated the terrain large orographic units.

Under tensioned compression, stretching, and folding the largest relief forms have been generated-such as the Greater and Lesser Caucasus mountain structures, Talysh Mts., and the Kur-Araz Trough. In other words, these intensive and complicated processes that occur in the Earth crust deep-seated layers have governed the generation of contemporary terrain outlines, typical for morphotectonic structure of the Alpine-Himalayan orogenic belt. Close interrelationship between morphotectonic structure and terrain shapes determined the formation of morphological structures of heterogeneous origin, typical for this orogenic zone, heterochronous, mainly folded-block, block and faulted ones, with general Caucasian, anti-Caucasian, and diagonal striking direction.

It should be noted that many specialists upholding various geotectonic views with regard to the Earth crust development have studied general regularities of Azerbaijan morphostructural configuration, reflecting the interaction between the Earth surface topography and its geological structure. Morphological structures were closely identified by Budagov et al. (1984, 1988), Budagov and Alizadeh (1995, 1998), Museibov (1968, 1973), Shirinov (1993, 1975, 1979, 1980), Shirinov and Dotduev (1976), Lilienberg (1985), Kuliyev (1977), Khalilov (1980, 1993), Gadjiev (1980, 1999), Mardanov (1980, 1981), Alizadeh (1984, 1987a, b, 1991, 1998, 1999). Recently, however, due to enhanced assertions of plate tectonics role in the Earth crust development and development of theoretical foundations of the contemporary terrain generation as a result of plate horizontal shifts, the necessity to examine morphological structure formation from new positions, from the viewpoint of plate tectonics, has emerged.

After integration of information and morphotectonic interpretation of aerospace photographs we have delineated the following large morphostructural units in the Republic territory that differ in their origin, development history, and geomorphological features.

The Greater Caucasus vigorously upraised horst-anticlinal, folded-block-thrust, highly differentiated suture ridges are essentially composed of the Jurassic and Cretaceous sedimentary and volcanogenic-sedimentary rocks, and their margins are composed of the Paleogene—Neogene sedimentary deposits. It is a vigorously compressed mountain area, and tectonically is the Scythian Plate frontal part. It is the product of compression and folding at the Scythian Plate surface in the area of its collision with the Transcaucasian Plate. The Transcaucasian Plate, moving north- and northeastward along the Alazan (Ganykh) trough, plunges under the Scythian plate, i.e., the collision takes place.

Intensive compression of the Earth crust on the Earth's surface resulted in differentiated, multiple-elevation terrain. Earth crust was broken by heterogeneous and heretochronous tectonic fracture of the general Caucasian and anti-Caucasian strike directions (Fig. 2.7).

Retrospective analysis within the Greater Caucasus eastern part permitted us to separate four generation stages of morphostructure modern framework: (a) the Lower Jurassic tension stage; (b) the Middle and partially Upper Jurassic compression stage; (c) the Upper Jurassic—Eocene stage of tension stabilization; (d) the Oligocene—Holocene compression stage.

Taking into account the interrelations and interdependences of geological structure and terrain in the Greater Caucasus Azerbaijan sector, a number of general Caucasian direction morphostructures have been revealed (Fig. 2.8). The Samur-Devechi graben-syncline plain that preserved its former shape, and the sharply elevated Gusar monoclinic plain as well as other relatively elevated piedmont plains were generated in the contact zone of the Greater Caucasus and the Scythian Plate, under weak folding and differentiation conditions. These monoclinally shaped morphostructures contact the Greater Caucasus mountain structure due to the Kryz-Siazan abyssal fracture, and within their borders, height difference amounts to 2000 m. Within these plains a number of lower-order morphostructures is prominent (the Telebi-Gainarja horst-anticlinal low-hill terrain, the Zeikhur depression, the Yalama-Khudat elevation, etc.).

The Greater Caucasus Mts. of folded-block-thrust origin that experienced geotectonic stress were generated in the suture zone, which consists of several large morphostructures of general Caucasian direction.

Horst-anticlinal Tufan (watershed) ridges generated in the compression axial zone, along the Main Caucasian abyssal fracture and most intensively raised hypsometrically are the most pronounced in terrain. This mountain system is composed of the Jurassic and partially Cretaceous sedimentary deposits and plunges northwestward to southeastward. These sedimentary rock complexes, relatively unstable against weathering are exposed to intensive splitting. Some morphostructures of the Main Dividing Ridge southern slope are sharply lowered along the abyssal fractures and are, therefore, partially reflected in contemporary relief.

The Alpine-Himalayan suture zone of interplate contacts with horst ridge and graben intermontane and intramontane kettle systems. Central part of the Alpine-Himalayan suture zone of interplate contacts with young, intensively elevated, block-ruptured ridge systems, and intensively lowered graben depression systems of kettles.

(1) I. The Greater Caucasus folded-block-thrust, intensively elevated, suture, horst-anticlinal, intensively dissected mountain structure.

(2) I.1. The Sumur-Devechi marginal, graben-synclinal, superimposed, poorly dissected plain. (3) I.2. The Gusar, marginal, horst-monoclinic, highly elevated, dissected plain. (4) I.3. The Sudur horst-synclinal, residual, intensively

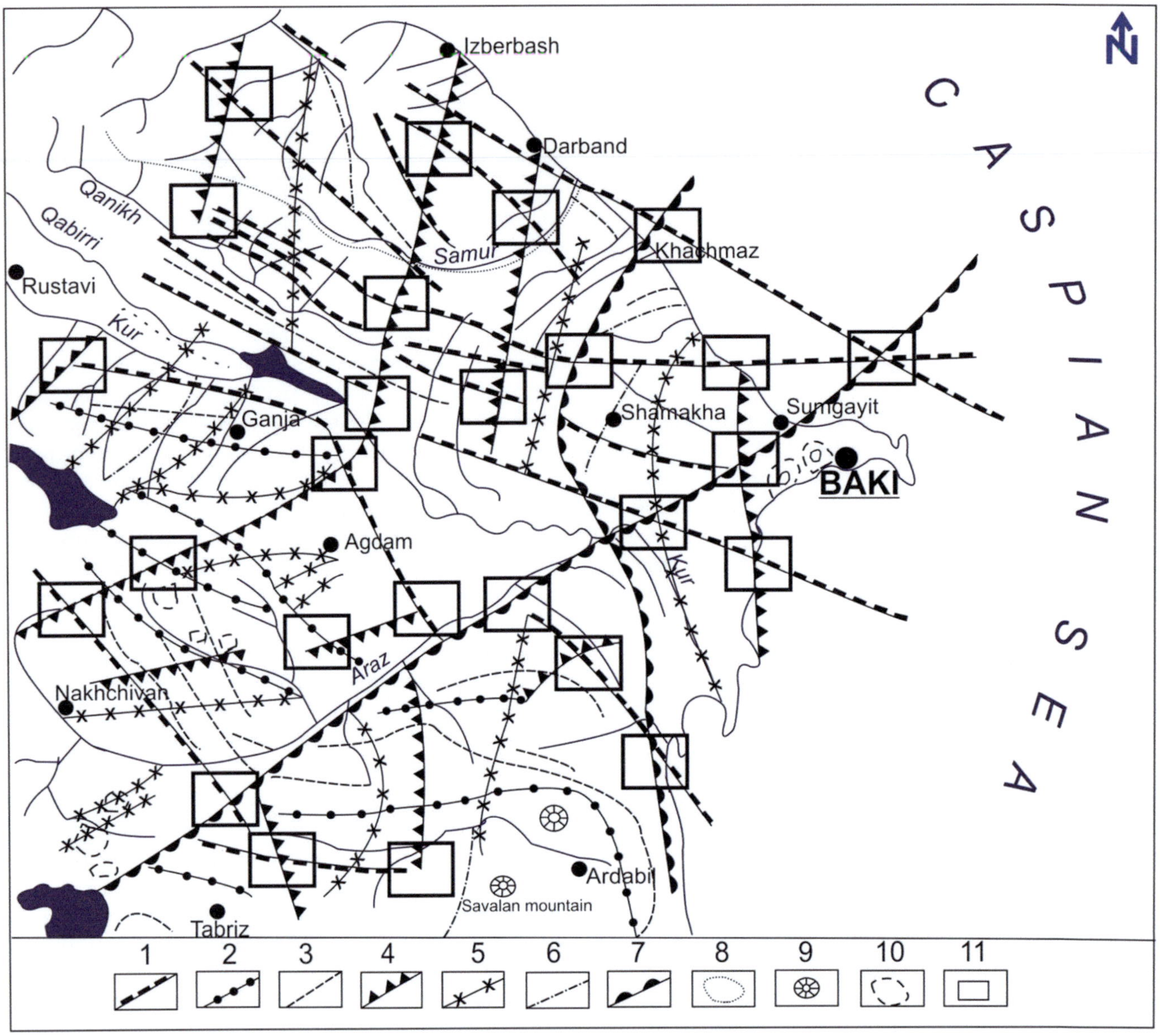

Fig. 2.7 Map-scheme of morphotectonic intensity of the Eastern Caucasus and adjacent territories (author: E.K. Alizadeh). Lineaments that correspond to faults (thrusts and overthrusts) of longitudinal (Lesser Caucasus) direction: (*1*) regional deep faults that limit large longitudinal folding and block steps, (*2*) local faults corresponding to the boundaries of longitudinal folding and block morphological structures, (*3*) faults that determine the details of morphological structure. Lineaments that correspond to faults (*strike slips*) of the transverse (anti-Caucasus) direction, (*4*) regional deep faults that limit transverse megablocks, (*5*) local faults that correspond to the boundaries of transverse block segments, (*6*) faults, (*7*) large interregional diagonal volcanic centers, (*8*) contour of the circular object of Samur, (*9*) great volcanic centers, (*10*) ring structures (tectonic and volcanogenic), (*11*) geodynamically stressed fields

elevated, intensively dissected ridge. (5) I.4. The Tengi-Beshbarmagh sub-fractural (suture), rear, horst-anticlinal, intensively elevated, highly dissected ridge. (6) I.5. The Shakhdagh-Khizi, superimposed, horst-synclinal, intensively elevated, highly differentiated, complicated by large tectonic nappes, dissected mountain massifs and plateaus. (7) I.6. The Tufan (water dividing) central, most elevated, horst-anticlinal, highly dissected ridge. (8) I.7. Ridges, intramontane ketttles, valleys of the Zagatala-Govdagh synclinal zone foretrough. (9) I.8. The Vandam (Nialdag) front, horst-anticlinal, averagely elevated, partially plunged, dissected ridge. (10) I.9. Ridges, ranges, valleys, kettles of the Shamakhy-Gobustan marginal graben-synclinal dissected low-hill terrain.

The Greater Caucasus transversal (anti-Caucasian) block morphostructural segments.

(11) I.A. The Belakan-Kakh relatively lowered; (12) I.B. The Sheki-Gabala elevated; (13) I.C. The Babadagh-Dibrar

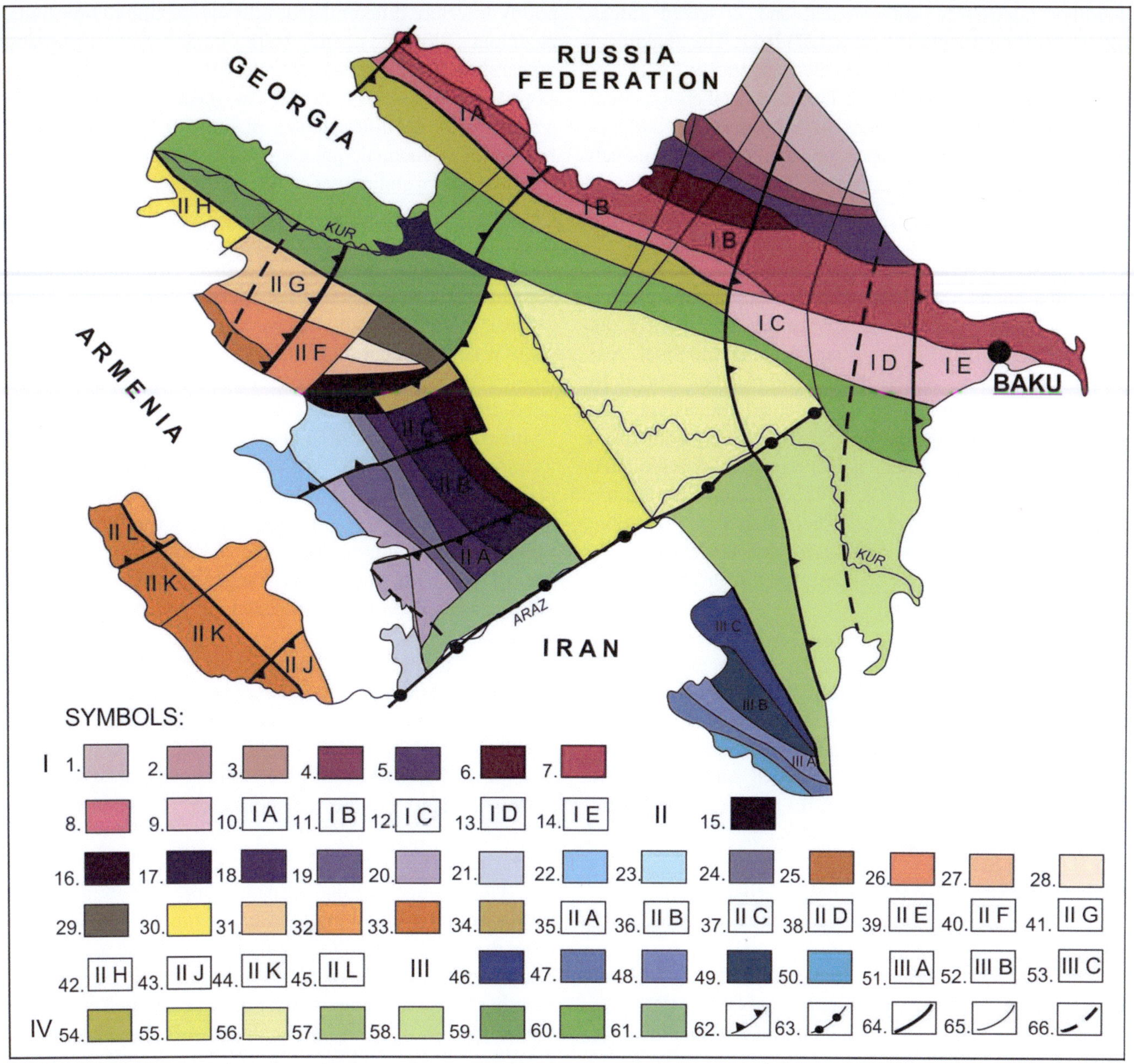

Fig. 2.8 Map of Azerbaijan morphostructures (constructed on the basis of space images analysis) (author: E.K. Alizadeh)

elevated; (14) I.D. The Maraza-Siazan lowered; (15) I.E. The Absheron most lowered.

(16) II. The Lesser Caucasus folded-block-thrust, intensively elevated, suture, horst-anticlinal, predominantly volcanogenic, averagely dissected mountain structure.

(17) II.1. The Murovdagh horst-anticlinal, articulated, suture, thrust, highly elevated, intensively dissected ridge. (18) II.2. The Agdam horst-anticlinal, rear, low-hill, averagely dissected ridges and ranges. (19) II.3. The Aterk (Umudlu)-Khojavend rear, graben-synclinal, en echelon located kettles and plateaus. (20) II.4. The Garabagh suture, partially ophiolite, highly elevated, highly differentiated, horst-anticlinal, highly dissected, asymmetric ridge. (21) II.5. The Lachin near-fracture, horst-anticlinal, highly broken ridges and plateaus. (22) II.6. The Gochaz rear, superimposed, synclinal, highly differentiated kettles, valleys, ranges and plateaus. (23) II.7. The Gafan (Bartaz-Oskyulyum) horst-anticlinal highly differentiated and dissected ranges and ridges. (24) II.8. The Garabagh volcanogenic, highly elevated, horst, poorly sloped, poorly dissected highland. (25) II.9. The Kelbajar transversely superimposed, volcanogenic trough (graben-synclinal) with intensively dissected ridges, kettles and valleys. (26) II.10. The Sarybaba horst-synclinal, suture, highly mylonitized,

ophiolite, intensively differentiated and dissected ridges, plateaus and intramontane kettles. (27) II.11. The Shakhdagh horst-synclinal, intensively elevated, averagely dissected ridge. (28) II.12. The Bashkend-Dastafyur graben-synclinal, front, lowered, dissected series of kettles, ridges and plateaus. (29) II.13. The Kyapaz horst-synclinal, intensively elevated, averagely dissected ridge. (30) II.14. The Goygol horst-anticlinal, averagely dissected elevation (plateau and ridges). (31) II.15. The Agjakend block-synclinal, front, poorly dissected series of superimposed kettles, ranges and monoclinal plains. (32) II.16. The Gazakh graben-synclinal transversely superimposed, relatively lowered, poorly dissected kettle. (33) II.17. The Shamkir horst-anticlinal, rear, dissected elevation—ridges, ranges, and plateaus. (34) II.18. The Daralagyoz-Zangazur horst-anticlinal, intensively elevated, highly dissected range. (35) II.19. The Middle Araz graben-synclinal, marginal, superimposed, differentiated, intramontane kettle. (36) II.20. The Touragchay suture, sub-thrust, intensively differentiated, ophiolite, block-synclinal, dissected ranges, ridges and kettles.

The Lesser Caucasus transversal (anti-Caucasian) block morphostructural segments.

(37) II.A. The Bazarchay-Garasu lowered; (38) II.B. The Ishykhly-Kirss elevated; (39) II.C. The Terterchay relatively lowered; (40) II.D. The Zod-Murovdagh highly elevated; (41) II.E. The Ganja-Murovdagh relatively intensively elevated; (42) II.F. The Ganja-Shamkir elevated; (43) II.G. The Shamkir-Tovuzchay relatively lowered; (44) II.H. The Tovuzchay-Afstafa lowered; (45) II.I. The Ordubad intensively elevated; (46) II.J. The Shakhbuz-Julfa elevated; (47) II.K. the Sharur lowered.

(48) III. The Talysh-Bogrovdagh folded-block-thrust, marginal, slightly elevated, horst-anticlinal, averagely dissected ridge.

(49) III.1. The Burovar rear, horst-anticlinal, relatively poorly elevated, poorly dissected ridge. (50) III.2. The Deman-Gosmalyan graben-synclinal and anticlinal, narrow series of intramontane kettles and valleys. (51) III.3. The Peshtasar horst-anticlinal, volcanogenic, elevated, averagely dissected ridge. (52) III.4. The Yardymly graben-synclinal, superimposed, intramontane differentiated kettle. (53) III.5. The proper Talysh horst-anticlinal, volcanogenic, relatively intensively elevated, highly dissected ridge.

The Talysh-Bogrovdagh Ridge transversal block morphostructural segments.

(54) III.A. The Astara-Gyzyurdu poorly elevated; (55) III. B. The Gyzyurdu-Kemyurkei relatively intensively elevated; (56) III.C. The Vilyashcahi-Bolgarchay lowered.

(57) IV. The Kur folded-block, graben-synclinal, greatly lowered, intermountain, differentiated series of kettles.

(58) IV.1. The near-Lesser Caucasian graben-synclinal, relatively poorly elevated, proluvial, alluvial, differentiated, sloped plain. (59) IV.2. The Garabagh graben-synclinal poorly lowered, differentiated, hilly-range, alluvial-proluvial, piedmont sloped plain. (60) IV.3. The Kyurdamir graben-synclinal, median, lowered, alluvial-proluvial, poorly sloped lowland. (61) IV.4. The near-Talysh piedmont, graben-synclinal, differentiated plain. (62) IV.5. The Lower Kur graben-synclinal, greatly lowered lowland. (63) IV.6. Kettles, valleys and ranges of the Alazan (Ganykh)-Agrichay marginal graben-syncline. (64) IV.7. the Jeiranchyol-Ajinour-Alyat horst-anticlinal, frontal, suture, low-mountain ridges and ranges. (65) IV.8. The Lower Araz intermountain, superimposed, graben-synclinal, sloped plain.

(66) IV.9. Abyssal transversal ruptured dislocations. (67) IV.10. Regional diagonal fracture. (68) IV.11. Abyssal longitudinal ruptured dislocations. (69) IV.12. Local ruptured dislocations. (70) IV.13. Ruptured dislocations.

Intensively elevated, horst-synclinal, complicated by tectonic nappes, highly dissected Shakhdagh-Khizi Mts., plateaus and valleys are prominent at the Greater Caucasus northeastern slope. These morphostructures with complicated interior configuration unite the Shakhdagh-Gyzylkai massif, the Yerfi monoclinal range, the Khizi, Sokhyub, Alayaz, etc., depressions elevated to various hypsometric heights.

During the Lower Cretaceous age, under the Greater Caucasus vigorous southwestward compression conditions, subaqeous gliding of organogenic and marbleized rocks of the "Sudur" origin that are abundant in its northeastern part, took place. As a result, a gravitational-tectonic nappe generated in lieu of the contemporary Shakhdagh-Budug plateau.

During subsequent periods, particularly after the Sarmatian (Upper Miocene) age, crucial differentiation in tectonic movement intensity along the Shakhdagh, Siazan and some other depressions of anti-Caucasian direction takes place. As a result of these movements, a certain section of this large morphostructure, situated northwestward of the Velvelechay transversal fault, and particularly, of the Shakhdagh-Gyzylkai massif, being compressed on all sides, is squeezed out and rises to 3500–4000 m height. Southeastward of the Velvelechay, the terrain heights and differentiation decrease, and here, the Shakhdagh-Khizi zone morphostructures, via the Germian abyssal fracture, encounter the Zagatala-Govdagh zone morphostructures that serves as the principal morphotectonic units of the Eastern Caucasus southern and southeastern slopes.

At the Greater Caucasus northeastern slope, at the boundary line of orogenic zone and the Scythian Plate, along the Siazan abyssal fracture, southeast of the Garachay River, the Tengi-Beshbarmagh horst-anticlinal (monoclinal) ridges have developed, composed essentially of the Cretaceous and Jurassic sedimentary thicknesses.

The ridge southern slopes, violently compressed by the Siazan fracture in the northeast, and by the Garabulagh fracture in the southwest, complicated by fractured ruptures, are very steep. It resulted from subsidence of the horst-anticline northwestern wing along the Siazan abyssal fracture. The Lateral Ridge subsides intensively southeast of Beshbarmak Mt.

Specific trait of the Greater Caucasus southern slope morphostructures increased clarity of their eastward reflection in topography. Structurally complex relief of this area west of the Damirapanchay River is prominent for its high steepness. The area exposed to intensive folding, south of the Tufan (water dividing) Ridge, displays mountains and intramontane kettles of the Zagatala-Govdagh horst-syncline that are mostly composed of the Cretaceous rocks. This large morphostructural unit is north delimitated by the Malkamud-Germian morphotectonic thrust, and South delimitated by the Zangi-Geradil morphotectonic thrust. Due to drastic vertical compression, the horst-anticlinal ranges and graben-synclinal depressions are formed in its western part. East of the Damiraparanchay, morphotectonic units are relatively well defined in the relief. Here, such morphostructures are prominent as the Babadagh horst-anticlinal, the Govdagh horst-synclinal, the Guzducha-Alatash horst-anticlinal ridges, as well as, the Demirchilyar, Astrakhan (Gyzmeidan) and additional synclinal, occasionally superimposed, kettles.

It should be noted that it is in this zone that the Zagatala-Govdagh morphotectonic unit, covering the Greater Caucasus water divide part, partially transfers to the northeastern slope of this mountain structure. East of the Girdymanchay-Velvelechay belt, starting from the West Caucasian transversal suture zone, differentiation of horizontal and vertical tectonic movements increases, and morphostructure intensive subsidence takes place. The Zagatala-Govdagh morphotectonic step played the role of autochthon for thrust nappes abundant at the southeastern slope. The well-known Astrakhan thrust nappe was generated here, owing to horizontal shift of more ancient (Cretaceous) rock systems onto the younger (Miocene—Quaternary) deposits along the Zangi-Geradil rupture. The Vandam horst-anticlinal ridge belt composed of the Middle Jurassic volcanogenic and volcanogenic-sedimentary rocks is prominent at the Greater Caucasus orogenic zone margin. Its western part is poorly pronounced in contemporary relief due to subsidence along the North Alazan (Ganykh) abyssal fracture. The Nialdagh horst-anticlinal lateral ridge, the Lagich graben-synclinal kettle and alternative morphostructures of relatively lower order stand out in the Vandam zone. Intensively dissected and highly complicated by contemporary landslides ridges and intramontane depressions, located in the Vandamchay-Akhsuchay interfluve are well pronounced in topography. The Vandam morphostructure sharply subsides at the town of Shamakhy meridian, and eastward is not morphologically pronounced in contemporary terrain.

Morphostructural analysis of terrain and of morphotectonic structure of the Greater Caucasus orogen serving as the interface between the Scythian and the Transcaucasian plates, shows that compression and tension intensities are also evident in anti-Caucasian direction, and as a result, morphostructural segments of the same direction have generated in this mountain system. Within the Greater Caucasus eastern part, the most dramatic morphotectonic and morphological differences have generated to west and east of the Western Caucasian morphostructural node, with its center in the Girdymanchay-Velvelechay fractured dislocation.

With due account for the relief morphotectonic and morphological peculiarities, resulted from the investigations conducted, and from the literature data analysis, the following relatively large transversal morphostructures are delineated within the Greater Caucasus: (a) the Belakan-Sheki relatively lowered; (b) the Sheki-Gabaly most elevated; (c) the Babadagh-Dibrar elevated; (d) the Maraza-Siazan lowered; (e) the Absheron most lowered one.

The Lesser Caucasus mountain structure, whose southeastern part is situated within Azerbaijan, has been developed in the Alpine-Himalayan zone, under contact and collision conditions of the Anatolian-Iranian microplate and the Transcaucasian microplate.

It should be emphasized that the Lesser Caucasian mountain system that underwent a complicated and prolonged way of morphotectonic development consists of a number of drastically differing by their origin large morphotectonic units. The Lesser Caucasus mountain structure composed by rocks of different composition (volcanogenic, volcanogenic-sedimentary, sedimentary, etc.) and age (Lower Paleozoic—Quaternary) according to its peculiar features, intensity of morphotectonic development and morphostructural genesis is subdivided into three large parts: (a) The Lok-Garabagh horst-anticlinal ridges generated in the highly compressed zone of the Transcaucasian plate southern margin between the Murovdagh-Garabagh and the near-Lesser Caucasian abyssal fractures; (b) the Tourachay-Sarybaby morphostructural zone, generated—starting from the Lower Cretaceous age—as a result of closure, and subsequently experiencing the folding of the Tutkhun rift that was confined to the Mesotethys Ocean northern part; (c) The Miskhan-Gafan zone morphostructures (ridges, highland, kettles) generated by folding and block formation at the site of the Anatolian-Iranian plate collision with the Transcaucasian plate.

Within the Lesser Caucasian mountain belt, foundation formation, and development for the contemporary morphostructures can be subdivided into two significant stages: (1) The subduction period—foundation formation for large pra-morphostructures possessing the Earth crust of oceanic and continental types (the Middle Jurassic—Upper Cretaceous ages).

(2) The collision period—formation of contemporary morphostructural carcass (the Upper Cretaceous—Quaternary age).

At the first stage, compression processes and general relief elevation typical for the Hercynian period were replaced by divergence and subsidence of the Lesser Caucasus great morphostructures.

At that period, the Miskhan-Gafan zone of the Anatolian-Iranian plate that encompasses frontal shelf zone separated from the Lok-Garabagh great morphostructure. The eugeosynclinal system of the Tutkhun Rift with the earth crust of oceanic type has formed in the tension zone. The relatively elevated Miskhan-Gafan and Lok-Garabagh zones were expressed by arc-like islands, whereas the Tutkhun rift looked like a narrow trough valley of 2000–3000 m depth that was exposed to continuous subsidence.

Development and formation of the Lesser Caucasus contemporary morphotectonic structure and laying the foundations for contemporary morphostructures start in the second half of this stage. During the Lower Cretaceous (Albian) age, horizontal movements changed their directions, and the Lok-Garabagh and Miskhan-Gafan zones interaction started, which resulted in subsequent closing of the Lesser Caucasus Tutkhun rift. During the period of vigorous compressions, folding, mutual thrust of heterogeneous rocks of the Earth crust oceanic and continental types, linear and areal lava eruptions, the oceanic rocks have compressed, separated from their roots, and underwent horizontal transfer at large distances. It resulted in the formation of the Sarybaba and Touragachay morphostructures of ophiolite composition.

Contemporary carcass of the Lesser Caucasus morphostructures is the product of collision that began, essentially, in the Upper Cretaceous period and is continuing until now. Thus, after compression of oceanic rift zone is over, the Anatolian-Iranian and the Transcaucasian plates finally collide, and the Miskhan-Gafan microplate starts plunging under the Lok-Garabagh zone. Orogenesis characterized by intensive magmatism, folding horizontal, and vertical movements has enhanced, starting from the Upper Eocene age, and folded-block, block, volcanic suture morphostructures of tectonic, and nappe origin were generated. Taking into account all morphotectonic development regularities within the Lesser Caucasus Azerbaijan part, the following morphostructural units can be separated in the collision zone of the Anatolian-Iranian and the Transcaucasian plates.

Among the Lesser Caucasus morphostructures exposed to vigorous compression and folding are the Murovdagh horst-anticlinal, suture, elevated, intensively dissected ridges and the suture, partially ophiolite Garabagh-Sarybaba elevated and highly differentiated horst-anticlinal and horst-synclinal highly dissected, asymmetrical ridges, and plateaus. The Sarybaba and the Touragachay morphostructures situated in this area and associated with the Murovdagh and Garabagh abyssal fractures are confined to the area where the compression and closing of the oceanic type Earth crust took place, therefore, ophiolites have been found among their constituting rocks.

The Lachin morphostructure and the morphostructures located southeastward of the Lachin-Bashlybel abyssal fracture were generated at the Miskhan-Gafan microplate marginal part and were exposed to intensive compression and folding (the Gochaz, Garabagh highlands, the Zangazur, Daralagyoz, Bartaz, Askyulyum ridges, etc.).

Heterogeneous morphostructures (Agdam, Aterk (Umudly), Khojavend, Shakhdagh, Bashkend-Dastafyur, Kyapaz, Goygel, Shamkir, etc.) situated northward of the Murovdagh Ridge and eastward of the Garabagh Ridge, were formed in the collision area in the Transcaucasian microplate southeastern part.

It should be marked that tectonic faults of anti-Caucasian direction that began their development in the Upper Cretaceous period (at the time of collision), played important role in the formation of the Lesser Caucasus contemporary topography. During the highest collision increase period (the Eocene—Pliocene and the Miocene—Quaternary periods), Earth crust dissection on this direction enhanced and consequently, development of endogenous and exogenous processes in the terrain resulted in the predominance of anti-Caucasian elements. In this manner folded-block, block morphostructures and morphostructural segments (map) were generated, typical for this trend and prominent in contemporary geomorphological features. For instance, the Kelbajar and the Gazakh graben-synclinal troughs, the relatively lowered Bazarchay, Garasu, Terterchay, Shamkir-Tovuz, Sharur, and elevated Ishykhly-Kirss, Ganjachay-Shamkir, Ordubad, etc., morphostructural segments.

The Talysh Mts., located at the southeast of Azerbaijan, are especially prominent among the Azerbaijan morphostructures by their morphotectonic characteristics. The reason is that this mountain system was generated in another morphotectonic zone, where the Iranian and Transcaucasian microplates of the Earth crust continental type collide with the South Caspian rift (graben) of the Earth crust of suboceanic type. Here, during the periods of collision and subduction enhancement (Upper Cretaceous, Paleogene—Eocene), intensive linear and areal volcanism, compression, and folding were observed. The Talysh horst-anticlinal suture ridge, limited by the pre-Talysh abyssal fracture, was a constituent part of the Talysh-Bogrovdagh-Elburs suture mountain system and generated at the Iranian microplate northern margin. This mountain system composed of volcanogenic and volcanogenic-sedimentary rock complexes

Table 2.1 Geomorphological zoning (after Budagov 1993a, b)

Province	Region	Subregion	District
1	2	3	4
The Crimean-Caucasian Mountain Land			
Piedmont depressions and uplands	**I** Samur-Devechi (Shabran)		1. Samur-Devechi lowland 2. Gusar sloping plain
The Greater Caucasus	**II** Eastern Greater Caucasus	II_1 Southern slope	3. Zagatala 4. Sheki 5. Lagich
		II_2 Southeastern slope	6. Bazardyuzyu 7. Shakhnabad-Khizi 8. Shakhdagh 9. Tengi-Beshbarmag 10. Dyubrar 11. Shamakhy 12. Gobustan 13. Absheron
South Caucasian depression	**III** Kur trough	III_1 Ganykh-Airichay valley	14. Ganykh 15. Ismailly
		III_2 Jeiranchel-Ajinour submountain region	16. Jeiranchel 17. Ajinour 18. Lengebiz-Alyat 19. Kharamy
		III_3 Kur-Araz trough	20. Shirvan plain 21. Southeastern Shirvan 22. Near-Kur plain 23. Mughan plain 24. Salyan plain
		III_4 Pre-Lesser Caucasus sloping plain	25. Ganja-Gazakh plain 26. Garabagh plain 27. Mil plain
Western Asian Highlands			
The Lesser Caucasus	**IV** Outer western ridges	IV_1 Northeastern slope	28. Agstafa 29. Shamkir 30. Dashkesan-Agjakend 31. Shakhdagh 32. Murovdagh
		IV_2 Eastern slope	33. Lower Araz
		IV_3 Southwestern slope	34. Kelbajar 35. Mykhteken 36. Kharmin
	V Inner Lesser Caucasus	V_1 Garabagh volcanic upland	37. Gyzylbogaz-Ishygly 38. Alagel-Chalbayir 39. East Goycha 40. Yazyn
		V_2 Near-Araz ridges	41. Daralagez 42. Zangazur 43. Bargushad
		V_3 Nakhchivan (Middle Araz) trough	44. Sharur 45. Negram 46. Ordubad
	VI Talysh Mts.		47. Lenkaran lowland 48. Burovar ridge 49. Yardymly 50. Peshtasar ridge 51. Zuvand 52. Talysh

Fig. 2.9 Geomorphological zoning of Azerbaijan territory

created elevated to 2000–2400 and 1000–1200 m, respectively, heights, the intensively compressed, dissected Peshtasar, Talysh, Burovar horst-anticlinal ridges and the Yardymly and Deman-Gosmalyan intramontane graben-synclinal kettles that separate them and are generated in relatively lowered areas.

The Kur intermontane graben-synclinal trough plays important role in the development of contemporary morphostructural carcass within Azerbaijan and in the creation of general morphological layout of contemporary terrain. In the Kur Depression northern part, in the collision zone of the Transcaucasian and Scythian plates, along the North Alazan (Ganykh) and Zangi-Geradil fractures, the Ganykh-Airichay and the Shamakhy-Gobustan morphostructures have been formed that are exposed to intensive subsidence and compression. Eastward of the Girdamanchay River valley, in West Caspian zone of tectonic faults, and southward of the Zagatala-Govdagh microstructure, the Shamakhy-Gobustan morphostructural step has developed that is relatively weakly compressed. It underwent subsidence to 1.5–2 km along the South Nialdagh and Zangi-Geradil abyssal fractures. Plateaus, ranges, hills, and kettles of nappe-thrust-folded genesis are prominent in the relatively poorly differentiated Shamakhy-Gobustan zone. Relatively southward of this zone the terrain is complicated by the Jeiranchyol-Ajinour-Alyat anticlinal low-mountain ridges and ranges that northward encircle the Transcaucasian microplate. Taking into account contemporary peculiarities of relief and foundation development and based on the electric resistivity log and other geophysical data analysis we can distinguish the Garabagh, Kyurdamir, near-Talysh, Lower Kur graben-synclinal, sharply lowered, and poorly dissected plains within the Kur trough.

Summarizing the above said, we note that we made an attempt to explain certain formation regularities in contemporary morphostructural configuration of Azerbaijan area from the viewpoint of "global plate tectonics" theory, on the basis of combined geophysical data analysis. However, the work was not intended for the clarification of all aspects of origin and specific development features of morphostructure in the region under study. Undoubtedly, in-depth study will further contribute to the development of these problematic issues of contemporary geomorphology.

2.4 Geomorphological Zoning

The proposed geomorphological zoning scheme (Budagov et al. 2011) is based on the principles of terrain morphostructural analysis and area dissection into variously ranked morphostructures—in view of their age, structural and developmental peculiarities, intensity, sign and manifestation manner of recent tectonic movements, depositional nature (for subsidence regions), erosional dissection, and the complex of genetically similar morphoscultpures complicating the morphostructural configuration (Table 2.1; Fig. 2.9).

References

Abasov, M. A. (1955). Geomorphology of the north-western part of Nakhchivan Republic. Transactions *of the Institute of Geography, Azerbaijan Academy of Sciences, 5* (in Russian).

Abasov, M. A. (1970). *Geomorphology of Nakhchivan Autonomic Republic*. Baku: Elm (in Russian).

Akhundov, C. A. (1978). *Drainage of sediments of mountainous rivers of Azerbaijan*. Baku: Elm. (in Russian).

Alizadeh, E. K. (1982). Morphostructural analysis of relief of the Astrakhan nappe series distribution. *Izvestiya of Azerbaijan Academy of Sciences, Series: Earth Sciences,* 3, 98–104 (in Russian).

Alizadeh, E. K. (1984). Morphostructural analysis of the southern slope of the south-eastern Caucasus with application of space images decoding. Ph.D. thesis, Institute of Geography, Baku (in Russian).

Alizadeh, E. K. (1987a). Structural-geomorphological analysis of space images of the south-eastern Caucasus. *Izvestiya of Azerbaijan Academy of Sciences, Series: Earth Sciences,* 5, 19–25 (in Russian).

Alizadeh, E. K. (1987b). Detecting a "skeleton" of morphotectonic blocks of the eastern part of the Lesser Caucasus and adjacent areas by space image analysis. *Geomorphology, 4*, 51–57. (in Russian).

Alizadeh, E.K., 1991. Identification of morphotectonic framework of the eastern part of the Lesser Caucasus and adjacent areas using space images. *Geomorphology*, No. 4, 51-57 (in Russian).

Alizadeh, E.K., 1998. *Morphostructural Structure of Mountainous Regions of Azerbaijan and Adjacent Areas*. Elm, Baku (In Russian).

Alizadeh, E.K., 1999. A retrospective analysis of forming and morphostructure development in the eastern part of the Lesser Caucasus. Proceed. of Baku University, No. 1, 202-213.

Alizadeh, E. K. (2007). Stable development of mountainous geosystems in conditions of strengthening morphodynamic tenseness (on example of Azerbaijan). *Herald of the Vladikavkaz Scientific. Center of the Russian Academy of Sciences, VII*(3), 19–24 (in Russian).

Alizadeh, E. K. (2011). Analysis of exodynamic stress in the southern slope of the Greater Caucasus associated with estimation of geomorphological risk. *Izvestiya of Azerbaijan Academy of Sciences,* 4, 62–69 (in Russian).

Alizadeh, E. K., & Tarikhazer, S. A. (2014). Dynamics of change in mountain ecogeosystems under the influence of natural-destructive phenomena (on example of the Greater Caucasus). In *Proceedings of "From the Caspian to Mediterranean: Environmental Change and Human Response during the Quaternary"* (pp. 11–13). Baku: Nafta-Press.

Alizadeh, E. K., Tarikhazer, S. A., & Gamidova, Z. A. (2014). Investigation of rock-slide danger within the Azerbaijanian part of the Greater Caucasus with the aim to revealing ecomorphological environments. *Geopolitics and Ecogeodynamics of Regions* (Simferopol), *10*(1), 266–273 (in Russian).

Antonov, B. A. (1972). Modern tectonics and evolution of morphostructures in Azerbaijan. In *Materials of International Geography of Congress in Moscow (Geomorphology and Paleogeography)*, Nedra, Moscow (in Russian).

Antonov, B. A. (1993). General characteristics of relief. *Relief of Azerbaijan* (pp. 108–111). Baku: Elm.

Budagov, B. A. (1959). On the modern tectonic movements of the south-eastern Caucasus. *Doklady Academy of Science in Azerbaijan Republic,* 3, 56–59 (in Russian).

Budagov, B. A. (1960). Vertical zonation of the relief forms of the Greater Caucasus within Azerbaijan Republic. *Transmision of GO*, 77–82 (in Russian).

Budagov, B. A. (1961). On the relation of modern tectonic movements with mudflow generation on the southern slope of the Greater Caucasus. *Doklady Academy of Science in Azerbaijan Republic, 17* (4), 21–27. (in Russian).

Budagov, B. A. (1962). On some problems of geomorphology of generations mudflows (on example of the southern slope of the Greater Caucasus). In *Transmission of V All-Union Conference on Studying Mudflows*, Baku, pp. 79–86 (in Russian).

Budagov, B. A. (1964). On the character of modern tectonic movements in the area of Shakhdagh massif in light of find of Upper Sarmatian fauna. *Doklady Academy of Science in Azerbaijan Republic, 155*(2), 7–13 (in Russian).

Budagov, B. A. (1965). Modern and ancient glaciations of the Azerbaijanian part of the greater Caucasus. *Academy of Science Azerbaijan*, Baku (in Russian).

Budagov, B. A. (1969). *Geomorphology of the southern slope of the Greater Caucasus*. Baku: Elm (in Russian).

Budagov, B. A. (1973). *Geomorphology and modern tectonics of the South-Eastern Caucasus*. Baku: Elm (in Russian).

Budagov, B. A. (1976). The main stages of the relief development in the Azerbaijanian part of the Greater Caucasus. In: *Problems of relief development and landscapes of Azerbaijan Republic. Transaction Institute for the Geography*, Elm, Baku, Vol. 6, pp. 3–11 (in Russian).

Budagov, B. A. (1993a). Level-glacial morpho-sculpture. In: *Relief of Azerbaijan* (pp. 20–22). Baku: Elm (in Russian).

Budagov, B. A. (1993b). Fluvial morpho-sculpture. In: *Relief of Azerbaijan* (pp. 28–35). Baku: Elm (in Russian).

Budagov, B. A., & Alizadeh, E. K. (1995). Complex methods of mapping of morphostructures the zones of interpolate contiguity of the Eastern Caucasus on the basis of interpretation of space photos materials. In *Proceedings of the Cartographic Conference of 10th General Assembly of the ISA*. Barcelona, 158–163.

Budagov, B. A., & Alizadeh, E. K. (1998). Peculiarities of forming and differentiation of Azerbaijan's morphostructures. *Geomorphology, 2*, 53–59 (in Russian).

Budagov, B. A., Khalilov, G. A., & Guluzadeh, V. A. (1993). Gravity morpho-sculpture. *Relief of Azerbaijan* (pp. 22–28). Baku: Elm.

Budagov, B. A., Mikailov, A. A., Aliyev, A. C., & Alizadeh, E. K. (1984). Morphostructural peculiarities of the area of Basgal nappe distribution. *Doklady Academy of Science of Azerbaijan Republic*, 1 (in Russian).

Budagov, B. A., Mikailov, A. A., & Alizadeh, E. K. (1988). Peculiarities of morphostructures of the eastern part of the Greater Caucasus by the use of space images analysis. *Geomorphology, 4*, 35–43. (in Russian).

Budagov, B. A., Shirinov, N. S., & Tarynverdiyev, H. K. (2011). *Paleogeomorphology of Azerbaijan*. Baku: Elm (in Russian).

Dumitrashko, N. V. (1962). Caucasus. In: *Mountainous countries of the European part of the USSR and Caucasus* (pp. 122–137). Moscow: Nauka, Moscow (in Russian).

Dumitrashko, N. V., Milanovsky, E. E., Antonov, B. A., Astakhov, N. E., Balyan, S. P., Liliyenberg, D. A., et al. (1968). Morphostructural peculiarities of the Caucasus. *Izvestiya of USSR Academy of Sciences, Series: Earth Sciences*, 6, 56–67 (in Russian).

Gadjiev, V. D. (1980). *Peculiarities of structure and evolution of terrain relief in areas of Cenozoic magmatism of Prearaz region (within the territory of Azerbaijan)*. Ph.D. thesis, Institute of Geography, Baku (in Russian).

Gadjiev, V. D. (1999). *Paleomorphology of Areas of Meso-Cenozoic Volcanism of Nakhchivan and Talysh*. Baku (in Russian).

Guluzadeh, V. A. (1982). *Morphostructures of north-eastern part of the lesser Caucasus, peculiarities of their structure and evolution (interfluve of Dzegamchay and Kurakchay)*. Ph.D. thesis, Institute of Geography, Baku (in Russian).

Kerimov, G. A. (1982). *Morphometric and morphostructural analysis of Gobustan's relief*. Ph.D. thesis, Institute of Geography, Baku.

Khalilov, G. A. (1975). On the linkage between the hydrographic net with disjunctive tectonics and its application for searching economic deposits (on example of interfluve of Touzchay and Chrami of the north-eastern slope of the Lesser Caucasus). *Doklady Academy of Science in Azerbaijan, 31*(3), 31–34 (in Russian).

Khalilov, G. A. (1980). *Morphostructural analysis of north-western part of the lesser Caucasus (within Azerbaijan)*. Ph.D. thesis, Baku (in Russian).

Khalilov, G. A. (1999). *Morphostructures of the eastern part of the Lesser Caucasus*. Baku (in Russian).

Kuliyev, R. Y. (1977). Experience in classification of morphostructures of orogenic regions (on example of the eastern part of the Lesser Caucasus). *Scientific Notes of the Azerbaijan State of University, Series: Geography and Geology*, 14–19 (in Russian).

Lilienberg, D. A. (1953). Levelling surfaces of the south-eastern Caucasus. In *Transactions of IV Conference on Problems on Geomorphology Transcaucasian*, No. 14, pp. 52–57 (in Russian).

Lilienberg, D. A. (1955). On the geomorphology of mud volcanoes of south-eastern Caucasus. In *Transactions of the Institute of Geography, USSR Academy of Sciences (materials on geomorphology and paleogeography of the USSR)* (No. 14, pp. 61–64). Moscow (in Russian).

Lilienberg, D. A. (1957). On the levelling surfaces of the south-eastern Caucasus. In *Transaction of IV Geomorphology Conference on Investment of the Caucasus and Transcaucasian*, Yerevan, pp. 11–14 (in Russian).

Lilienberg, D. A. (1958). Levelling surfaces of the south-eastern Caucasus. In Transactions *of the Institute of Geography, USSR Academy of Sciences (materials on geomorphology and paleogeography of the USSR)* (Vol. 74, pp. 36–42). Moscow (in Russian).

Lilienberg, D. A. (1959). Peculiarities of relief and modern tectonic movements of the Eastern Caucasus. In: *Geographical Reports* (pp. 37–42). Moscow: Nedra (in Russian).

Lilienberg, D. A. (1985). Modern geodynamics of Alpine orogenic belt of southern Europe. *Geomorphology, 4*, 4–9. (in Russian).

Mardanov, I. E. (1980). Main peculiarities of evolution of morpho-sculptures of the south-eastern Caucasus. *Izvestiya Academy of Sciences, Series: Earth*, 18–23 (in Russian).

Mardanov, I. E. (1981). Genetic classification of morphostructures of the Greater Caucasus. *Izvestiya Academy of Sciences, Series: Earth,* 2, 7–13 (in Russian).

Milanovsky, E. E. (1960). On the signatures of Upper Pliocene glaciations in the mountainous part of the Central Caucasus. *Doklady Academy of Science in USSR, 30*, 77–83. (in Russian).

Museibov, M. A. (1968). Regularities of Azerbaijan's morphostructures. *Scientific Notes of the Azerbaijan State University*, Series: *Geography* and *Geology*, Baku, 2, 9–14 (in Russian).

Museibov, M. A. (1973). Morphostructures of Azerbaijan. Scientific Notes of the Azerbaijan State University, Series: Geography and Geology, 4, 3–8 (in Russian).

Shirinov, N. S. (1965). *Geomorphology of the Absheron oil & gas region*. Academy of Sciences in Azerbaijan, Baku (in Russian).

Shirinov, N. S. (1973). *Geomorphological structure of Kur-Araz depression: Morpho-sculptures*. Baku: Elm (in Russian).

Shirinov, N. S. (1975). *Modern tectonics and development of relief of Kur-Araz depression*. Baku: Elm (in Russian).

Shirinov, N. S. (1979). Morphostructural analysis of Azerbaijan's relief. *Izvestiya of Azerbaijan Academy of Sciences, Series: Earth Sciences,* 6, 35–42 (in Russian).

Shirinov, N. S. (1980). Morphostructural analysis of relief of Azerbaijan. *Izvestiya of Azerbaijan Academy of Sciences, Series: Earth Sciences,* 3, 18–26 (in Russian).

Shirinov, N. S. (1993). Absheron peninsula. In: *Relief of Azerbaijan* (pp. 164–167). Baku: Elm (in Russian).

Shirinov, N. S., & Dotduev, S. I. (1976). Some problems of geomorphology of Sakhdagh massif and adjacent areas of the Greater Caucasus. *Scientific Notes of the Azerbaijan State University, Series: Geography and Geology,* 5, 17–20 (in Russian).

Vekilov, B. G. (1956). Quaternary deposits of Precaspian region of Azerbaijan. Transactions *of the Institute of Geography, Azerbaijan Academy of Sciences, 17* (in Russian).

3 Stratigraphy and Lithology

The geological structure of Azerbaijan includes complex sedimentary, metamorphic and magmatic formations representing wide stratigraphic diapason from Cambrian through recent eras. The majority of deposits are well-characterized faunistically. Simultaneously, the same age deposits undergo serious lithofacial changes in space based on the different circumstances of their formation (Khalilov 1965; Azizbekov et al. 1972; Agayev 1990; Aliyev et al. 2005b; Ismail-Zadeh et al. 2005; Alizadeh et al. 2007; Mamedov et al. 2008a; Babayev et al. 2015).

3.1 Precambrian—Cambrian

The ancient metamorphic complexes of Azerbaijan had previously been studied by Sh.A. Azizbekov, P.S. Bernstein, T. Ab. Hasanov, K.N. Paffenholtz, M.I. Rustamov, A.Sh. Shikhalibeyli, I.I. Sitkovskiy, G.M. Smirnov, V.V. Weber, and others. Within the country's boundaries, the Baikalian basement protrudes in the core of the Shamkir elevation in the Lok-Garabagh zone of the Lesser Caucasus (Azizbekov 1961; Gasanov 1985; Mamedov et al. 2008a). There, in the Asrikchay River valley, the metamorphic shale rocks outcrop with faunistic characteristics that are similar to Precambrian and Cambrian deposits of the Dzirul and Lok massifs in neighboring Georgia. Higher in the succession, the crystalline shale rocks are replaced by phyllites containing thin interbeds of quartzitic sandstone and lenticular interbeds of marble. The 170-m-thick series is transgressively overlapped by Lower Jurassic (Hettangian-Sinemurian) basal conglomerates, sandstones, and tuffites.

Within the boundaries of the Zangazur elevation of the Araz megazone, the "Pazmara" metamorphic complex and its agmatites are distinguished—a block of xenolith situated in the body of the Megri-Ordubad batholith, represented by gneiss-slates, amphibolites, metabasites, and metakeratophyres (Azizbekov 1961; Mamedov et al. 2008a).

Finally, the rubble and the blocks of Precambrian mylonites, gneiss-slates and binary gneiss are detected in Upper Jurassic tectonic breccias and conglomerates from the southern slope of the Greater Caucasus (Shikhalibeyli 1956).

The ancient metamorphic rocks of Azerbaijan are well-compared to Precambrian complexes of the adjoining regions of the Southern Caucasus and Iran.

3.2 Paleozoic

Besides a couple of isolated outcrops of Middle and Upper Carboniferous volcanic bodies detected in the Shamkir elevation (Lok-Garabagh zone of the Lesser Caucasus), the Paleozoic complex is revealed only within the Sharur-Julfa elevation of the Araz megazone (Nakhchivan Republic). Moreover, deep drilling operations implemented in north-eastern Azerbaijan (Aghzybirchala well) discovered the Upper Paleozoic plagioporphyres at a depth of 4700 m inside the pericratonic settling line of Scythian-Turanian epi-Hercynian platform (Azizbekov et al. 1972; Aliyev et al. 2005b; Mamedov et al. 2008a).

3.2.1 Devonian

The Devonian system is represented by the entire spectrum of its divisions and stages. It forms a terrigenous-carbonate formation occurring within the Sharur segment of the Sharur-Julfa elevation (Azizbekov 1961; Mamedov et al. 2008a). This complex does not come to a surface here, but it does protrude on the right river-bank of Araz, transgressively overlying the Ordovician-Silurian volcano-sedimentary formations.

In general, the Devonian section is represented by an alternation of various kinds of limestone, quartz sandstones, quartz rocks, marl and clay slates, and argillites.

Devonian terrigenous-carbonate deposits have been previously studied and described by G. Abich, H.I. Aliyev, V. A. Aristov, V. Arkhipov, G. Artgaber, Sh.A. Azizbekov, A. E. Baghyrov, N.A. Balashov, P. Bonnet, V.P. Felix, F.

A.A. Alizadeh et al., *Geosciences of Azerbaijan*, Regional Geology Reviews,
DOI: 10.1007/978-3-319-27395-2_3

Frekh, I.A. Grechishnikova, T.G. Gadjiev, A.A. Kadirov, P. Khalatov, G.P. Kornev, Y.S. Levitskiy, E.Kh. Madatov, A.B. Mammadov, K.N. Paffenholtz, N.V. Pashaly, M.A. Rjonsnitskaya, K.O. Rostovtsev, M.I. Rustamov, M.A. Sadykhov, G. Tsulukidze, N.N. Yakovlev, V.F. Zakharov, M.B. Zeynalov, and others.

3.2.1.1 Lower Devonian

In the northwest of the Araz zone, Lower Devonian formations are distinguished in the Dahna-Validagh area, represented by the Lower Lockhovian, Middle Praghian and Upper Emsian stages with carbonate-terrigenous facies.

Formations of the Lockhovian stage do not outcrop. However, in the 1950s, an approximately 700-m-thick series of Lockhovian graphitized limestone, quartz sandstones, quartz siltstones, and argillites had been discovered in the bottom section of deep well drilled on the left river-bank of Araz (at the southeastern foot of Mt. Dahna). The middle part of the well's geological section contained 300-m-thick layers of graphitized limestone and quartz sandstones of the Praghian stage. Finally, an approximately 400-m-thick alternation of Emsian limestone, quartz rocks, and argillites can be detected in the upper section.

Built by a 153-m-thick alternation of diverse limestone, siltstones, quartz rocks, and argillites, the upper part of the Lower Devonian section outcrops in the crest of the Validagh anticline, situated in the southern foothill of the same-named mountain. This section with a buried bottom was characterized as the Upper Emsian *Sarajly suite* (Mamedov 1979, 1980a), distinguished within the volume of *Arduspirifer extensus* and *Uncinulus keltibericus* zones. The suite is divided into the following two subcomponents:

(1) The lower subsuite is built by a 25-m-thick layer of grey and dark-grey sandy limestone in the bottom section and by a 40-m-thick layer of argillites with siltstone interbeds in the top. Faunistically, the *Arduspirifer extensus* is distinguished in the volume of brachiopod zone. Besides brachiopods, it also contains remnants of various muscoids, crinoids, rugose, conodonts, and sea ore.

(2) The bottom section of the upper subsuite is represented by an 11-m-thick layer of dark-grey limestone, while higher successions are represented by 77-m-thick black and violaceous argillites, siltstones, clayey limestone and light-grey quartz rocks. Faunistically, the *Uncinulus keltibericus* is distinguished in the volume of brachiopod zone. Besides brachiopods, it also contains remnants of various muscoids, crinoids, rugose, conodonts, tabulate, stromatoporata, dread nautilus, ostracods, pelecypoda, and foraminifers.

3.2.1.2 Middle Devonian

Represented by Lower Eifelian and Upper Givetian stages (terrigenous-carbonate facies), the Middle Devonian formations are detected in the northwest of the Araz zone (river basins of Chapanchay, Jahannamdarasi, Baghyrsagdara and Arpachay as well as Dahna-Validagh area).

The Eifelian stage is represented by limestone, limey-quartz sandstones, dolomites, and clayey and marly schists. It is divided into lower (Sharur suite), middle (Validagh and Daznik suites), and upper (Gurdgapysy suite) substages distinguished in the volumes of *Zdimir pseudobaschkiricus velidagensis—Megastrophia yralensis, Arduspirifer intermedius, Spinospirifer araxicus—Emanuella takwanensis, Spinatrypa aspera—Undispirifer rzonsnitzkajae, Gruenewaldtria latilinguis* and *Devonogyra begljari-Subrensselandia amygdalina* brachiopod zones.

The ***Sharur suite***, named by Mamedov (1980b), is found in the Dahna-Validagh area. It is built by a 197.5-m-thick alternation of grey and dark grey sandy-argillaceous and organic limestone with siltstones, quartzites, schists, and argillites. Faunistically, the suite is distinguished by the volumes of a *Zdimir pseudobaschkiricus velidagensis-Megastrophia yralensis* brachiopod zone. Besides brachiopods, it also contains remnants of various muscoids, crinoids, rugose, conodonts, tabulata, stromatoporata and pelecypoda.

The ***Validagh suite***, named by Mamedov (1980b), is found on the left bank of the Araz River (Dahna-Validagh and Garaburun areas) and the left shore of the Arpachay reservoir (Dolanansu area). From 60 to 210 m thick, the suite is composed of dark grey massif organic limestone with interlayers of quartzite, sandstone, dolomite and siltstone. Faunistically, it is distinguished by the volume of *Arduspirifer intermedius* brachiopod zone. Besides brachiopods, this section contains remnants of various muscoids, crinoids, rugose, conodonts, tamnoporides, tentaculites, conicons, trilobites, tabulata, stromatoporata and pelecypoda.

The ***Danzik suite***, named by Rzhonsnitskaya (1948) is a stratotype situated on the left shore of the Arpachay reservoir in the entry point of the Payadara Brook. The suite's openings are also detected in the Dahna-Validagh, Garaburun, Dashburun and Kalafa Agil areas. The complex is built by dark-grey and black limestone alternated with light-grey sandstones, siltstones, argillites and quartz rocks. Depending on dislocation, the suite's total thickness varies between 50 and 140 m. Faunistically, it is distinguished in the volumes of *Spinospirifer araxicus-Emanuella takwanensis* brachiopod zone. Besides brachiopods, the suite's section contains remnants of various crinoids, rugose, conodonts, tamnoporides, tentaculites, tabulata, stromatoporata, and pelecypoda, among others.

The ***Gurdgapysy suite***, named by Rjonsnitskaya (1948) (see also Mamedov 1962) is a stratotype situated in the Dahna-Validagh area (Gurd Gapysy pass). Separate outcrops have also been detected in the Garaburun and Dashburun areas as well as around Arpachay reservoir (Dolanansu, Payadara and Gumushlu areas). The suite is divided into three subsuites: (1) a lower subsuite mainly built by limestone; (2) a middle subsuite represented by the alternation of limestone, argillites, and siltstones; and (3) an upper subsuite distinguished by the alternation of limestone, siltstones and sandstones. The suite's thickness varies between 85 and 350 m. Faunistically, it is distinguished by the volumes of *Spinatrupa (Isospinatrupa) aspera aspera-Undispirifer izonsnitzkajae, Gruenewaldtia latilinguis* and *Devonogyra begliari-Subresselandia amygdalina* brachiopod zone. Besides brachiopods, the suite's section contains remnants of various muscoids, crinoids, rugose, conodonts, tentaculites, trilobites, tabulata, stromatoporata, dread nautilus, ostracods, gastropods and pelecypoda.

The Givetian stage manifests itself as carbonate facies that conformably (and sometimes transgressively) overlap the Eifelian stage's formations. Represented by heavy-bedded coral limestone alternating with sandy limestone and calcareous sandstones, the stage is divided into lower (Sadarak suite) and upper (Arpachay suite) substages distinguished by the volumes of *Stringocephalus burtini, Crurithyris inflatus, Spinocyrtia transcaucasica* and *Undispirifer undiferus—Emanuella pseudovolhynica* brachiopod zones.

The ***Sadarak suite***, named by Rjonsnitskaya (1948) (see also Mamedov 1962), is a stratotype situated in the vicinities of the Sadarak settlement (southern slope of the Ujubiz Mt.), with various outcrops also detected in the Dahna-Validagh area, the Kalafa Agil area in the upper reaches of the Baghirsagdara River, as well as around the Arpachay reservoir (Dolanansu, Payadara, Devaolan, Gumushludara and Yayjydarasi areas). Represented by dark grey massive and heavy-bedded limestone, the suite is 110–335 m thick depending on the sections. Faunistically, it presented by the volumes of *Stringocepnalus burtini* and *Crutithyris inflatus* brachiopod zones. Besides brachiopods, the suite's section contains remnants of various muscoids, crinoids, rugose, tentaculites, tabulata, stromatoporata, gastropods, ostracods, foraminifers and pelecypoda.

The ***Arpachay suite***, named by Mamedov (1962), is a stratotype situated in the vicinities of Gumushlu village, with various outcrops also detected in the Garaburun-Ujubiz and Dahna-Validagh areas, Kalafa Agil area in the upper reaches of Baghyrsagdara River, as well as around Arpachay reservoir (Dolanansu, Payadara, Devaolan, Gumushludara and Yayjydarasi areas). The suite is divided into two components: (1) a lower subsuite mainly built by massive limestone and (2) an upper subsuite represented by an alternation of limestone, argillites, sandstones and siltstones. The suite's total thickness varies between 117 and 200 m depending on the outcrop. Known for rich brachiopod and coral fauna, the suite is faunistically distinguished by the volumes of *Spinocyrtia transcaucasia* and *Undispirifer undiferus-Emanuella pseudovolhynica* brachiopod zones. Besides brachiopods, the suite's section contains remnants of various muscoids, crinoids, rugose, conodonts, tabulata, stromatoporata, ostracods and pelecypoda.

3.2.1.3 Upper Devonian

Represented by Lower Frasnian and Famennian stages with terrigenous-carbonate facies, the Upper Devonian formations are revealed in the northwestern part of the Araz zone, in the river basins of Chapanchay, Jahannamdarasi and Baghyrsagdara.

The **Frasnian stage** is represented by clayey schists, fakes, crystalline (sometimes organogenic) limestone, and rarely quartz rocks. It is divided into lower (Kalafaagil suite), middle (Baghyrsagdara suite) and upper (Yayjy suite) substages distinguished in the discovered paleontological material: *Cyrtospirifer disjunctus elegans* and *Cyphoterorhynchus koraghensis—Cyrtospirifer subarchiaci* brachiopod zones.

The ***Kalafaagil suite***, named by Mamedov (1981) (see also Mamedov et al. 2008a), is a stratotype situated on the right bank of Baghyrsagdara River's upper reaches near the ruins of Kalafa Agil village. The suite's outcrops are also detected in the vicinities of the Sadarak settlement (Ujubiz Mt.), river basins of Jahannamdarasi and Baghyrsagdara, as well as around (Dolanansu, Payadara, Ardyjlydara, Devaolan, Gumushludara and Yayjydarasi areas) and to the south (Gorangalasu area, etc.) of the Arpachay reservoir. Built by an alternation of yellowish-grey limestone, white clayey schists, sandstones, siltstones, and quartz rocks, the suite is 20–220 m thick. Faunistically, it is distinguished in the volumes of the *Cyrtospirifer disjunctus elegans* brachiopod zone. Besides brachiopods, the suite's section contains remnants of various crinoids, rugose, conodonts, and tabulata.

The ***Baghyrsagdara suite***, named by Mamedov (1981) (see also Mamedov et al. 2008a), is a stratotype situated in the upper reaches of Baghyrsagdara River. Various outcrops are also detected around the Sadarak settlement (Ujubiz Mt.), in the river basins of Jahannamdarasi and Baghyrsagdara, as well as around (Dolanansu, Payadara, Ardyjlydara, Deveolen, Gumushludara and Yayjydarasi areas) and to the south (Gorangalasy area, etc.) of the Arpachay reservoir.

Built by dark grey limestone, black clayey schists, and rarely yellowish-grey sandstone and quartz rocks, the suite's thickness varies between 190 and 205 m depending on the outcrops. Faunistically, it is distinguished in the volumes of the *Cyphoterorhynchus koraghensis-Cyrtospirifer subarchiaci* brachiopod zone. Besides brachiopods, the suite's section contains remnants of various rugose, conodonts, tabulata, stromatoporata, tentaculites, and sea ore.

The ***Yayjy suite***, named by Rzhonsnitskaya and Mamedov (1988), has a boundary established according to the disappearance of the *Cyphoterorhynchus koraghensis-Cyrtospirifer subarchiaci* zone's fauna-containing limestone from the section of Frasnian stage. The suite's stratotype is situated in sources of the Yayjydarasi River, with various outcrops detected also in the Jahannamdarasi and Baghyrsagdara river basins as well as around (Dolanansu, Payadara, Ardyjlydara, Devaolan, Gumushludara and Yayjydarasi areas) and to the south (Gorangalasy area, etc.) of the Arpachay reservoir. Represented by and alternation of dark-grey and grey argillites, quartz rocks, limestone, sandstones, and siltstones, the suite's maximal thickness reaches 154 m.

The **Famennian stage** is represented by clayey schists, sandstones, limestone and quartz rocks and divided into lower (Gabagdagh suite), middle (Payadara suite) and upper (Munhbalaoghlu suite) substages distinguished in the volume of *Mesoplica meisteri, Cyrtiopsis orbelianus—Cyrtiopsis davidsoni famennianus, Dmitria semioni, Cyrtospirifer pamiricus-Enchondrospirifer ghorensis, Paurogastroderhinchus nalivkini* and *Spinocarinifera niger-Hamilingella goergesi—Sphenospira julii* brachiopod zones.

The ***Gabagdagh suite***, named by Mamedov (1981) (see also Mamedov et al. 2008a), has a lower boundary established by appearance of *Cyrtospirifer asiaticus* Brice and its' accompanying fauna in the relevant section. The suite's stratotype is situated on the slopes of Gabagdagh Mt., with various outcrops detected also in the Jahannamdarasi and Baghyrsagdara river basins as well as around (Dolanansu, Payadara, Ardyjlydara, Devaolan, Gumushludara and Yayjydarasi areas) and to the south (Gorangalasy area, etc.) of the Arpachay reservoir.

The complex is divided into two components: (1) a lower subsuite mainly built by quartz rocks, clayey schists, and partly by limestone and siltstones; and (2) an upper subsuite composed of limestone, clayey schists and quartz rocks. In various outcrops, the suite's total thickness varies between 140 and 410 m. Faunistically, it is distinguished in the volumes of *Mesoplica meisteri* and *Cyrtiopsis orbelianus—Cyrtiopsis davidsoni famenniana* brachiopod zones. Besides brachiopods, the suite's section contains remnants of various foraminifers, conodonts, pelecypoda, and sea ore.

The ***Payadara suite***, named by Mamedov (1981) (see also Mamedov et al. 2008a), is a stratotype situated in the upper reaches of the Payadara Brook, with various outcrops detected also in the Jahannamdarasi and Baghyrsagdara river basins; around (Dolanansu, Ardyjlydara, Devaolan, Gumushludara and Yayjydarasi areas), to the south (Gorangalasy area, etc.) and to the west (Gabagdagh area) of the Arpachay reservoir; and in several other places. The suite is divided into two components: (1) a lower subsuite built predominantly by limestone, quartz rocks, argillites and sandstones; and (2) an upper subsuite composed of limestone, argillites, siltstones, sandstones and quartz rocks. In various outcrops, the suite's total thickness varies between 130 and 170 m. Faunistically, it is distinguished in the volume of *Dmitria seminoi* and *Cyrtospirifer pamiricus-Enchondrospirifer ghorensis* brachiopod zones. Besides brachiopods, the suite's section contains remnants of various corals, conodonts, foraminifers and ostracods.

The ***Munhbalaoghlu suite***, named by Mamedov (1981) (see also Mamedov et al. 2008a), is a stratotype situated on the eastern and southeastern slopes of the Munhbalaoghlu Mt. with parastratotype detected in the vicinities of Ashaghy Yayji village. The suite's outcrops are also detected in the river basins of Jahannamdarasi and Baghyrsagdara; around (Dolanansu, Ardyjlydara, Devaolan, Gumushludara and Yayjydarasi areas), to the south (Gorangalasy area, etc.) and to the west (Gabagdagh area) of the Arpachay reservoir; and in several other places. The complex is divided into two components: (1) a lower subsuite built predominantly by clayey schists and siltstones; and (2) an upper subsuite represented by limestone, clayey schists, sandstones and quartz rocks. The suite's total thickness varies between 80 and 220 m. Faunistically, it is distinguished in *Paurogastroderhynchus nalivkini* and *Shinocarinifera niger-Hamlingella goergesi-Sphenospira julii* brachiopod zones. Besides brachiopods, the suite's section contains remnants of various muscoids, crinoids, rugose, conodonts, trilobites, ostracods and clams.

3.3 Carboniferous

In Azerbaijan, Carboniferous deposits have developed in the Araz zone and on the northern slope of the Lesser Caucasus in the upper reaches of the Asrikchay and Zayamchay rivers.

In the Araz zone, such deposits are represented by Lower Carboniferous marine carbonate facies revealed within the Sharur segment of the Sharur-Julfa elevation (basins of the Arpachay and Baghyrsagdara rivers, Bozdagh area). Relevant outcrops have been studied and described by G. Abich, H.I. Aliyev, Sh.A. Azizbekov, P.L. Bezrukov, P. Bonnet, O. L. Eynor, V.T. Felix, F. Frekh, I.A. Grechishnikova, E.S. Levitskiy, K.I. Lisitsin, A.B. Mammadov, V.I. Meller, A.D. Miklukho-Maklay, K.N. Paffenholtz, A.M. Sadykhov, A.A. Stoyanov, O.I. Tumanskiy, N.N. Yakovlev, etc.

Within the boundaries of the Lesser Caucasus, the Carboniferous system is represented by its middle and upper sections, which outcrop locally within the Lok-Garabagh zone of Shamkir elevation in the Zayamchay-Asrikchay interfluve. Outcrops have been studied by A.A. Bayramov, T.G. Gadjiev, H.M. Hasanov, T.Ab. Gasanov, and other researchers.

3.3.1 Lower Carboniferous

Revealed in the northwestern Araz zone (river basins of Arpachay and Baghyrsagdara as well as in the Mt. Bozdagh area), the Lower Carboniferous deposits are represented by Lower Tournaisian, Middle Visean and Upper Serpukhovian stages of terrigenous-carbonate facies (Azizbekov 1961; Azizbekov et al. 1972; Alizadeh et al. 2007; Babayev et al. 2015).

The **Tournasian stage** is built by an alternation of limestone, quartz rocks and clayey schists, which conformably overlies the Munhbalaoghlu suite's Upper Devonian (Famennian) formations. The stage is represented at the Gorangalasy suite and distinguished by the volume of *Rhipidomella micheilini-Unispirifer ussiensis* and *Marginatia deruptoides-Marginatia burlingtonensis-Cleiothyridina obmaxima* zones.

The ***Gorangalasy suite***, named by Mamedov (1980c), is a stratotype situated on the left river-bank of Arpachay at the northern slope of the Gorangalasy Mt. Its outcrops are also detected in the Munhbalaoghlu, Gabagdagh, Shahbulagh and Gunnut areas, as well as in the vicinities of Karki village. The suite is established by an alternation of coral-containing limestone with segmental interlayers of dark-grey clayey schists, thin calcareous sandstones and quartz rocks. The complex is divided into two components:

(1) A 50- to 90-m-thick lower subsuite represented mainly by sandy limestone and more rarely by clayey and sandy schists and quartz rocks, which is faunistically distinguished by *Rhipidomella micheilini-Unispirifer ussiensis* brachiopod zone. Besides brachiopods, its section contains remnants of various rugose, pelecypoda, conodonts and ostracods.
(2) A 20- to 75-m-thick upper subsuite built by an alternation of quartz rocks, sandstones and sandy limestone with thin layers of dark-grey clayey schists, which is faunistically distinguished by *Marginatia deruptoides-Marginatia burlingtonensis-Cleiothyridina obmaxima* brachiopod zone. Besides brachiopods, this section contains remnants of various rugose, pelecypoda, crinoids, and foraminifers.

The **Visean stage** is built of sandstones and limestone and represented by the Diza suite, distinguished by *Spirifer striatus-Composita verkhotomica* and *Gigantoproductus giganteus* zones.

The ***Diza suite***, named by Mamedov (1981), is a stratotype situated on the left bank of the Arpachay River on the northern slope of the Gorangalasi Mt. The outcrops are also detected in the Dahna-Validagh area at the slopes of the Bozdagh (Bozaghyl) elevation. The suite is divided into two components:

(1) A 15- to 30-m-thick lower subsuite represented by dark-grey, black, partly bituminized, sandy, heavy-bedded limestone, which is faunistically distinguished by the volume of *Spirifer striatus-Composita verkhotomica* brachiopod zone. Besides brachiopods, its section contains remnants of various rugose, pelecypoda, crinoids, gastropods and foraminifers.
(2) A 25- to 105-m-thick upper subsuite built by an alternation of yellowish-grey and dark-grey limestone with sandstones, which is faunistically distinguished by a *Gigantoproductus giganteus* brachiopod zone. Besides brachiopods, its section contains remnants of various rugose and gastropods.

The **Serpukhov stage**, represented by Bozaghyl suite (Alizadeh et al. 2007), is built by massive limestone with rich brachiopod and rugose fauna and detected only within the slope of the Bozdagh (Bozaghyl) elevation in the Dahna-Validagh area. The stage is characterized by the following fauna complexes: *Cliothyridina aff. obmaxima* (Mc.Chesny), *Cliothyridina* sp., *Stafella sphaerica* (Abich.), *Echinaria* sp., *Dibunopyllum aff. Turbinatum* M'Coy. Depending on the section, the stage's thickness varies between 50 and 65 m.

3.3.2 Middle-Upper Carboniferous

Paleontologically characterized Middle and Upper Carboniferous deposits are revealed as an uplifted tectonic block situated in the northern slope of the Lesser Caucasus within the body of the Shamkir high of the Lok-Garabagh zone. Protruding in the upper reaches of Zayamchay and extending for 7–8 km further to the northwest (towards the Asrikchay River), the 20–25 km^2 block manifests itself as a tectonic wedge (its northern and southern sides are limited by transtensional faults). Carboniferous deposits of this area are known as Lazylar suite described in Gasanov (1985). This suite is represented by a 170-m-thick tuffaceous-sedimentary complex divided into two subsuites according to its lithological peculiarities.

The lower volcanomictous subsuite manifests itself as a homogenous pack of the interstratified effusive with the coherent tuffs of average-to-main composition. Also detected here were small portions of the thin interlayers and lenses of tuffaceous siltstones and clayey schists. The latter formations contain many Middle Carboniferous molds of *Annularia sphenophylloides* (Zenk), *A. microphylla* Souv. and *Calomites cf. multiramus* Woiss. Represented by a homogenous dragfolded tuff of average composition, the upper subsuite had presumably formed during the Upper Carboniferous period.

3.4 Permian

The Permian system within the Araz zone is found in the geological section of both northwestern Sharur (Araz River basin) and southeastern Julfa (Julfa Canyon) segments of the Sharur-Julfa high. The system is represented by an upper division mainly built by coral-foraminifer, brachiopod-coral, and bituminized limestone. The Permian outcrops have been studied and characterized by G. Abich, H.I. Aliyev, Q. Artgaber, Sh.A. Azizbekov, P. Bonnet, I.O. Chediya, G.A. Dutkevich, O. Eynor, F. Frex, T.A. Grunt, Y.A. Khalilov, G. S. Kropachova, G.V. Kotlyar, E.Y. Leven, A.B. Mammadov, A.D. Miklukho-Maklay, Y. Moysisovich, F.A. Mustafayev, K.N. Paffenholtz, G.P. Pronina, K.O. Rostovtsev, M.A. Rjonsnitskaya, V.Y. Rujentsov, A.M. Sadikhov, T.G. Sarychev, H.G. Shenk, A.A. Shevyrev, A.A. Stoyanov, O.G. Tumanskaya, N.N. Yakovlev and Y.D. Zakharov.

3.4.1 Lower Permian

Lower Permian deposits do not outcrop in the territory of the Nakhchivan Autonomous Republic (AR). However, the Upper Bolorian regiostage is represented by the Develi suite's limestone outcropping in the neighboring areas of Armenia. It is presumed that the Lower Permian formations may be buried somewhere in the Nakhchivan AR's geological section.

3.4.2 Upper Permian

The Kubergandi, Murgab, Midia, Julfa and Darasham regiostages of the Upper Permian formations are represented in the geological section of the Araz megazone. These formations are built mainly by bituminized foraminifer-algal limestone as well as an alternation of limestone and argillites in the higher successions. Basal conglomerates of the material complex transgressively overlie deeply eroded Upper Devonian and Lower Carboniferous rocks and get overlapped by Lower Triassic and sometimes Upper Cretaceous and Middle Eocene formations (Paffengoltz 1940, 1959; Azizbekov 1961; Permian System 1966; Azizbekov et al. 1972; Leven 1975; Alizadeh et al. 2007).

The **Kubergandi regiostage** corresponds to the lower division of the Upper Permian deposits. Being built by an alternation of foraminifer-algal limestone with their brecciate types and sometimes dolomitic interlayers, the regiostage is represented in the lower segment of the Gyzylgaya suite, distinguished by the volume of the *Armenina-Misellina (Misellina) ovalis* zone.

The **Murgab regiostage** is the second-from-the-bottom Upper Permian regiostage. It is constituted by an alternation of siliceous-bitumen, sometimes schist-like, algal-detritic limestone with heavy-bedded sandy types. The regiostage is represented by the upper segment of the Gyzylgaya suite, distinguished by the volume of *Neoschwagerina simplex* zone, as well as by the Shahbulag suite distinguished by the volume of the *Neoschwagerina craticulifera* zone.

The ***Gyzylgaya suite***, named by Mamedov (1981) (see also Mamedov et al. 2008a), is a stratotype situated in the Gyzylgaya Mt. (Sharur segment of the Sharur-Julfa elevation), with various outcrops detected also in the Dahna-Validagh (slopes of Bozdagh (Bozaghyl) high) and Baghyrsagdara areas, basin of the Arpachay River (Ardyj, Zarlidara, Danzik, Mehridarasi, Dolanansu, Gumushludara, Devaolan, Munhbalaoghlu, Gorangalasy, Yayjydarasi, Khyrmanjyg and Baysal areas), and several other places. The suite is built by dark-grey, medium- and heavy-bedded foraminifer-algal limestone (with their brecciate types) and sometimes dolomitic interlayers. The suite overlaps the Upper Devonian and the Lower Carboniferous formations with basal conglomerates and bauxitic limestone retained in its basement. Depending on the section, the suite's thickness varies between 80 and 250 m. Faunistically, it is distinguished by *Armenina-Misellina* (Misellina) *ovalis* and *Neoschwagerina simplex* fusulinid zones. Besides fusulinids, the suite's section also contains remnants of various tetracorals, conodonts, and cancellinas.

The ***Shahbulag suite***, named by Mamedov (1981) (see also Mamedov et al. 2008a), is a stratotype situated in the vicinities of the Shahbulag village (Sharur segment of the Sharur-Julfa high), with various outcrops detected also in the Baghyrsagdara and Kechaltapa areas, Arpachay river basin (Ardyj, Mehridarasi, Dolanansu, Davaolan, Gorangalasy, Yayjydarasi, Gabagyal and Baysal areas) and several other places. From 50 to 200 m thickness (depending on the outcrop), the suite is built by an alternation of dark-grey and grey, thin-, medium- and heavy-bedded siliceous-bitumen, sometimes schist-like algal-detritic limestone, with heavy-bedded sandy types. Faunistically, the suite is distinguished by the *Neoschwagerina craticulifera* zone or *Eopolydiexodina persica* and *Pseudofuzulina padangensis-Chusenella shengi* fusulinid zones. Besides fusulinids, this section also contains remnants of various tetracorals and brachiopods.

The **Midia regiostage** is the third-from-the-bottom Upper Permian stage built by an alternation of clayey-siliceous, bituminized, and schist-like algal-detritic limestone. It is represented by Deveolen and Baysal suites distinguished by *Neoschwagerina margaritae* and *Yabeina-Lepidolina* zones.

The ***Davaolan suite,*** named by Mamedov (1981) (see also Mamedov et al. 2008a), is a stratotype situated in the source of the Davaolandarasi Brook (Sharur segment of the Sharur-Julfa high), with various outcrops detected in Ardyj, Zarlidara, Mehridarasi, Gumushludara, Davaolan, Munhbalaoghlu, Yayjydarasi, Khyrmanjyg, Gabagyal, Baysal and Kechaltapa and several other areas. The suite is built by an alternation of grey and light-grey, massive and heavy-bedded clayey-siliceous and algal-detritic limestone. Depending on the outcrop, the suite's thickness varies between 70 and 240 m. Faunistically, the complex is distinguished by *Yangchienia thompsoni* and *Chusenelia abichi* local fusulinid zones. Besides fusulinids, the suite's section also contains remnants of various foraminifers, tetracorals and brachiopods.

The ***Baysal suite***, named by Mamedov (1981) (see also Mamedov et al. 2008a), is a stratotype situated on the southwestern slope of the Baysal Mt. (Sharur segment of Sharur-Julfa elevation), with various outcrops also detected along the slopes of the Gabagyal, Khyrmanjyg and Ardyj Mts., in the vicinities of the Shahbulag village as well as in the Davaolandarasi ravine. Within the Julfa Canyon (Julfa segment of the Sharur-Julfa high), the suite constitutes a basement for the outcropping Permian formations. The suite is built by dark-grey and black heavily bituminized algal-foraminifer and detritic limestone. Depending on the outcrop, the suite's thickness varies between 60 and 220 m. Faunistically, the suite is distinguished by *Baisalina puichra, Hemigordiopsis orientalis, Midiella irregulariformis-Orthotetina azarjani* and *Pseudodunbarula arpaensis-Araxilevis intermedius* local foraminifer zones. Besides foraminifers, the suite's section also contains remnants of various fusulinids, tetracorals, brachiopods and conodonts.

The **Julfa regiostage** is the fourth-from-the-bottom Upper Permian stage built by an alternation of argillites and limestone. It is represented by Akhura and Julfa suites distinguished by *Araxoceras latum* and *Vedioceras ventroplanum* zones.

The ***Akhura suite*** (named by Leven (1975) stratotype is situated on the southern slope of Mt. Baysal, 1.5 km to the north of Akhura Village (Sharur segment of the Sharur-Julfa high), with various outcrops detected also along the slopes of the Gabagyal, Khyrmanjyg, Ardyj and Kechaldagh mountains. Built by an alternation of multicolored argillites and limestone, the 20- to 50-m-thick suite is conformably overlaid by the Lower Triassic Garabaghlar suite. Faunistically, it is distinguished by *Araxoceras latissimum* and *Vedioceras ventrosulcatum* localammonoidea zones in the bottom, and *Phisonites triangulus, İranites transcaucasus, Dzhulfites spinonsus, Shevyrevites shevyrevi, Paratirolites kitti* localammonoidea zones in the top. Besides ammonoideas, the suite's section also contains remnants of various foraminifers, brachiopods and conodonts.

The ***Julfa suite,*** named by A.D. Miklukho-Maklay (Alizadeh et al. 2007; Mamedov et al. 2008a), is a stratotype situated to the north of the Darasham II railway station (Julfa Canyon within the Julfa segment of the Sharur-Julfa high), with outcrops detected only around Darasham I and Darasham II stations. A 40- to 50-m-thick suite is built by an alternation of greenish and dark-grey schist-like argillites with light-grey and yellowish pelitomorphic limestone. The complex conformably overlies the Baysal suite's limestone and gets conformably overlapped by Darasham suite's formations. Faunistically, the suite is distinguished by the volume of *Pseudodunbarula arpaensis-Araxilevis intermedius* local foraminifer zone in the bottom, *Araxoceras latissimum* local ammonoidea zone in the middle, and *Vedioceras ventrosulcatum* local ammonoidea zone in the top. Besides foraminifers and ammonoideas, the suite's section also contains remnants of various tetracorals, brachiopods, and conodonts.

The **Darasham regiostage** is the topmost Upper Permian regiostage built by argillites with scattered interlayers of concretionary limestone. The regiostage is represented by a same-named suite distinguished by *Phisonites triangulus, İranites transcaucasus, Dzhulfites spinonsus, Shevyrevites shevyrevi, Paratirolites kitti,* and *Pleuronodoceras occidentale* zones.

The ***Darasham suite*** named by Kotlyar and Ganelin (1983) stratotype is situated to the north of the Darasham II railway station (Julfa Canyon within the Julfa segment of the Sharur-Julfa high), with outcrops detected only around the Darasham I and Darasham II stations. The 30- to 35-m thick suite is built by multicolored argillites with scattered interlayers of concretionary limestone. The complex conformably overlies the Julfa suite's argillites and gets conformably overlapped by the Garabaghlar suite's Lower Triassic formations. Faunistically, the suite is distinguished by *Phisonites triangulus, İranites transcaucasus, Dzhulfites spinonsus, Shevyrevites shevyrevi, Paratirolites kitti,* and *Pleuronodoceras occidentale* local ammonoidea zone. Besides ammonoideas, the suite's section also contains remnants of various foraminifers, brachiopods and conodonts.

3.5 Mesozoic

The Mesozoic section in Azerbaijan is composed of a set of sedimentary, volcanic-sedimentary and volcanogenic rocks. Within the Greater Caucasus, the marine Jurassic and

Cretaceous deposits play the main role in the Mesozoic section. The volcanogenic-sedimentary rocks occur within the Lesser Caucasus and Kur Depression (Azizbekov et al. 1972; Ismail-Zadeh et al. 2005).

3.5.1 Triassic

Within the territory of Azerbaijan, the Triassic system is presented in the Araz zone, where all three Triassic divisions are represented as carbonate formations within the geological sections of both the northwestern (Arpachay River basin) and southeastern (Julfa canyon) parts of the Sharur-Julfa zone (Azizbekov 1961; Azizbekov et al. 1972; Khalilov 1978; Aliyev et al. 2005b; Alizadeh et al. 2007; Babayev et al. 2015). System's outcrops have been studied and described by G.V. Abich, H.I. Aliyev, G.I. Allahverdiyev, G. Artgaber, Sh.A. Azizbekov, P. Bonnet, F. Frekh, T.A. Grunt, T.A. Hasanov, L.D. Kiparisova, G.P. Kornev, E. Moysisovich, F.A. Mustafayev, K.N. Paffenholtz, K.O. Rostovtsev, M.I. Rustamov, A.M. Sadykhov, T.G. Sarychev, A.A. Shevyrev, V.N. Shimanskiy, A.A. Stoyanov, etc.

3.5.1.1 Lower Triassic

Within the geological section of the Araz megazone, the Lower Triassic period is presented through its Induan and upper Olenekian stages, represented by mainly limestone containing the Garabaghlar suite.

Built predominantly by limestone, the Induan stage is detected in the lower part of the Garabaghlar suite. The Olenekian stage is composed by limestone and partly by dolomites, and located in the upper part of the Garabaghlar suite.

The ***Garabaghlar suite,*** named by K.O. Rostovtsev (Aliyev et al. 2005b; Alizadeh et al. 2007), is a stratotype situated 3 km to the northeast of the Garabaghlar village (southeastern margin of the Sharur segment of the Sharur-Julfa high). Its major outcrops are tracked in the Gabaglychay River basin, in vicinities of the Tananam, Garabaghlar, and Chalkhangala villages. Minor outcrops are detected on the western slope of Mt. Gabagyal, in the upper reaches of the Davaolandarasi Brook and Baghyrsagdara River, to the south of Mt. Ardyj, and in the Mehridarasi area. The suite's formations are also presented in the geological section of the northeastern limb of the Darasham anticline (Nehram valley of the south-eastern Julfa segment). The suite with nonevident disconformities overlies the Upper Permian formations of the Akhura and Darasham suites and gets conformably overlapped by the Ardaghy suite.

The Garabaghlar suite is divided into three components: (1) the lower subsuite confined to the Indian and Lower Olenekian substages and represented by mainly thin-bedded fucoid limestone, (2) the middle subsuite confined to a lower segment of the Upper Olenekian substage and built by heavy-bedded massive limestone, (3) the upper subsuite confined to the Upper Olenekian substage and mainly represented by dolomitic limestone. Depending on the outcrops, the suite's overall thickness varies between 260 and 420 m. This section stands out for its rich fauna of bivalves, ammonoideas and conodonts, including the following: *Glaraia clarai* Emm., *Cl. stachei* Bitt., *Cl. Aurita* Hauer, *Cl. of. exclamata* Spath., *Cl. ef. dalpiasi* Leon., *Cl. intermedia* Bitt., *Cl. orbicularıs* Pichth., *Cl. Tredintina* Bittn., *Proptychites discoides* Waag., *Ophiceras (Lythophioeras) facuntala* Diener, *O. medium* Griesd., *Eumorphotis venetiana* Hauer., *E. cf. inaequicostata* (Ben.)., *E. cf. multiphormis* (Bittn.), *Anadontophora fassaensis* (Wissm.), *A. canalensis* (Cats), *Entolium microtis* Bittn., *Neoschizodus laevigatus* (Ziet.), *Eumorphotis venetiana* Hauer., *E. reticulata* (Richt.), *E. spinicostata* Witt., *E. costata* Witt., *E. cf. hinnitidae* Bitth., *Pseudocorbula cf. nuculiformis* (Zenk.), *P. cf. plana* (Hoh.), *Neoschizodus cf. ovatus* (Goldf.), *Anadontophora aff. isocardiodes* French., *Anohignathodus parvus* Koz. et Pjat., *Lsaroioella Isaroicus* Kozur, etc.

3.5.1.2 Middle Triassic

Within the geological section of the Araz zone, the Lower Triassic is presented through its lower Anisian stage. This stage is represented by the Ardaghy suite containing brecciate limestone and siliceous dolomites.

The ***Ardaghy suite***, named by H.I. Aliyev (Alizadeh et al. 2007), is a stratotype situated on the northern slope of the Ardaghy high in the upper reaches of the Gabaglychay River (south-eastern margin of the Sharur segment of the Sharur-Julfa high). Various outcrops are also found in the south-western slopes of the Hinadali, Garovultapa and Bazakli mountains. The suite is built by an alternation of dark- and greenish-grey, heavy-bedded, brecciate dolomitic limestone with clayey sandstones and calcareous clays. Depending on the outcrops, the suite's thickness varies between 50 and 150 m. Its section contains fauna of bivalves, ammonoideas and conodonts, including the following: *Megalodon aff. rimosus* Mstr., *Myophoriopis cf. nuculaeformis* Zenk., *Eumorphotis inaeguicostata* Ben., *E. venetiana* (Rauor), *E. of. hinnitidea* Ditt., *Anadontophora aff. brevieformis* Spath., *A. canalensis* Cat., *Omphaloptychia* sp., *Pleuromya* sp., *Pleuromia* sp., etc.

3.5.1.3 Upper Triassic

Within the geological section of the Araz zone, the Lower Triassic is presented through its lower Carnian and (presumably) middle Norian stages, represented by the Tananam suite containing dolomites and limestone.

The **Carnian stage** is represented by the lower part of the Tananam suite built by dolomites and dolomitic limestone unconformably overlying the Lower Triassic Anisian limestone and dolomites. The **Norian stage*** is represented by the upper part of Tananam suite built by massive dolomites.

The ***Tananam suite*** (named by K.O. Rostovtsev) (Aliyev et al. 2005b; Alizadeh et al. 2007) corresponds to the Carnian and Norian* stages. Its stratotype is situated to the northeast of the Tananam village (the southeastern margin of the Sharur segment of the Sharur-Julfa high). The suite's major outcrops are widely developed in the Gabaglychay and Jahrichay river basins, as well as in the vicinities of the Chalkhangala, Garabaghlar, Tananam, Akhura and Havush villages. The outcrops are also detected on the peaks and slopes of the Ardaghy, Hinadali, Remlar, Garovultapa, Anabad-Gadik, Garagush and Bazakli mountains. The suite's formations are also presented in the geological section of the northeastern limb of the Darasham anticline (Nehram valley of the south-eastern Julfa segment). The suite transgressively overlies the Lower (Garabaghlar suite) and Upper (Ardaghy suite) Triassic formations, with its eroded roof unconformably overlapped by various layers of Jurassic and Cretaceous formations.

The suite is divided into two components: (1) the Carnian lower subsuite, which is composed of grey, light- and dark-grey, sometimes brownish-grey, heavy-bedded and massive, fractured, ferruginous, lightly siliceous hollow-bored dolomites, dolomitic limestone and thin-bedded fucoid limestone; and (2) the Norian* upper subsuite, which is composed of grey massive dolomites. Depending on the outcrops, the suite's overall thickness varies between 400 and 800 m. Its section includes remnants of rare *Undularia cf. scalata, Omphaloptycha* sp., *Pleuromya* sp. fauna.

3.5.2 Jurassic

Outcrops of Jurassic deposits occupy a vast territory in the priaxial parts and slopes of the Greater and Lesser Caucasus ridges. Moreover, these deposits are being exposed by wells in the Gusar-Devechi region and Kur Depression. Within the Azerbaijanian part of the Greater Caucasus, the first studies of the Jurassic complex were done by G.V. Abich at the end of the nineteenth century, and then continued in the twentieth century by V.B. Agayev, G.A. Akhmadov, A.A. Alizadeh, K.I. Bogdanovich, R.G. Babayev, D.V. Drobyshev, N.M. Gasymova, L.A. Grechishkin, V.A. Grossgeim, M.I. Gutman, G.M. Huseynov, B.M. Isayev, T.N. Kangarli, V.Y. Khain, G.Y. Krimholz, D.D. Mazanov, M.F. Mirchink, S.S. Mustafayev, K.N. Paffenholtz, I.F. Pustovalov, V. P. Rengarten, N.N. Rostovtsev, A.N. Shardanov, A.Sh. Shikhalibeyli, A.M. Shurygin, A.N. Solovkin, N.V. Tuschev, N.B. Vassoyevich, A.V. Vikhert, V.V. Weber, etc.

In the Lesser Caucasus, the Jurassic deposits have been studied by K.O. Abdulgasymzadeh, G.A. Aliyev, Ch.G. Aliyeva, A.Z. Abdullayev, R.N. Abdullayev, G.V. Abich, G.I. Allahverdiyev, R.B. Askerov, Sh.A. Azizbekov, R.G. Babayev, A.A. Bayramov, P.S. Bernstein, V.V. Bogachov, P. Bonnet, M. A. Kashkay, A.A. Gasymzadeh, T.A. Hasanov, T.Ab. Hasanov, B.P. Juze, Y.V. Karyakin, V.Y. Khain, A.G. Khalilov, L.K. Konyushevskiy, G.Y. Krimholz, L.N. Leontyev, F.A.Mustafayev, K.N. Paffenholtz, M.R. Rostovtsev, N.M. Salimkhanov, M.I.Rustamov, A.Sh. Shikhalibeyli, A.N. Sitkovskiy, V.I. Slavin, A.N. Solovkin, V.V. Tikhomirov, M.I. Varentsov, etc.

3.5.2.1 Lower Jurassic

Greater Caucasus

In the Greater Caucasus, Lower Jurassic complexes are composed of marine (often deep marine) sediments (up to 4000 m). These deposits occur in the northwest part of the Tufan zone (basins of the Mazymchay, Balakanchay, Katekhchay and Talachay rivers) and are represented by Pliensbachian and Toarcian stages (Khain 1947; Shikhalibeyli 1956; Agayev 1966, 1990, Azizbekov et al. 1972; Khalilov 1978; Aliyev et al. 2005b; Alizadeh et al. 2007).

Pliensbachian stage. The Pliensbachian deposits occur at the base of the Mazymchay, Belakanchay and Katekhchay rivers and correspond to the Belakan suite, represented by clay shales, aleurolites and sandstones (with a thickness up to 3600 m). These deposits are subdivided into lower and upper substages.

Toarcian stage. Deposits of the Toarcian stage occur in the Tufan zone. A typical section of Toarcian deposits can be observed in the upstreams of the Mazymchay River. It is represented by a series of clay shales, aleurolites and sandstones, separating there into the Gubakh suite (800–1600 m). These deposits are characterized by ammonites, *Dactylioceras* sp. and *Dumortieria bleicheri* Ben.

Lesser Caucasus

Within the Azerbaijani part of the Lesser Caucasus, the Jurassic complex is widely developed and outcropped on the northeastern slope, southeastern plunge, and partly within the southwestern slope (Araz zone) of a mountain-fold structure (Paffengoltz 1940, 1959; Azizbekov 1961; Gasanov 1961, 1967, 1973, 1985; Abdulkasumzadeh 1963; Shikhalibeyli 1964–1967; Abdulkasumzadeh et al. 1972; Azizbekov et al. 1972; Khalilov 1978; Abdulkasumzadeh 1988; Shikhalibeyli 1994; Aliyev et al. 2005b; Alizadeh et al. 2007; Babayev et al. 2015). Meanwhile, the Lower Jurassic deposits are developed to a lesser degree. In

* Determination of Norian stage here is under question.

the north, such formations outcrop within the boundaries of the Shamkir elevation (Asrikchay River basin). In the south, their individual spots partly superpose the Triassic carbonates of the Sharur-Julfa elevation of the Araz zone (in vicinities of Chalkhangala village and Nehram valley).

The Lower Jurassic deposits are transgressively deposited on the Paleozoic metamorphic shales and are covered by volcanogenic formations of Lower Bajocian. The Lower Jurassic deposits are represented by all stages and marine facies.

Hettangian stage. The deposits of the Hettangian stage are exposed in the regions of the Boyuk Gyshlag, Shamlyg, Chatakh and other villages. They are represented by basal conglomerates, lenses of sandstones, metamorphic shales, argillites, quartz porphyrites, and others (to 300 m). These deposits correspond to the Shamlyg suite.

Sinemurian stage. The deposits of Sinemurian occur in the upstream of the Asrikchay River; in the midstream of the Akhynja River; and in the areas of Shamlyg, Chatakh, Boyuk Gyshlag, Safarli and Garadash villages. They are represented by sandstones in the upper part passing into tuffstone (80 m). Stratigraphically, this stage is equal to the Boyukgyshlag suite and contains ammonite fauna—*Arnioceras objectum* (Fucini), *Echioceras declivis* (Tr.), and *Arietites* sp.

Pliensbachian stage. This stage occurs in the upstream of the Asrikchay River in a core of the narrow anticlinal fold composed by thick-bedded shales. The most complete sections of the Pliensbachian deposits are in the area of Boyuk Gyshlag Cheshmali, Mollalar and Safarli villages (from 5 up to 26 m), where ammonite fauna *Arietticeras algovianum* Opp. and *Liporoceras ex gr. henleyi* (Sow) have been found. Pliensbachian, Toarian and Lower Aalenian deposits are united here in the Safarli suite.

Toarcian stage. This stage widely occurs and outcrops in the upstreams of the Asrikchay River, as well as in the midstreams and upstreams of the Akhynja River. Lithologically, it is represented by shales with 5- to 10-cm lenses of marls and limestone (60 m). In sections from Cheshmali and Goyaly villages, the following fauna is found: *Grammoceras saemanni* Dum., *Pseudoggrammoceras fallaciosum* Bayle, and *Dumortieria tabulata* Buckm.

In the Araz zone, the volcanogenic Nehram suite (270 m) can be related on its stratigraphic position to the Lower Jurassic.

3.5.2.2 Middle Jurassic

Greater Caucasus

Deposits of the Middle Jurassic within the Greater Caucasus occur in the Vandam, Zagatala-Govdagh, Tufan, Guton-Gonagkend, Shahdagh-Khizi, Sudur, and Tahirjal zones (Khain 1947; Khain et al. 1955; Shikhalibeyli 1956; Khain and Shardanov 1957; Agayev 1966, 1990; Azizbekov et al. 1972; Khalilov 1978; Alizadeh et al. 2007; Aliyev et al. 2005a, b; Kangarli et al. 2013).

Aalenian stage. These deposits widely occur in the Tufan, Guton-Gonagkend, Shahdagh-Khizi, Tahirjal zones, and Duruja subzone of the Zagatala-Govdagh zone and are represented mainly by alternations of dark-grey and black clay shales, argillites and sandstones with interbedding of carbonate rocks. The deposits compose the Garkhun suite, including the Lower Siderite, Atashgaya and Upper Siderite units (the thickness of this suite reaches 2600 m). A typical section of the Garkhun suite is located at the interfluves of the Garachay River and contains *Planammotorecas planinsigne* Vacek. In the buried structures of the southeastern part of the Greater Caucasus, the Aalenian section is exposed by wells in the Beimdagh-Tughchay and Keshchay areas.

Bajocian stage. Deposits of the Bajocian stage occur in structures of the Tufan, Guton-Gonagkend, Shahdagh-Khizi, Tahirjal and Vandam zones. The Lower Bajocian deposits are lithologically identical to the Upper Aalenian ones and, together with them, form the Jimi suite, where argillites and aleurolites are dominant (up to 450 m). The Jimi suite is overlapped by the sandstones of the Khynalyg suite (700–1000 m). In the Gusar-Devechi region, these deposits are exposed by wells in the Yalama, Khachmaz and Khudat areas. Within the Vandam zone, the Bajocian formations outcrop at separate places in the volcanic facies ("porphyry Jurassic") and compose the Emberchay suite.

Bathonian stage. The Bathonian deposits in the Greater Caucasus do not widely occur and are exposed only in its southern slopes (Kurmukchay River basin) and north (basins of the Aghchay, Garachay and Babachay rivers). Lithologically, they are represented by a series of alternating shales, argillites, aleurolites and sandstones (up to 450 m) and are separated into the Keyvan suite. Their age is dated by foraminifera determined in the basins of the Babachay and Garachay rivers. In the buried structures of the Caspian coast, the Bathonian deposits are exposed by exploration wells in the Yalama, Khachmaz and Khudat areas.

Callovian stage. The outcrops of the Callovian are observed in the south areas of the Tufan zone and northwest areas of the Zagatala-Govdagh zone. They form the Zemchay suite, composed by flysh alternations of shales, siltstones and sandstones (up to 350 m). The fauna *Partchiceras subobtusum* Kud. and *Sigaloceras calloviensis* Sow were revealed there. Within the Shahdagh-Khizi zone (basin of the Gilgilchay River), the Callovian stage (J_2k_2 + J_3o_1) is represented mainly by conglomerates, gritstones, sandstones and argilliters (150–200 m), which compose the Molt suite (Kangarli et al. 2013). Within northeastern Azerbaijan, deposits of the Callovian stage (J_2k_2 + J_3o_1) can be observed in the Sudur zone. They are represented mainly by schists and sandstones (50–60 m) and compose the Tahirjal suite (Kangarli et al. 2013).

Kur Depression

Within the Kur Depression, the Jurassic deposits are deeply subsided and can be discovered by drilling and geophysical methods (Aliyulla and Babayev 1984; Bagirzadeh et al. 1987; Alizadeh et al. 2000; Eppelbaum and Khesin 2012). A number of intervals of nondeposition or erosion were revealed (confirmed by angular unconformities) in the sections of these deposits. Jurassic deposits have been revealed in the Jarly area and Saatly super deep boreholes (SD-I and OP-I) in the Saatly-Kurdamir buried uplift. An andesite-basalt complex of volcanogenic rocks discovered here at a depth greater than 3529 m and distinguished as a Muradkhanly suite may be related to the Lower and Middle Jurassic (see Fig. 5.21 in this Volume; integrated physical-geological model of the Saatly superdeep borehole area is presented in Chap. 5 of Volume II). Fauna of radiolarians, revealed later at a depth of 6560 m in interbedding of tuffosilts, indicates the age of these rocks as in the second half of the Lower Jurassic to the beginning of the Middle Jurassic. The upper part of the Muradkhanly suite consists of porphyrites and andesits with a thickness of 755 m. Analysis of samples withdrawn from the Saatly Superdeep well over the Muradkhanly suite shows that there is a thick series of carbonate rocks (Saatly series). The lower part of this series (int. 3529–3445 m), related to the Sarygol suite, is represented by crystallic oolite—small fragmental limestone with interbeds of tuff and spilite—and contains single foraminifera: *Nubeculinella* sp., *Marsonella* sp. and *Trochammina* sp. The age of the Sarygol suite conditionally can be considered as Callovian-Lower Oxfordian (int. 3445–3207 m). This stage is represented by organogenic-fragmental, oolitic-fine fragmental limestone with interbeds of spilite (10 m). In the upper part of the carbonate series (Jarly suite) composed of coral limestone with *Microsolena foliosa* Ron., the Upper Oxfordian age has been determined. Also here were found *Nubeculinella* sp., *Dentalina* sp., *Nodosaria* sp., and *Turrispirillina* sp., etc.

Deposits of the Kimmeridgian stage (Kurdamir suite) in the well SD-1 (int. 3207–3150 m) are represented by facies identical to the Upper Oxfordian and contain massive colonial corals: *Heliocoenia* (*Hexaheliocoenia*) *orbignyi* Ron., *Microsolena* sp., *Lingulina* sp., etc. In well OP-I in this interval (int. 3207–3157 m) are found volcanogenic formations with interbeds of breccia limestone, which contain numerous sponge spicules in the upper part.

In numerous drilled wells in the Kur Depression, faunistically dated deposits of the Tithonian stage were not found.

Lesser Caucasus

Lithologically, Middle Jurassic complexes are represented by volcanogenic, pyroclastic and sedimentary rocks with a thickness of about 4 km. Middle Jurassic associations compose structures of the Lok-Garabagh, Goycha-Hakeri, Gafan and Araz zones. The Aalenian stage is presented in the north-eastern area of the Lesser Caucasus.

Aalenian stage outcrops in the Asrikchay River basin are represented by the lower division and built by sand-clay schists and argillites containing *Leioceras cg. Opalium* Rein., *Catulloceras aratum* Buckm., *Calliphyltoceras semseyi* Ptinz., *Hammatoceras subinsique* Opp. A thickness of these deposits varies between 80 and 120 m. They are transgressively overlapped by Lower Bajocian volcanic formations.

Within the boundaries of the Araz zone, the Aalenian stage is represented by yellowish-grey medium-grained and gravelitic quartz sandstones (2–35 m). Stratigraphically, it is equal to the Chalkhangala suite (Alizadeh et al. 2007; Babayev et al. 2015) and contains pelecypod fauna—*Pholadomya, Mytiloides cinctus* Goldf., etc.

Bajocian stage. Deposits of the Bajocian stage are presented by volcanogenic pyroclastic rocks and cropping out the northeast slope (basins of Asrikchay, Zayamchay, Shamkirchay, Ganjachay, Terter, Khachynchay and other rivers) and southeastern settling (basins of the Hakeri and Okhchuchay rivers) of the Lesser Caucasus. These deposits are divided into two parts according to their lithofacial characteristics. In the first part, Lower Bajocian deposits (Zayam suite) contain lava and pyroclastic rock associations are represented by diabase, pyroxene and plagioclase porphyrites, volcanic breccias, agglomeration tuffs and tuff-conglomerates. In the second part, the Upper Bajocian (Gyzylja suite) is built by liparite-dacite porphyrites and associated pyroclastic formations (tuffs, tuff-breccias, tuff-conglomerates and tuff-sandstones). In the upstreams of the Asrikchay River, Bajocian pyroclastic rocks over the thin thickness (2 m) basal conglomerates are transgressively bedded on the deposits of different stages of the Lower Jurassic. In the interfluve of the Tovuzchay-Kurakchay rivers in the vulcanites of the Upper Bajocian have been found *Parkinsonia subarietis Wetz.* and *Oppelia subradiata* (Sow). In the northern Garabagh in the area of the Shahmansur village in the breccia-like acid tuffs, *Dinolyticeras zivagovi* Besn., *Pseudophylloceras kudernatschi* (Hauer), etc. were revealed.

In the Araz zone, Bajocian deposits are transgressively bedded on Aalenian and defined in the Gulfa Canyon and in the area of Chalkhangala, Bilava, and some others villages. The Lower Bajocian is represented by carbonated sandstones and shales (40–80 m) containing *Otoites contractus* (Sow.), *Stephanoceras humphriesianum* (Sow.), *Stemmatoceras subcoranatum* (Quenst), and other ammonites. The Upper Bajocian stage lithologicaly consists of interbedding argillites, limestone and marls (45–100 m) and faunistically is characterized by *Parantiana garantiana* (Orb.), *Parkinsonia* (Sow.), etc. Bajocian deposits correspond to the lower and middle sections of the Babek suite.

Bathonian stage. Deposits of the Bathonian stage crop out in the northeast slope (basins of Tovuzchay, Ganjachay, Terter, Khachynchay, Domychay and other rivers) and southeast settling (basins of Hakeri and Okhchuchay rivers) of the Lesser Caucasus. The base of Bathonian deposits is penetrated by the top of the Upper Bajocian liparite-dacite porphyrits and their tuffs. In the regions where Bathonian deposits are gradually passing into Callovian in the upper contact of Bathonian, the typical Callovian fauna is found. In a lithological context, the Bathonian is divided into a low-middle substage (600 m) and is represented mainly by the porphyrites, tuffobreccia, gritstone, tuffoconglomerates and tuffites. The upper substage (300 m) is represented by tuffstone, argillites, clay sandstone, and clays (Kurakchay suite). In Dashkasan, Aghdara, and Sarybaba synclinal bowing, the Bathonian deposits on ammonite fauna are divided into three substages. The lower substage contains *Oxycerites limosa* (Buckm) and *Oecotraustes formosus* Ark.; the middle substage contains *Proecotraustes splendes* Ark. and *Cadomites zlatarskiil.* Steph.; and the upper substage contains *Oxycerites aspidoides* (Opp) and *Oecotraustes maubcugeil.* Steph., etc.

In the Araz zone, the Bathonian deposits (upper section of the Babek suite) are presented by clay series (15–90 m). In some places, they are covered by Cretaceous basal conglomerates; in other places, they are covered by carbonated sandstones of Callovian. The Bathonian deposits are characterized by the following ammonites: *Oxycerites limosa* (Buckm.), *Oecotraustes genicularis* (Waag.) and *O. formosus* Ark., etc.

Callovian stage. In the northeastern areas of the Lesser Caucasus, Callovian deposits occur in the basins of the Akhynjachay, Asrikchay, Shamkirchay, Ganjachay, Terter, and Khachynchay rivers. The Callovian stage transgressively—and in some places, concordantly (Dashkasan synclinal bowing)—lies on the Bathonian and is concordantly covered by the Lower Oxfordian sediments. Lithologically, it consists of the alternation of argillites, tuffstones, shales, and conglomerates (up to 300 m). Together with the Oxfordian, the Callovian comprises the Kapaz suite. Due to ammonite fauna, it is divided into three substages. The lower is characterized by *Macroceohalitis* (Scholoth.) and *Chof-Erymnoceras coronatum* Br.; the middle has *Reineckeia anceps* (Rein.) and *Erymnoceras coronatum* Br.; and the upper has *Peltoceras athleta* (Phill.), etc.

In the Araz zone, the Callovian deposits (Anabadgadik suite) consists of carbonated sandstones (200 m) containing *Procerites funatus* (Opp.) and *Hecticoceras bunuloides* Kil. Here, these deposits compose the Chalkhangala suite. Callovian deposits are transgressively overlaid by the Upper Cretaceous.

3.5.2.3 Upper Jurassic

Greater Caucasus

Deposits of the Upper Jurassic in the Greater Caucasus occur in the limbs of the Tufan, Zagatala-Govdagh, Guton-Gonagkend, Shahdagh-Khizi, Sudur and Vandam zones (Khain 1947; Khain et al. 1955; Shikhalibeyli 1956; Khain and Akhmedbeyli 1957; Khain and Shardanov 1957; Agayev 1966, 1990; Azizbekov et al. 1972; Khalilov 1978; Aliyev et al. 2005a; Alizadeh et al. 2007; Kangarli et al. 2013).

Oxfordian stage. Deposits of the Oxfordian stage are observed in the southern areas of the Tufan zone and northwest areas of the Zagatala-Govdagh zone in the interfluve of the Katekhchay and Jimichay rivers. Lithologically, they are represented by interbedding of argillites, shales, sandstones, limestone and marls (300–800 m). In sandstones of the areas of Ilisu and Dashbulagh *bivalvia Chlamys viminea* Sow. and *Astarte ovata* Smith, etc., were found. In the northeast regions, isolated outcrops of the Middle-Upper Oxfordian, which is represented by carbonate rocks, are traced in the line stretching from the Tangaalty, Takya, Chyrag and Guleh villages as far as the Beshbarmagh area. Within the Sudur zone (basin of the Tahirjalchay River), deposits of the Oxfordian deposits (J_3o_2) are represented mainly by evaporates and sandstones (60–85 m) and compose the Gushgala suite (Kangarli et al., 2013). Within the Guton-Gonagkend zone (basin of the Gilgilchay River), the Oxfordian stage (J_3o_2) is represented mainly by gritstones, sandstones, limestone and argilliters (up to 650 m), which compose the Khaltan suite.

Kimmeridgian stage. Deposits of the Kimmeridgian stage on the south slope of the Greater Caucasus are represented by terrigenous-carbonate rocks. In its northeast part, these deposits (together with the Middle-Upper Oxfordian and lower Tithonian deposits) form united reefogenous facies. In the Zagatala-Govdagh zone in the region of Ilisu village and around the upstreams of the Kurmukchay and Shinchay rivers, the Kimmeridgian deposits (up to 500 m) consist of sandy limestone in the lower parts and finely-bedded limestone, marls, shales and sandstones in the upper parts; they contain *Perisphinctes garneri* Font. and *Calliphylloceras koshi* (Opp.), etc. Reefogenous facies in the northeast regions form easily recognized outcrops near the Sohub, Chyrag and Takya villages. Upward, the limestone section becomes massive and recrystallized (98 m). These deposits are dated by the coral fauna related to Kimmeridgian and Tithonian.

Tithonian stage. Deposits of the Tithonian stage in the Greater Caucasus are characterized by sharp facies variability. In the Zagatala-Govdagh zone, the Tithonian stage is represented by argillites, limestone and sandstone; the Sudur zone has sandy-argillaceous deposits; and the Shahdagh and Tanga-Beshbarmagh areas exclusively have biogenic breccias composed of like limestone and dolomite clasts (the Shahdagh facies). A typical section of the Tithonian is located near Gumbashi on the left bank of the Garachay River. Here, it is composed of heavy-bedded limestone, siltstones and shales (up to 430 m) and contains gastropods *Pentaptyxis staszycii typica* (Zitt.) and *Diptyxis conoidea* (Pet). A standard section of the Shahdagh facies is located on the south cliff of the Shahdagh Mt. The lower part of this section is composed of grey, coarse-bedded reefogenous dolomitized limestone (225 m), whereas the upper part has pink coarse-bedded breccia-like limestone and dolomites with beds of breccia and conglomerates occurring (nearly 550 m).

Kimmeridgian and Tithonian stages. These deposits are cropped out in the north slope of the Greater Caucasus in the Sudur and Guton-Gonagkend zones (Kangarli et al. 2013). Within the first zone, they are represented by limestone and dolomites (up to 500 m) and compose the Gukhur suite. Within the Guton-Gonagkend zone, these deposits are represented by terrigenous-carbonate rocks (rude flysh) and compose the Gyzylgazma suite (up to 250 m).

Within the Vandam zone, Upper Jurassic (Upper-Oxfordian-Tithonian) deposits are represented by the Sheki suite (Agayev 1990; Aliyev et al. 2005b; Alizadeh et al. 2007) composed of carbonate-terrigenous flish (more than 100 m).

Lesser Caucasus

Within the Lesser Caucasus, Upper Jurassic deposits occur in the basins of Tovuzchay, Asrikchay, Zayamchay, Shamkirchay, Ganjachay, Terter, Khachynchay, and Okhchuchay and some other rivers and are composed of terrigenous, pyroclastic and (rarely) volcanogenic rocks. The Upper Jurassic deposits in the Lesser Caucasus contain different kinds of ore and non-ore natural resources (ferrum ore, alunite, gypsum, etc.).

Oxfordian stage. Lithologically, this stage very often consists of terrigenous (rarely of carbonate) rocks (thickness is about 500 m). In some structures, based on ammonite fauna, the Oxfordian deposits are divided into three substages. The lower substage contains *Griniceras renggeri* (Opp.) and *Sower-byceras protortisulcatum* (Pomp.); the middle has *Perisphinctes* (Sow.) and *Gregoriceras transversarium* (Quenst.); and the upper has *Taramelliceras hauffianum* (Opp.) and *Ochetoceras marentianum* (Orb.).

Kimmeridgian stage. This stage consists of carbonate-terrigenous and volcanogenic rocks represented by tuffoconglomerates, tuffobreccia, tuffstone, limestone, and marls (up to 800 m). The lower Kimmeridgian is conformably underbedded by Upper Oxfordian and unconformably covered by Kimmeridgian which, in its turn, is unconformably covered by Tithonian. Analysis of ammonite fauna at this stage enabled a division into two substages. The lower stage contains *Streblites tenuilobatus frotho* (Opp.) and *Taremelliceras externnodosum* (Dorn), whereas the upper stage contains *Hybonoceras beckeri* (Neum.) and *Ptychylloceras ptychoicum* (Quenst.).

Tithonian stage. This stage is represented by carbonate deposits with interbedding of tuffaceous limestone, tuffite, and conglomerate (about 335 m). They are unconformably bedded on the volcanogenic rocks of Upper Kimmeridgian and covered by Lower Cretaceous deposits. Tithonian deposits have corals, mollusks and brachiopods. Based on this fauna, Tithonian sediments were divided into substages. The lower stage contains *Subplanites contiquus* (Cat.), *Haploceras carachtheis* (Zeuschn.), *Perisphinctes zittely* Siem., *Middle-Upper-Phaneroptyxis rugifera* Zitt. and *Thecosmilia moraviensis* Ogil. In the Lower Tithonian, Gushchular suite can be seen, with Shamkirchay suite in the Middle-Upper substage.

3.5.3 Cretaceous

3.5.3.1 Lower Cretaceous

Greater Caucasus

Within the Azerbaijani part of the Greater Caucasus, the Cretaceous complex is represented by its full section, which had been discovered in the Sudur, Shahdagh-Khizi, Zagatala-Govdagh, and Vandam structural zones (Khain 1947; Shikhalibeyli 1956; Akhmedov 1957; Khain and Akhmedbeyli 1957; Khain and Shardanov 1957; Grossgeim and Khain 1963; Khalilov 1965, 1978; Alizadeh 1966; Isayev and Grigoryantz 1968; Azizbekov et al. 1972; Mirchink and Shurygin 1972; Cretaceous Dep. 1980; Cretaceous Fauna 1988; Khalilov and Aliyev 1985; Aliyev et al. 2005a, b; Ismail-Zadeh et al. 2005; Alizadeh et al. 2007).

Berriasian stage. The lower part of the Babadagh suite corresponds to Berriasian on the southeastern margin of the Greater Caucasus. It contains two ammonite zones, corresponding to substages of the lower *Spiticeras obliquilobatum* and the upper *Fauriella boissieri*. The outcrops of Berriasian in the Northeastern Caucasus can be observed in the Sudur, Shahdagh-Khizi and Zagatala-Govdagh zones. In the Shahdagh and Sudur areas (the sections of Shahdagh and Tahirjalchay), the Berriasian-Valanginian is expressed by carbonate facies, consisting of (70–100 m) dolomitic limestone. Facies of zoogene dolomitic limestone of Berriasian crop out also in the Tanga and Sohub sections where the following fauna was revealed: *Euthymiceras transfigurabilis* (Bogosl.), *Salinea soloviensis* (Pcel.), *Trochoptygmatis neisatzensis* (Fogdt.), etc. Southward, the lithological content of Berriasian is changing to terrigenous rocks. Within the Khizi zone (Ugakh, Dahnachay and Atachay sections), Berriasian with basal conglomerates transgressively covers the different horizons of Middle and Upper Jurassic and, in the neighbouring Gonagkend, by piedmont where Mt. Kalovudagh comes into tectonic contact, grouped in packages of flyschoids. There are ammonites, *Spiticeras obliquilobatus*, in the Middle Jurassic shales.

Here, the Lower Berriasian is represented by a series (72 m) of alternating grey compact marls, breccia limestone, siltstones, argillites, clays and small-medium grained sandstone. These rocks contain *Holephylloceras tauricum* (Ret.) and *Berriasella calisto* (d'Orb.), as well as foraminifera and radiolarians. The Upper Berriasian is represented by a rhythmical alternation (40 m) of grey and dark-grey thin-layered carbonate argillites and clays, light-grey compact marls, light-grey limestone and (rarely) sandstone containing ammonites: *Neocomites cf. occitanicus* (Pict.), *Pseudosubpanites ponticus* (Ret.), foraminifera and radiolarins. In the Tanga-Beshbarmag structural high and Nardaran area, Berriasian is represented by a series (96 m) of alternating carbonate sandstone, sandy and organogenic limestone, marls, carbonate argillites and clays grouped in packets of rude flysch. The separate beds of conglomerate (up to 2–3 m) are *Lamellaptychus atatscaicus* A.Khal., *Symphithyris neocomiensis* (d'Orb.), *foraminifera-Rhizammina indivisa* Brady, *Glomospira gordialis* Park., *Marginulina micra* Tairov and *radiolarians-Cenosphaera kizilkazmaensis* Kh. Aliyev, *Cenellipsis giganteus* Rust, *Cenodiscaella nummulitica* Kh. Aliyev and some others. In the Zagatala-Govdagh zone, the Berriasian stage crops out in the western area of the Kheybarichay River and in the upstreams of the Girdimanchay River, where it is represented by a thick series (up to 170 m) of carbonate-terrigenous flysch consisting dark-grey carbonate argillites, clays, limestone, marls, sandstone and (rarely) by gritstone. There are rare examples in clays: *Lamellaptychus beyrichi moravica* (Mug. et Fill), *L. dardery* (Colom.), *foraminifera Bigenerina gracilis* Anton, *Glomospirella gaultina* (Berth.), and radiolarian *Cenosphaera kizilkazmaensis* Kh. Aliyev, *C. mocropora* Rust., *Cenodiscsella nummulitica* Kh. Aliyev. These Cretaceous deposits occur in all structures of the Zagatala-Govdagh and Vandam zones on the southern slope of the Greater Caucasus.

In the above-mentioned zones, Cretaceous deposits are poorly faunistically characterized, which strongly complicates any application of known stratigraphy methods. The outcrops of Berriasian-Valanginian are fixed in the basins of the Girdimanchay, Akhokhchay, Dashaghylchay, Shinchay, Kishchay, Mukhakhchay, Katekhchay, and Balakanchay rivers. Here, Berriasian together with volcanogenic forms the whole series (400–700 m) recognized in the Kepuch suite, which is bedded on the Sheki suite of the Upper Jurassic and consists of flysch alternations of organogenic-clastic sandy and pellitomorphic limestone, marly argillites and rare sandstone. In the base and upper part of the Akhokhchay River, the following foraminifera species are revealed: *Glomosplirella gaultina* (Berthelin), *Glomospira subcharoides* Chal., *Glangulina lacrima* Reuss, *Marssonella neocomica* (Chalil.) and others.

Valanginian stage. In the southeastern Caucasus, the deposits of the Valanginian stage are developed in the same zones that underlie the Berriasian rocks, which gradually pass upward into Valanginian deposits. The upper portion of the Babadagh suite corresponds to this stage and is represented by light-grey carbonate-terrigenous flyschoid of the Gaytar suite. Faunistically, the Valanginian is divided into Lower and Upper substages corresponding to ammonite zones: the *Olostephanus drumensis* (lower) and *Lamellaptychus didayi* (upper) zones. The Valanginian deposits in the Shahdagh and Sudur zones in Mt. Shahdagh, the Tahirjalchay River, and the Tanga area are represented as a series (up to 100 m) of grey (compact in some places) siliceous, organogenic-clastic, dolomitic limestones with *Belbekella coralina neocomiensis* Jac. et Fall. In the western and southwestern areas of the Khizi zone, Kalovudagh, Dahnachay, Ugakh and some other districts this stage consists of alternations of shale beds, sandstones, limestones and marls, with a total thickness up to 200 m. In the Kalovudagh Mt., Lower Valanginian is represented by a series (40 m) of alternating dark-grey, grey carbonate argillite and clays, marls and grey limestones containing *Lamellaptychus didayi* (Coq), *Conobelus conicus* (Blain), *Pseudobelus bipartitus* (Blainv.) as well as foraminifera and radiolarians.

The Upper Valanginian is represented at a low (30 m) level by interbedding dark-grey, grey carbonate argillites and clays, grey marls and in some places—by limestones and conglomerates (5–10 m); the upper portion (40 m) has dark grey, carbonate clays with interbeds of grey fragmental marls. In clays, together with the above-mentioned cephalopods, the following fauna is revealed: *Duvalia binervia* (Rasp.) and Upper Valanginian association of foraminifera: *Vaginulina arguta* Reuss, *V. recta* Reuss, *V. renngarteni* Samysch., *Dentalina nana* Reuss and others. Towards the southeast in the vicinity of Mt. Beshbarmag in the Nardaran sections of the Keshchay and Atachay rivers, the role of clastic rocks-sandstones, gritstones, and conglomerates increases. In Nardaran, the conglomerates are grouped into packages up to 6–10 m in thickness and consist of the fragments of boulders in blocks of the Upper Jurassic, differing in size and content. The other components of rude flysch (up to 30 m) are silty marls, organogenic sandy and oolite limestones, carbonate sandstones and gritstones with thin interbeds of carbonate clays with radiolarians:

Conosphaera kizilkazmaensis Kh. Aliyev, *dictyomitra ordinaria* Kh. Aliyev, *Xitus clivosa* and others.

The deposits of the Valanginian stage in the eastern segment of the Zagatala-Govdagh zone are very similar with those of the Khizi zone. In the upstream of the basins of the Girdimanchay and Pirsaatchay rivers, they are presented by a thick (to 500 m) carbonate-marl-argillite series with rich microfauna: *Epistominata canariforma* (Chalil.), *Caudryina* Porosh, *Cenodiscaella nummulitica* Kh. Aliyev, *Cenellipsis gigante* Kh. Aliyev, and others. On the southern slope of the Greater Caucasus in the Vandamian zone, the Berriasian-Valanginian series on the eastern part (Akhokhchay River) consists of sandy organogenic-fragmental and pelitomorphic limestones, sandstones, conglomerates and carbonate breccia. Towards the west in the basins of the Dashaghylchay, Kishchay, and Shinchay in the Berriasian-Valanginian section, the role of limestones and breccia is decreased. At the Bashdashaghyl River, these deposits are represented by siliceous limestones, shales, and argillites with rare interbeds of sandstones (300 m); they have *Lamellaptychus mortilleti londa* (Trauth) and poor remains of foraminifera.

Hauterivian stage. The deposits of the Hauterivian stage in the southeastern part of the Caucasus occur everywhere except the northern slope (Sudur zone and Shahdagh-Gyzylgaya massif) and northeastern area of the Tanga-Beshbarmag structural high, where the Hauterivian is represented by carbonate facies. In all other zones, it is represented mainly by shaly facies. In the southeastern part of the Greater Caucasus, the upper, major area of the Gaytar suite (to 700 m) corresponds to the Hauterivian stage, the lower area of which belongs to Valanginian. Due to faunal content, the Hauterivian stage is divided into 2 substages. The lower substage corresponds to the local zone of *Lyticpceras regale*, whereas the upper one corresponds to *Speetoniceras auerbachi*. In the Shahdagh-Gyzylgaya massif and Sudur zones, the Hauterivian stage (up to 100 m) with underlying Valanginian limestones are distinguished by the appearance of beds of argillites, clays, and sandstones in the section. *Spitoniceras cf. inostranzewi* Kar., *Pleurotomaria daghestanica* Anth., and *Loptha rectangularis* Roem, etc. were found there. Lithologically, in all studied sections of the Khizi zone and in the southern flank of the Tanga-Beshbarmag structural high, the Hauterivian deposits are identical. They are mainly composed of dark-grey shales with interbeds of sandstones, limestones and marls with common thickness from 300 m (Kalovudagh Mt.) to 700 m (Sitalchay River).

A more typical section of Hauterivian is observed in the Mt. Kalovudagh, where it is divided into two substages. The Lower Hauterivian is represented by a series (95 m) of dark-grey, grey carbonate argillite and clays with rare interbeds of thin-layered marls and carbonate sandstones with fauna of foraminifera: *Haplophragmoides infracretaceous* Mjati., *H. voccontianus* Moull., *Pseudocyclammina lituus* (Jok.), *Lenticulina muensteri* (Roem), and *Darbyella irregularis* Samych, etc. In the Ugakh, Dahnachay, and Gyzylgazmachay sections in the Lower Hauterivian, were revealed: *Lyticoceras regale* (Pavi.), *Phyllopachyceras katschience* Drush., *Duvalia binervia* (Rasp.), etc. The Upper Hauterivian can be related to the series (150 m) of dark-grey, carbonate clays with interbeds of carbonate sandstones and marls containing *Haplophragmoides concavus* (Champ.), *Dantalina legumen* (Reuss), *Lenticulina ouachensis* Sigal, and others. In the eastern segment of the Zagatala-Govdagh zone, the Hauterivian deposits are similar to those of the Khizi zone. In the most western area of Gobustan, in the Pirsaatchay River basin (near Zarat-Kheybari village) and Girdimanchay River, the Hauterivian is represented by dark-grey carbonate clays and argillites with frequent interbeds of sandstones, limestones and rarely gritstones. The remains of belemnites were discovered here: *Duvalia binervia* (Rasp.), *Hibolites longior* Schwetz and foraminifera. The Gyrkhbulag suite in the Vandam zone consists of alternation of shales with interbeds of sandstones and limestones (60–300 m). In the Akhokhchay section, poor microfauna is discovered, including *Lenticulina caltarica* (Agal) and *Anommalinasp*, etc.

Barremian stage. In the southeastern Caucasus, Barremian deposits occur in all tectonic zones and everywhere gradually overlie the Hauterivian deposits. The Khalchay suite corresponds to the Barremian stage, which is represented by clay facies, except the Shahdagh-Gyzylgaya massif, where these deposits are represented by carbonate rocks with interbeds of shales. These deposits are divided into two substages: the Lower Barremian corresponds to the zone of *Holcodiscus caullaudi*, whereas the Upper Barremian corresponds to *Phyllopachceras ectocostatum*. In the Shahdagh-Gyzylgaya massif and Sudur zone of Barremian (90–180 m), it is represented mainly by light-grey limestones, with beds of clays in the middle part of the section. There, *Nautilus neocomiensis* d'Orb., *Nucula planata* Desh., and *Lopha rectangularis* Roem are revealed.

In the Khizi trough and Zagatala-Govdagh zone, the Barremian is represented by monotonous clayey facies. The Barremian section (335 m) of Mt. Kalovudagh is typical for these zones, where rare interbedding of sandy limestones and marls and blocks of Jurassic limestones (up to 2 m) can be observed. Here are *Dorocidaris urcustensis* Neb., *Mesohibolites of varians* (Schwetz.), and others. Around Mt. Beshbarmag, the interbedding of gritstone and conglomerates was traced. The thickness gradually increases up to 600 m (Dahnachay River), 620 m (Atachay River), and 885 m (Gyzylgazmachay River). In this section, the typical Barremian ammonites are observed: *Phyllopachyceras infundibulub* (d'Orb), *Ph. segne* Drush, *Ph. eichwaldi* (Kar,), and *Barremites charrierianus* (d'Orb). In the eastern area of the Zagatala-Govdagh zone near the Altyaghaj village

(215 m), Mt. Dibrar (415 m), Chikilchay River (425 m), and the upstream of Tudarchay (400 m) and Gadisu (more 425 m) rivers, the Barremian stage consists of grey carbonate clays with alternation of limestone marls and calcite and contain *Phyllopachieceras katschiense* Drush. In this area inside the Barremian sediments, the large blocks of the Upper Jurassic limestones are observed. In the Nearcaspian-Guba area, the Barremian deposits are discovered by structural mapping and exploration wells. In the Yalama area of the Gusar-Devechi zone, a part of Barremian deposits (up to 350 m) is represented by greenish-grey argillites and sandstones with alternation of limestones, with *Lima cotaldina* d'Orb and *Trigonia* sp. Barremian in Girdimanchay-Pirsaatchay water contact near the Zarat-Kheybari village consists of shales with an alternation of marls, sandstones and limestones that contains *Phyllopachyceras infundibulum* (d'Orb), *Ph. eichwaldi* Kar etc. Farther toward the northwest in the Zagatala-Govdagh zone in the Barremian section appear interbeds of clays and argillites with the thicknesses from 175 to 386 m. In the Vandamian zone, Barremian corresponds to the Gandob suite (150–450 m), consisting of an alternation of light-grey limestone marls, sandstones, argillites and gritstones.

Aptian stage. The deposits of the Aptian stage crop out within the Shahdagh-Khizi and Zagatala-Govdagh zones and are discovered by exploration wells. Lithologically, Aptian deposits in all regions are comparatively similar ones. The Septarian horizon and Khanagah suite correspond to the Aptian stage, which is faunistically subdivided into: (1) the Lower Aptian, *Dufronoya furcata*; (2) the Middle Aptian, *Neohibolites inflexsus*; and (3) the Upper Aptian, *Acanthoplites multispinatus*. In the Khizi trough (the section of Mt. Kalovudagh), the lower Aptian (40 m) is represented by greenish-grey, strong carbonate clays with interbeds of marls and sandstones with *Ammodiscus spirillinaformis* Tair., *Bigenerina reopfax* Tair, and *Bifarina aptica* Tair. The Middle-Upper Aptian (45 m) is presented by reddish-brown carbonate shales with interbeds of marls, containing *Neohibolites montanus* Ak. Alizadeh and *N. cairicus* Natz. The more complete sections of Aptian are known in the southeastern sections in the basin of the Atachay River (e.g., Tugchay River, Gariban village and the Beimdagh Mt.), where shales contain abundant remains of belemnites: *Neohibolites ewaldi* (Stromb), *N. clava* Stoll, *N. montanus* Ak. Alizadeh, *N. inflexus* Stoll, and *N. cairicus* Natz. In the Gusar-Devechi zone in the Yalama-Khudat area, the Aptian is discovered by drilling wells to a depth of 300 m. Here, they are represented by clays and sandstones with *Trigonia longa* Agal., *Nucula impressa* Sow., *Hedbergella aptica* (Agal.), and *Globospirillina bulloides* (Agal). The clay facies of Aptian with rich remains of macro- and microfauna are observed in the eastern segment of the Zagatala-Govdagh zone in the basin of the Jarkhachichay River (198 m), Tudarchay River (191 m), Gadisu River (212 m) and in Mt. Dibrar (152 m). Here, the Lower Aptian is represented by greenish-grey clays with marly septaria, containing a great number of belemnites: *Neohibolites ewaldi* (Stromb), *N. monranus* Ak. Alizadeh, *N. compressus* Ak. Alizadeh, *Mesohibolites abkhasiensis* Krimh and foraminifera. The Middle and Upper Aptian consist of grey, red, greenish and yellowish carbonate shales and together with the above-mentioned belemnites which contain the Middle-Upper Aptian species as well: *N. wollemanni* Stoll. and *N. Lesser* Stoll. Towards the northwest in sections of Zarat-Kheybari (150 m) and Girdimanchay-Pirsaatchay (240 m) appears an alternation of sandstones and sandy limestones. In the southern slope in the Zagatala-Govdagh zone, the Aptian (60–300 m) is represented by flysch facies consisting of a rythmic alternation of clays, marls and limestones, usually without organic remains. The Kohnadakhar suite (Lower Aptian) and the Aghbulagh suite (Middle-Upper Aptian) correspond to the Aptian stage in the Vandam zone. In the basin of the Girdimanchay River, the Kohnadakhar suite is represented by series (105 m) of siltstones, limestones, sandstones and shales. The Aghbulagh suite is represented by multicolored deposits (to 50 m), with an alternation of carbonate sandstones, sandy limestones and siliceous argillites.

Albian stage. In the southeastern Caucasus, the Albian deposits occur in the same zones as the Aptian deposits and are usually concordantly lying on them. In the central area of the Khizi trough and in the eastern segment of the Zagatala-Govdagh zone, the Albian section is complete and expressed by all three substages.

In the area of the Gonagkend village, the Lower Albian (28 m) is represented by clays with interbeds of sandstones and shists. The Middle Albian (30) is represented by the same rocks and contains *Neohibolites minimus* (List.), *N. pinguis* Stoll, and foraminifera. The Upper Albian (Mamedov 1962) consists of flyschoid alternation of clays, sandstones and marls and is characterized by a complex of *belemnites Neohibolites stylioides* Renng., *N. subtilis* Krimh., *auceline Aucellina aptiensis* (d'Orb.), *A. pavlovi* Sok., *A. renngarteni* Sok., *A. pompeckli* Pavi, as well foraminifera and radiolarians. In the Gusar-Devechi zone, Albian deposits (65 m), according to drilling wells in the Yalama and Khudat areas, are represented by flysch alternations of grey, compact clays, sandstones, siltstones and marls. In the lower portion of the section, corresponding to the horizon of Kyulyulyu sandstones, *Hoplites dentatus* (Sow.), *H. impressa* Sow. and *Inoceramus concentrcus* Park, were found. The upper portion of the section corresponding to the Aucelline horizon contains *Neohibolites stylioides* Renng. and *Aucellina ef nassibianzi* Sok. In the central part of the eastern segment of the Zagatala-Govdagh zone, the Lower Albian (25–60 m) and Middle Albian (50–60 m) are represented by shales with interbeds of sandstones; the Upper Albian (50–60 m) is represented by shales, marls, sandstones and limestones.

In sections of Mt. Dibrar and in the area of Takhta-Yaylag, all three Albian substages are found—the presence of which can be proved by the discovery of belemnites, foraminifera and radiolarians. In sections of the Girdimanchay River, the Albian deposits are subdivided into two parts: the Lower Albian (85–90 m) is represented by clays and limestones with *Globuligerina tardita* (Anton.) and *Ticinella gaultina* (Moroz), whereas the Upper Albian (30 m) is represented by clay-sandy deposits with remains of foraminifera from the Hedbergella group. Within the Vandamian zone, the main part of the Albian section (from 200 to 600 m) consists of the tuffaceous rocks. The deposits of the Lower-Middle Albian are from the Lahij suite, whereas the Upper Albian is from the Duylun suite. A more complete Albian section can be observed in the Sulutchay and Girdimanchay rivers, where it is represented by alternations of argillites, tuffstones, tuffs and (rarely) by siliceous limestones, where poor remains of foraminifera can be revealed: *Gyroidinoides aff. nitidus* Reuss., *Gavelinella* sp., and *Osangularia* sp.

Lesser Caucasus

Within the Azerbaijani part of the Lesser Caucasus, Cretaceous deposits are widely represented in the structure of the Lok-Garabagh, Goycha-Hakeri and Gafan zones (Paffengoltz 1940, 1959; Aliyev et al. 1958; Khalilov 1959; Alizadeh 1966; Mamedzade 1967; Agalarova and Mamedova 1967; Shikhalibeyli 1964–1967; Khalafova 1969; Azizbekov et al. 1972; Khalilov et al. 1974; Khalilov 1978; Gasanov 1985;Khalilov and Aliyev 1985; Cretaceous Fauna 1988; Shikhalibeyli 1994; Alizadeh et al. 2007). They also form the lower structural stage of theOrdubad trough of the Araz zone (Azizbekov 1961; Khalafova 1969; Azizbekov et al. 1972; Khalilov 1978; Khalilov and Aliyev 1985; Aliyev etal. 2005a; Alizadeh et al. 2007; Babayev et al. 2015).

Lower Cretaceous deposits are found in the Lok-Garabagh, Goycha-Hakeri and Gafan structural zones; they usually are characterized by thedifferentiation of deposits and uncompleteness of sections. Regional stratigraphic schemes of the Lower Cretaceous in the Lesser Caucasus wereprepared on the basis of the prolonged research efforts of Ak.A. Alizadeh, A.G. Khalilov, V. P. Rengarten, and E. Sh. Shikhalibeyli, etc.

Berriasian-Valanginian stages. Taking into account the complex structure of these stages, the boundary between these stages in the LesserCaucasus usually is considered together. They are not widely presented and usually occur in the central zone. In the whole zone of their arealextent, the Berriasian-Valanginian deposits are lithologically represented by carbonate facies with tuffogenous admixture. More complete andpaleontologically characterized sections are found within the Sarybaba and Toragaychay troughs in the Goycha-Hakeri zone and the Gochazsynclinal bowing in the Gafan zone. In Mt. Sarybaba, Berriasian-Valanginian deposits are represented by limestones, often by silicified differences(230 m), and transgressively lie on Bathonian-Callovian rocks. They contain Berriassella callisto (d'Orb), Lamellaptychus beyrichi (Opp.), L. mortilleti (Pict. et Lor), and L. didayi (Coq.), etc. These deposits are traced in a narrow and interrupted strip far to the north and within the Toragaychaysynclinal bowing (in Mts. Susuzlug and Bogdadagh) form a mountainous series (160 m) that contains crystal, sandy and (rarely) gristone andsilicified limestones. They are linked with the underlying Tithonian by a gradual transition. In the middle part of the series, there have beendetermined Fauriella boissieri (Pict.), Lamellaptychus beyrichi (Opp), and L. didayi (Coq), etc. In the Gochas synclinorium, a typical section of thecarbonate series of Berriasian-Valanginian is observed in the basin of the Yaghlydara River, near Lachyn town and in Mt. Gyzyldash. In the lowerpart of the carbonate series (140 m), which contains silicified and (rarely) sandy limestones with interbeds of tuffs and tuff-sandstones, the Berriasianfauna complex was found: Lamellaptychus beyrichi (Opp), L. lamellosus (Park), L. mortilleti (P. et. L.), Duvalialata (Blainv.), and D. lata constricta(Uhl.), etc. The upper part (100 m), corresponding to the Valanginian stage, is represented by coarse crystallic and sandy-flinty limestones, whichcontain Lamellaptychus angullicostatus symphysocostata Tranthand Duvalia lata (Blain).

Hauterivian stage. Deposits of this stage occur in the Goycha-Hakeri and Gafan zones and are represented by the carbonate facies. In the centralpart of the Lesser Caucasus, the Hauterivian deposits occur widely and take part in the construction of the main tectonic units. In the extremenorthern parts of the central zone (head waters of Takagayachay areas), the Hauterivian (60–100 m) is found in a continued section of theTithonian-Barremian carbonate series composed of strong silicified sandstones, here and there with gritstone limestones. In the aforementioneddeposits were discovered Lamellaptychus angulicostatus (Pict. et Lor), L. angulicostatus cristobalensis (O.Conn.) and Hibolites longior Schwetz.

In these sections, deposits (up to 200 m) are represented by thin-layered, silicified limestones with theadmixtures of tuffogenous materials. Lamellaptychus angulicostatus (Pict. et Lor.), L. noricus (Wkl.) and Hibolites subfusiformis (Rasp) were foundthere.

In the Hochaz synclinal downwarping the Hauterivian deposits compose tops of Gyzylgaya, Topaghaj, Dibatli, Gyzyldash and Lachindagh Mts.; everywhere it is linked with the Valanginian gradual transition.

Barremian stage. Deposits of this stage occur within the Aghdam, Khachynchay-Khojavand and Toragaychay structural subzones. In the Aghdam anticlinal dome, Barremian deposits compose its northeastern flank and are exposed near Garakend (202 m), Abdal Gulably (201 m), Gasymly (85 m), Shelli (96 m) and some other villages. In these sections, Barremian is represented by light-grey organogenous-fragmental crystallic limestones with rare interbeds of tuffogenous rocks, which transgressively (together with the basal conglomerate at the base) lie on volcanogenic-sedimentary Kimmeridgian rocks. Age of limestone is dated by the presence of complex of brachiopods and echinoderms *Cymatoceras pseudoelegans* (d'Orb.), *Barremites difficilis* (d'Orb), *B. biassalensis* (Kar.), *Duvalia grasiana* (Duv.-Jouve), *Hibolites subfussiformis* (Rarp.), *Oxytoma cornueliana* (d'Orb), *Cyclothyris castellanensis* (Jac. et.Fall.), and *Cidaris punctaissima* Agass, etc. Within the Khachynchay-Khojavand trough, Barremian deposits as isolated outcrops can be traced near Dilagarda (174 m), Gajar (160 m), the Yukhary Syznag (30 m) and Shusha plateau (35 m). In these sections, Barremian lies on the surface of Oxfordian and is represented by limestone monotonous series that contain mainly rich brachiopod fauna. Here, *Moutonithyris moutoniana* (d'Orb.), *Symphythyris neocomiensis* (d'Orb.), *Cyclothyris lata* (Jac. et. Fall.), *Loriolithyris russilensis* (Lor.), and some others were determined. In the northeast flank of the Gafan zone, Barremian deposits together with Lower Aptian deposits compose the coarse-fragmental-organogenous-carbonate series within the layers of volcanogenous-sedimentary rocks. In the southern part closer to the left bank of the Araz River in Vejnali (340 m), Bartaz (190 m), Seyidlar (187 m), etc., the Barremian section is dominated by carbonate rocks-organogenous fragmental limestone and marls.

Volcanogenic-sedimentary formations have a lesser contribution. In the limestone and marls, an abundant fauna was found: *Diozoptyxis coquandi* (d'Orb), *Pterotriginia caudata* (Agass.) *Neithea atava* Roem., *Sellithyris sella* (Sow.), *Heteraster rengarteni* Par. and *H. magnus* Por. In the northern parts, in the interfluve of the Araz and Bazarchay rivers in the Aliguluushaghy, Davutlu and Shahverdilar sections, the Barremian composition with the volcanogenic-sedimentary components dominates over the carbonate material. The latter is represented by reef limestones, which form different types of organogenous constructions (biostomes and bioherms). A rich complex of reef-consrtructing sclerenhymas and also large foraminifers, phipidistia, gastropods and myarian mollusks are found there. Their age is determined by *Protetragonites karakaschi* Drush., *Siliesites seranonsis* (Uhl.), *Barremites strettostoma* (Uhl.), *Monopleura urgonensis* Math., *M. sulcata* Math., and *Felixigyra duncani* Prever. A thickness of Barremian is about 450 m. In the northeastern part of the Gochaz Depression, Upper Barremian (22 m) is represented by carbonate tuff-sandstones and sandy limestones with *Silesites seranonsis* (d'Orb) and *Costisdiscus recticostatus* (d'Orb).

Aptian stage. Deposits of the Aptian stage occur in the northeastern part of the Gafan zone and in the central part of the Goycha-Hakeri zone. The Aptian stage is the most complete stratigraphically and well characterized faunistically in the Gochaz synclinal bowing depression (Dolanlar village) in the Gafanchay and Araz rivers, as well as in the Mt. Gartyz (e.g., Guyudara-Khashtab), and is represented here by all three substages.

The Lower Aptian (360 m) is represented by argillites, limestones, marls, sandstones, tuff-sandstones, and (rarely) gritstones with typical fauna: *Deshayesites weissi* (Neum. et Uhl.), *D. dechy Pap.*, *Duvalia grasiana* (Duv.-Jouv.), and *Neohibolites ewaldi* (Stromb.).

The Middle Aptian transgressively lies on the upper horizons of the Lower Aptian and in some places (e.g., Chardaghly village) on Barremian or older stages. Lithologically, it is represented (nearly 900 m) by tuff-sandstones, tuff-conglomerates, marls, argillites and rarely by tuffs and limestones. There are found *Salfeldiella hofmanni* Sim., *S. guettardi gelmiensis* Renng., *Epicheiloniceras martini* d'Orb., *Acanthoplites subpeltoceratoides* Sin., *A. aschiltaensis* Antn., gastropods, pelecypods, and brachiopods, etc.

The Upper Aptian deposits do not widely occur and are traced near Saraly-Khashtab village over marls of the Middle Aptian, which conformably lie on a series (35 m) of marls, clays and sandstones with interbeds of limestones with typical clansenian fauna: *Acanthoplites multispinatus* Anth., *A. aschiltaensis* Anth., *Neohibolites andrusovi* Natz, and *Epiaster kemali* Veber. In the central zone—the Sarybaba trough—Aptian deposits do not widely occur and are found near Mt. Sarybaba and Gyshlag village. Here, the Aptian is represented by a series (50 m) of cross-bedded, coarse-grained sandstones and tuffaceous sandstones, tuffo-gritstones and limestones with *Neohibolites ewaldi* (Stromb.), *Aucellina caucasica* (Buch.) and *A. nassibianzi* Sok.

Albian stage. Albian deposits are represented by the Middle and Upper substages only. Faunistically characterized deposits of the Lower Albian in the Lesser Caucasus are not known. Within the Lok-Garabagh zone, Albian deposits (in a narrow interrupted strip in the form of separated isles) are traced from the northwest as far as the southeast and outcrop in some sections of the Gazakh, Aghjakend, and Aghdara synclinal bowing. The Albian deposits here (up to 150 m) are represented by tuff-sandstones, marls and (rarely) by sandstones; they contain *Neohibolites minimus* (List.), *N. stylioides* Renng., *Exogyra caucasica* Mordv., *Apcellina aptiensis* (d'Orb.), and *Ostrea papyracea* Sinz., etc. which characterize the whole of the Middle and Upper Albian stages. In the Goycha-Hakeri zone, Albian deposits occur more widely and are found in most sections of Lower

Cretaceous in the Sarybaba, Toraghaychay and Khachynchay-Khojavand troughs. In this zone, the Albian stage is better characterized faunistically in the Mt. Kechaldagh. Here, the Middle Albian (120 m) is represented by tuff-sandstones, limestones sandstones and argillites with a typical complex of mollusks: *Hoplites hexagonalis* Lupp., *Turrilites catenatus* d'Orb., *Neohibolites minimus* (List.), *N. stylioides* Renng. and *Aucellina aptiensis* (d'Orb).

The Upper Albian (205 m) is represented by alternations of marls, argillites, and tuff-sandstones with leading species of *Pervinquieria inflata* (Sow.), *P. fussicostata* Spath, *Neohibolites stylioides* Renng. and *Actinoceramus sulcatus* (Park). Albian deposits within the Gafan zone are found in all main sections of the Gochaz trough, where they transgressively overlap different horizons of the Upper Aptian, Barremian and Hauterivian. Typical for this zone is the Dolanlar section, where the Middle Albian (130 m) consists of alternations of tuff-sandstones, argillaceous, sandstones and sand with *Puzosia cf. mayoriana* (d'Orb), *Actinoceramus sulcatus* (Park) and *Inoceramus concentricus* Park. The Upper Albian (46 m) is composed of tuff-gritstones, tuff-sandstones and sandy argillites with fauna of ammonites: *Hysteroceras orbignyi* Spath., *H. carinatum* Spath., *Kossamatella agassiziana* (Pict.) and *Puzosia planulata* (Sow).

3.5.3.2 Upper Cretaceous

Upper Cretaceous deposits occur more widely than those of the Lower Cretaceous and are presented in all tectonic structures. The section of the Upper Cretaceous here is characterized by a large vertical thickness of the stratigraphic column, by rich mollusk complex, seaurchins foraminifers, radiolarians, coccolithophorids and other organisms. The lower (Cenomanian–Turonian) and upper (Upper Santonian–Maastrichtian) parts of this section are formed by carbonate-terrigenous and carbonate deposits, respectively. The middle part (Coniacian–Lower Santonian) consists of volcanogenic-sedimentary and volcanogenic formations.

Greater Caucasus

The Upper Cretaceous deposits of the southeastern end of the Greater Caucasus are characterized by monotonous lithofacies and richness of ammonite, belemnite, inoceram, echinoderms, corals foraminifera, ostracoda, radiolarians remains, etc. In Cenomanian, Campanian and Maastrichtian deposits, the spore-pollen complexes can be observed. The scheme of Upper Cretaceous deposits division in the region has been developed on the basis of the paleontologic-stratigraphical analysis by Alizadeh (1966), Khalilov and Aliyev (1985), and some other scientists. The outcrops of the Upper Cretaceous are confined to the same structural-facial zones as in the Lower Cretaceous. In the zone of the northern slope, namely in the observed area (Sudur and Shahdagh-Khizi zones), the normal marine and flysch-like formations mainly have accumulated; the southern area of region (the eastern segment of the Zagatala-Govdagh zone) has mainly flysch deposits. In the Sudur zone, the Upper Cretaceous is bedded non-corcondantly on the Lower Cretaceous and is represented by terrigenous-carbonate sediments with poor thickness. In the Shahdagh and Tanga-Beshbarmag massifs, the Upper Cretaceous deposits are represented mainly by carbonate rocks and are characterized by the availability of breaks in the sections. In the Zagatala-Govdagh zone, the section of the Upper Cretaceous can be characterized by a lack of breaks and a significant thickness (up to 2200 m) of flysch deposits with a prevalence of shaly rocks.

Cenomanian stage (Kemishdagh suite). The deposits of this stage cropped out in the Khizi and Govdagh troughs and are discovered in the Gusar-Devechi area. In many sections (Kalovudagh, Dibrar and others), the transition from Albian deposits to Cenomanian is gradual. In the Gusar-Devechi area, the Cenomanian deposits are discovered at depth less than 2200 m and have lithological likeness with the Upper Albian substage; they are represented by marls with alternations of sandy clays and clayey sandstones with mixed fauna of the Upper Albian-Cenomanian *Neohibolites stylioides* Renng., *Inoceramus cf. scalprum* Boehm, and *Aucellina cf. nassibianzi* Sok, etc. In Tanga-Beshbarmag anticlinal bowing, Cenomanian deposits are discovered by exploration wells in the Shuraabad area, where they consist of dark-grey, greenish-grey argillite and clays, fine-grained sandstones, limestones and conglomerates (5590 m), with rare interbeds of tuffs and carbonate tuffites. Here, *Inoceramus cf. scalprum* Boehm, *Neohibolites ultimus* d'Orb., and *Thamanninella appenninica* (Renz), etc., are found. In the Shahdagh-Khizi zone, isolated outcrops of Cenomanian deposits can be observed in Budug syncline in the Chulgazidagh Mt., where they are transgressively bedded on washed-out horizons of Albian and covered nonconcordantly by conglomerates of the Upper Turonian. They are represented by carbonate-terrigenous flysch (up to 50 m) consisting of grey argillite, clays with interbeds of limestones, carbonate sandstones, and small shingle conglomerates. *Neolibolites ultimus* Orb., *N. subtilis Krimh*., and in the lower part of section, complexes of foraminifera are detected here.

In the Khizi trough in Mt. Kalovudagh, the Lower Cenomanian (20–22 m) is represented by alternations of grey and dark-grey fragmental, carbonate argillites and clays; organogenic-fragmental, sandy limestones with interbeds of marls; and thick-plated carbonate sandstones with rare interbeds of carbonate gritstone. Here are *Neolibolites ultimus* d'Orb, *N. subtilis* Krimh., as well as foraminifera and radiolarians. The Middle and Upper Cenomanian (20 m) are represented by alternations of small shingle conglomerates, gritstones and carbonate sandstones with thin interbeds of sandy clays. There are *foraminifera Nodosaria obscura* Reuss., *Dentalina lorneiana* d'Orb., *Frondicularia guadrigana* Pern., *Vaginulina truncata* Reuss and *radiolarians Cenodiscus cenomanicus* Kh. Aliyev and others. In the eastern segment of the Zagatala-Govdagh zone, the Cenomanian deposits are close to the same age of Khizi trough deposits based on lithological content, differing from them by distinct, large thickness (300–330 m) and strict reduction of small shingle conglomerates. A fuller section of Cenomanian deposits occurs in the Mt. Dibrar, where lithologically it is represented by a sandy-clayey carbonate series (up to 300 m) and is characterized by rich complexes such as benthos and planktonic foraminifera, radiolarians and spore pollens. In the same facies, Cenomanian is represented in the neighborhood of Takhta-Yaylag (158 m), where rich fauna of foraminifera and radiolarians can be discovered. In a section of the Tudarchay River (115 m) in Cenomanian, *Neohibolites ultimus* (d'Orb) were found. In northeastern area of the zone in the basin of the Pirsaatchay River near the Zarat-Kheybari village, the thickness of Cenomanian significantly reduces (65–70 m) and is represented by a series of limestones, argillites, clays and marls containing *Inoceramus pictus* Sow., *Neohibolites ultimus* (d'Orb), and microfossils. In the southeastern direction in the Sarydashchay River valley, the Cenomanian is represented by interbedding of grey, whilish-grey limestones, thin-plated sandstones and marls with thin interbeds of carbonate argillites and clays, where *Parahibolites tourtiae* (Weign.) and complex of foraminifera have been found. In the Gadisu River basin, a series (90 m) of white grey clays, marls, sandstones and limestones corresponds to Cenomanian. They contain a Cenomanian complex of foraminifera: *Hedbergella infracretacea* (Glassen.), *H. globigerinellinoides* (Subb.), *Heterohelix globulosa* (Ehrenb.), *H. cenomanica* (Agal), etc.

Turonian stage (Zorat horizon and the lower part of the Kemchi suite). Deposits of the Turonian stage concordantly are bedded on Cenomanian and lithologically represented in two differing facies. The Lower Turonian, corresponds to the Zorat horizon and is represented by series of limestones, marls, sandstones, carbonate argillites and clays with interbeds of combustible shales. The Upper Turonian is represented by carbonate facies. Due to their lithological likeness with Coniacian deposits and poor paleontological remains, these stages are considered together. In the zone of the northern slope, the natural outcrops of the Turonian deposits are unknown. Their upper part (to 200 m) is drilled by wells in the Yalama area of the Gusar-Devechi zone, where it is represented by light-grey marls and organogenic-fragmental limestones with interbeds of sandstones containing *Inoceramus lamarki* Park. These deposits are completely represented in the Khizi trough and Mt. Dibrar area of the Zagatala-Govdagh zone. In the section of Mt. Kalovudagh, the deposits of the Lower Turonian are represented by interbedding (to 30 m) of compact, grey plate marls, limestones, argillites, dark-grey and (in some places) black clays with *Inoceramus labiatus* Scholth. The Lower Turonian complex has foraminifera—*Helvetoglobotruncana helvetica* (Boll), *Gyroidinoides nitidus* Reuss, and *Whiteinella holli* (Hagn.) in shales—as well as mass accumulation of radiolarians: *Alievium superbum* Pess., *Dictyomitra striata* Lipm., *Amphipyndax atocki* (Camp. et Clark), etc. The Upper Turonian is represented by a series (17 m) of grey, dark-grey argillites and clays, limestones, marls, conglomerates and sandstones, containing a typical complex of foraminifera *Gavelinella montiliformis* (Berth.), *Heterohelix globulosa* (Ehrenb.), *H. glibifera* Reuss., etc. In the Govdagh zone, the Turonian concordantly occurs on the Cenomanian and is represented by facies of typical terrigenous-carbonate flysch. In Mt. Dibrar, the Turonian deposits are divided into two substages. The Lower Turonian consists of alternations (26–28 m) of greenish-grey, reddish-brown carbonate argillites and clays, light-grey layer compact limestones and sandstones, with occasional interbeds of purple-violet clays with *Inoceramus labiatus* Scholoth. The remains of foraminifera and radiolarians can be observed in shales.

The Upper Turonian is represented by rhythmical alternations (47–49 m) of greenish-grey, brown argillites and clays, marls, light-grey crystallic limestones transforming in gritstones and fine shingle conglomerates. In shales are observed *Foraminifera Gavelinella moniliformis* (Berth.), *Heterohelix globulosa* (Ehrenb.), *H. globifera* Reuss and *radiolarians Gongylothorax annulatus* Dum., *G. verbeeki* (Tan Sin Hok), etc. Turonian deposits are falling out in some places (the Gozluchay River) and stretch as narrow stripes to the west and crop out in basins of the Chikilchay, Pirsaatchay and Girdimanchay rivers. In the most westerly area of Gobustan, in the neighbourhood of the Zarat-Kheybari village, Turonian deposits (77 m) are represented by rhythmic alternations of many coloured (in some places, cross-layered) limestones, marls, carbonate argillites and clays containing *Globotruncana renzi* Gand., *Gl. lapparenti* Brotz., *heterohelix globifera* Reuss and radiolarians. In such facies, Turonian stage is represented in the Sarydashchay River basin where *Inoceramus lamarki* Park and abundant remains of radiolarians *Lithostrobus turritella* Lipman and foraminifera *Stensioina praeexculpta* (Kell), *Globotruncana linneiana* (d'Orb), *Bolivinita eouvigeriniformus* Kell are observed. Towards the east, Turonian deposits are registered in the neighbourhood of the Takhta-Yaylag area and along the right bank of the Ambizlar River. In both sections, the Lower Turonian (up to 40 m) is characterized by flysch alternations, light-grey carbonate argillites and clays with a complex of microfossils *Eggerellina subsphaerica* (Reuss), *Globotruncana lapparenti* Brotz, *radiolarians Sethocapsa microacanthos* Squin., *Amphipyndax stocki* (Campb. et Clark), etc. In the Upper Turonian (40 m), in addition to the flysch components, the beds contain compact marls with *Globotruncana linneiana* (d'Orb), *Gl. imbricata* Morn., and *Heterohelix globulosa* (Ehrenb.), etc.

Coniacian stage (the upper part of the Kemchi suite). The deposits of the Coniacian stage concordantly cover the Turonian deposits and are represented by shaly-marlous and carbonate facies. Faunistically, the stage is characterized by a rich complex of foraminifera and radiolarians, and as well by the presence of rare inocerama in the lower part. A series of marls and limestones, rarely interbedding with clays and containing a typical complex of foraminifera, was discovered by wells (Yalama-Khudat area). This series corresponds to the northern slope area. In the Sudur zone and Shahdagh area in the natural outcrops, Coniacian deposits are unknown. Within the Khizi trough, Coniacian deposits have been preserved from Pre-Santonian washout in the Budug incline, where they are represented by carbonate-terrigenous flysch, which consists of grey-pinkish argillites and clays, sandy limestones, and marls with rare interbeds of fine-shingle conglomerates and contain the typical complexes of foraminifera. In the Mt. Kalovudagh section, the Coniacian stage is presented by both substages.

The Lower Coniacian (45 m) is represented by light-grey argillites, clays, marls and limestones, which contain a complex of microfauna *Gyroidinoides depressus* (Alth.), *Globorotalites multiseptus* (Brotz.), *Globotruncana angusticarinata* Gand., *Genodiscus cenomanicus* Kh. Aliyev., *Gryptamphorella conara* Dum., and *Dictyomitra* Zitt, etc. In the base of the Upper Coniacian (15 m), massive sandstones appear with small shingle conglomerates and on the roof-interbeds of conglomerates. The following can be observed here: *Gavellinella praeinfrasantonica* (Mjatl.), *Rugoglobigerina ordinaria* (Subb.), *Striataella striata* (Ehrend.), *Gongylothorax verbeeki* (Tan Sin Hok), and *Theocapsomma teren* For., etc. In the Govdagh zone, the deposits of the Coniacian stage concordantly occur on the rocks of the Upper Turonian and are represented by flysch facies. In the basin of the Pirsaatchay River near the Zarat-Kheybari village, to the Lower Coniacian corresponds a series (40 m) of alternations of grey and dark-grey compact carbonate argillites and clays, grey layered marlous limestones and marls containing *Inoceramus koeneni* Mull., *Inderformis* Meek., and foraminifera as well.

The Upper Coniacian is represented by a series (75 m) of rhythmic alternations of grey and dark-grey carbonate argillites and clays, various-grained limestones, and marls with rare interbeds of leaf-like sandstones. In shales, the following foraminifera have been found: *Tesserraella pseudotessera* (Cushm.), *Globotruncanaglobigerinoides* Brotz., and *Gl. tricarinata* (Quer.) etc. The same facies of Coniacian with foraminifera can be observed in the basin of the Sarydashchay River, where the thickness of deposits reduces to 60 m. In a section of Mt. Dibrar, the Lower Coniacian can be seen as flysch formations (75 m) represented by light-grey, chocolate-brown, and greenish-grey carbonate argillites and clays, marls, limestones and sandstones. In shales, abundant remains are observed of foraminifera. including *Globorotalites hangensis* Vass., *Whiteinella holli* (Hagn.), *Globotruncana lapparenti* Brotz. and *Heterohelix globulosa* (Ehrenb.). The Upper Coniacian is represented in the same facies as the Lower Coniacian, but the thickness reduces almost by a factor of two. The shales contain *Globotruncanella chalilovi* (Alij.), *Rugoglobigerina ordinaria* (Subb.), and *Cryptamphorella conara* Dum., etc.

Toward the south in the neighbourhood of the Takhta-Yaylag area, the Lower Coniacian (70 m) is also represented by carbonate facies with microfauna *Globotruncana lapparenti* Brotz., *Gl. linneiana* (d'Orb), *Gl. angustica-rinata* Gand., *Dictyomitra multicostata* Zitt., and *Amphipyndax enesseffi* For. The Upper Coniacian (40 m) contains foraminifera *Rugoglobigerina ordinaria* (Subb.), *Globotruncana subbotinae* Alij, and *Tesseraella pseudotessera* (Cushm). Here can be observed the abundant remains of radiolarians: *Porodiscus cretaceous* Lipm., *Orbiculiforma monticelloensis* Pess., and *Gonhylothorax verbeeki* (Tan Sin Hok). In the basin of the Gadisu River, the Coniacian stage (50 m) is represented by rhythmic alternations of chocolate-brown and greenish-grey argillites and clays, light-grey limestones and marls, and rarely, microconglomerates. Among them were revealed *Inoceramus subquadratus* Schlut and foraminifera.

Santonian stage (the lower part of the Yunusdagh suite). Santonian deposits concordantly occur on the Coniacian and are represented by terrigenous-carbonate facies with a prevalence of clayey rocks. They are paleontologically characterized mainly by foraminifera and rare finds of belemnite and inocerum, dating to the Upper Santonian age. The Santonian section in the zone of the northern slope (the wells in Yalama, Khudat) is represented by interbedding of clayey limestones and marls with typical microfauna. In the Sudur, Shahdagh and Tanga-Beshbarmag areas, these deposits are unknown. Santonian deposits widely occur in the Khizi trough, where they are bedded on the deposits of the Upper Coniacian and in some places on the different horizons of the Lower Cretaceous. In the section of Mt. Kalovudagh, the Santonian stage is represented by a series (52 m) of light-grey and grey massive overcrystallized limestones with alternation of fine-grained conglomerates. This stage contains *Globotruncana tricarinata* (Quer.), *Globotruncanella chalilorvi* (Alij.), *Stensionia exsculpta* (Reuss), *Gyroidinoides turgidus* (Hag.), *Orbiculiforma monticelloensis* Pess., *Sethocapsa microacanthos* Squin. and

Tricolocapsa dispar (Tan Sin Hok). In the Zagatala–Govdagh zone, the Santonian deposits are bedded concordantly on the deposits of the Coniacian stage and lithologically differ by increasing the shale thickness and prevalence of brown colors. In Mt. Dibrar, the Lower Santonian is represented by a series (82 m) of rhythmic alternations of reddish-brown and greenish-grey strong carbonate argillites and clays, with interbedding of grey and greenish-grey carbonate sandstones and limestones and, rarely, fine-grained conglomerates and grit-stones. In the series are *Inoceramus inconstans* Wods foraminifera and radiolarians. The series (77 m) of rhythmic alternationh of greenish-grey and pinkish-brown, compact (in some places, leaf-like), sandy clays with interbeds of grey limestone correspond to the Upper Santonian. The complex of microfossils *Stensioina exsculpta* (Reuss), *Gyroidinoides turgious* (Hag.). *Orbiculiforma monticelloensis* Pess., *Sethocapsa microacanthos* Squin., and *Tricolopsa dispar* Tan Sin Hok can be observed in shales. In sections of the Zarat-Kheybari (50 m) and Takhta-Yaylag (64 m), the Upper Santonian lithologically is almost monotype and is represented by interbedding of argillites, clays, marls and limestones with prevalence of clays, in which the same foraminifera can be observed as in Mt. Dibrar. In the southeast, the outcrops of the Santonian deposits can be observed in the basins of Gadisu (80 m), where they are represented by red carbonate argillites and clays with alternations of white-grey marls with *Cavelinella infrasantonica* (Bolakhm.), *Globotruncana ventricosa* White., Gl. concavata (Brotd.), etc.

Campanian stage (the upper part of the Yunusdagh suite and lower part of the Aghburun suite). The deposits of the Campanian stage occur in all troughs and are represented in the same facies as the Lower Santonian deposits, but their thickness is significantly increased (up to 215 m). In the Campanian, belemnites, pelecypods, and foraminifera widely occur, on the basis of which the substages are distinguished. In the Sudur zone and Tanga-Beshbarmag structural high, the Lower Campanian deposits are transgressive, with basal conglomerates on the base of the rocks of the Upper Barremian. In sections of wells on the northern slope, they concordantly cover the Upper Santonian and are represented by a weak series of multicolored clays, clayey limestones and marls containing microfauna remains. Here, the Upper Campanian consists of argillites, clays, marls and sandy limestones with rare interbeds of fine-grained conglomerates and the typical fauna of ammonites, foraminifera, belemnites, pelecypods and ostracoda. In the Yalama and Khudat areas, drilling data indicate that the Campanian stage (more 200 m) is represented by light-grey detritus limestones, marls and sandstones and contain *Inoceramus desipients* (Zitt.), *In. balticus* Boehm., and the remains of foraminifera. In the Shahdagh-Gyzylgaya massif, at the base of the Upper Cretaceous section, the deposits of the Lower Campanian are bedded transgressively and non-concordantly cover the different horizons of Barremian and Aptian. In Mt. Shahdagh, the Lower Campanian (6 m) is represented by shales with rare interbeds of limestones and marls, containing an early Campanian complex of foraminifera: *Ataxophragmium crassum* (d'Orb.), *Gavelinella menneri* (Kel.), *Globotruncana arca* (Cushm.), *G. rugosa* (*marie*), etc. The Upper Campanian corresponds to the thickness (144 m) of interbedding of limestones, marls and clays containing rich fauna of belemnite: *Belemnitella mucronata senior* Now., *B. langei langei* Jel., *B. conica* Arkh., *Inoceramus balticus* Brehm., *In. regularis* (d'Orb)., and foraminifera as well. In the Khizi trough, in the base of the Campanian, conglomerates are also observed, which cover the Santonian deposits and different horizons of the Cretaceous.

In Mt. Kalovudagh, the Lower Campanian (30 m) is represented by a series of pelitomorphic limestones, sandstones, marls and clays with *Belemnitella mucronata senior* Now, *Inoceramus balticus* Roehm. and *Liostrea incurva* Nilss. The Upper Campanian (34 m) is represented by interbedding of sandy limestones, marls and clays with remains of foraminifera *Globotruncana arca* (Cushm), *Gl. fornicata* (Plumm.), *Globotruncanella stuarti* (Lapp) and *radiolarians Gongylothorax verbeki* (Tan Sin Hok), *Tricolocapsa granti* Campb. et Clark, *Cryptamophorella conara* etc. The Campanian deposits widely occur in the eastern segment of the Zagatala-Govdagh zone and have similarities with those of the Khizi trough, differing by the prevalence of shaly rocks. The Lower Campanian deposits in Mt. Dibrar have fine-shingle conglomerates in the base (0.4 m) and are represented by a series (52 m) of alternations of white-grey, greenish-grey and deep-red carbonate argillites and clays with rapid interbeds of marls, carbonate sandstones and compact limestones. Here were found *Inoceramus balticus Boehm* and abundant remains of foraminifera.

The Campanian stage is represented by flysch alternation (50 m) of greenish-grey strong corbominiferous clays, plated sandstones, cross and wave-layer marls and limestones, containing foraminifera *Stensiona pommerana* Brotz., *Globotruncana conica* White, *Gl. arca* (Cushm), *Gl. caliciformis* (Lapp) and radiolarians *Coccodiscus holmesi* Camph. et Clark, *Gongylothorax verbeeki* (Tan Sin Hok), *Theocapsoma ancus* For., etc. Toward the south in the Campanian (Takhta-Yaylag area and Chikilchay River basin), the thickness of shaly beds significantly increases, in which *Belemnitella mucronata* Link., *B. mucronats senior* Now. and *B. mucronata parva* Najd. were revealed. In the Zagatala-Govdagh zone, the lithological content of Campanian deposits (up to 80 m) are close to Mt. Dibrar facies and crop out in the basin of Pirsaatchay in the neighbourhood of the Zarat village, where *Globotruncana linneiana* (d'Orb), *Heterohelix globulosa* (Shrenb.), *Tesseraella pseudotessers* (Cushm.), *Schackoina multispinata* (Cushm. et Wick). were found in shales. Outcrops of Campanian deposits can be observed in the Shykhandagh Mt., where Lower Campanian is represented by a series (80 m) of alternations of grey, greenish-grey, and sometimes red-brown carbonate argillites and clays, as well as white-grey and pink marls and fine-grained sandstones. *Belemnitella mucronata* Link. and *Inoceramus agdjakendensis* Aliyev were found here. The Upper Campanian (60 m) is represented by the same facies as in the Lower Campanian and contains *Belemnitella mucronata profunda Naid., B. mucronata senior Now. B. conica Arkhang* and others.

Maastrichtian stage (the upper part of the Aghburun suite). The deposits of the Maastrichtian stage concordantly are bedded on Campanian. Faunistically, they are characterized by ammonites, belemnites, inoceramus and foraminifera. In the area of the northern slope, the Maastrichtian deposits are discovered by wells (Yalama-Khudat area) and are represented by a thickness (about 200 m) of grey limestones with interbeds of marls. Here, *Diplomoceras* sp., *Pachydiscus* sp., *Pecten spathulatus* Roem., *Magna pulmilis* Sow., and the complex of foraminifera in them were identified. Weak conglomerates of the Lower Maastrichtian remain in the Sudur zone. The shingles of these conglomerates contain Aptian-Albian mollusk fauna. In the Tanga-Beshbarmag structural high in the Tanga ravine, the Lower Maastrichtian is represented by a series (67 m) of alternation of grey sandy limestones and dark-grey carbonate clays with rare interbeds of small-shingle conglomerates. In the upper part, the clays are substituted by sandy marls containing *Inoceramus tenginiccus* R. Aliyev, *In. caucasicus cabardinica* Dobrov, *In. caucasicusmonlifera* Dobrov., and *In. caucasicus ravni* Dobrov.

In the southeastern zone, Maastrichtian deposits can be observed along the coast of the Caspian Sea near the Shuraabad village, where they are represented by light-grey marls (260 m thickness) and marlous clays containing *Globotruncana arca* (Cushm.), *Gl. conoca* White, *Gl. Contusa* (Cushm.) and *Racemiguembelina fructicosa* (Egger). In the Shahdagh-Khizi zone, Maastrichtian deposits, as a result of recent erosion processes, have been preserved in some sections of the Khizi trough. In the Shahdagh Mt., the section of the Lower Maastrichtian (25 m) consists of organofragmental limestone with interbeds of clays, sandstones and marls containing foraminifera *Planoglobulina acervulinoides* (Egger) and *Pseudotextularia varians* Rzehak, etc. In the Budug syncline (Mt. Chulgazidagh), the Maastrichtian consists of shaly-carbonate thickness with interbeds of marls, sandstones and fine-shingle conglomerates. The Lower Maastrichtian is dated according to determined paleontological features: *Stegaster chalmasi* Seunes, *Inoceramus caucasicus ravni* Dobr., *Jn. Caucasicus monlifera* Dobr., etc.; the Upper Maastrichtian is characterized by belemnites *Belemnella arkhangelskiy* Najd. and *B. pontica* Najd. These deposits contain the typical complexes of foraminifera and ostracoda, as well as spore-pollen complex. In Mt. Kalovudagh, the deposits of the Maastrichtian stage are represented only by the lower substage, consisting of a series (35 m) of limestones, marls, and clays with interbeds of sandstone. In the upper part of the series are *Inoceramus caucasicus varni* Dobr. and *In. tegulatus* Hag. The rich association of foraminifera and radiolarians has been revealed as well in marls.

In the eastern segment of the Zagatala-Govdagh zone, the deposits of Maastrichtian are widely developed and connected with the Campanian by gradual transition, with lithological characterization by an increase of terrigenous material content. In the Dibrar Mt., the Maastrichtian stage (70 m) is represented by flysch alternation of grey and greenish conglomerates, characterized by the presence of *Stensioina exculpa* (Reuss), *Globotruncana lapparenti* Brotz., and *Gl. contusa* (Cushm.). The Mt. Dibrar flysch facies of Maastrichtian can be observed in sections of the Chikilchay (Takhta-Yaylag section) and Tudarchay (Ambizlar syncline) basins, differing by some increase of sandy limestone interbeds. There are *Pseudovalvulineria alievi* Ali., *Globotrunca linneina* (d'Orb) and *Gl. arca* (Cuchm.). The complete section of Maastrichtian crops out in the area of Diyally village, where the Lower Maastrichtian (85 m) is represented by interbedding of marls, limestone and shales in which *Inoceramus balticus* Boehm., *In. ismailliensis* M. Aliyev et R.Aliyev, and *Stegaster chalmas* Seun, etc. were found. The Upper Maastrichtian (50 m) deposits occur in the same facies and that the Lower Maastrichian; however, in the top of sections, fine-grained conglomerates are observed. Here, *Gyroidinoides turdinus* (Has.), *Globutruncana arca* (Cushm.), *Gl. linneiana* (d'Orb) and *Abathomphalus mayaroensis* (Bolli) were recognized.

Kur Depression

Within the Kur Depression, Cretaceous deposits are deeply subsided and discovered by geophysical field analysis and wells (Agabekov and Mamedov 1961; Mamedov 1973; Aliyulla and Babayev 1984; Bagirzadeh et al. 1987; Alizadeh et al. 2000). Especially Cretaceous carbonate and volcanogenic rocks are here associated with commercial oil and gas accumulations. In the Cretaceous deposits sections, the number of intervals of nondeposition or erosion was confirmed by angular unconformities. In this context, some problems with the stratigraphy of deeply occurring Mesozoic hydrocarbon deposits have not been solved yet. A Cretaceous complex of deposits was revealed in the Jarly, Amirarkh, Muradkhanly, Sor-Sor, Zardab, Saatly areas and Kur-Gabyrry (Kur-Iori) interfluve at depths greater than 2500 m. The oldest Cretaceous horizons were discovered by

Superdeep wells SD-I and OP-I drilled on the Saatly-Kurdamir buried uplift (see also Chap. 5 of Volume II).

In numerous drilled wells in the Kur Depression, faunistically dated deposits of the Berriasian and Valanginian stages were not found. Deposits of the Hauterivian-Barremian in the Middle Kur Depression (Jarly area, wells 1 and 4) are represented in the lower part by andesite porphyrites; in the upper section, these are replaced by tuffs, sandstone, tuffites, argillites, clays and (rarely) by limestone, with total thickness up to 1000 m. Single foraminifers (*Lenticulina* sp., *Trocholina* sp., etc.) were discovered in this area. Barremian-Lower Aptian deposits in wells SD-I and OP-I transgressively lie on the eroded surface of the Kimmeridgian and are compose various carbonate rocks with *Complexastraea* sp. The Aptian deposits outcropped in the Jarly filed are represented by tuffaceous sandstones, tuffs, tuffaceous siltstones and siltstones (100–140 m). In the Saatly area, the Upper Aptian deposits together with Albian deposits form the Sor-Sor suite, which is composed of carbonate rocks and is transgressively overlapped by the Sarmatian deposits. The Albian deposits in other areas of the Kur Depression have not been determined. The lower stages of the Upper Cretaceous (the Cenomanian, Turonian and Coniacian stages) within the Kur Depression are found in single wells (Sor-Sor and Muradkhanly areas) and are represented mainly by tuffogenous-sedimentary formations (the total thickness of which is up to 450 m). Usually, they are poor in faunal remnants. Their age is determined by the stratigraphic position and lithofacies. In the studied wells, these deposits, as a rule, lie under faunistically dated deposits of the Santonian stage. Santonian stage deposits within the Kur Depression have a peculiar set of types of thin component-rocks. In the Jarly area (well 1, int. 3370–3637 m), they are represented by limestone and dolomites with interbeds of sharply dependent sandstone and tuffaceous sandstone. Within the same interval, Santonian deposits in the Zardab, Amirarkh and Muradkhanly areas are represented by tuffogenous-sedimentary complex of rocks-tuffosandstones, argillites, limestone, tuffs and marls. *Globotruncana linneiana* (d'Orb.), *Gl. concavata* Brotz., *Gl. lapparenti* Brotz., *Rugoglobigerina ordinaria* (Subb.) etc., were found there. The Campanian-Maastrichtian in the Middle Kur Depression, the were found in a majority of wells. Campanian deposits in the Muradkhanly (well 7, int. 3782–3855 m) and Amirarkh (well 3, int. 4990–5100 m) areas are composed of a series of limestone interbedded with marls. They contain *Globotruncana arca* (Cuchm.), *Rugoglobigerina ordinaria* (Subb.), *Striataella striata* (Ehrenb.), etc. The Maastrichtian stage in the Muradkhanly (well 7, int. 3715–4235 m) and Amirarkh (well 2, int. 5210–5250 m) fields is represented by a series of white dense limestone with rare interbeds of marls, argillites and clays. They are characterized by various species of foraminifers: *Globotruncana contusa* (Cushm.), *Racemiguembelina fructicosa* (Egger), *Planoglobulina acervulinoides* (Egger), *Bolivinoides draco* (Marsson), etc.

Lesser Caucasus

Cenomanian stage. Cenomanian deposits are represented stratigraphically completely only in the central depressions and are exposed in certain areas in the foothills of the Lesser Caucasus. Within the Lok-Garabagh zone, certain outcrops of weak Cenomanian deposits are found in the Gazakh and Aghdara depressions. They transgressively (60 m) lie on the Albian where they are composed of tuffogenous sandstones, sandy marls, limestone and argillites. Near the Kichik Garabay and Sugovushan villages, these deposits are represented by *Oligoptyxis ornata* Pcel., *Exogyra haliotides* Sow., *E. olisponensis* Sharpe and *Nautilus munieri* Choff. Cenomanian deposits widely occur in the Goycha-Hakeri zone, where the argillite-sandy-gritstone series, without considerable lithological change, is traced in an uninterrupted line from the northeast edge of the Toragaychay trough as far as the southwest edge. In addition, the maximum thickness (up to 400 m) of these deposits is observed in the north-east and central areas of this zone. Here, faunistically characterized Cenomanian deposits are found in the Umudlu, Aghdaban, Elbekdash, Meydanchay, Galaboynu, Jomard, Pichanis, Chovdar and Ningi sections. One of the most typical facies complexes is a section in the Aghdaban River where the Cenomanian stage is mostly completely presented. The Lower Cenomanian is represented by a series (60 m) of interbeds of dense sandy carbonate clays and argillites with rare interbeds of marls. In this series were discovered the Lower Cenomanian cefalopods *Pizosia dachumiensis* (Sim., Bas. et Sar), *Mantelliceras cf. man-telli* (Sow.), *Neohibolites ultimus* d'Orb, and typical complex of foraminifers. The Middle Cenomanian (128 m) is represented by a series of clays and argillites that contain *Mantelliceras cf. mantelli* (Sow), *Hypholites crassifalcatus* Sim. and foraminifers *Gyroidinides nitidus* (Rouss.), *Hedbergella infracretacea* (Glaessn.), *Praeglobotruncana stephani agdavanesis* Alij.

The Upper Cenomanian (150 m) is composed of shales, argillites, sandstone and (rarely) gritstones with *Acanthoceras cf. rhotamagense* Defr., *Amphidonta columba plicatula* Lam and rich foraminifera complex. Within the Gafan zone, Cenomanian deposits are stratigraphically completely represented only in the axial part of the Gochaz trough; in the southwestern walling they are absent. Here they conformably lie on the underlying Albian deposits and are represented by a series (180 m) of limestone, clays, argillites and (rarely) by marls and gritstones. In these rocks were discovered *Hedbergella brittonensis* Loeb. et Tapp., *Praeglobotruncana stephani* (Gand.), *Lenticulina incrassata* (Maric), and also mollusks *Amphibonta columba plicatula* Lam. and *Neithea aequicostata* Lam relating to Cenomanian.

Turonian stage. Deposits of the Turonian stage in the Lesser Caucasus occur very narrowly. They are absent completely in the foothill areas of central depressions and in the Araz zone. Within the Gochaz trough near the Dalidaghsu River and Sheylanly village, the Turonian stage conformably lies on the Cenomanian and is represented by a series (130 m) of alternating black schistose argillites, black marls and grey sandy limestones, and rarely by sandstones and clays. There were found *Stensioina praexculpta* (Kell.), *Helvetoglobotruncana helvetica* (Bolii), *Globotruncana sigali* Reich, *G. globigerinoides* Brotz., etc. Related to the Turonian stage is the lower part of the flinty diabasic series, which is exposed along the whole territory of the Sarybaba and Toragaychay troughs in the Goycha-Hakeri zone. Among this series are observed silicified argillites and radiolarites (near Jomard, Arikli, Garasaggal and some other villages), which contain rich remnants of radiolaria *Cenodiscus* Kh. Aliyev, *Holocryptocanium barbui* Dum., *H. tuberculatum* Dum, *Theocorys antique* Ror. and *Cyrtophormis azerbaijanensis* Abbasov et Kasin.

Turonian deposits in the marine and continental facies with rich gastropods, pelecypods, foraminifers and seaweeds are found within the Araz zone, where they transgressively lie in the Jurassic and older rocks. In the section between the Payiz and Chalkhangala villages and Nehram canyon, Turonian deposits are represented by Lower and Upper sections. The Lower Turonian (Arkhaj suite) is found in volcanogenic facies, whereas the Upper Turonian (Jahrichay suite) has argillaceous-argillite and marl facies (Alizadeh et al. 2007) with a typical Turonian complex of foraminifers: *Bolivinopsis praelonga* (Reuss), *Valvulineria lenticula* (Reuss.), *Helvettoglobotruncana helvetica* (Bolli) and *Heterohelix obtusa* (Agal.) Continental formations of the Turonian are traced in the basin of the Arpachay River, where in the series (80 m) of marls, clays and sandstones are found *charophytes seaweeds Atopochara multivolvis* Peck., *A. submultivolvis* Romasch. and *Raskyella caucasica* Romasch.

Coniacian stage. Deposits of this stage widely occur and are represented by sedimentary and volcanogenic-sedimentary rocks. Within the Lok-Garabagh zone in the most distant northeastern parts of the Gazakh depression, the Coniacian deposits (120 m) in Gushchu-Ayrym and Yukhary Oksuzlu villages are represented by alternations of limestone, sandstone, tuff-sandstone and marls with *Plesioptygmatis bicinata* Bronn., *Caprinula robusta* Renng. and *Inoceramus koeneni* Muell. In the Aghjakend and Aghdara synclinal bowings, deposits of the Coniacian transgressively (together with the basal conglomerates in the base) lie on different horizons of the Jurassic, Albian and Cenomanian. In a typical section of the Gulustan section down the Injachay River, the Coniacian stage is represented by both substages. In the inner depressions of the Goycha-Hakeri zone, within the central parts of the Sarybaba and Toragaychay synclinorium, the Coniacian stage together with the Lower Santonian compose non-segmented volcanogenic-sedimentary series (nearly 800 m) represented by different porphyric covers, their tuffs, tuff-breccia and layers of flinty argillites, radiolarites and other silicates. A rich complex of radiolarians was found in different sections, (Gylychly, Nadirkhanly, Jomard, Fingya, etc.)—*Alievium praegallowayi* Pess., *A. superbum* (Squin.), *Thanarla veneta* (Squin.), *Dictyomitra forquata* For., *D. multicostata* Zittel, *D. inornata* Bolii., *Globotruncanella chalilovi* (Alij.), *Pseudovalvulineria monterelensis* Maic., etc.—indicating the age of these deposits as Coniacian–Early Santonian.

In marginal parts of the above-mentioned depressions (e.g., Mt.Galaboynu), the Coniacian is represented by sedimentary facies and consists of grey carbonate argillites, shales and sandy limestone (200–280 m). There were discovered rich macrofauna *Actaeonella cf. subovum* Pcel., *Trigoncarca quadrans* Renng., *Inoceramus cf. cordiformis* Sow., *In. crassus* Petz., etc. In the southwestern flank of the Gochaz trough (Saraly-Khashtab and Ashaghy Amazli sections), only Upper Coniacian is identified, which usually transgressively lies on Lower Cretaceous limestone and is represented by a series (88 m) of clays, sandstones, and limestone. There were found *Haustator kuruistanensis* Pcel., *Actaeonella gracilis* Pcel., *Inoceramus inconstans* Woods and *In. crassus* Pets. In the northwestern flank of the Gochaz trough, in the section of the Dalidaghsu River, the Coniacian deposits (90 m) conformably lie on the Turonian stage and are represented by argillites, sandstones, marls and (rarely) by fine-gravel conglomerates with *Actaeonella gracilis* Pcel. and *Protocardia hillana* Sow.

Coniacian deposits in the Araz zone are found in the Arpachay, Jahrichay and Ordubadchay river basins and the Nehram canyon and are represented by the Buzgov suite (Alizadeh et al. 2007) in the sandy-argilaceous and carbonate facies with *Haustator subnodosa* Pcel., *Inoceramus subquadratus* Schlut., *In. glatzia* (Fleg.) and by the complex of foraminifers *Globotruncana andustricardinata* and *Striataella santonica* (Agal.).

Santonian stage. Deposits of the Santonian stage occur very widely and usually with the underlying Coniacian deposits. Analysis of the lithofacial peculiarities and fauna species enabled to reliably distinguis the Lower and Upper substages. In the north-east part of the Lok-Garabagh zone, the Lower Santonian is represented by a thick (up to 600 m) volcanogenic-sedimentary series; in the southeast part, the thickness of the Lower Cantonian decreases. The most typical is the section of the Injachay River, near the Gulustan village, where the Lower Santonian is expressed by three different lithological levels. The lower part of the section (54 m) is represented by grey stratified and often dry sandstone, clays and marls; the middle part has argillaceous facies (50 m); whereas the upper section has carbonate-argillaceous facies. In these deposits, a rich complex of foraminifers with *Gaudryina pyramidata* Cushm., *Discorbis anellus* Alij., and *Globotruncana subbotinae* Alij was found. Within the Gochaz trough, the Lower Santonian is represented by a thick suite (up to 700 m) of tuff-conglomerates, tuff-sandstone, porphyrites and tuffs containing *Protocardia hillana* Sow, *Trigonia tatevensis* Mordv. and *Tr. scabra* Lam. The Upper Santonian in the Lesser Caucasus is represented by marl-limestone facies. In the Gulustan section, the Upper Santonian consists of a series of marls, limestone and marlaceous clays, which in the upper parts transit into the pelitomorphic limestone and contain *Globotruncana djaffarovi* Chal., *Gl. subarca* Alij., *Globotruncanella chalilovi* (Alij.), *Striatella striata* (Eht.) and *Pseudotextularia plummerae* (Loett.). In the central depressions of the Goycha-Hakeri zone near the middle stream of the Terter River, the Upper Santonian lies on the flinty-volcanogenic rocks and is represented by a series (80 m) of thick-bedded limestone and marls with veins of calcite. The following were discovered there: *Inoceramus balticus* Boehm., *In. decipiens* Zittel, and foraminifers *Globotruncana linneiana* (d'Orb), *Gl. ventricosa* White and *Stensioina exsculpta* (Reuss). In the central part of the Gochaz trough in the middle stream of the Hakeri River (Lachyn town), the Upper Santonian is represented by the calcareous-sandy facies (56 m) with *Neithea regularis* Schol. and *N. quinquicostata* Lam.

The Santonian stage in the Araz zone was found in the Arpachay, Jahrichay and Ordubadchay rivers basins and the Julfa Canyon. It is represented by the Kotam suite (Alizadeh 2012) and consists of alternations of clays, argillites, limestones, marls and sandstones with *Radiolites galloprovincialis* Math., *Inoceramus subquadratus* Schlut and *In. lobatus* Mull.

Campanian stage. Deposits of the Campanian stage in the Lesser Caucasus occur in all main tectonic structures and are expressed in some areas by carbonate facies. In the Lower Campanian, there appear volcanogenic formations. In the whole extension of the Lok-Garabagh zone, the Campanian deposits conformably lie on the Upper Santonian and are represented by a monotonous series of the carbonate rocks. The Campanian stage in the Gulustan village is divided into two substages. The Lower Campanian is entirely composed of limestone (90 m) with *Gl. stuarti* (Lapp.) and *Gl. rosseta* Cars. The Upper Campanian is composed of a series (90 m) of limestone; sandy shales containing *Globotruncana arca* (Cushm.), *Gl. caliciformis* (Lapp.), *Gl. conica* White and *Gl. linneiana* (d'Orb) occur there, as well as radiolarian *Patellula verteroensis* Pess., *Pseudoaulophacus parguaerensis* Pess. and *Amphipyndax stocki* (Camp. et Clark). Campanian deposits in the Goycha-Hakeri zone widely occur in the left bank of the Terter River, in the Tutgun River basin (Galaboynu and Jomard sections) and near the Seyidlar village. Lithologically, they are represented by a series (220 m) of interbedding of pelitomorphic limestones and marls. They contain *Inoceramus balticus* Boehm, *In. regularis* d'Orb. and *Galeola papillosa* Klein, and also complexes of foraminifera. To the south within the Khachynchay-Khojavand zone, Campanian deposits contain carbonate series and volcanogenic facies. The Campanian volcanogenic series (up to 200 m) is represented mainly by basins, porphyrites, dolerites and their pyroclasts, alternating in some places with grey limestones. There were discovered the early Campanian foraminifers *Stensioina exsculpta* (Reuss.), *Globotruncana arca* (Cushm.) and *Gl. lapparenti* Brotz. The carbonate series (up to 500 m) covers the Upper Campanian and is represented by bedded, sometimes platy, limestone with beds of marls, clays and sandstone. *Pachydiscus lvyi* Gross., *P. icenicus* Scharpe., *Inoceramus regularis* d'Orb. and *In. balticus* Boehm were discovered there.

In the Araz zone, deposits of the Campanian stage were found in the basins of the Arpachay, Jahrichay and Ordubadchay rivers and the Julfa Canyon. They are represented by the Garmachatag suite (Alizadeh et al. 2007) and consist of alternations of limestone, marls and sandstone (80–300 m). The stage is characterized by the presence of ammonites, inocerams, sea-urchins and foraminifers; the leading species include *Inoceramus azerbajanensis* Aliyev., *In. sagensis* Owen. and *Micraster* Stoll.

Maastrichtian stage. The Maastrichtian stage in all zones of the Lesser Caucasus is represented by carbonate facies. In the northeastern parts of the Lok-Garabagh zone (Gazakh trough), they do not widely occur, are limited in thickness (up to 45 m), and are represented by crystallic organogenous-fragmental limestones. To the south, in the central part of the north-east foothills of the Lesser Caucasus, the carbonate rocks of the Maastrichtian stage occupy larger territories and play a considerable role in the construction of the Aghjakend and Aghdara synclinal bowing. As a rule, they conformably cover rocks of the Campanian and differ in the amount of argillaceous-sandy material and sometimes pebbles. In this zone, a typical section of the Maastrichtian is observed in the Garachay River basin near the Boru village. The Lower Maastrichtian deposits are here represented by a series (165 m) of white and light-grey sandy limestone with interbeds of coarse-grained carbonate sandstone. Recognized remnants of mollusks and sea-urchins are found in the Lower Maastrichtian:

Inoceramus balticus Boehm., *Seunaaster lamberti* Charl., *Homoeaster tunetanus* Pom., as well as a complex of foraminifers. The Upper Maastrichtian deposits (60 m) are represented by sandy limestone and marls with remnants of sea-urchins: *Echinocorys ovata* Leske, *E. pyramidata* (Portl.), *Pullenia dampelae* Dain. and *P. kelleri* (Subb.) On the left bank of the Terter River, the Maastrichtian is expressed by a series (420 m) of dark-grey polymorphous limestone and marls, with *Inoceramus regularis* d'Orb. and *Echinocorys arnaudi* Seum.

Maastrichtian deposits in the Araz zone are exposed in the basins of the Arpachay, Jahrichay and Ordubadchay rivers and represented in the Pirchay suite (Alizadeh et al. 2007). For sea urchins and foraminifers both stages of the Maastrichtian have been determined here. The Lower Maastrichtian (132 m) is represented by limestone, marls and argillites with a typical complex of foraminifers with the *Globotuncana contuse*. The Upper Maastrichtian (150 m) is represented by sandy and pelitomorphic limestone, marls, clays and sandstone with *Rasemiguembelina fructicosa* and *Abathmphalus mayaroensis*. It should be noted that Maastrichtian deposits are conformably overlaid by Paleocene carbonate rocks in many areas of the Caucasus.

Talysh

Within the Mountainous Talysh, Cretaceous deposits are widely represented in the structure of the Astara zone (Azizbekov et al. 1979) and are presented in the Upper Cretaceous carbonate facies. Such deposits include small outcrops of limestone outcropping from under the tuffaceous-sedimentary Paleogene rocks in the river valleys of Istisuchay and Tangaru. Besides bedding outcrops, the Upper Cretaceous limestone are detected in the forms of tectonic rocks protruding along the faulting zones. The Santonian-Maastrichtian age of the limestone is established by the presence of typical fauna, such as *İnoceramus* sp., *Ostrea* sp., *Textulariidea, Globotruncana, Globigerina, Zithathamninium, Globotruncana aff. stuarti Lapparent, Globotruncana conica rohite,* as well as crinoideans, bruozoans, and algae.

3.6 Cenozoic

3.6.1 Paleogene

Paleogene deposits in Azerbaijan are represented by different types of sedimentary, volcanogenic-sedimentary and volcanogenic rocks of a large thickness and take part in the structure of large tectonic units of the Greater and Lesser Caucasus, Kur Depression and Mountainous Talysh (Azizbekov et al. 1972; Khalilov 1978; Aliyev et al. 2005a; Alizadeh et al. 2007).

Tracing the distribution of Paleogene deposits in Azerbaijan and clarification of their stratigraphic position are especially urgent to solve the problems of stratigraphy and the correlation of Paleogene deposits of the Mediterranean-Alpine region.

Later, detailed studies of the Greater Caucasus Paleogene complex were undertaken by D.A. Agalarova, G.A. Akhmadov, A.H. Aliyev, K.A. Alizadeh, A.A. Alizadeh, V.V. Bogachev, E.Y. Dmitriyev, S.F. Fyodorov, D.V. Golubyatnikov, I.M. Gubkin, D.I. Jafarov, V.Y. Khain, D.M. Khalilov, A.M. Lednyov, M.F. Mirchink, A.G. Pominova, I.I. Potapov, I.F. Pustovalov, A.L. Putkaradze, S.G. Salayev, A. Sh. Shikhalibeyli, V.V. Tikhomirov, N.B. Vassoyevich, V. V. Weber, and V.P. Yasenev, etc. The aforementioned scientists had divided the complex into a number of suites and horizons. Similar studies had been implemented in the Lesser Caucasus and Talysh by R.N. Abdullayev, G.V. Abich, A.G. Aliyev, H.I. Aliyev, A.A. Alizadeh, G.A. Alizadeh, G.I. Allahverdiyev, G. Artgaber, I.N. Aslanova, Sh. A. Azizbekov, Sh.A. Babayev, M.A. Bagmanov, A.E. Baghyrov, A.A. Bayramov, V.V. Bogachev, P. Bonnet, Ye. N. Yemelyanova, A.A. Florenskiy, F. Frech, M.A. Kashkay, T.G. Gajiyev, A.D. Ismailzadeh, B.P. Juze, V.E. Khain, D. M. Khalilov, D.M. Khalilov, M.N. Mammadov, T.A. Mammadov, B.F. Meffert, Sh.F. Mekhtiyev, F.A. Mustafayev, L.A. Nechayev, N.Sh. Nijeradze, K.N. Paffenholtz, V. P. Rengarten, M.I. Rustamov, A.N. Ryabinin, A.Sh. Shikhalibeyli, I.N. Sitkovskiy, G. Tsulukidze, Y.K. Ustiyev, M.M. Valiyev, and V.A. Varentsov, etc. Implemented research has resulted in the development of a detailed stratigraphic scheme of Azerbaijan's Paleogene formations.

3.6.1.1 Paleocene

Greater Caucasus

Paleocene deposits are widely developed in the southeastern part of the Greater Caucasus, where they have contributed to the structure of the Gusar-Devechi (Siyazan monocline), Sudur, Shahdagh-Khizi, Zagatala-Govdagh, Vandam and Shamakhy-Gobustan structural zones (Potapov 1954; Akhmedov 1957; Khain and Akhmedbeyli 1957; Azizbekov et al. 1972; Mirchink and Shurygin 1972; Khalilov 1978; Bagmanov 1980; Alizadeh et al. 1989, 2007; Aliyev et al. 2005a).

Deposits of the ***Danian stage*** occur in the Ilkhidagh suite and are represented by grey and dark-grey clays, marls, marlaceous clays, and sandstone with intervals of limestone with *Globoconusa dauojergensis* Bronn., *Acarinina schachdagica* Chal., *Cruciplacolithus tenuis* (Str.), and *Chiasmolithus danicus* (Brotz.), etc. The thickness of these deposits ranges from 90 to 450 m. Danian deposits have been determined in the Budug syncline (Zeyid village), Shuraabad area, Yunusdagh ridge, Aghdara, Khilmilli, Chukhuryurd, Goturdagh and in some other areas.

Deposits of the ***Selandian stage*** occur in the lower part of the Sumgayit suite and are represented by red-brown clays and grey marls with *Globorotalia angulata* (White), *G. compressa* (Plumm) and *Fasciculithus tympaniformis* Hay et Mohl., *Chiasmolithus bidens* (Bram. et Sull.). The thickness of these deposits ranges between 40 and 140 m. They have been found in Zeyid and Shuraabad sections, Yunusdagh ridge, Goturdagh and Aghdere areas, etc.

Deposits of the ***Tanethian stage*** occur in the upper parts of the Sumgayit suite and are represented by brick-red non-carbonate gritstones, clays, sandstone, and conglomerates with thickness of 40–180 m. Here were discovered *Phizammina indivisa* Brady, *Ammodiscus incertus* d'Orb., *Acarina subsphaerica* (Subb.), *Cenosphaera caucasica* Boriss., *Discoaster gemmeus* Strad., *Marthasterites bramletteu* Br. et Strad., etc. These deposits have been determined in the Zeyid, Gazma-Gryz, Rustov, Shuraabad, Nardaran, Goturdagh and Aghdara areas.

Kur Depression

The outcomes of geophysical exploration (Bagirzadeh et al. 1987; Aliyev et al. 2005c; Mamedov et al. 2008b; Eppelbaum and Khesin 2012) presume the presence of a Paleocene sedimentary complex in the Middle and Lower Kur depressions. This assumption was confirmed by the results of deep drilling within the boundaries of the Middle Kur Depression (Agabekov and Mamedov 1961; Azizbekov et al. 1972; Mamedov 1973; Rasulov and Mamedzade 1978; Alizadeh et al. 2007), where two large sedimentary Paleogene basins were discovered. They are confined to the interfluve of Kur and Gabyrry rivers and to the east from the Ganja meridian in the deep Yevlakh-Agjabedi trough.

Deposits of the ***Danian stage*** in the interfluve of the Kur and Gabyrry rivers are exposed in the Tovuz-Gazakh and Khatynly areas. They conformably lie there on the Upper Cretaceous limestone and are represented by marls and marlaceous limestone with interbeds of clays (the thickness of the deposits reaches 100 m). *Globigerina varianta* Subb., *G. edita* Subb., *Subbotina triloculinoides* (Plumm.), and *Globorotalia compressa caucasica* Chal., etc., were discovered there. Deposits of the Tanethian stage are exposed in the same areas and are composed of marly sandy clays, the thickness of which is up to 100 m.

Within the Yevlakh-Agjabedi trough, the Paleocene deposits are exposed in Mt. Duzdagh, as well as the Amirarkh and Muradkhanly areas. The Danian stage is composed of limestone, marls and clays (20–40 m) with *Globigerina quadritriloculinoides* Chal., *G. edita* Subb., *Acarinina shachdagica* Chal., and *Marcalius inversus* Defl., etc. Selandian deposits are exposed in the same areas and are represented by clays, marls with interbeds of sandstone, limestone with *Globorotalia angulata* (White), *Gaedita* Subb., and *Fasciculithus tympaniformis Hay et* Mohl, etc. Tanethian deposits are exposed in the Duzdagh, Amirarkh, Muradkhanly and Mil areas and are represented by clays and marls with interbeds of limestone (25–95 m), with *Acarinina subsphaerica* Subb., *A. acarinata* Subb., *Discoaster multiradiatus* Bram. et Ried, etc.

3.6.1.2 Lesser Caucasus

Within the Azerbaijani part of the Lesser Caucasus, Paleocene formations transgressively overly the Cretaceous complex in some submerged areas of the Lok-Garabagh and Goycha-Hakeri structural zones (Shikhalibeyli 1964–1967; Azizbekov et al. 1972; Shikhalibeyli et al. 1976; Khalilov 1978; Paffengoltz 1979; Bagmanov 1980; Gasanov 1983; Alizadeh et al. 1989, 2007; Aliyev et al. 2005a) and outcrop on the southeastern limb of the Araz zone's Ordubad trough (Azizbekov 1961; Azizbekov et al. 1972; Khalilov 1978; Bagmanov 1980; Alizadeh et al. 1989, 2007; Aliyev et al. 2005a; Babayev et al. 2015).

In the northwest foothills of the Lesser Caucasus, Danian deposits do not widely occur. Their outcrops are observed in the Gazakh (Dash Salahly village), Aghjakend (Dozular, Shorbulag, Boru villages), Aghdara (Godakburun, Tazakhachyn, Ortakend villages), and Khachynchay-Khojavand (Sardarkend, Farrukh, Pirlar villages) troughs and represented by limestone, marls, and marlaceous clays (5–94 m) with *Globigerina pseudobulloides* Plumm., *G. fringa* Subb., *G. quadrata* White, *Acarinina inconstans* (Subb.), *Hercoglossa danica* Schloth., *Homoeaster abichi* and *Cyclaster danicus* Schlut. Selandian deposits crop out in the Aghjakend (Dozular, Shorbulag, Boru, Gulustan, etc.) trough and are represented by carbonate clays, limestone, and marls with *Globorotalia angulata* (White), *Bolivina subincrassata* Chal., *Stensioina cucasica* (Subb.) (thickness of the deposits ranges within 20–100 m). Tanethian deposits are represented by marls and clays with interbeds of carbonate sandstones; they occur in the Gazakh (Dadavan), Aghjakend (Dozular, Boru, Injachay, Gulustan etc.), Aghdara (Aghdara, Sugovushan villages), and Khachynchay-Khojavand (Khanabad village) troughs. The thickness of the deposits ranges within 40–140 m. The deposits contain *Nummulites fraasi* de la H., *N. deserti* de la H., *Cardita minutula* Roman., *Acarinina subsphaerica* (Subb.), *Globigerina velascoensis* Cushm., *Discoaster multiradiatus* Br. et Ried., etc.

In the central part of the Lesser Caucasus, deposits of the Danian stage are almost absent. In the area where the left tributary of the Bulanlyg River merges into the Terter River (in the axial line of the Sarybaba trough), limestone and marls with *Globorotalia pseudobulloides* (Plumm.) and *Subbotina triloculinoides* (Plumm.) were identified (the thickness of the deposits is 165 m). Selandian deposits have been also determined. They are represented by a series of argillaceous limestone and marls (the thickness of which is 214–250 m) with *Subbotina triloculinoides* (Plumm.) and *Globorotalia angulata* (White). In the southern part of the Shahdagh ridge in the northwest margin of the Sarybaba trough (Guneypaya, Yalkend and Jamilli villages), the thickness of the Lower Paleocene argillaceous clays, argillites, siltstones, sandstone, and limestone reaches 375 m. They contain *Globorotalia angulata* (White), *Bulimina inflata* Seg., and *Bolivina subincrassata* Chal., etc.

The ***Upper Paleocene*** deposits occur in the Kalbajar superimposed trough and in the southeastern part of the Sarybaba trough (e.g., Kalafalik, Faraj, Nadirkhanly and Bashlybel villages). They are represented by argillites and limestone with interbeds of sandstone, with a total thickness of 500–600 m. They contain *laster abchasicus* (Schwetz), *Nummulites fraasi* de la H., *N. solitarius* de la H., *Discocyclina seunesii* (Douv.), *Stensioina caucasica* (Subb.), and *Globorotalia nana* Chal., etc. In the southern part of the Shahdagh ridge (Goygol and Jamilli areas), the Upper Paleocene deposits are represented by argillaceous shales, argillites and limestone with thickness ranging within 50–200 m. They contain *Stensioina whitei* Mor. and *Globorotalia nana* Chal., etc.

In the Araz zone of the Lesser Caucasus, Paleocene deposits widely occur in the Ordubad trough in the Turkesh-Badamli and Ordubad-Yayjy areas. Deposits of the Danian stage are represented by the Aza suite (Babayev et al. 2015) and consist of carbonate clays, gritstones and conglomerates (the thickness of the deposits varies within 100–300 m). They contain *Globoconusa daubjergensis* (Bronn.), *Globigerina schachdagica* Chal., *Acarinina schachdagica* Chal., *A.inconstans* Subb., *Coccolithus eopelagicus* Br. et Ried., *Toweis emineus* (Bram. et Sull.), etc. Deposits of the Selandian stage are represented by Turkesh suite (Babayev et al. 2015) that overlap the Danian ones; they consist of carbonate clays, marls, sandstone, gritstones, with *Globorotalia angulata* (White), *Globigerina varianta* Subb., *Stensioina whitei* Moroz., etc. Their thickness is in the range of 100–230 m. Deposits of the Tanethian stage are represented by the Ordubad suite (Babayev et al. 2015) and are lithologically similiar to the Selandian stage. Their thickness varies within 60–370 m. They contain *Acarinina subsphaerica* (Subb.), *Globigerina velascoensis* Cushm., *Discoaster multiradiatus* Br. et Ried., *Nummulites fraasi* de la H., *N. deserti* de la H., *Operculina heberti* Muen.-Chalm., *Discocylina seunesi* Douv., *Pitar pavlovi* Arkhang., and *Goniopora rudis* (Reuss), etc.

Talysh

Paleocene deposits in the Talysh Mountain (Astara suite) are represented by argillites, aleurotuffites, marls, siltstones, limestone, tuff-sandstones with *Discocyclina* sp., *Globorotalia angulata* (White.), *Acarinina* (Subb.), *Globigerina nana* Chal., etc. They occur in the upper interfluve of the Astarachay and Lakarchay rivers, in the southeastern part of the Astara anticlinal dome (Problems of the Talysh Geology 1958; Bagmanov 1963; Azizbekov et al. 1972, 1979; Khalilov 1978; Alizadeh et al. 2007). Their thickness varies within 960–1270 m.

3.6.1.3 Eocene

Greater Caucasus

The Eocene complex conformably overbuilds the Paleogene formations in the southeastern plunge of a mountain-fold structure. It partly contributes to the structure of the Gusar-Devechi (Siyazan monocline), Sudur, Shahdagh-Khizi and Vandam zones, having maximal development within the boundaries of Zagatala-Govdagh and Shamakhy-Gobustan zones (Potapov 1954; Akhmedov 1957; Azizbekov et al. 1972; Mirchink and Shurygin 1972; Khalilov 1978; Agabekov et al. 1982; Bagmanov 1980; Cretaceous Fauna 1988; Alizadeh et al. 1989, 2007; Aliyev et al. 2005a).

Within the southeastern part of the Greater Caucasus, the Govun (Koun) suite is found, with subsuites corresponding to Eocene. The Lower Eocene deposits (Ypresian stage) are in the lower section of the Govun suite and are represented by marls and marlaceous clays with interbeds of bentonite sandstone and limestone; their thickness is 60–120 m. They conformably lie on the Upper Paleocene deposits with *Globorotalia subbotinae* Moroz., *G. aragonensis* Nutt., *G. caucacica* Glaessn., *Cenosphera grossheimi* Boriss., *Marthasterites tribrachiatus* Braml. et Ried., etc. The upper section of the Middle Eocene (Barthonian stage) is represented by marl bituminous shales with interbeds of sandstone, the thickness of which is 80–290 m. They are traced in Zeyid, Gazma-Gryz, Nardaran, Shahdagh, Aghdara, and Khilmilli sections. These deposits contain *Globigerina turkmenica* Chal., *G. praebulloides* Blow., *Nummulites badensis* Hant., and *Reticulofenestra umbilica* (Levin), etc. Deposits of the Upper Eocene (Priabonian stage) refer to the Upper Govunian. These deposits are represented by marlaceous clays and sandstone with interbeds of marls with fucoids. Their thickness varies from 100 m to 200 m. They include *Globigerinatheka tropicalis* (Bakn. et Blow), *G. index* (Finla), *Globigerina corpulenta* Subb., *Bolivina antegressa* Subb., *Discoaster barbadiensis* Tan Sin Hoc, and *Chiasmolithus oamuruensis* (Defl.), etc. These deposits are traced in Gazma-Gryz, Yalama, Shuraabad, Yunusdagh ridge, Aghdara, Khilmilli, etc.

Kur Depression

The presence of the Eocene complex in the deep structure of the Kur Depression has been established by geophysical exploration and deep drilling materials (Agabekov and Mamedov 1961; Azizbekov et al. 1972; Mamedov 1973; Rasulov and Mamedzade 1978; Agabekov et al. 1982; Bagirzadeh et al. 1987; Aliyev et al. 2005a; Alizadeh et al. 2007).

In the interfluve of the Kur and Gabyrry rivers, Ypresian deposits are exposed in Tovuz-Gazakh, Khatynly, Mammadtapa, Gyragkasaman and Sajdagh areas. They are represented by carbonate clays, sandstone, and marls (rarely, by tuffs and limestone), with *Globorotalia subbotinae* Moroz., *G. aragonensis* Nutt., *Nummulites praelucasi* Douv., *N. pernotus* Schaub, *N. burdigalensisde* la H., *N. spileccensis* Muen-Chalm., *Operculina parva* Douv., *Discoaster binodosus* Mart., and *D. lodoensis* Mar., etc. Middle Eocene deposits occur in the same areas, as well as in the Damirtapa-Udabno, Koyrukkeylan, and Gurzundagh areas. They are nearly alike in different stages and occur with interbeds of marls, limestone, and tuffs—the thickness of which is 70–250 m. They contain *Acarinina bullbrooki* (Bolli), *A. rotundimarginata* Subb., *Globigerina turkmenica* Chal., *Reticulofenestra umbilica* (Levin), *Nannotetrina fulgence* Stradn., etc. Upper Eocene deposits have been studied according to data obtained from wells in Damirtapa-Udabno, Mammadtapa, Sajdagh and Khatynly areas. They are represented by clays with interbeds of sandstone, with thickness varying within 70–320 m. They contain *Globigerina corpulenta* Subb., *G. officinalis* Subb., *Nummulites prestwichianus* Jon., *N. variolarius* Lam., *N. orbignyi* Gah., *N. rectus* Curry, *Chiasmolithus oamarunsis* Defl., *Sphenolithus pseudoradians* Br. et Wilc., etc.

The Eocene deposits in the Yevlakh-Agjabedi trough are exposed in the Duzdagh, Amirarkh, Shirvanly-Barda, Zardab, Muradkhanly, Shiringum and Mil areas. Unlike the younger Eocene formations, Lower Eocene deposits are exposed only in the Duzdagh, Amirarh, Shirvanly-Barda, and Muradkhanly areas. They are represented by marls, shales, and siltstones with interbeds of limestone—the thickness of which is 100–200 m. They contain *Globorotalia subbotinae* Moroz., *G. caucasica* Glaessn. and *Marthasterites tribrachiatus* Br. et Ried. The Middle Eocene deposits are represented by clays with interbeds of marls, limestone, and sandstone—the thickness of which is 50–460 m. They contain *Acarinina rotundimarginata* Subb., *Globigerina turkmenica* Chal., *Nannotetrina fulgence* Str., *Reticulofernestra umbilica* (Levin), etc. The Upper Eocene deposits are represented by sandy carbonate with interbeds of marls, limestone and tuffs—the thickness of which is 120–140 m. They contain *Globigerina corpulenta* Subb., *G. officinalis* Subb., *Istmolithus recurvus* Defl., *Sphenolithus pseudoradians* Br. et Wil., etc.

Lesser Caucasus

Within the Azerbaijani part of the Lesser Caucasus, Eocene formations either continue in the section of the Paleocene or transgressively overlie the Upper Cretaceous formations at the immersed areas of the Lok-Garabagh and Goycha-Hakeri structural zones (Shikhalibeyli 1964–1967; Gasanov 1968; Azizbekov et al. 1972; Khalilov 1978; Paffengoltz 1979; Gasanov 1983, 1985; Alizadeh et al. 1989, 2007; Aliyev et al. 2005a). They also form a major part of the outcropping section of the Araz zone's Ordubad trough (Azizbekov 1961; Azizbekov et al. 1972; Khalilov 1978; Alizadeh et al. 1989, 2007; Aliyev et al. 2005a; Babayev et al. 2015).

Deposits of the Lower Eocene in the Gazakh trough have been observed near the Musakoy village in volcanogenic-sedimentary facies (limestone, sandstone, and shales dolerites). Their thickness is 220 m. They contain large amounts of *Nummulites praelucasi* Douv., *N. burdigalensisde la H.*, *N. inkermanensis* Schaub, *N. leupoldi* Schaub, *N. subramondi* de la H., *Operculina parva* Douv., *Pitar lamberti* Desh., etc. In the Aghjakend trough, the Lower Eocene deposits are represented by sandstone, clays, alevrolites, and limestone (with their thickness varying between 10 and 230 m). They contain *Nummilites planulatus* (Lam), *N. exilis* Douv., *N. spileccensis* Muen-Chalm., *Globorotalia subbotinae* Mor., *G. marginodentata*, *Discoaster binodosus* Mart., *D. lodoensis*, *Cardita minutula* Rom., and *Tellina denudata* Desh., etc.

Deposits of the Lutetian stage consequently lie on the deposits of the Lower Eocene. They are represented by marls and marly clays, with interbeds of sandstone and a thickness of 20–120 m. They contain *Nummulites irregularis* Desh., *N. anomalis* de la H., *N. anomalis* de la H., *Acarinina bullbrooki* (Bolli), *A. rotundimarginata* Subb., *Turritella imbricataria* Lmk., etc.

Deposits of the Bartonian stage lie stratigraphically higher comparing with the previous stage and are represented by clays and sandstone, with a thickness varying from 10 to 140 m. They contain *Nummulites irregularis* Desh., *N. murchi-soni* Ruet., *N. atacius* Leum., *N. anomalis* de la H., *Acarinina bullbrooki* Chal., *G. praebulloidess* Blow, and *Reticulofenestra hampdensis* Edwards, etc. Deposits of the Upper Eocene are represented by marls, clays, sandstone, and argillites—the thickness of which is 150–200 m. They contain *Nummullites budensis* Hantk., *N. bouillei* de la H., *N. stellatus* Rov., *Bolivina nobilis* Hantk., *Globigerina corpulenta* Subb., *Nucula peregrine* Desh., *Nuculana alexeevi* Mir. et Jark, *Chlamus solea* Desh, etc. Outcrops of the Lower Eocene are located in the central part of the Lesser

Caucasus within the Kalbajar superimposed trough, the Shahdagh depression and the southeastern subsidence of the Gochaz trough. Layers of the Ypresian stage in the Kalbajar trough are exposed in the northeastern flank of the Sarybulagdagh fold, along the north flank of the Mykhtokan-Dalidagh syncline and in the northeast flank of the Garakhach brachy-anticline. They are represented by limestone, sandstone, argillites and marls—the thickness of which is up to 200 m. They contain *Heterohelix subglabra dosularensis* Chal., *Nodellum velascoense* (Cushm.) *Nummulites silvanus* Schaub, *N. akkuurdanedsis* Nemk., *N. praemurchisoni* Nemk. et Barkh, *N. leupoldi* Schaub, *N. distance* Desh., and *Operculina gigantea* May-Eim., etc.

Lower Eocene deposits in the Shahdagh trough are exposed on the south slope of the Murguzdagh ridge; at the heads of the Polad, Tarsa, Akhynjachay and Shamkirchay rivers; and to the south from Chobandagh. In the central part of the Shahdagh ridge, Ypresian deposits are represented by tuffogenous-sedimentary series (400 m) and limestone (320 m) series. The tuffogenous-sedimentary series (near the Novoivanovka village and Sariyal Mt.) is represented by tuff-sandstone, lens gritstone organogenous limestone, sandstone, tuff-conglomerates, andesites and interbeds of tuff breccia. Limestone contain *Nummulites praelucasi* Douv., *N. pustulosus* Douv., *N. praemurchisoni* Nemk. et Barkh, *N. pernotus* Schaub, *N. planulatus* (Lam.), *Discocyclina sella* d'Arch., etc. The limestone series of the Lower Eocene is observed both in the north (near Alagollar, Garaarkhach, Chobandagh, and Gojadagh) and on the southern slope of the Shahdagh ridge (from Ardanish to Jil villages). This series is represented by brecciated multilayered limestone, which nonconformably lie on the carbonate deposits of Campanian-Maastrichtian. They contain *Nummulites praelucasi* Douv., *N. murchisoni* Ruet., *N. planulatus* (Lam.), *N. irregularis* Desh., *N. partschi* de la H., *N. distans* Desh., *Discocyclina pratti* (Mich.), etc. In the southern part of the Shahdagh ridge, the Lower Eocene is represented by terrigenous and terrigenous-carbonate facies with a thickness of 80–140 m. They contain *Nummulites partschi* de la H., *N. planulatus* (Lam.) and *N. praelucasi* Douv. The Lower Eocene deposits in the form of a narrow strip are traced in the southeastern subsidence of the Gochaz trough from Gubadly region as far as Aliguluushagy village, and also near the Ishygly and Charali villages. They are represented by limestone, argillites, shales and sandstone (5–45 m), which non-conformably overlap the Campanian-Maastritchtian limestone. They contain *Globigerina nana* Chal., *G. aequa* Cushm and *Renz*, *Acarinina acarinata* Subb., etc.

Deposits of the Lutetian stage are represented by limestone, shales, sandstone and siltstones. They are observed in the Kalbajar superimposed trough (220 m), as well as the Gochaz and Gafan subzones of the Gafan zone (260 m). In the central part of the Shahdagh depression, these deposits are represented by a thick volcanogenic-sedimentary complex (1000 m). *Nummulites distans* Desh., *N. pratti* d'Arch et H., *N. murchisoni* Ruet., *N. irregularis* Desh., *Discocyclina pratti* (Mich.), *N. rotularis* Desh., *Acarinina bullbrooki* Bolli, and *A. otundimarginata* Subb, etc., were discovered there.

Deposits of the Bartonian stage are represented by a thick volcanogenic series (the Kalbajar superimposed trough) and a volcanogenic-sedimentary complex (the Shahdagh trough, 1800 m). They contain *Nummulites gizehensis* Forsk., *N. ataticus N. incrassatus* de la H., etc.

Deposits of the Priabonian stage are represented by conglomerates, breccia, limestones, argillites and marls in the Kalbajar superimposed trough (250 m) and in the southeastern subsidence of the Hochaz and Gafan subzones of the Gafan zone (160 m). They contain *Nummulites orbignyi* Gal., *Globigerina corpulenta* Subb., *Diastoma constellata* Lmk., etc.

In the Araz zone of the Lesser Caucasus, the Ypresian stage is represented by the Aylis, Arysu, Gilanchay and Kalaki suites (Babayev et al. 2015). The Aylis, Arysu and Gilanchay suites (lower and middle parts of the stage) mainly consist of sedimentary rocks: alternating clays, argillites, sandstones and aleurolites (up to 1500 m). The Kalaki suite (upper part of the stage) is represented by volcanogenic facies (more than 1000 m). They are observed in the basin of the Nakhchivanchay River and on the southwestern slopes of the Zangazur ridge. They contain *Nummulites spileccensis* Muen.-Chalm., *N. bolcensis* Muen-Chalm., *N. aquitanicus* Ben., *N. partschi* de la H., *Globorotalia subbotinae* Mor., *G. aragonensis* Nutt., *G. marginodentata* Subb., *Discoaster barbadiensis* Tan Sin Hok, *Marthasterites tribrachiatus* (Br. et Ried), and *Alveolina bosci* (Defr.), etc.

Deposits of the Lutetian stage are represented by the Kyukyuchay and Bilav suites (lower and middle parts of the stage) and the lower subsuite of the Paradash suite by (Babayev et al. 2015). In the northeast bank of the Ordubad trough, Lutetian deposits are represented by clays, sandstones, tuff-sandstones, and limestone (420 m); in the northeast flank of the Sharur-Julfa anticlinal dome, there are basal conglomerates, limestone, sandstones, and clays (100–200 m). The central and the southeastern part of the Ordubad trough has volcanogenic-sedimentary (80–900 m), tuff-conglomerate series (80–800 m) and the lower parts of the sedimentary series (50–400 m). They contain *Nummulites laevigatus* (Brug.), *N. uroniensis* Heim, *N. polygyratus* Desh., *Assilina laxispira* (de la H.), *A. exponensis* (Sow), *Acarinina bullbrooki* (Bolli), *Globigerapsis subconglobatus micra* (Schutz.), *Discoaster barbadiensis* Tan Sin Hok, *Spondylus radula*, and *Pseudoamussium corneum* So, etc.

Deposits of the Bartonian and Priabonian stages widely occur in the Paradash, Nursu and Turkesh depressions.

Deposits of the Bartonian stage represented by middle substages of the Paradash suite (Babayev et al. 2015) consist of alternations of limestone, sandstone and clays (400–600 m). They contain *Nummulites brongniarti* d'Arch. et H., *N. acutus* (Sow), *N. variolarius* (Lmk.), *N. millecaput* Boub., *N. perforatus* (Montf.), *N. gizehensis* Forsk., *Assilina exponens* (Sow), *Operculina alpine* Douv., *Truncorotaloides rohri* Bronn. et Berm., *Globigerina pseudocorpulenta* Chal., *G. azerbaidjanica* Chal., *Discoaster saipanensis* Br. et Ried., *Coccolithus eopelagicus* (Br. et Ried), *Variamusslum fallax* Korob., and *Diastoma costellatum* Lmk, etc. Deposits of the Priabonian stage are represented by the upper substage of the Paradash suite (Babayev et al. 2015), which consists of alternations of argillites, clays, sandstones, conglomerates, and nummulitic limestones with a thickness of 30–40 m. The deposits contain *Nummulites garnieri* Bouss., *N. fabianii* Prev., *N. budensis* Hantk., *N. concinnus* Jarz., *N. rectus* Curry, *Operculina alpina* Douv., *Rotalia octocamerata* Chal., *Bolivina antegressa* Subb., and *Globigerina corpulanta* Subb., etc.

Talysh

Eocene formations are widely developed in the Talysh zone, being represented by sedimentary and volcanic-sedimentary variations (Problems of the Talysh Geology 1958; Bagmanov 1963; Azizbekov et al. 1972, 1979; Khalilov 1978; Alizadeh et al. 2007). Ypresian and Lutetian deposits are represented by the Gosmalian suite; they transgressively occur with sharp angular unconformity and basal conglomerate in the base of the Paleocene deposits. They are represented by sedimentary-tuffogenous series of leucite-sanidine tuffs of the volcanic breccia and conglomerates of andesite-basalts and basalts, lavas and pyroclasts of picrite-trachybasalts and leucite basalts (880–1330 m), spread in the Gosmalian depression and Astara rise. In the upper part, these deposits contain *Nummulites laeviugatus* (Brug.), *N. uroniensis* Heim, *N. gallensis H.*, *N. brongniarti* d'Arch. et H., *N. millecaput* Boub. and *Asterocyclina stella* (Guemb.).

Bartonian deposits are represented by the Nesli suite, which transgressively occur with the basal conglomerates in the base above the Lutetian layers. They are represented by conglomerates, tuff-sandstone, shales, marls, tuffs and limestone (300–800 m). They contain *Nummulites prestwichianus* Jon., *Operculina alpina* Douv., *Globigerina turkmenica* Chal., *G. ouachitaensis* Howe et Wall., *Valvatia lamarckii* Desh., and *Chlamys parisiensis* (Orb.), etc.

Priabonian deposits are represented by lavas and pyroclasts of porphyritic trachyandesite-basalts and leucite trachyandesites (Peshtasar suite—900–1000 m), which change above to tuffs, tuff-sandstone, tuff-siltstones and shales (Arkivan suite—up to 1000 m) in the Lerik-Yardymly and Buravar zones. They contain *Nummulites chavannesi* de la H., *N. incrassatus* de la H., *Globigerinoides conglobatus* Brady, *Rotalia aff. similis* Chal., and *R. aff. cubanensis* Chal., etc.

3.6.1.4 Oligocene

Greater Caucasus

Oligocene formations conformably continue in the Eocene section on the southeastern plunge of a mountain-fold structure. They contribute to the structure of the Gusar-Devechi (Siyazan monocline), Sudur, Shahdagh-Khizi, Zagatala-Govdagh, Vandam and Shamakhy-Gobustan zones (Alizadeh 1945; Potapov 1954; Khain and Akhmedbeyli 1957; Khain and Shardanov 1957; Salayev 1961; Azizbekov et al. 1972; Mirchink and Shurygin 1972; Khalilov 1978; Aliyev et al. 2005a; Alizadeh et al. 2007). Deposits of the Maykop series (except for the uppermost Lower Miocene layers) in the southeastern part of the region are related to the Oligocene period. In the Shahdagh-Khizi (Gazma-Gryz village) and the Gusar-Devechi (Khanagah and Rustov villages) depressions, Oligocene deposits do not widely occur and are represented by non-carbonaceus shaly clays with a powdery coating of jarosite, concretions of siderite and fish remnants (with a total thickness of about 100 m). In the Gusar-Devechi area, Oligocene is represented by marlaceous clays with jarosite and sandy clays with interbeds of sands (with thickness up to 500 m). In the Shamakhy-Gobustan and Zagatala-Govdagh zones, the Maykop deposits widely occur and are represented by argillaceous and sandy-argillaceous facies. Their thickness is up to 500–1150 m. The Maykop deposits are poor with microfauna. They contain *Globigerina ouchitaaensis* (Howe et Wall.), *Bolivina nobilis* Hantk., *Nonion curviseptum* Subb., *N. pseudomartkobi* Chal., *Cibicides lobatulus* (W. et J.), etc. At the same time, here were discovered numerous remnants of fish Lepidopus *meletta priacanthus* and remnants of Cedroxylon trees.

Kur Depression

Oligocene formations outcrop only within the Kur-Gabyrry interfluve (Georgian border). However, according to data from deep drilling and geophysical research, they have developed quite widely, having contributed to a buried structure of the Middle and Lower Kur depressions (Alizadeh 1945; Agabekov and Mamedov 1961; Azizbekov et al. 1972; Mamedov 1973; Rasulov and Mamedzade 1978; Bagirzadeh et al. 1987; Aliyev et al. 2005a; Alizadeh et al. 2007; Mamedov et al. 2008b). In the interfluve of the Kur and Gabyrry rivers, the Lower Maykop deposits in the lower section are represented by a series of sandy-argillaceous alternations (100–1000 m) with the *Caucasina schischkinkayae* (Sam.), *Chiloguembelina gracilina* and *Coccolithus subdistichus* Roth et Hay. The upper section of the Lower Maykop is presented by sandy shaly clays and sandstones (700–900 m) with *Rotalia unduluta* Chal. and *Neobulimina elongata* (Orb.). Upper Maykop is represented by a series of sandy-argillaceous alternations with *Virgulinella aequale* Kusn., *V. pertusa* (Reuss), *V. poiliensis* Kuzn., Bolvina, etc. Lower Maykop in the lower part of the Yevlakh-Agjabedi Depression, is represented by shaly clays with interbeds of sandstone (up to 1000 m). They contain *Globigerina tumbely* Chal., *G. bulloides* Orb. and *Coccolithus subdistichus* Roth et Hay, which above changed by alternations of sandy and limestone shaly clays (300–900 m). Upper Maykop is represented by sandy-argillaceous facies (up to 300 m).

Lesser Caucasus

Within the Lesser Caucasus, the Oligocene formations continue in the Eocene's section and outcrop in separate windows of the Middle Kur Depression in the northeastern wing of the Lok-Garabagh zone, on the southeastern plunge of a mountain-fold structure on the left bank of the Araz River, and within the Araz structural zone (Alizadeh 1945, 1968; Azizbekov 1961; Shikhalibeyli 1964–1967; Azizbekov et al. 1972; Khalilov 1978; Alizadeh et al. 1980, 2007; Babayev et al. 2015).

Oligocene deposits in the northeastern foothills of the Lesser Caucasus are represented by facies of the Maykop series and occur very widely. Lower Maykop in the lower parts is represented by alternations of marly sandy shaly clays and sandstone (600–700 m) with *Caucasina schkinskae oligocaenica* Chal., *Rotalia zeivensis* Chal., *R. diverosepta* Chal., *Nonion pseudomartkobi* Chal., *Arca sulcocosta* Nyst, *Leda perovalis* Koen., *Thyasira vara* Korob., *Planorbella*, *Pontocypris oligocaenica* Zal., *Disopontocypris oligocaenica* (Zal), etc. Upper parts of the Lower Maykop are presented by alternations of sands, sandstone, conglomerates and shaly clays (300–1000 m) with *Glycymeris obovatus* Lmk., *Cordiopsus incrassate* Sow., *Ostrea callifera* Lmk., *Lentidium elongatum* (Sandb.), *Neritina fluveatilis* Zin., etc. Deposits of the Upper Maykop are represented by alternations of shaly clays, sands and sandstone (485–525 m) with *Cardium* sp., *Pecten* sp., *Quinqueloculina acneriana* (Orb.), *G. gracilis* (Karr.), etc.

In the southeastern subsidence of the Lesser Caucasus, Oligocene is found on the left bank of the low flow of the Bazarchay River, along the southwestern wing of the Gochaz trough, in the Gubatly area, and in the left bank of the Hakeri River. Oligocene deposits in the lower parts are represented by conglomerates, sandstone, gritstone argillites, siltstones and limestone (920 m) with *Lentidium donatciforme* (Nyst), *Corbula sokolovi slussarevi* Merkl., *Arca sandbergery* Desh., *Loxoconcha tunicata* Mand., *Cytherura placida* Mand., *Nonionella azerbaidjanica plana* Chal., etc. Higher horizons of the Oligocene deposits are exposed near the Gubadly and Khudafarin villages. The deposits are represented by argillites, limestone, sandstone and conglomerates (300–800 m) with *Glycymeris obovatus* Lmk., *Lentidium lamberti* Cossm., *Cordiopsis incrassata* (Sow.), *Turritella planispira* Nyst, etc.

Oligocene deposits in the Araz zone are presented by the Darydagh (Lower Oligocene) and Abragunus (Upper Oligocene) suites (Babayev et al. 2015), which occur in the east and northeast parts of the Nehram plateau, down the flow of the Alinjachay River, in the region of Darydagh Mt., etc. These deposits are represented by tuff-breccia, tuff-gritstones, tuff-sandstone, tuff-conglomerates and tuffs (1000–1100 m).

Talysh

In the Talysh zone, the sedimentary formations of the Oligocene transgressively overlie the Eocene volcanic-sedimentary complex (Alizadeh 1945; Problems of the Talysh Geology 1958; Bagmanov 1963; Azizbekov et al. 1972, 1979; Khalilov 1978; Aliyev et al. 2005b; Alizadeh et al. 2007). They contribute to the structure of the Lerik-Yardymly, Buravar and Jalilabad zones. The lower parts of Oligocene (Tilakend and Pirembel suites) are represented by carbonate and shaly clays and sandstone with *Nummulites* sp., *Globigerina officinalis* Subb., *Rotalia mexicana* Nutt., *Traxinus juglandina* Sap., *Pteris cadigera* S., etc. The Upper Oligocene deposits (the Shishnavar suite) are represented by thin layered shaly clays, alternating with sandstone with a powder coating of jarosite. Between the basins of the Vilashchay and Bolgarchay rivers, an increased presence of rudaceous rocks with single foraminiferas, small fragments of bones and scales of fishes were observed. The total thickness of these deposits varies from 670 to 2400 m.

3.6.2 Neogene

3.6.2.1 Miocene

The Miocene deposits of Azerbaijan widely occur in intermontane and premountain areas, as well as in the adjoining aquatorium of the Caspian Sea.

During the almost 150-year history, the Miocene deposits were studied by a number of scientists: G.V. Abich, M.G. Agabekov, G.A. Akhmadov, H.I. Aliyev, A.A. Alizadeh, Ak.A. Alizadeh, G.A. Alizadeh, G.I. Allahverdiyev, N.I. Andrusov, G. Artgaber, E.Z. Atayeva, A.I. Azizbekova, Sh. A. Azizbekov, V.V. Bogachev, P. Bonnet, R.A. Fataliyev, D.V. G.M. Gasymova, Golubyatnikov, I.M. Gubkin, D.V. Hajiyev, B.P. Juze, V.Y. Khain, D.M. Khalilov, A.N. Kudryavtsev, V.P. Kutsev,, L.N. Leontyev, L.D. Mamedova, A.V. Mammadov, Sh.F. Mekhtiyev, E.Z. Movlazadeh, F.A. Mustafayev, F.F. Osvald, K.N. Paffenholtz, I.I.

Potapov, I.A. Preobrajenskiy, M.T. Pronina, I.F. Pustovalov, M.I. Rustamov, S.G. Salayev, G.M. Sultanov, V.V. Tikhomirov, M.I. Varentsov, A.G. Voroshilova, V.V. Weber, and others.

Lower Miocene

The deposits of the Lower Miocene in Azerbaijan are related to the Maykop series and represent the Sakaraulian and Kotsakhurian regiostages. Conformably continuing in the Oligocene's section, they contribute to the geological structure of the southeastern plunge of the Greater Caucasus, Kur Depression, Lesser Caucasus and Talysh (Potapov 1954; Akhmedov 1957; Problems of the Talysh Geology 1958; Agabekov and Mamedov 1961; Azizbekov 1961; Azizbekov et al. 1972, 1979; Mirchink and Shurygin 1972; Mamedov 1973; Khalilov 1978; Rasulov and Mamedzade 1978; Alizadeh et al. 1980, 1986, 2000, 2007; Babayev et al. 2015).

The ***Sakaraulian regiostage*** (50 m) is the lower part of the Upper Maykop. It is represented by clayey and clay-sandy lithofacies (Lesser Caucasus and Talysh) and contains macro- and microfauna (beds with the *Neobulimina elongata leninabadensis*).

The ***Kotsakhurian regiostage*** (130 m) is represented by shaly lithofacies (non-carbonate shales) and is lacking in microfauna (Gazakh-Mammadtapa, Tumarkhanly, Khudafarin-Bahmanly). It contains fish and plant remains.

The Lower Miocene is the lower part of the salt-bearing, grey-colored series (carbonate, terrigenous), with a thickness of 70–160 m (Khachaparakh, Makhmud and others). It has macrofauna (Planorbarius, Lithophaga, etc.) in the Nakhchivan Depression of the Araz zone (Khalkhal and Garagala suites).

Middle Miocene

The deposits of the Middle Miocene represent Tarkhanian, Chokrakian, Karaganian and Konkian regiostages. On the earth's surface, they are mostly developed inside the boundaries of the Shamakhy-Gobustan zone within the Greater Caucasus. They also outcrop in the Gusar-Devechi zone (Siyazan monocline), Kur-Gabyrry interfluve, Nakhchivan superimposed trough and Talysh (Potapov 1954; Akhmedov 1957; Problems of the Talysh Geology 1958; Agabekov and Mamedov 1961; Azizbekov 1961; Azizbekov et al. 1972, 1979; Mirchink and Shurygin 1972; Mamedov 1973; Khalilov 1978; Rasulov and Mamedzade 1978; Alizadeh et al. 1980, 1986, 2000; Aliyev et al. 2005b; Babayev et al. 2015).

Tarkhanian regiostage. The deposits of the Tarkhanian regiostage in Azerbaijan are represented by clayey and sandy-clayey lithofacies. Their natural outcrops are of narrow occurrence and are chiefly defined from well samples by the availability of foraminifera *Globigerina tarchanesis*.

The Tarkhanian regiostage is represented by marls and shales in the Gusar-Devechi trough (Shuraabad, Tangaalty, Kurkachidagh sea, etc.), Shamakhy-Gobustan zone and Absheron (Shorbulag, Balakhany, Gyrmaki, etc.). Tarkhanian deposits are hardly separated in drilled wells from the Chokrakian deposits due to the weak thickness of the Tarkhanian and frequent absence of microfauna in cores. This regiostage was recognized in the Kurdakhany, Sulutapa, Gijaki-Akhtarma, Shargi Sundi, and Cheyildagh areas due to the availability of typical foraminifera (*Globigerina tarchanensis* and *Bolivina tarchanensis*). It is represented by shaly-marl deposits with a thickness of 4–30 m.

Besides microfauna, the *Spiratella andrusovi tschokrakensis* has been found in Tarkhanian deposits within the Kur and Gabyrry interfluve in Mammadtapa, Molladagh, Gushkuna, and some other areas, as well as around Mt. Gabagtapa.

Tarkhanian deposits are bedded transgressively on the grey-coloured series in the Nakhchivan Depression (Araz zone) and are represented by volcanic (Chashyrdagh suite) and sedimentary (Dizazina and Arazin suites) rocks. Here were found *Turritella vermicularis*, *T. strangulata*, etc.

Tarkhanian deposits are bedded transgressively on the grey-coloured series in the Nakhchivan Depression (Araz zone). Here were found *Turritella vermicularis*, *T. strangulata*, etc.

Within the Talysh zone, Tarkhanian deposits (up to 80 m) crop out on the right bank of the Bolgarchay River. They are represented by clays with sand interbeds, sandstones, marls and conglomerates.

Chokrakian regiostage. The deposits of the Chokrakian regiostage of Azerbaijan are represented by sandy-clayey and clayey lithofacies. This regiostage is represented by clays (with *Spiratella* and *russovi tschokrakensis*, *Loripes dujardine*, etc.) with thin interbeds of sands (50–340 m) in the Gusar-Devechi region.

In the Shamakhy-Gobustan zone, the Chokrakian regiostage is represented by clayey lithofacies in the north (Kemchi, Boyanata, etc.) and by sandy-clayey in the south (Umbaki, Ajiveli, Cheyildagh, etc.). The first lithofacies is represented by clays with dolomite interbeds of Spiratella. Mollusks of Arca, Ervilia, Gibbula, Ceritrium genera, foraminifera, Ostacoda, and otholites are found in sandy-shaly lithofacies (up to 500 m).

The Chokrakian regiostage is represented by shaly lithofacies on Absheron (Uchtapa, Govun, Jorat, Fatmayi areas, etc.). It is expressed by clays with interbeds of dolomites and marls, except for the southern part of the western Absheron, where the facies have been developed.

Chokrakian deposits have been revealed as a result of drilling in the Gushkhana, Garadagh and Shorbulag areas of Absheron, with observed thickness of 60 to 300 m. They have also been found in the Gyrgyshlag and Donguzdug areas in Gobustan. Their thickness varies from 198 to 451 m in the Duvanni area.

In the southeastern margin of the Lesser Caucasus (Jabrayil region), Chokrakian deposits are represented by a series of interbeddings of clays, sands and sandstones with inclusions of pebbles. There exists a rich mollusk fauna: *Spiratella andrussovi tschokrakensis*, *Loripes dujardini*, *Cultellus probus*, etc. The thickness of these deposits is about 140 m.

Within the Nakhchivan Depression of the Araz zone, Chokrakian deposits are conformably laid on turritel layers and expressed by agrillaceous-gritstone-sandy rocks with a thickness of 42–100 m (Garadash suite). In sandstones in the eastern part of this depression, *Sandbergenia acicularis* was found. In the clays of its deeper central part, *Spiratella andrussovi tschokrakensis* was discovered.

Within the Talysh zone, Chokrakian deposits are concordantly bedded in Tarkhanian deposits. They are represented by alternations of a series (up to 400 m) of sandstones with interbeds of sandy clays and marls. Numerous remains of *Spiratella andrussovi tschokrakensis* mollusks; *Ostrea digitalina*, *Nuculara pella* and *Ervilia pusilla;* as well as foraminifera, Ostracoda, otholites, bones and scales of fish were found here.

Karaganian regiostage deposits in Azerbaijan are mainly composed of grey clays with interbeds of sands, sandstone, dolomites and marls. In the Gusar-Devechi Depression, Karaganian deposits outcrop in the basins of the Valvalachay, Jagajugchay, Turyanchay and Gilgilchay rivers. They are determined by well data in the Yalama, Shuraabad, Zeyva and Saadan-Amirkhanly areas, and are composed of *Spaniodontella gentilis*, *otolithes-Rhombuscories* and single foraminiferas.

In the Absheron-Gobustan region, deposits of the Karaganian regiostage compose the lower pack of the Diatom suite. In the Shamakhy-Gobustan zone, this deposit is represented in the north (Mayash, Goradil, Siyaki, etc.) by clays with interbeds of dolomites and marls. The south (Umbaki, Ajiveli, Ilkhichi etc.) has sandy-agrillaceous deposits with a thickness of 350 m. In the Absheron zone, the Karaganian regiostage is represented by agrillaceous and agrillaceous-sandy rocks with dolomites. Their thickness is 130 m. These deposits are cropped out by wells in the Gyrmaki, Binagadi and Shorbulag areas. In all wells, Karaganian deposits are characterized by typical fragmental thick clays with the remains of otolithes (*Rhombus cories*, *Rh. cories binagadinica*), as well as the vertebras, teeth, bones of fishes and embryos of *Spaniodontella gontiles*.

In the interior part of the Middle Kur Depression, deposits of the Karaganian regiostage developed in the northwestern part of the Kur and Gabyrry rivers' interfluve. In the Sajdagh, Chobandagh, and Armudlu areas these deposits are represented by alternations of grey clays (202 m) and grey marls (110 m).

Exposures of the Karaganian deposits in the southeastern margin of the Lesser Caucasus (Jabrayil region) are represented by brown and red clays and sandstone (with a total thickness of 80 m) with *Spaniodontella*, *Rhombus cories*, *Quinquelloculina consobrina*, etc.

In the Nakhchivan Depression (Araz zone), deposits of the Karaganian regiostage (Payiz suite) developed in the eastern and central part (in Jahri and Chalkhangala villages) and are represented in the lower part of the Karaganian regiostage by reddish-brown clays, siltstones and sandstone with interbeds of gypsum, anhydrite, and rock salt. Their thickness is 85 m. In the central part of the depression, these layers are represented by gypsum clays and (rarely) by sandstone, siltstones and marls. Their thickness is 350–400 m. Foladic layers of aleurolite and sandy littoral formations with a thickness of 60–350 m are very well developed in the eastern part of this depression. They are characterized by the mollusk fauna *Clithon pictus*, *Barnea ujratamica*, and *B. kubanica*, as well as by the earth's mollusks: *Microstele cf. caucasica*, *Caucasotachea cf. kubanica*, *Chondrula*, etc.

In the Talysh zone, natural outcrops occur in the valley of the Talachay River and are represented mainly by argillaceous and sandy rocks (thickness of 25 m). In the Novruzali, Tumarkhanly and Jalayir areas, Karaganian deposits are determined according to the presence of the following paleontological remains in the cores: *Spaniodontella gentilis*, *Nonion miocenicus*, *Leptocythere stabilis*, *otolithes of fishes —Rhombus corius*, *Rh. corius miocenicus*, etc.

Konkian regiostage deposits in Azerbaijan occur in the same places as Karaganian deposits. There exist some difficulties in separating these deposits from the Lower Sarmatian and Karaganian, which is why Karaganian and Konkian deposits are usually examined together.

In the territory of the Gusar-Devechi Depression, Konkian deposits do not widely occur and developed only in the southeastern part of this depression. They have be detected by drilling in the Gyzylburun, Saadan-Amirkhanly, Zaghly and Zeyva areas. They are expressed in the agrillaceous facies and by fauna *Mactra basteroti konkensis*, *Dorsanum duplicatum*, *Elphidium kudakoensis*, *Trigla konkensis*, etc.

In the Absheron-Gobustan region, Konkian deposits compose the second pack of the Diatom suite. In the northern part of the Shamakhy-Gobustan zone (Mayash, Goradil and Siyaki), the Konkian deposits are represented in the agrillaceous facies with a number of otolithes of fishes, *Trigla konkensis*. To the south, increasing sandy material is observed (e.g., Cheyildagh, Umbaki and Arzani) with mollusk fauna (*Spiratella konkensis foraminifera—Elphidium kudakoense* etc.), otolithes (*Trigla konkensis*), and ostracoda

(*Paracytherois gracilis*, etc.). In the upper part of the sections with light-grey foliaceous clays, there exist diatomic *sea-weed-Coscinodiscus radiatus*, *C. oculus*, and *Cocconeis scutelum*, etc. The Konkian deposits within the Absheron occur mainly in its western part (Guzdak, Gobu, etc.) and are represented by agrillaceous facies; the clays contain foraminifera (*Elphidium kudakoense*, *Bulimina konkensis*, etc.) and otolithes (*Rhombus cories miocenicus*, etc.). In the wells drilled in the Gyrmaki valley, the Pirallakhy Island, Fatmayi, Binagadi, Atashgah, and Shorbulag, Konkian deposits are expressed by shales with *Trigla konkensis*.

The existence of Karaganian-Konkian deposits was determined lately as a result of drilling in the Garadagh, Utalgi, Kahnizdagh, Duvanny, Duvanny-west, Solakhay, Sangachaldeniz and Duvanny deniz areas. Here, they compose the lower part of the Diatom suite and are expressed by grey sandy clays with interbeds of sands, sandstone and marls.

In the interfluve of the Kur and Gabyrry rivers, the Konkian regiostage has the same characteristics (Armudlu, Mammadtapa and Khatynly). In the interior part of the Lower Kur Depression, the Konkian deposits are represented in the west by sandy sediments and in the east by argillaceous sediments with *Spaniodontella gentilis* and *S. tapesoides*. Their averaged thickness is 75 m.

Konkian deposits in the southeastern margin of the Lesser Caucasus (Jabrayil region) do not widely occur. Here, they are expressed by the alternation of brown-yellow and grey-brown sandstone with interbeds of grey-brown clays (40 m), with *Ervilia trigonula* and *Pholas* (*Barnea*) *ustjurtensis*.

In the Nakhchivan Depression (Araz zone), deposits of the Konkian regiostage (Tumbul suite) developed in the eastern and central parts (in Jahri, Boyukduz, Khok and Chalkhangala villages). They are represented by clays, aleurolites and sandstone (up to 360 m) with Pholas *ex gr. bogatshovi* Ossip., *Bornea oft.sinzovi* Ossip., *B. pseudoustjurtensis* Bog., *B. ustjurtensis* Ossip., *B. ujramatica* Andrus., *B. tumbulica* Azizb., *Theodoxus pictus* Ferussac, *Microstele cf. caucasica* Stekl., *Caucasotachea cf. kubanica* Stekl., *Chondrula* (*Mastus*) *tumbulica* Azizb., *Ch.* (*Mastus*) alizadei Azizb. and *Helix nachishewanicus* Azizb.

Konkian deposits in the Mountainous Talysh are cropped out in the Talachay River and conformably lie on the Karaganian deposits. Konkian deposits are represented by grey clays with interbeds of sandstone and marls (70–80 m). Sandy-agrillaceous deposits discovered in wells contain foraminifera, such as *Elphidium kodakoense*, *Articulina konkensis* and *otolithes-Trigla konkensis*.

Upper Miocene

Within Azerbaijan's boundaries, the Upper Miocene formations are most developed within the southeastern plunge of the Greater Caucasus, Talysh Mts. and Gusar-Devechi zone, and Kur and Nakhchivan superimposed troughs (Sultanov 1953, 1960; Potapov 1954; Akhmedov 1957; Problems of the Talysh Geology 1958; Agabekov and Mamedov 1961; Azizbekov 1961; Shirinov and Bazhenov 1962; Vekilov 1962; Azizbekov et al. 1972, 1979; Mirchink and Shurygin 1972; Mamedov 1973; Alizadeh 1974; Khalilov 1978; Rasulov and Mamedzade 1978; Alizadeh et al. 1980, 1986, 2000, 2007; Babayev et al. 2015).

The deposits of the Upper Miocene are characterized by a wide occurrence and are represented by sediments of the Sarmatian and Meothic regiostages. Those of the Sarmatian regiostage have mainly a marine character and divide most of the regions into three subregiostages. In the southeastern part of the Greater Caucasus, Sarmatian deposits are exposed almost uninterruptedly along the southern edge of the Gusar-Devechi, Sudur and Shamakhy-Gobustan zones. The Sarmatian sediments in the Gusar-Devechi Depression are represented by all of its subregiostages and widely occur in the basins of the Gilgilchay, Valvalachay, Kamalchay, Garachay and Gudiyalchay rivers. The Lower Sarmatian subregiostages from the Valvalachay River to the northwest are represented in the shallow-water littoral facies and contain a rich association of the mollusk fauna *Ervilia pusilla dissita* and *Gibbula subalatro* (Chakhmagdagh). To the east from the river, the agrillaceous facies have the foraminifera fauna *Porosononion markobi* and *Nonion bogdanowiczi*. Their thickness is 150 m.

The Middle Sarmatian deposits, according to the foraminifera fauna, are subdivided into two zones: the zone of nonionides (Porosonion and subgranosus) and the zone of elfidiides (e.g., Parellina regina). These deposits have mainly shallow-water characteristics (basins of the Garachay, Gudiyalchay and Valvalachay rivers). Their thickness is about 360 m and they are widely represented faunistically. The Nearcaspian type of lithofacies (Gilgilchay) is expressed by shales with rare interbeds of sand and sandstone containing *Paphia naviculata*, *Mactra urupica* and *Hydrobia elongata*. Their thickness is 91 m. The Guba type of lithofacies is expressed by clays, sands, sandstone, and coquina with *Musculus sarmaticus*, *Cerastoderma fittani* and *Calliostoma sarmates*. This lithofacies is especially rich faunistically near the Chakhmagdagh area. Its thickness is 102 m. The Upper Sarmatian subregiostage is subdivided into two horizons according to its lithological and faunal characteristics: the lower is the agrillaceous horizon (Rostov) and the upper is the sandy-agrillaceous (Kherson). These deposits contain *Mactra caspia*, *M. bulgarica*, *M. crassicolis* and *M. nalivkini*. Their thickness is 250–300 m.

Sarmat deposits in the Shamakhy-Gobustan and Absheron zones are represented by the lower and middle subregiostages and are related to the third pack of the

Diatom suite. In the Shamakhy region (Meysari, Khynysly and Gyz-Galasy), these deposits have analogous complexes of mollusk fauna and foraminifera, as in the Gusar-Devechi zone. In the Absheron zone, the Lower and the Middle Sarmatian subregiostages have been determined in the northwestern, northern and central parts of the peninsula (e.g., Uchtapa, Kasmalydagh, Orjandagh and Fatmayi). They are expressed by yellowish-grey and grey-brown clays with marl interbeds. Faunistically, Sarmat has very poor representation, being characterized by the presence of a great number of *Algae-Ovulites-Ovulites sarmaticus* calci, *otolithes-Gadidarum minusculus* fish, and *Gobius sarmatus*. For sections expressed in the sandstone-agrillaceous lithofacies (except the mentioned fish and vegetative remnants), the presence of foraminifera, such as *Elphidium macellum* and *Porosonion granosus*, is typical. The thickness of the deposits reaches 100 m. Sarmatian deposits (with the presence of otolithes and ovulites) were determined in Absheron peninsula in the Gala, Zira, Mashtagha-Buzovna, Gyrmaki, Binagadi, and Garadagh areas. The thickness of the deposits here reaches 300 m.

In the Kur and Gabyrry interfluve and within the inner zone of the Middle Kur Depression, the Sarmatian deposits are cropped out in the Eldaroyughu, Mammadtapa, Molladagh, Palantokan, Gurzundagh, Alajyg, Yaylajyg and Aghtakhtatapa areas and were also exposed by drilling in the Armudlu, Molladagh, Sajdagh, Gurzundagh, Mammadtapa, Damirtapa-Udabno, and Khatynly areas. The Lower Sarmatian deposits are outcropped mainly in the western part of the region (Mammadtapa) and are represented by shales (containing *Abra reflexa* and *Ervilia pusilla dissata*) and sandy deposits (containing rich fauna genus of *Abra*, *Ervilia* and *Gibbula* with foraminifera fauna, ostracoda and otolithes). The Middle Sarmatian subregiostage in the mentioned regions is represented by *cryptomarctra* layers, where the faunistic complex in the grey clays consists of mollusks (e.g., *Cryptomactra pesanseris*) and foraminifera genus (*Quinqueloculina*, *Miliolinella*, *Dogielina*, *Articulina*), as well as ovulites, otolithes and ostracoda. The horizon containing typical Middle Sarmat fauna is presented by sandstones and limestones with rich mollusk fauna genus of *Paphia*, *Mactra* and Cerastoderma. The Upper Sarmatian deposits occur in the eastern part of the region (Eldaroyugu, Boyuk Palantokan and Aghtakhtatapa) and are represented by marine (Rostov and Kherson horizons) and continental deposits (Eldar suite). The Rostov horizon (argillaceous rocks of 200 m) contains mollusks *Mactra podolica* and *Solen subfragilis*. The Kherson sandy horizon (400 m) contains a rich complex of hipporion fauna (hipparions, antelope, giraffe, seals, hyaena, etc.). The Eldar suite (350 m) is structured by alternations of clays, sands and sandstone with thin layers of gypsum. It contains a rich complex of freshwater mollusks of genus *Unio*, *Melanopsis* and Helix. The bones from antelope, ostrich and hipparion were found.

The Sarmatian deposits (Duzdagh suite) within the Nakhchivan Depression (Araz zone) are subdivided into the Lower-Middle Sarmatian and the Upper Sarmatian subregiostages. Lower Sarmatian deposits are outcropped in the eastern and central parts of the trough, where they are expressed by aleurolite-sandy layers (60–80 m) with the *Pirenella disjuncta*, *P. picta mitralis*, etc. In the central part of the trough, the deposits of the Lower Sarmatian subregiostage (350 m) and the lower part of the Middle Sarmatian subregiostage (140 m) are represented by an alternation of sandstone, siltstones, clays and marls with the *Sphaerium sarmaticum* and *Pisidium schakhtachticum*. The Upper Sarmatian subregiostage (140–150 m) contains the deposits of rock salt (5–10 m) containing *Mactra caspia* and *M. bulgarica*.

Within the Talysh zone, the Sarmatian deposits occur in the northeastern and northwestern parts and are represented by all three subregiostages. The Lower Sarmatian sub-regiostage (200 m) is represented by grey shales with rare interbeds of sandstone and marls (Tumarkhanly, Aghdash, etc. villages) with a poor fauna of *Musculus sarmaticus*, *Ervilia pusilla dissita*, and *Cerastoderma plicatum*, etc.; foraminifera of *genera Elphidium*, *Non-ion*, and *Porosononion*; ostracoda and otolithes. The Middle Sarmatian subregiostage (about 600 m) is represented by shales, limestone and clays with interbeds of sandstone and sand with the mollusk fauna *Musculus sarmaticus*, *Paphia naviculata, and Mactra urupica*, *M. vitaliana*. The Upper Sarmatian subregiostage has been determined by drilling near the Tumarkhanly, Garmali, Uzuntapa (Novogolovka) and some other villages. The last subregiostage is presented by thick grey shales with interbeds of sandstone containing *Mactra podolica*, *Solen subfragilis* and *Porosononion granosus*, dating to the Rostov horizon. Their thickness is 120 m.

Meotian regiostage. The Meotic deposits within Azerbaijan occur in the Gusar-Devechi, Shamakhy-Gobustan and Absheron regions and in the regions of the internal zone of the Middle Kur Depression. In the Gusar-Devechi zone, the Meotian regiostage is expressed by breccia-shaped dolomites with thickness of 25 m (Gilgilchay River). Near the Zaghly village, the olitolic limestone with thickness of 70 m was adjusted by the Meotian deposits. According to drilling data in the Shuraabad-deniz, Yashma-deniz and Kurkachidagh-deniz, Meotic sediments were adjusted according to the occurrence of fish otolithes (*Sparidarum meoticus*, etc.), ostracoda (*Loxoconcha meotica*, etc.), foraminifera (*Quinqueloculina seminulum*, etc., *diatoms algae*) and also larva shells of myarian mollusks.

In the Gobustan region and in the Absheron peninsula, the Meotian deposits compose the fourth (upper) pack of the

Diatom suite. The base facies of Meotian are represented by shales and argillaceous clays (the latest are more typical) with interbeds of dolomites, breccia and volcanic ashes with remnants of diatom algae. The Lower Meotian deposits are known in the central and northern Gobustan (Mayash, Siyaki, Burgut) and in the Shamakhy region in the south slope of the Pirdaraki Mt., where *Mytilaster incrassatus*, *Abra tellinoides* and *Ervilia minuta* were found in the Early Meotic sandstone. The Upper Meotian deposits (Osmanbeyli, Adnaly and Bakla villages) are represented by sandstone and limestone containing *Tirricaspia cf. jalpuhensis* and *Rissoa cf. trochus*. The thickness of the deposits reaches 30 m.

In central Gobustan, Meotian deposits were adjusted in boreholes drilled in the Gijaki, Sheytanud, Ajiveli, Rahim and Arzani-Glych areas. Here, the Meotian regiostage is represented by sandstone-argillaceous lithofacies and contains *diatoms algae Coscinodiscus gigas* E. and *Grammatophora arens* Ktz., foraminifera (*Quinquelocu-lina seminulum*, *Q. simplex*, etc.), ostracoda (*Leptocythere maeotica* and *Loxoconcha maeotica*), otolithes (e.g., *Percidarum sigmolinoides*), shallow clays of myarian and Gastropodas mollusks.

The Meotian regiostage in the Absheron area is represented by clays and argillaceous shales containing fish and vegetative remnants. The thickness of the deposits is 70–150 m.

In the Kur and Gabyrry interfluve of the Middle Kur Depression, Meotian deposits are represented by continental facies and compose the lower part of the Shirak suite in Georgia.

The Meotian deposits (Gulshanabad suite) within the Nakhchivan Depression of the Araz zone are outcropped in the central parts of the trough (near the Paiz village), where they are expressed by clay layers (25–30 m) with *Quingueloculina aff. Seminulum meotica* Gerke, etc.

Pontian regiostage. The deposits of the regiostage in Azerbaijan are represented by marine and continental formations. These deposits widely occur in the southeastern submergence of the Greater Caucasus within the Absheron peninsula (Atashgah area, near Gobu and Binagadi villages, Zigilpiri-shor, Masazyr, Kopruk-shor and Keyraki-shor sections), in the Shamakhy-Gobustan area, and in the Kur Depression (southeastern Shirvan, Agsu and Goychay towns). The deposits have been discovered by drilling in the islands of Absheron and Baku archipelago, as well as in the Gusar-Devechi Depression. The marine facies, widely developed in Eastern Azerbaijan, is represented by coastal shallow-water and relatively deep-water sediments. Continental formations occur in the central and western regions of Azerbaijan.

The marine deposits of the Pontian regiostage are bedded transgressively everywhere on the older deposits of Miocene (Absheron peninsula, Gobustan and Shamakhy regions), Oligocene (Shamakhy-Gobustan area and interfluve of Aghsu-Goychay) and the Upper Cretaceous (Shamakhy-Gobustan and Gusar-Devechi zones). They are also transgressively covered by deposits of the Balakhany and Akchagyl regiostage.

The more complete sections of the lower substage of the Pontian regiostage are known in the Maraza plateau (north-western Gobustan) in area of the Sundi ravine (more 600 m), as well as in the Nabur, Khilmilli and Absheron peninsulas (corresponding section near the Jorat village consists of 155 m), which survived washout of the Middle Pontian transgression. Analysis of the paleontological species of the Black Sea and Caspian area fauna allowed to apply for the identification of the lower substage of the Azeri Pontian a combination of Odessa and Yevpatory beds.

The Lower Pontian subregiostage deposits within the Absheron peninsula are lithologically represented by dark-grey, shaly clays, brown clays interbeds (21 m), containing small, sharp-ribbed *Cardium*, *Limnocardium*, and some species of the *Paradacna abichi R. Hoern genus*. Lithological content and fauna forms indicate their deep-water origin. The Middle Pontian deposits (*Shamakhy suite*) of the Absheron peninsula are more developed than the Lower Pontian ones; they are represented by alternations (79 m) of dark-grey and light-brown rocks exposed within interbeds of silts and volcanic ash. These deposits are faunistically characterized by the representatives of *Gastropods Valenciennius-V. annulatus* Rouss., *V. revolutus* Bog., *V. orientalis* Bog., and also *Paradacna abichi* R. Hoern. In addition, the following forms of mollusks and Ostracoda were found: *Didacna lutrae* Andrus., *D. schemachinica* Andrus., *D. meisserensis* Andrus., *Dreissena meissarensis* Andrus, *Melanopsis dianaeformis* Andrus., *Zagrabica rugosa* Andrus., *Pontoniella acuminata* (Zal.), *P. loezyi* (Zal.) and *Xestoleberis lutrae* (Schn.). The Upper Pontian subregiostage (*Babajan suite*) is represented on the Absheron peninsula by grey, dark-grey, and greenish-grey compact clays, as well as interbeds of shaly limestones and clayey rocks. The thickness varies from 35 to 55 m in different sections (Alajalar-Jorat, Gyrmaki, etc.). The Pontian deposits are nonconcordantly underbedded by clays of diatomic beds and are covered by sands of the lower part of the Gyrmaki suite of the Mahsuldar gat (Productive Series).

Within the Shamakhy-Gobustan zone, Pontian deposits are expressed lithologically by sandstone-limestone-clay facies and developed in the northwestern direction of region in the Shamakhy-Meysari ridge. They are represented by classic sections of Meysari and Khynysly, which were cut out by the deep Khynysly and Meysari ravines. In addition, the Pontian deposits cropped out in vicinity of the Madrasa, Kalakhana, Charkhan, Nuydu, Goylar, Gushchu and Gubaly villages. The lower subregiostages (130 m) are characterized by not-large complex of small-size, weak, salt-water mollusk and Ostracoda fauna: *Congeria panticapaea* Andrus., *C. novorossica* Sinz., *Dreissena simplex* Barb., *Monodacna pseudocatillus* Barb., *Chione* (*Parvivenus*) *wildhalmi* Sinz., *Abra* (*Syn-desmya*) *tellinoides* Sinz., *Trachyleberis pontica* Liv. and *Caspiolla dorsoarcuata* (Zal.). The dominating fauna of the middle subregiostage (*Shamakhy suite*—115 m) is represented by the following species: *Paradacna abichi* R. Hoern., *Valenciennius annulatus* Rouss., *Val. revolutus* Bog., *Chartoconcha bayerni* R. Hoern., *Didacna schemachinica* Andrus. and *D. meissarensis* Andrus. The upper subregiostages (Babajan suite, 350 m) are represented by the following dominating forms: *Cardium negativum* Andrus., *Monodacna babadjanica* Andrus., *Didacna depereti* Andrus., *D. laskareni* Andrus., *D. Pirsahatica* Andrus., *Prosodacna schirvanica* Andrus. and *Limnocardium nargiavagicum* Ebers.

The Pontian deposits cropped out in the valley of the Gudiyalchay River, in the vicinity of the Kusnat village of the Gusar-Devechi Depression. Lithologically, they are represented by a series (300 m) of dark-grey, bluish-grey striped clays with interbeds of sands and sandstones, non-concordantly bedded on the Cretaceous deposits.

The Pontian deposits within the Middle Kur Depression are represented by continental facies; in the Ajinohur foothills and Shirak desert, there are mainly by river and delta formations. The remains of representatives of fresh-water mollusks and also fauna of vertebral animals (*tooth Dinotherium*) have been found in these deposits.

3.6.2.2 Pliocene

The Pliocene deposits of Azerbaijan have been studied by G. V. Abikh, M.V. Abramovich, M.G. Agabekov, D.A. Aghalarova, G.A. Akhmadov, A.G. Aliyev, A.A. Alizadeh, K.A. Alizadeh, N.I. Andrusov, A.K. Babayev, Yu. P. Bajenov, L.F. Batsevich, V.B. Baturin, V.V. Bogachov, Z.K. Gadirova, D.V. Golubyatnikov, V.I. Gorin, I.M. Gubkin, V.G. Idrisov, G.A. Ismailov, D.I. Jafarov, Z.K. Kadirova, D.M. Khalilov, A.V. Mammadov, Sh.F. Mehtiyev, I.S. Mustafayev, I.I. Potapov, I.F. Pustovalov, Putkaradze, N.I. Sorokin, Kh.M. Sheydayeva-Guliyeva, A.Sh. Shikhlinskiy, F.A. Shirinov, S.G. Simonovich, V.A. Sulin, A.D. Sultanov, B.I. Sultanov, E.A. Tagiyev, G.G. Tsulukidze, N.I. Usheykin, B.G. Vekilov, P.Y. Volarovich, V.V. Weber, B. P. Yasenev, and others.

The Pliocene deposits of Azerbaijan widely occur in the southeastern part of the Greater Caucasus, in the Absheron peninsula, the Gusar-Devechi and Kur Depression, the Araz and Kalbajar zones of the Lesser Caucasus, as well as in the adjoining aquatorium of the Caspian Sea (Potapov 1954; Akhmedov 1957; Agabekov and Mamedov 1961; Azizbekov 1961; Shirinov and Bazhenov 1962; Shikhlinskiy 1967; Alizadeh 1969; Azizbekov et al. 1972; Mirchink and Shurygin 1972; Mamedov 1973; Khalilov 1978; Alizadeh et al. 2000, 2007; Babayev et al. 2015). The deposits of the Pliocene in Azerbaijan represent the lower (Zanclean) and upper (Piacentian) stages.

Lower Pliocene

The deposits of the Lower Pliocene in Azerbaijan represent the ***Balakhany regiostage***. The marine facies of this regiostage (Mahsuldar gat or Productive Series) of Azerbaijan strongly differ by hydrocarbon distribution (see also Chapter II of the second volume, "Economic Deposits and Applied Geophysics"). The depositions of Mahsuldar gat (MG) are widely developed and have a great thickness in the Absheron peninsula, Lower Kur and Shamakhy-Gobustan (especially in Jeyrankechmaz area of the South Gobustan) depressions. These deposits occur also in the Gusar-Devechi zone and Ajinohur region of the Middle Kur trough. Outside Azerbaijan, analogies of these deposits have developed in Turkmenistan (Red Rocks series), Georgia (tops of the Shirak suite), and the Northeastern Caucasus (Terek suite). The stratigraphic position of MG, based on the studies of N. I. Andrusov, D.V. Golubyatnikov and I.M. Gubkin (Azizbekov et al. 1972; Alizadeh et al. 2007), was defined between underlying the Pontian and overlapping Akchagyl regiostages. The most difficulty arose in paleontological correlations due to the lack of typical fossils. The stratigraphy of the MG (and its subdivision into separate suites and horizons) was defined on the basis of the total lithological composition. Five complexes (sandstone, sandstone-argillaceous, clay, shingle, rock debris-loam and sandstone-loam) had been distinguished by Potapov (1954) in the MG of the Absheron region according to the lithological peculiarities. According to mineralogical data, the stratigraphic scheme of the MG of Gobustan was suggested by A.G. Aliyev and G.A. Akhmadov (Azizbekov et al. 1972; Alizadeh et al. 2007). Because the common stratigraphic scheme has not been developed yet, we give here the MG stratigraphy for the various facies zones: I—Absheron type of sediments (the Absheron peninsula, Absheron archipelago); II—Lower Kur type of sediments (the Lower Kur depression and adjacent part of Baku archipelago); III—Gobustan type of sediments (Jeirankechmez depression and the part of Baku Archipelago); IV—Nearcaspian—Guba type of sediments; and V—Ajinohur type of sediments.

The Absheron Type of Sediments

Within the Absheron peninsula and the adjacent aquatorium of the Caspian Sea, MG is studied both on outcrops (Gyrmaki and Yasamal valleys, Atashgah, Shabandagh, etc.) and on large numbers of wells' sections located in both onshore and offshore areas. MG is overlapped by deposits of

Akchagyl regiostage (for which the presence of black clays in the base is typical). In the lowermost interval of MG, *Monodacna babadjanica*, *Didacna lascarevi*, *Limnocardium* sp., and fish remnants have been recorded. A sharp increase of the MG thickness has been observed not only in the southern, eastern and southeastern directions, but also in the southwest in the direction of Garadagh and Lokbatan areas.

Seven sedimentary rhythms of the MG deposits were distinguished by I.I. Potapov for all Absheron regions: I rhythm—Gala suite (GaS); II rhythm—Pre-Gyrmaki and Gyrmaki suites (PGS and GS); III rhythm—Post-Gyrmaki clay and sandy suites (PGCS and PGSS); IV rhythm—"clayes under the VIII horizon," IX and X horizons of the Balakhany suite and the Fasila suite; V rhythm—series IV, V, VI, VII and VIII horizons of the Balakhani suite; VI rhythm—middle and top of Sabunchu suite; and VII rhythm—Surakhany suite. The stratigraphical unit characteristics of MG for the Absheron region are given below.

Gala suite (GaS). These deposits were studied from numerous wells, which exposed this suite in different areas of the Absheron peninsula and adjacent aquatorium of the Caspian Sea. The deposits of GaS were exposed by the wells for the first time in the southeastern pericline of the Gala fold and then in the eastern limb of the Surakhany and Garachukhur-Zykh anticlines, within the south pericline on the Bibi-Heybat fold and in the areas of the Absheron archipelago (Pirallakhy Island, Chilov Island, Gurgan-deniz and Neft Dashlary).

Recently, GaS was exposed by onshore wells in the Turkan and Zira areas, Bina-Hovsan syncline, as well as offshore in Gunashli, Hazi Aslanov, Palchyg Pilpilesi, Khali, Janub, Gum-deniz areas; on western limb of the Darvin uplift; the southern periclinal part of the Mardakan-deniz structure; and the southern limbs of the Garadagh and Lokbatan folds of southwestern Absheron. GaS generally consists of sandy-argillaceous formations. Until 1968–1969, this suite had been divided into 3 subsuites (from the bottom to top): GaS-3, GaS-2 and GaS-1. GaS-2 is more sandy, GaS-3 is generally composed of clays, and GaS-1 is represented by alternation of clays and sands with the predominance of the first ones. GaS-3 has the greatest thickness (150–160 m). The thickness of GaS-1 varies from 75 to 80 m. The thickness of GaS-2 varies from 30 to 80 m (in the northwest) to 125 m (in the southwest).

Lately, with the appearance of new drilling data in the southeastern part of Absheron area, new subsuites (GaS-4, GaS-5 and GaS-6) have been distinguished in the lower part of the Gala suite. In the Janub, Chilov, Palchyg Pilpilesi, and Hazi Aslanov areas, as well as the Pirallakhy Island and Gurgan-deniz of Absheron archipelago, GaS-5 and GaS-6 were analyzed and described. The regular increase of thickness of GaS in the eastern and southern directions has been determined. The maximum thickness of this suite is marked in areas of the Absheron archipelago. In well 7 in the Hazi Aslanov area, the exposed part of the GaS section is more than 670 m; in well 900 in the Neft Dashlary area, it is more than 570 m; and in well 1 in the Oguz area, it is more than 460 m. In Palchyg Pilpilesi and Janub areas, the full thickness of GaS varies from 328 to 423 m. In the southwest of the Absheron peninsula (in Garadagh-deniz area), the thickness of GaS is 30 m (well 236); in the Lokbatan-deniz area, the thickness is 45–65 m (well 844).

Pre-Gyrmaki suite (PGS). The deposits of this suite within the Absheron area outcrop in the Gyrmaki valley, Mt. Gyrmaki, and Chilov Island. In all other parts of the area, this suite has been studied by the drilling of numerous wells. The PGS occurs in a wider area than GaS and is distinguished by changes in the lithological composition. The section of this suite is represented by light-grey and grey quartz sands with inclusions of black grit and large rounded grains of quartz. The sands are poorly sorted, medium and coarse grained, rarely fine-grained, and often cemented. Based on the appearance of sand intervals, PGS is subdivided into five separate horizons (from the bottom to top): PGS-5, PGS-4, PGS-3, PGS-2, PGS-1. This zone is absent in western and northwestern Absheron, where PGS has been found in Garadagh and Lokbatan structures and has thickness of 30–70 m. In the Atashgah-Shabandagh zone, this PGS also has a small thickness. The maximum thickness the PGS is revealed in the Hazi Aslanov (192 m), Neft Dashlary (174 m), Janub (162 m), Turkan (163 m) and Zira (142 m) areas.

Gyrmaki suite (GS). The deposits of GS occur in many areas of the Absheron peninsula (Sarynja-Gulbakht, Shongar, Garaheybat, Shorbulag and others), and also on arches of some uplifts of the northern and northeastern Absheron (Binagadi and Buzovna-Mashtagha areas). The deposits of GS outcrop in the Gyrmaki valley, Mt. Gyrmaki and Chilov Island. In other areas, they have been studied by drilling. GS is usually represented as a thick series of alternating thin sands, sandstone, siltstones and clays with persistent thickness and lithological composition; the suite section becomes sandier toward the bottom. In some areas of northern and northwestern Absheron (as well as in the separate areas of eastern Absheron—Buzovna, Khali), the Gyrmaki suite occurs on the eroded surface of older stratigraphical units (Pontian, Diatomian, Maykop and Govun). In northwestern Absheron, the suite occurs on the deposits of the Pontian regiostage and Diatom suite, with angular inconformity of 10°–15°. In the GS deposits relics of Pontian microfauna were found: ostracoda [*Cythere cellula* Liv., *C. praebacuana* Liv., *C. olivina* Liv., *C. liven-tali* Schn., *Loxoconcha djafaroff* Schn.) and foraminifera (*Nonion ex gr.granosa* d Orb., *Rotalia beccarii* (Lin)]. The maximum thickness of GS is recorded in the Palchyg Pilpilesi area (>400 m), as well as in the Hazi Aslanov, Chilov Island and Bakhar areas (>300 m). In the central and eastern Absheron, the thickness of GS is

200–290 m (Yasamal valley, Balakhany, Surakhany, Garachukhur, Gala, Turkan, Zira and other areas). A similar thickness for GS was found in Bibi-Heybat and in the southern limbs of the Garadagh and Lokbatan uplifts.

Post-Gyrmaki sandy suite (PGSS). The deposits of this suite are cropped out in the Gyrmaki valley, to the east of the Fatmayi salt lake, in Shabandagh high, in raised parts of the uplifts of Chilov Island and Neft Dashlary. These deposits were exposed by drilling in different parts of the Absheron area. The suite is represented by a series of grey and white fine-, medium- and coarse-grained sands; lens-shaped sandstone; and rare interbeds of clays. In the northwestern part of the Guzdak syncline in the Gara-Heybat, Shongar and Sarynja-Gulbakht areas, as well as in the Govun and Yunusdagh anticline zones, PGSS is absent. The thickness of PGSS varies from 15–25 m to 60–70 m within the Absheron area, but it often varies from 30 to 50 m. The ostracoda species have been discovered in rocks of this suite: *Cythere olivina* Liv., *C. cellula* Liv., *Hemycythere pontica* Liv. and *Cytherideа torosa littoralis* Brady.

Post-Gyrmaki clay suite (PGCS). This suite, according to the rock composition, looks like the GaS and GS; however, it contains greater amounts of shale. The interbeds of fine-grained sands are generally grouped in the lower parts of the sections. In the western and southwestern directions in the PGCS section, many interbeds of sands gradually disappear and the suite becomes almost clay. Faunistically, this suite is represented only by *Cythere cellula* Liv and *C. torosa litoralis* Brady. In the strip occurring over the Kurdakhany, Fatmayi, Binagadi, Balakhany, Garachukhur, Hovsan, Gyrmaki and other areas, the thickness of the suite is 155 m. To the west (Sulutapa, Lokbatan and Puta) and to the east (Gala, Turkan, Shah Dili, Neft Dashlary and others), the thickness of the suite gradually increases and reaches the maximum values of 120–200 m. The greatest thickness of PGCS is marked in the east of the Oguz (up to 200 m), Hazi Aslanov (>160 m), Shah Dili (156 m) and Zira (135–140 m) areas, as well as southwest of Lokbatan (160 m) and Garadagh (>200 m) areas.

The ***Fasila suite*** is a basis horizon of the upper section of MG. This suite is characterized by the availability of a great amount of coarse material in the section (gritstones, conglomerates, pebbles, coarse-grained sands, sandstone and gravels). The Fasila suite in central Absheron is composed of coarse sands and sandstone, gradually enriched by clay and silt interbeds and becomes an oil-bearing one. The greatest thickness of the suite is revealed in the east of the Absheron peninsula (140–250 m) and offshore (Bakhar, Janub, Janub-2 and Gunashli). Onshore, the maximum thickness of this suite is revealed in the Hovsan and Shah Dili areas. In some areas of the peninsula (Kurdakhany, Buzovna and Yasamal valley), the thickness of the suite does not exceed 80–100 m. In the Fasila suite, *Helix* and *Cytherideа torosa littoralis* Brady have been found. In the redeposited rocks, the following fauna has been found: fragments of *Inoceramus, sea-urchin and Terebratura Cretaceous, Spaniodontella, Cryptomacra Liocene, Cardium and Dreissena* Pont, and *Low Permian* Pusulinids.

The ***Balakhany suite*** is thicker in the upper section of the MG. The lower part of the suite is predominately sand and sandstones (70–80 %). However, in the upper part, clay plays an essential role in the interbeds. For areas of eastern and central Absheron, the section of the Balakhany suite in the Balakhany and Sabunchu areas is very typical. In the west, south and southeast directions, the argillization of the Balakhany suite can be observed. The thickness of the suite varies within 250 to 700–800 m, reaching a maximum of 900–1000 m in Oguz, Shah Dili and Janub-2 offshore areas. In the Balakhany suite, *Cytheria torosa littoralis* Brady is found.

Sabunchu suite. This suite has a very typical clayey section and regional stability of lithological composition of separate horizons. The transition zone from Sabunchu to Surakhany suite is very sharp. Therefore, their contact produces reliable data for the comparison of well sections. The quantity of the sand levels of Sabunchu suite is observed in the southeast and west directions. Microfaunistically, the Sabunchu suite is characterized by *llyocypris gibba, I. bradyi, Darvinela aurea* and others. The thickness of this suite varies from 185 to 450–520 m (Zira, Bakhar, Gum-deniz and other areas).

The ***Surakhany suite*** is the first from the top. It is thicker and has the largest areal extension in the MG section. The Surakhany suite has a clayey character and differs by a wide range of thickness alternations from 60 to 100 m (Kurdakhany, Binagady and Balakhany areas) to 1400–1800 m and more (Shah Dili, Zira, Bakhar, Janub-2, Sayyar and other areas). The Surakhany suite is represented by alternations of fine- and close-grained sands and clays with significant predominance of the latter ones. In the west and southwest directions, the clayness of the Surakhany suite increases sharply; the sandy layers (clearly presented in eastern Absheron) are completely replaced by clays. In the Surakhany suite, freshwater fauna have been found, including *Unio*, *Planorbis*, *Limnaea* and Melania, as well as ostracods *Cypridopsis nidua*, *Cypris salinus*, *Cyprinotus incongruens* and others.

The Lower Kur Type of Sediments

The deposits of MG within the Lower Kur Depression and Baku archipelago have a wide areal extension and large thickness. They crop out in the northeastern flange and partially in the inner part of the depression in the arch of

Baba Sanan, Boyuk and Kichik Harami, Kalamaddin, Mishovdagh and Bandovan. In the Baku archipelago, MG partially crops out on the arched uplifts under the modern formations of the Caspian Sea (Alat-deniz, Hamamdagh-deniz, Sangi-Mughan, Yanan-Tava and others). Taking into account the sharp alternation of thicknesses and lithological characteristics of deposits in the area, G.A. Ismailov, V.G. Idrisov and E.A. Tagiyev (Azizbekov et al. 1972; Alizadeh et al. 2007) suggested a new scheme of the MG section separation in the Lower Kur Depression and Baku archipelago. They divided the MG section into three facial-lithological complexes: Langabiz-Kalamaddin, Alat-Pirsaat and Garachala (Kurovdagh)-Neftchala.

The **Langabiz-Kalamaddin complex** is represented by Garanohur, Langabiz and Kalamaddin suites.

The ***Garanohur suite*** is cropped out with a large thickness in the Garanohur gorge, where its exposed part is 602 m and is represented by clays with a weak parting of conglomerate, gritstone and sandstone. In the Dashmardan area (232 m), this suite is represented by alternations of clays and pervious variations.

The ***Langabiz suite*** is represented by alternations of dense clays, sandstone and sands. In the Dashmardan area, the sandier part of the section occurs in the interval of 1680–2730 m.

The ***Kalamaddin suite*** in Garanohur, Ovchulu and Ortabulag gorges is represented by dense, cloddy clays with separate parting of dense sands, sandstone and gritstones. In the Baridash gorge and Kalamaddin and Dashmardan areas, the quantity of sandy variations is increased. The thickness of the suite is 480–873 m. N.G. Akatov and S.N. Alekseichik found *Planorbis* sp., *Helix* sp. and *Pura* sp. in the tops of the Langabiz suite (Azizbekov et al. 1972).

Alat-Pirsaat complex. The area of development of this complex occurs over most of the Lower Kur Depression, confined by the Alat ridge in the north, northeast, and southwest; it passes in the northeast and southeast directions of Galmas (the northwestern periclinal) and Kursanga (the deep submersion of the northeastern limb). This line offshore occurs over all the structures of the Baku archipelago between Alat-deniz to the north and Atashgah to the south. This complex is revealed by multiple wells and is reaches over 5 km in thickness. Analogues of the following suites developed here: GaS (30–60 m), PGS (35–70 m), GS (270–760 m), PGSS (35–83 m), PGCS (150–268 m), Fasila suite (60–120 m), Balakhany suite (710–1070 m), as well as Hamamdagh (420–800 m), Dashgil (310–750 m) and Goturdagh (590–880 m), which finalize the upper section of MG. Meanwhile, the Hamamdagh suite corresponds to Sabunchu, while Dashgil and Goturdagh suites correspond to the Surakhany suite of the Absheron region.

Garachala (Kurovdagh)-Neftchala complex. Within the development zone of this complex, the maximal thicknesses of MG's section (2800–3500 m) is revealed in Neftchala, Khylly, Durovdagh, Kurovdagh and other areas. The section is mainly represented by an alternations of sands, sandstone and clays, with the latter prevailing. Meanwhile, the lower division's section is revealed by individual wells drilled in Mishovdagh, Kurovdagh, Garabaghly, Durovdagh, Khylly and Neftchala areas. Serious claying is present in the section, often with the replacement of sands by clays and vice versa, which complicates a fractional breakdown of MG. However, a correlation of well's sections by different areas has allowed to identify analogues of all suites with the Absheron type of deposits even within the complex composition. Analogues of the following suites developed in this area: GaS (32 m), PGS (60–70 m), GS (370–380 m), PGSS (30–55 m), PGCS (115–230 m), Fasila suite (100–120 m), as well as Garachala or Kurovdagh (330–350 m), Neftchala (420–960 m) and Baba Sanan (820–1350 m) suites. Meanwhile, the Garachala suite corresponds to the Balakhany; the Neftchala suite corresponds to the Surakhany suites of the Absheron complex. Paleontological species such as *Planorbis.* sp., *Helix* sp., and some others have been found in the Baba Sanan suite.

The Gobustan Type of Sediments

MG deposits in the Gobustan are mainly extended in the Jeirankechmez depression and transform into the areas of central Gobustan. On the northern flange of the Jeirankechmez depression (Donguzdug-Gyrgyshlag line), in separate regions of central Gobustan (Boyuk Siyaki, Kichik Siyaki and Mayash) and in the southern part of the Sundi-Maraza depression, the half-continental and continental analogies of the tops of MG (with a thickness of 200–400 m) are developed. On the other parts of this depression, MG has been studied by drilling. Different schemes of PS separation have been suggested. Analyses indicate that the Absheron scheme of separation of MG may also be utilized in the Gobustan.

The lowest stratigraphical unit of MG (Gala suite) in Gobustan has not been found yet. In all wells exposing the lower section of MG, the analogy of PGS occurred in the depositions of the Pontian regiostage. PGS is composed of conglomerates in the lower part and calcareous sandstone and siltstones in the upper part. Its thickness is 15–30 m in the Miajik and Kahnizdagh areas and up to 50 m in a zone of the Alat ridge (Goturdagh, Dashgil and others). The PGS suite is overlaid by a clayey interval that is 160–250 m thick —GS's analogue. Staying higher in the succession, the PGSS is represented by alternations of sands and clays. Its thickness is 37–60 m (Miajik and Kahnizdagh areas).

The upper section of the MG in Gobustan begins by the Balakhany suite, which is represented by alternations of sands, sandstones and clays. Five sandy horizons with thicknesses from 85 to 115 m can be recognized in this suite. The thickness of the Balakhany suite is 500–900 m. The Sabunchu suite is represented by clays with rare interbeds of sands. The thickness of the suite is 450–600 m. The Surakhany suite is the most representative stratigraphical unit of the MG in Gobustan. Its lower part consists of clays, with a thickness of 400 m; the upper part is represented by steady alternations of sandy-clayey formations. The greates thickness of the Surakhany suite (up to 1900 m) is observed in the northeastern flank of the Kahnizdagh fold. The total thickness of the MG in Gobustan varies from 200–450 to 3900 m.

The redeposited Pontian fauna of Ostracoda has been found in the MG section: *Paracypria acuminate* Zal, *Cythere praebacuana* Liv., etc. In the Cheildagh and Rahim areas, the mollusk fauna *Planorbis* and *Melania rhodensis* Buk. have been found.

Nearcaspian-Guba Type of Sediments

MG in the region is represented by two facies: rudaceous-pebble (Guba facies) in west-northwest region and sandy-loamy (Gyzylburun facies) in the east-southeast region. A thickness of the MG significantly decreases from south-southeast and reaches zero in southern Dagestan (where these deposits completely pinched out). The MG is composed of poorly sorted sands and pebbles in the north-northwest zone of the region. The reddish shales can be found in the right tributary of the Gusarchay River in the low part of the section; they play the role of cement. The MG contains gritstones and pebbles in the west and northwest between the Aghchay and Dustahirchay rivers. The thick interbeds (25–30 m) of dark-grey colored shales, shaly sands and rare sandstones are found in the area of the Dustahirchay and Tahirjalchay rivers. The interbedding of volcanic ash (0.25 m) is discovered in the upper part of section.

Thin interbeds of sands and loose sandstone (1–5 m) are found in clay intervals. The upper part of section is characterized by the occurrence of thick beds of sandstone, whereas the lower part has interbeds of gritstones. Redeposited Cretaceous, Miocene and (typical for Pontian) Ostracoda *Xestolebres lutrae* Schn., *Cythere praebacuana* Liv., *Paracypris loezyi* Zal were found in the lower part of the section.

A great amount of different species are found in the PS section in Gilgilchay (probably, corresponding to the upper section of the MG—mainly Cretaceous, Paleogene and Sarmatian foraminifera and ostracods, among them *Nonion* sp., *Cibicides ex gr. granosa* (d'Orb.), *Joxoconcha eichwaldi* Liv., *Cytheridea torosa lithoralis* (Braty).

Ajinohur Type of Sediments

Shirinov and Bazhenov (1962) described MG deposits occurring in the western margin of Ajinohur, including valley of the Ganykh (Alazani) River. These authors supposed also the MG occurrence to the west of the mentioned river.

Within the Ajinohur area, the more complete section of MG crops out in the right bank of the Girdimanchay River, near the Bilistan village. The number of outcrops is developed in the eastern part of the Ajinohur area; they extend from the north to south.

The lower portion of the Girdiman suite (with a thickness of about 500 m) consists of alternations of conglomerates, different-grained sandstone and sandy clays. The upper portion has loamy gypsum containing the Aghdara suite with a thickness of 300 m; it consists of clays with interbeds of conglomerates in the southern section of the Girdimanchay River. In the northern section of the Girdimanchay River, the clays transfer into loams, which contain a great amount of admixtures of sands and small pebbles; sandstone are substituted by gritstones.

There are the beds of poorly-cemented conglomerates in the lower part of the section. According to Shirinov and Bazhenov (1962), loamy "gypsum-ferrous" (Aghdara) and sandy-pebble (Girdiman) suites are analogies of the Kalamaddin and Langabiz suites.

It should be mentioned that the MG section was revealed in Ajinohur in a well-drilled area southward from the lake of the same name. The section is represented by alternations of clays, sands and sandstone, with a total thickness of about 1000 m. In the cores of the Dashuz ridge of the MG, deposits of paleontological species *Helix* and *Planorbis* were found.

Volcanogenic facies of the Balakhany regiostage have been developed in the highlands of the Lesser Caucasus. The complex of volcanogenic formations (from 10 m up to 300 m), consisting of andezites, dacites, liparilo-dacites, tuffs and tufobreccia, was revealed within the Sarybaba and Kalbajar troughs. This complex is called the Ishygly suite. Within the Ordubad trough (Araz zone), the volcanogenic Lower Pliocene is called the Bichanak suite (800 m).

Upper Pliocene

Azerbaijan's Upper Pliocene deposits (Piacentian stage) are represented by the ***Akchagyl regiostage***. The deposits of the Akchagyl age were discovered by N.I. Andrusov, who studied them for more than 15 years (1895–1912). He contributed significantly to the study of Neogene deposits not only of Azerbaijan, but also of the whole Caspian region. After the research of N.I. Andrusov in the beginning of the twentieth century, studies of the Akchagyl regiostage of Azerbaijan were conducted by D.V. Aghalarova, L.I. Aliyeva, A.A. Alizadeh, G.A. Alizadeh, V.V. Bogachov, Z. K. Gadirova, D.V. Golubyatnikov, I.M. Gubkin, Z.K. Kadirova, V.P. Kolesnikov, S.A. Kovalevskiy, and others. These investigators studied the fauna and stratigraphy of Akchagyl deposits, clarified the stratigraphic position of Akchagyl and justified it in the common scale of Neogene of the Caspian region.

At present (since 1991), the Akchagyl regiostage is the single member of Upper Pliocene in the stratigraphic scale of Cenozoic of the Caspianregion. Akchagyl deposits have great significance in the geological structure of Azerbaijan. They widely occur in the Azerbaijanian part of the Lesserand Greater Caucasus, in the Kur Depression and in the Azerbaijan sector of the Caspian Sea.

Within Azerbaijan, the Akchagyl deposits (up to 1000 m) with basal conglomerate at the base occur on the different subdivisions of the Neogene,Paleogene, Cretaceous and Jurassic. They are represented in marine, continental and (partially) volcanogenic facies.

The marine deposits of Akchagyl overlap by almost solidly covering the depression zones of Azerbaijan. Lithologically, they are represented by interbedding of grey, dark-grey, and blue-grey weakly sandy clays; fine- and medium-grained sands, sandstones, and shell rocks; interbeds ofvolcanic ash; and partially, mudvolcanic breccia (Gobustan, Baku archipelago). They are faunistically characterized by the representatives of marinemollusks (*Cerastoderma, Avicardium, Mactra, Avimactra, Crytomactra, Potamides* (*Pirinella*), etc.), foraminifera (*Cassidulina, Cidicides, Bolivina*, etc.),and ostracoda (*Candona, Candonella, Leptocythere, Limnocythere, Loxoconcha*, etc.).

A faunistically well-founded subdivision of Akchagyl in Azerbaijan is linked with the name of great investigators of Akchagyl, such as A.A. Alizadeh,G.A. Alizadeh and V. P. Kolesnikov. They subdivided the deposits of the Akchagyl regiostage on three subregiostages—Lower, Middle and Upper.

According to the subdivision scheme of V.P. Kolesnikov and G.A. Alizadeh (Azizbekov et al. 1972 ; Alizadeh et al. 2007), the Lower Akchagylcorresponds to the Lower and Middle Akchagyl; the Middle and Upper correspond to the Upper Akchagyl of Alizadeh's (1969) scheme. Accordingto these schemes, the borders between the subregiostages do not coincide. Therefore, for a long time, the definition of these borders has had adisputable character. In the Caspian region, some investigators had accepted the division scheme of V.P. Kolesnikov and G.A. Alizadeh,with another group accepting A.A. Alizadeh's division scheme. Nevesskaya and Trubukhin (1984) developed a modified scheme of atwo-member subdivision of Akchagyl of the Caspian region. This scheme was developed on the basis of the paleomagnetic characteristics ofdeposits and the development of marine mollusk fauna.

According to this scheme, the lower border of the Akchagyl regiostage can be defined by the appearance in sections of poor marine fauna with the mollusks (*Mactra, Cerastoderma, Potamides* (Pirenella)), foraminifera (*Cibicides, Cassidulina*), and ostracoda (*Lymnocythere intermedia,Loxoconcha aktchagylica*); it corresponds to the border of the Gilbert/Gauss paleomagnetic epochs at 3.3 Ma. The most ancient dating of Akchagyldeposits is obtained using the Treck method of ash interlayers in the Yasamal valley: 3.34 ± 0.35 Ma. The border between Lower and Upper-Akchagyl is defined by the frontier of the Gauss-Matuyama paleomagnetic reversal (2.52 Ma) and also by the appearance of representatives ofendemic (rare-costate *Cardiid-Avicardium*), peculiar (*Mactriid–Avimactra, Cryptomactra*), and other marine mollusks.

The upper border of Akchagyl with the Absheron regiostage (Eopleistocene) can be observed by the disappearance of almost all representatives ofmarine mollusk fauna and the appearance of salt-water mollusks: *Dreissena, Monodacna, Theodoxus* and others. The recently acceptedstratigraphic scale of the Cenozoic of the Eastern Paratethys—the upper border of Akchagyl—is the border between the Pleistocene and Pliocene.From the paleomagnetic data analysis, it follows that it may be dated near the Olduvei episode of the Matuyama paleomagnetic epoch.

The newly developed scheme of the subdivision of the Akchagyl in the Caspian region is more justified and simplified. Due to this novel scheme, a two-member subdivision of the Akchagyl regiostage deposits is accepted in Azerbaijan as the most optimal one.

Within the Nearcaspian-Guba zone, the Akchagyl deposits sections are observed in the Agchay, Gusarchay, Susaycay, Gudiyalchay, Jagajugchay, Valvalachay and other river valleys. In the southwest regions (Agchay, Gusarchay and Susaycay) of this zone, the Lower Akchagyl is represented with the thick layer of sands and sandstones, with layers of sandy clays, shingles and conglomerates. In thesoutheastern areas (Gudiyalchay, Jagajugchay and Valvalachay), the role of coarse deposits (shingles, conglomerates) is decreasing considerably, and sands and sandstone are becoming more small-grained. Sandy clays contain *Cerastoderma dombra, Mactra subcaspia* and *Clessiniolla sp*.The thickness varies from 60 m (Agchay) to 175 m (Gudiyalchay). The Upper Akchagyl contains *Mactra inostranzevi, M. venjucovi, Avicardium nikitini*and

Cerastoderma konschini. The thickness of the Upper Akchagyl varies from 150 m (Jagajugchay) to 240 m (Gudiyalchay).

In the ***Absheron periclinal sag***, within the Absheron Peninsula, the complete sections of Akchagyl are observed in the areas of the Korgoz (BakyGulaglary), Guzdak, and Garagush mountains in the Yasamal valley. They were discovered by wells drilled in all oil deposits in the peninsula andthe Caspian aquatorium (Absheron archipelago). The Lower Akchagyl is represented by slate clays, with less thick layers of volcanic ash and sand; it contains *Cerastoderma dombra, Mactra subcaspia* and *Clessiniolla intermedia*. The Upper Akchagyl is represented by sandy clays, with layers of sands, sandstone and shell-limestone (Guzdak); it contains *Avicardium nikitini* and *Mactra inostranzevi*. The thickness varies from 21 m (Yasamalvalley) to 50 m (Guzdak).

Within the Lower Kur Depression, Akchagyl deposits with a comparatively small thickness are observed in the structures of Kichik Harami (20 m),Boyuk Harami (35 m), Baba Sanan (106 m), and discovered by wells in the marine structures, such as Atashgah bank (35 m) and Chighil Island (80 m) in the Baku archipelago. The Akchagylian deposits with a great thickness are discovered in the parametric wells in the structures of Galmas(240 m), Baba Sanan (355 m), Saryjalar (365 m), Padar (496) and Daykend (640 m). In all structures of this depression, in the lower parts, theAkchagyl deposits are represented by leaf-clays; in some areas, they are represented by weak sandy clays with thin layers of sands and sandstone and numerous (to 16) layers of volcanic ash. The lower parts contain only foraminifers: *Cassidulina crassa* and *Cibicides lobatulus*. Inthe upper parts of the sections (sandy clays) were detected small shells of mollusks (*Mactra karabugasica, Cerastoderma vogdti, Potamides (Pirinella) caspius, Micromelania sp.*) and Upper Akchagylian ostrocodes (*Candona abichi* and *Loxoconcha eichwaldi*).

In the ***Gobustan zone***, Akchagyl deposits are observed within the Jeirankechmez depression in the areas of Cheyildagh (150 m) and Gushgaya(225 m) in the Alat range zone, in the Solakhay area (190 m) and in the sea continuation, and in the rows of the anticlinal belts of southeasternGobustan, such as Sangachal-deniz (45 m) and Khara-Zire Island (80 m), etc. The Lower Akchagyl in all sections of Gobustan (up to 100 m) isrepresented with leaf-like clays (in the lower parts) with thin layers (up to 22 m in Cheyildagh) of volcanic ash and packets of mud-volcanic brecciacontaining *Cerastoderma dombra* and Mactra subcaspia. The Upper Akchagyl (up to 130 m) consists of sandy clays and thin layers of shellsandstone;it contains mollusks such as *Avicardium nikitini, Cerastoderma konuschevski, Mactra pizum,* and *Potamides* (Pirinella) *jumudicum*.

In the ***Shamakhy region***, more complete and faunistically richly characterized sections were observed in Karkanj (350 m), Langabiz (162 m),Ortabulag (268 m) and Goylar (130 m) ravines, and in the valley of the Aghsu River (343 m). In the section of Karkanj, the Lower Akchagyl isrepresented with alternations of slate clays with thin layers of sands and volcanic ash; it contains *Cerastoderma dombra, Mactra subcaspia,* and*Clessiniolla sp*. The thickness reaches 190 m (Kerkenj). The Upper Akchagyl, in the lower part, is represented by weak sandy clays, sands,sandstone with shingle and shale-shells; these rocks contain mollusks: *Cerastoderma mirum, Avicardium radiferum, Avimactra aviculoides, Cryptomactra acutecarinata*, etc. In the upper part of the section in the oolitic shells, the following representatives of fresh water mollusks werefound: *Pyrgula, Alizadella, Micromelania, and Theodoxus*. The thickness of the Upper Akchagyl is up to 260 m (Kerkenj).

In the ***Ajinohur zone***, the thicker sections of the Akchagyl deposits are observed in the valleys of the Girdimanchay River (650–850 m), AljiganchayRiver (850 m), Turyanchay River (540 m), and in the Bozdagh Mt. (590 m), Kudbarekdagh Mt. (820 m), Gamigaya Mt. (615 m) and Dashuz village (800 m). Within the southern regions of Adjinohur, the characteristic sections of the marine deposits of Akchagyl (along exposure) were discoveredin the ranges of Bozdagh, Duzdagh and Garbi Gojashen. The Lower Akchagyl in the bearing section of Mt. Bozdagh (200 m) is represented byalternations of the thick layers of clays and thin layers of sand, sandstone and volcanic ash. Here were found mollusks, such as *Mactra subcaspia*and *Potamides* (Pirinella) caspius. The Upper Akchagyl, consisting of alternations of sandy clays with thin layers of volcanic ash and layers of sand,sandstone, shell-limestone, and shingle, contains characteristic Upper Akchagylian species of mollusks: *Avicardium dagestanicum, Avic.mainocaricum, Cerastoderma mirum, Mactra nazarlebi,* and *Cryptomactra acutacarinata*, etc. The upper part of the section (sandy clays: 390 m) contains mainly freshwater mollusks: *Melanoides costatus, Unio samuchicus*, etc.

As distinct from the south areas of Ajinohur, in south-east (Girdimanchay, Turyanchay and Aljiganchay rivers) and south-west (Dashuz, Gamigaya Mt., Kudbarekdagh Mt. and left bank of the Ganykh River) areas, the Akchagyl sections acquire coastal-continental character.

Here are observed the increase of conglomerates, the thickness of sandstone, shingles and conglomerates; and the appearance offreshwater and continental-fauna mollusks.

In the region between the ***Kur and Gabyrry rivers***, complete and faunistically richer sections of Akchagyl were discovered in the areas of anticlinalstructures Boyuk Palantokan (500–550 m), Gurzundagh (500–600 m), Koyrukkeylan (525 m), Molladagh (325 m) and Gushguna

(330 m). In thesesections, the Lower Akchagyl is represented by clays and layers of sands, sandstone, clay-shells and volcanic ash and contains *Mactra subcaspia,M. decora* and *Cerastoderma dombra*. The thickness varies from 113 m (Gushguna) to 205 m (Boyuk Palantokan). The Upper Akchagyl isrepresented by alternations of marine and coastal continental deposits. The lower part of the Upper Akchagyl in sections of Boyuk Palantokan,Kirzan, Gaflandara and Molladagh is represented by alternation of sandy clays, sands, sandstones, and shells and contains shells of the seamollusks *Mactra nazarlevi, Potamides* (Pirinella) *eldaricus, Avicardium cumli*, etc. The upper part of the Upper Akchagyl has a coastal-continental character. Here, in loams and sands were found freshwater (*Melanoides costatus*) and continental (*Pomatias* sp., *Helix* sp.) mollusks, as well as theremains of (Gushguna) *vertebrates Archidiscodon gromovi* and the eggshells of the ostrich *Gazella borbonica*, etc. Its thickness varies from 125 m(Gushguna) to 195 m (Boyuk Palantokan).

In the ***Ganja region*** (northwest foothills of the Lesser Caucasus), faunistically richly characterized sections of marine deposits are observed in theNaftalan region (to 300 m), near the spring Gazanbulagh (145 m), in Ajidara (120 m), in the villages of Gashalty (90 m) and Tapgaragoyunlu (78 m).In the bearing section of the Naftalan region, the largest part of the Lower Akchagyl is discovered by prospecting drilling. According to drilling data,the Lower Akchagyl is represented by clays with sand, sandstone, and shell–limestone. These rocks contain mollusks *Mactra subcaspia, Clessiniollasp., Microfauna, Cassudilina grassa, Leptocythere gubkini,* etc. The thickness of the Lower Akchagyl is 100–120 m. The outcropped LowerAkchagyl has a thickness of 27 m and is represented by the alternation of sandy clays with sand layers and volcanic ash (2 m). Here, the followingmollusks were found: *Mactra garabugazica, Ceras-toderma dombra, Potamides* (*Pirinella*) *caspius*, etc. The total thickness of the Lower Akchagyl inthe sections of Naftalan is accepted within the limits of 127–147 m. The lower part of the Upper Akchagyl is represented by the alternation of sandsand sandy clays with the layers of shell-limestone and volcanic ash; it contains the rich fauna of mollusks: *Mactra venjucovi, Avimactra aviculoides,Cryptomactra carinatacurvata, Cerastoderma sulini, Avicardium nikitini, Potamides* (Pirinella) *levis*, etc. The upper part of the Upper Akchagyl isrepresented by clays, sands, sandstones, and detritus limestone; the layer of microconglomerate is characterized by small shells of marine Mactrasubcaspia, Cerastoderma dombra, saltwater Dreissena, freshwater *Pyrgula Naphtalanica, Avardaria andrussovi* and other mollusks. The thickness ofthe Upper Akchagyl is 163 m. Incompletely but faunistically richly characterized sections of the Akchagyl are observed in Ajydara (118 m), byGazanbulagh spring (140 m), and in the areas Gashalty (90 m) and Tapgaragoyunlu (78 m).

Akchagyl deposits are revealed in numerous wells in the right bank of the Kur River—from the Gazakh region to Beylagan. These deposits are represented here in the lower part of section by clays, sands, sandstone, shingles and sometimes conglomerates. Along the whole section, sandy-shells containing mollusks were observed: *Cerastoderma dombra, Mactra subcaspia* and *M. garabugazica*.

Within the ***Lower Araz Depression*** (the southeast submersion of the Lesser Caucasus) in the Horadiz area, the Akchagyl deposits are representedby clays and sand layers, sandstones and volcanic ash. The thickness of the deposits in the Dashburun area is 600 m; in the southeast, it isgradually reduced (for example, in the western villages Ashaghy and Abdurrahmanly, the marine deposits of Akchagyl are not discovered by drilling).

On the foothills of ***Mt. Talysh***, Akchagyl deposits are discovered with by drilling in the Uzuntapa (Novogolovka) and Tumarkhanly areas. Here,Akchagyl deposits are represented by the considerably weak sandy clays and thick layers of sand containing *Mactra subcaspia, M. venukovi,Clessiniolla vexatilis*, foraminifers: *Cassidulina crassa, Rotalia beccari*, and ostrocodes *Loxoconcha laevigata*. The thickness of these deposits is220 m.

3.6.3 Quaternary

Quaternary deposits widely occur in Azerbaijan and play an important role in its geological structure. Stratification, fauna, lithology and other peculiarities of the Quaternary deposits have been continuously studied since end of the nineteenth century. It is necessary to note the significant contributions of the following scientists: M.G. Aghabekov, F.S. Akhmedbeyli, L.I., B.D. Aleskerov, Aliyeva, A.A. Alizadeh, G.A. Alizadeh, S.A. Alizadeh, N.I. Andrusov, B. A. Antonov, Y.P. Bazhenov, N. Barbot-de-Marni, V.V. Bogachov, B.A. Budagov, N.V. Dumitrashko, P.V. Fyodorov, V.D. Fyodorov, T.G. Gadiyev, D.V. Golubyatnikov, I.M. Gubkin, P.Ya. Guliyev, M.P. Hajiyev, V.E. Khain, V. P. Kolesnikov, S.A. Kovalevskiy, D.A. Liliyenberg, V.Y. Liventhal, A.V. Mamedov, D.V. Nalivkin, G.M. Pashaly, V. Y. Rujentsov, A.N. Shardanov, E.Sh. Shikhalibeyli, N.Sh. Shirinov, F.A. Shirinov, Kh. Shoegren, A.D. Sultanov, N.V. Sultanov, Kh.K. Tanryverdiyev, B.G. Vekilov, V.V. Weber, and others.

Lithologically, Quaternary deposits are represented by argillo-arenaceous formations, limestone, shingle conglomerates and interbeds of gypsum and volcanic ash (Kovalevsky 1936; Akhmedov 1957; Agabekov and Mamedov 1961; Shirinov and Bazhenov 1962; Sultanov 1964; Vekilov

1969; Mamedov et al. 1970; Azizbekov et al. 1972; Alizadeh 1973; Mamedov 1973; Alizadeh et al. 1978, 2000, 2007; Khalilov 1978; Alizadeh 1987; Babayev et al. 2015).

3.6.3.1 Eopleistocene

Eopleistocene within the Azerbaijan's territory is represented by ***Absheron regiostage***, deposits of which have widely developed in premountain and depression areas, as well as in the adjoining aquatorium of the Caspian Sea. The Absheron regiostage was first discovered in the Absheron peninsula by G. Shergen in 1891. In 1897, N.I. Andrusov justified it as the upper stage of the Pliocene (Alizadeh et al. 2007). Over the last few years, significant changes in the stratigraphical scale of the Cenozoic in the Caspian area were done. For example, the International Stratigraphic Commission accepted the Absheron regiostage within the Anthropogene and its lower section, the Eopleistocene. Thus, the lower border of the Anthropogene is accepted at the level of 1.7–1.8 Ma.

The deposits of the Absheron regiostage, according to data of faunal and lithofacial content, are separated into three subregiostages. Here, three complexes of ostracoda and mollusk fauna were recognized, which correspond to three regiostages of Absheron transgression development: lower, middle, and upper. Chronologically, the Lower Absheron subregiostage is characterized by the micro- and macrofauna: *Caspiola acronasuta* (Riv), *Rectocypris reniformis* (Schw), *Liventalina gracilis* (Liv), *Leptocythere nostrata* (Liv), *L. angusta* Klein., *L. saljanica* (Liv), *Monodacna laevigata* (Andrus), *M. catilloides* (Andrus), *Parapsheronica raricostata* (Sjoegr.), *Dreissena distincta* (Andrus) and *Lymnaea apsheronica* Andrus. The Middle Absheron subregiostage contains four complexes of ostracoda: *Absheron Caspiocypris candida* (Liv), *Loxoconcha kalickyi* Liv, *Leptocythere casusa* (Mark), *Azerbaijanella pirsahatica* (Liv), *Loxoconcha liventali* Scheid., *Leptocythere camelii* (Liv); *Herpetocypris diginitatis* Step., *Trachyleberis papillosa* (Schw.), *T. bailovi* (Liv), *Leptocythere malva* (Liv). Mollusk fauna are represented by *Apsheronia propinqia*, *Monodacna kabristonia* (Andrus), *M. beibatica* (Alz.), *Pseudocatilus caucasica* (Andrus), *Didacnomya caucasica* (Andrus) and *Dreissena eichwaldi* Andrus. The Upper Absheron subregiostage is represented by fauna of ostracoda: *Advenocypris kurovdagensis* Klein, *Leptocythere salebra* Kul., *L. rostrata* (Liv), *L. messeriensis* Step, *L. martha lingula* (Liv), mollusks—*Hyrcania hyrcana* (Andrus), *H. intermedia* (Eichw), *Apscheronia propinqua* (Eichw), *Monodacna minor* Andrus Psedocatillus, *catiloides* Andrus, etc.

As can be seen from the above-mentioned list, the faunal content is characterized by both typical saltwater and freshwater species. The Absheron deposits are represented by marine and continental facies. The area of their distribution more or less coincides with the area of Akchagyl formations occurrence. Within the Gusar-Devechi trough, the Absheron deposits are not completely represented. Here one can recognize two types of lithofacies: clay-sandy in the southeast and coastal-shallow continental in the northwest. Their thickness reaches 1000 m in the southeast. The deposits of the Absheron regiostage in the Absheron peninsula crop out in the area of the Yasamal valley, Korgoz (Baky Gulaghlary), Garagush, Sabunchu, "Baku stage" Mt. and in other places. The Lower Absheron deposits occurring in Absheron without noticeable concordance on the Akchagyl rocks consist of clays with interbeds of sands, marls, volcanic ash and gypsum. Their thickness is from 90 to 500 m in the southeastern areas. The Middle Absheron deposits occur correspondingly on the Lower Absheron rocks. They are represented by alternation of clays, sands, sandstones, limestones, rarely conglomerates and shingles with thin layers of gypsum interbeds and volcanic ash. The availability of large-grained and hard-fragmental material in lithofacies content shows the existence of coastal and coastal shallow-water conditions during the Middle Absheron period. The thickness of the Middle Absheron varies from 170 to 300 m. According to lithological content, the Upper Absheron deposits do not differ from the Middle Absheron.

The changeability of lithofacies in the Absheron regiostage was detected in some territories within the southeastern submergence of the Greater Caucasus. It is expressed by a gradual replacement of thick limestone by clayey-sandy formations in the southwest direction. Absheron deposits widely occur in the Gobustan and Pre-Kur lowland, in the ridges of Kichik Harami, Alat range, Cheyildagh Mt., near station Sangachal, Duvanny and other areas. In sections of the Lower Absheron occur series of large-grained sandstone (Kichik Harami) and clay rocks with interbeds of sandstone shells and ashes (Alat range). Their thickness ranges from 150 to 350 m. The Middle Absheron is represented by alternations of sandy grey clays with interbeds of shell limestone and volcanic ash. Its thickness is 350–450 m. The Upper Absheron subregiostage of an area of Gobustan is characterized by the domination of shallow-water and coastal-shallow water facies, which consist of limestone and sandstone, and contain clay in small amounts. This thickness is 400–500 m.

The sections of Absheron deposits in the Western Azerbaijan are classic both for Azerbaijan and for the whole southern part of the former Soviet Union. Their outcrops can be observed within the Bozdagh, Duzdagh and Garaja folds. Faunistically, they are represented by salt water fauna with freshwater and, in some places, by semi freshwater forms. They are bedded with basal conglomerates on the base of Akchagylian deposits and can be characterized by the presence of all three subregiostages in the section: low, middle and upper. The low subregiostage (90–180 m) consists of alternations of clays with interbeds of sands and sandstone. The middle subregiostage (100–380 m) consists of a clayey series with thick sandy interbeds. The upper subregiostage (up to 500 m) consists of interbedding of conglomerates, sandstone and ashes. Besides marine and coastal-marine facies, the Absheron deposits are represented also by continental facies developed within the Lower Araz Depression (so-called Araz suite) and Gafan zone (Hakeri suite). Here, freshwater gastropods were found (*Bythinia. aff. fenticulata*, *Pisidium amicum* Mull), as well as the land form *Planorbella* Helicela. The thickness of the first and second suites approximately reaches 250 and 200 m, respectively.

3.6.3.2 Pleistocene

The Pleistocene is represented by the Turkan beds, Baku horizon and Mingachevir beds (Lower Pleistocene), Khazar (Middle Pleistocene) and Khvalyn (Upper Oleistocene) horizons.

The ***Tyurkan beds*** do not widely occur in Azerbaijan and are characterized mainly by a freshwater complex of fauna, on the basis of which they were recognized in the independent stratigraphic subsections. Some investigators suggested to include the Turkan deposits in the content of Baku horizon as the lower subhorizon. However, the special freshwater nature of faunal content, in our opinion, shows the validity of this independent stratigraphic unit. In sections of Mt. Duzdagh and Garaja in the Western Azerbaijan, the Turkan deposits are represented by dark-grey and grayish-brown clays, silts, crossbedded sandstones and sands. Their thickness is about 35–40 m.

Baku horizon. At the present time, the Baku horizon is divided into lower and upper Baku subhorizons. The rocks of this age differ from the Turkan deposits by the presence of a great amount of sandy-silt hard-fragmental and shell limestone rocks. The faunal content is characterized by the presence of the following species and gender of saltwater mollusks and ostracoda: *Didacna parvula* Nal., *D. catillus* Eichw, *Dreissena polymortha* (Pal), *Dr. rostriformis* Desh., *Micromelania caspia* Eichw and *Theodoxus pallasi* Lindh. Here, three complexes of ostracoda are distinguished: (1) *Caspiocypris filona* (Schw.), *Loxoconcha endocapra* Shar., *Leptocythere martha* (Liv), *L. lunata* Step. and *L. aff. resupina* Step., (2) *Caspiola liventalina* (Evlach.), *Leptocythere medicata* Step., *L. pondoplicata* Step. and *Xestoleberis ementis* Mand, and (3) *Caspiola gracilis* (Liv), *Bacunella dorsorcuata* (Zal), *Loxoconcha gibboida* Liv. and *Leptocythere bakinica* Scheid. The Upper Baku subhorizon is represented by mollusks: *Didacna rudis* Nal., *D. carditoides* Andrus., *Dr. polymortha* (Pall.), *Dr. ponto-caspia* Andrus., *Clessiniola triton* (Eichw.), *ostracoda-Pseudoctenocypria asiatica* Schn., *Bacunella dorsuarcuata* (Zal.) and *Leptocythere aff. referata* Step. A stratotypical section of Baku horizon is the "Baku stage" Mt. in the Absheron peninsula. The presence of much faunal remains in the rocks of Baku age enables the separation of these deposits into two subhorizons. Lower subhorizon, situated on the basal conglomerate, consists of non-compact sandstones, sands with limestone beds, and greenish-grey sandy clays, passing to upper beds of loose sandstones and sands. The deposits of Lower Baku age, concordantly covered by the upper Baku subhorizon, are characterized by the presence of loose sandstones, shells, inter-bedding of clays and sands, and series of shells and sands. The total thickness is 60 m, with a maximum thickness of Baku deposits of 100–120 m in the Absheron peninsula (accumulated in the eastern territory where shallow water facies prevailed). In the southwest and northeast, the coastal facies were formed as mainly gravel-shingle and shell rocks.

In the Gobustan and Lower Kur Depression, the content and changeability of lithofacies of Baku age are the same as in the Absheron peninsula; that is, the fraction of rudaceous rocks and shells increases in the western and northern directions. The thickness varies from 10 to 240 m. The rocks of this period outcrop in sections of the Shykhygaya, Davalidagh, Kalamaddin, Kichik and Boyuk Harami, Garachala (Kurovdagh), Alat range and some others. In the Western Azerbaijan, the deposits of Baku age occur in ridges of Bozdagh Mt., Duzdagh Mt., Garaja, Gojashen and Godakboz areas. These rocks are characterized by more rudaceous nature in comparison to the Absheron peninsula. Here is revealed a tendency of increasing a role of more small-grained material from west to east. On the whole, the deposits (120–140 m) consist of grey, grayish-brown shales, coarse-grained sandstone including shingles, interbeds of sandstone, conglomerates and volcanic ash. In the Nearcaspian-Guba region, Baku rocks are represented by terraces with high presence of gypsum. Their thickness varies from 250 to 400 m. In Lenkaran lowlands, the deposits of Baku age are characterized by the domination of sandy-silt and clayey material. As a result of the determination of the absolute age of Baku deposits by different radiometric and paleomagnetic methods, the low border of this horizon is fixed on the mark of 0.6–0.4 Ma. Figure 3.1 shows the behavior of Baku lithofacies in the depression zones of Azerbaijan.

Mingachevir beds. Here, one can understand the deposits with *D. eulachia* Bog. described for the first time by V.V. Bogachov in the Mingachevir region. Later, P.V. Fyodorov recognized the analog beds with *D. eulachia* Bog in Turkmenia and called them *Urundjik beds*; he defined the independent position for these deposits. However, some investigators considered the Mingachevir beds in the content of the upper Baku subhorizon. In our view, as a result of research for the specific structure of shells of dominating species *Didacna eulachia* Bog., we can speak about the transition of this form containing the features of catiloid

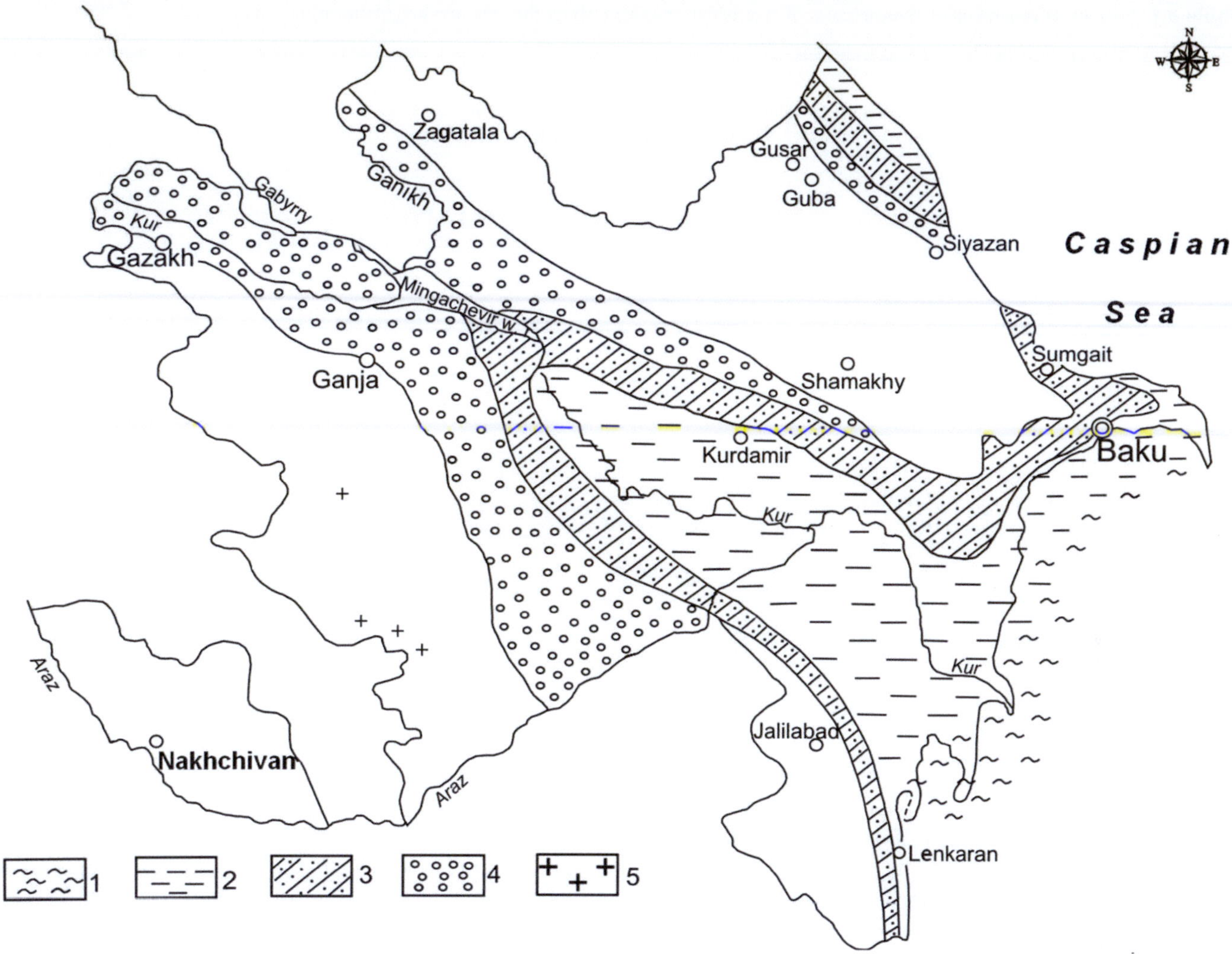

Fig. 3.1 Scheme of facies of Baku horizon deposits (after Alizadeh 1987): (*1*) clays, (*2*) predominantly clay-sandy sediments; (*3*) clay-sandy sediments interbedded with conglomerates and limestone, (*4*) continental coarse fragmental sediments; (*5*) volcanogenic rocks

(stretch) forms of Baku age to the trigonoid (triangula) shells of the *Khazar Didacna*. The peculiarities in *Didacna eulachia*, together with stratigraphic position of beds and occurrence of these species in the proper upper the Baku formation, confirm the concerns about validity of deposits with *D. eulachia* Bog. (Mingachevir beds). The rocks of the period are developed within the Absheron Peninsula, Gobustan, and also in the axes area of the Kur Depression. Lithologically, they are represented by coquinas, sands, gravel-shingle material and flank parts of folds-sandy clays. In these rocks, interbeds of volcanic ash (Duzdagh Mt.) can be observed.

The greatest thickness of 100 m was determined in Mt. Duzdagh. The following faunistic features (besides *Didacha eulachia* Bog) were found: *D. pravoslavlevi* Fed., *D. karelini* Fed., *D. mingetschaurica* Vekil; *Ostracoda-Loxoconcha liventali* Scheid., *L. endo-capra* Schar., *L. kalickyi* Lub., *Leptocythere periculosa* Step., *L. posteriobiplicata* Step., *L. pondoplicata* Step., *Paraleptocythere caspia* (Liv), *Trachileberis pseudoconvexa* (Liv) and others.

The ***continental generations of Lower Pleistocene*** gradually replace marine sediments toward the mountainous rims of depression zones, being represented by rocks of alluvial-proluvial, alluvial, volcanic and chemical origin.

Alluvial-proluvial deposits are widely distributed within the northeastern slope of the Greater Caucasus (Gusar suite), in zone of the Ajinohur foothills (Ushtal and Ivanovka suites), and in the strip at the northeastern foothills of the Lesser Caucasus.

They are composed of pebbles and conglomerates with interlayers and lenses of sand, loam and volcanic ash. The age of these deposits is determined by their interrelation with known age of river terraces, as well as by a comparison of the composition of volcanic ash horizons contained in the sections of these deposits, with those in the faunistically known marine sediments.

Table 3.1 Scheme of separation of river terraces of Azerbaijan (numbers show the absolute heights in m)

No.	Azerbaijani part of the Greater Caucasus	Lesser Caucasus		Talysh	Age
	According to B.A. Budagov	According to N.Sh. Shirinov and M.P. Gadjiev	According to N.Sh. Shirinov, P.Ya. Guliyev and H.K. Tanryverdiyev	According to B. A. Antonov	
I	1–3	2	2–6	0.8–1.3	Novocaspian (Holocene)
II	4–6	6		6–8	
III	10–12	10–12	8–15	12–16	Khvalyn (Late Pleistocene)
IV	16–18	18–22	20–27	18–22	
V	25–30	30–35	32–38	–	
VI	35–40	40–45	40–48	37–45	
VII	50–55	55–65	50–60	–	Khazar (Middle Pleistocene)
VIII	65–70	80–85	64–77	60–85	
IX	90–100	–	80–105	–	
X	–	110–115	110–128	100–125	
XI	110–125	–	135–140	–	
XII	150–170	140–160	148–160	150–155	
XII	180–200	180–200	176–210	–	Baku (Early Pleistocene)
XIV	220–240	–	217–240	200–220	
XV	250–280	240–260	–	250	
XVI	300	280–290	248–270	–	
XVII	–	300–330	290–350	–	Absheron
XVIII	–	350–360	–	360	
IX	–	400–420	–	–	
XX	–	480–500	–	–	

The thicknesses of these deposits reaches 90–100 m in the northeastern foothills of the Lesser Caucasus, 400–450 m in Ajinohur, and 150–200 m in the northeastern foothills of the Greater Caucasus.

Alluvial deposits are developed in the middle and upper reaches of the rivers of the Greater and Lesser Caucasus, Talysh, and compose a series of erosion-accumulative terraces.

Alluvial deposits are developed in the middle and upper reaches of the rivers of the Greater and Lesser Caucasus, Talysh and compose a series of erosion-accumulative terraces. The number and relative heights of these terraces differ in different valleys (Table 3.1).

The deposits composing these terraces consist of boulder gravels with sand, small gravel filler, and loam filler.

Within the Garabagh volcanic highland and the upper basin of the Terter River, this age relates to volcanic strata presented by dense, porous and vesicular basaltic andesite lava flows of gray, brown and black colors with a thickness from 10 to 80 m. These lava flows are characterized by columnar jointing passing up within each flow to clumpy or spherical. Between these separate flows occur layers of brown and black volcanic ash.

Khazar horizon. The deposits of this stage in the territory of Azerbaijan are expressed in marine and continental facies. Marine facies of this horizon are developed in the Kur-Araz lowland, in the Alat ridge, in the southeastern Gobustan, in the Absheron peninsula and the Precaspian region in northeastern Azerbaijan.

The Khazar horizon in marine facies is divided into the Lower Khazar (Gurgan) and the Upper Khazar deposits of Caspian area and their analogues for the Evksinsk basin. The Lower Khazar horizon can be observed within the Middle Pleistocene and the Upper Khazar, and within the Upper Pleistocene as a lower horizon. Faunistically, both subhorizons contain a large number of mollusks and ostracoda. The Lower Khazar subhorizon contains mollusks: *Didacna nalivkini* Wass., *D. pallasi* Prav., *D. paleotrigonoides* Fed., *D. subpyramidata* Prav., *D. kovalevskii* Bog., *D. lindleyi* Dasch., *D. charamica* Fed, etc.; ostracoda *Trachileberis pseudoconvexa* (Liv), *T. azerbaijanica* (Liv), *Xestoleberis ementis* Mand., etc. The Upper Khazar beds contain mollusks: *Didacna surachanica* Andrus., *D. delenda* Bog., *D. nalivkini* Wass., etc. In comparison with Baku deposits, the Lower Khazar rocks are characterized by domination of more large-grained (sandy-silts) and rudaceous material. The periclinal troughs were accumulated in Absheron both as the coastal-shelly sandy facies (in the western areas) and as deep water sandy-clay facies (in eastern areas). The analogical content of facies is typical for the Upper Khazar sedimentary complex. The Upper Khazar transgression is less than the Lower Khazar. In Gobustan, the deposits of both horizons in the western area consist of shingles conglomerates and large-grained sands. On the eastern territory dominate sandy-silts and clayey deposits. Numerous interbeds of cone breccia show the intensive mud volcanic activity. The clay formations are widely developed in the eastern area of Nearcaspian territory and Baku archipelago.

The Khazar deposits formed in the Nearcaspian lowland with terraces have a height up to 200 m. In western areas of the lowland, the Khazar formations are characterized by sandy material with shingle conglomerates, as in the Western Azerbaijan for the lower parts of sections of the Duzdah Mt. and Garaja. They consist of greenish-grey, yellowish, strong-sandy clays, sands, sandstone, shingles, gritstones and conglomerates. Continental analogues of Khazar marine deposits developed in Ajinohur area are Ivanovka and Post-Ivanovka suites, synchronizing correspondingly with the Lower Khazar formations. The age of Lower Khazar rocks by absolute dating is 254–300 kyr (the lower border) and 90 kyr ago (the upper border). The more real age dating for the lower part of the upper Khazar deposits, obtained from the thermoluminescene method (TL-method), is 91 ± 17 kyr ago. The upper part of the subhorizon was formed 30.7 ± 1.5 kyr ago (14S-method).

The Middle Pleistocene is expressed in continental facies in the most territory of Azerbaijan. It is represented by glacial, lacustrine, alluvial, alluvial-proluvial, deluvial-proluvial, colluvial and volcanogenic formations.

Glacial deposits are developed as separate spots in the origins of the Gusarchay, Goychay, Kishchay, Shinchay, Mukhakhchay, Balakanchay, and Mazymchay rivers within the Greater Caucasus and in the origins of the Terter, Hakeri, Gilanchay, and Nakhchivanchay rivers in the Lesser Caucasus. They are represented by boulder and blocky formations with loamy and sandy-gravelly filler and a maximum thickness up to 50–60 m. Their age is determined by the way of correlation with the river terraces of a known age.

Lacustrine deposits developed in the western part of the Ganja-Gazakh sloping plain along the right bank of the Kur River and presented by alternation of thin layers of clay and sand, among which occur bands of volcanic ash and pebbles lens. Color of the rocks is light gray and bluish-gray. Their thickness is 80–100 m. From the fauna, only freshwater mollusks *Sarbicula fluminalis* (Müll) were found.

Alluvial deposits are confined to the river valleys and compose erosion-accumulative and accumulative terraces with predominant relative heights within 60–200 m (see Table 3.1). These deposits are presented in the valleys of mountainous rivers by boulders, pebbles and gravel with silt filler, and in the valleys of lowland rivers by small pebbles, sand and loam. Their thickness sometimes reaches 90–100 m or more.

Alluvial-proluvial deposits occur in a wide strip border at the foothills of the north-eastern slope of the Lesser Caucasus and widely distributed in the middle stream of the Araz River (Nakhchivan trough), Ajinohur foothills and Gobustan. There are mainly deposits of cones rivers, temporary streams and ravines. They are mainly represented by pebbles. In the strip of the northeastern foothills of the Lesser Caucasus and Ajinohur area, these deposits contain numerous layers of volcanic ash which allow to correlate them with the Khazar marine horizon deposits containing the same volcanic ash. The thickness of these deposits in the Ajinohur foothills reaches 180–200 m, and in the northeastern foothills of the Lesser Caucasus is 100–150 m.

Talus-proluvial deposits developed at the foot of the northeastern slope of the Talysh Mts. and are presented by sandy loam, loam and clay. Near the foot, the rocks clogged by gravelly material and redeposited pebbles. The thickness of these deposits ranges from a few to 10–12 m, increasing to the foot of the ridge.

Colluvial deposits have a very limited distribution. They are developed in the form of small islands in the southeastern Gobustan on the eastern slopes of the Boyukdash and Kichikdash Mts. and are blocky scree of Absheron (Eopleistocene) limestone. In these clumps were found the famous Gobustan cave paintings (Environment of Azerbaijan 2010).

Volcanogenic deposits are developed in the Lesser Caucasus in the upper Hakeri River. They are represented by basaltic lava and andesibasalts, and sometimes breccia and ash layers. Their thickness, depending on the degree of erosion, reaches 70–80 m and more.

Lithofacies and thickness of Khazar deposits (Fig. 3.2) indicate a significant change in the conditions of sedimentation in the Middle Pleistocene, when the sea space (compared to the Baku time) was sharply reduced and the generation of coastal and continental deposits on the depression sides was formed.

Khvalyn horizon. These deposits in the territory of Azerbaijan are also presented by marine and continental facies (Fig. 3.3).

In the marine facies, they developed in the same depression areas that the Khazar deposits, but have a much smaller area of distribution. At the same time, these deposits are separated from the Khazar deposits almost everywhere by interruption and angular unconformity.

In natural sections, they almost always are represented by terraces. Analysis of available fauna clearly reveals two subhorizons: Lower Khvalyn and Upper Khvalyn. The leading fauna of the Lower Khvalyn subhorizon are *Didacna parallella* Bog., *D. praetrigonoides* Nal. et Anis., and for the Upper Khvalyn subhorizon are *Didacna praetrigonoides* Nal. et Anis and *D. trigonoides* Eichw. Khvalyn horizon completes the Upper Pleistocene stage. The age of Lower Khvalyn deposits is defined by the range of 42 ± 5 kyr (14S-method). The Upper Khvalyn subhorizon is estimated as 14.6 + 18.5 kyr (TL-method) and 12.8 + 9.6 kyr (14S-method).

The Lower Khvalyn deposits are most widespread in the Kur Depression. They go far to the west, skirting from the

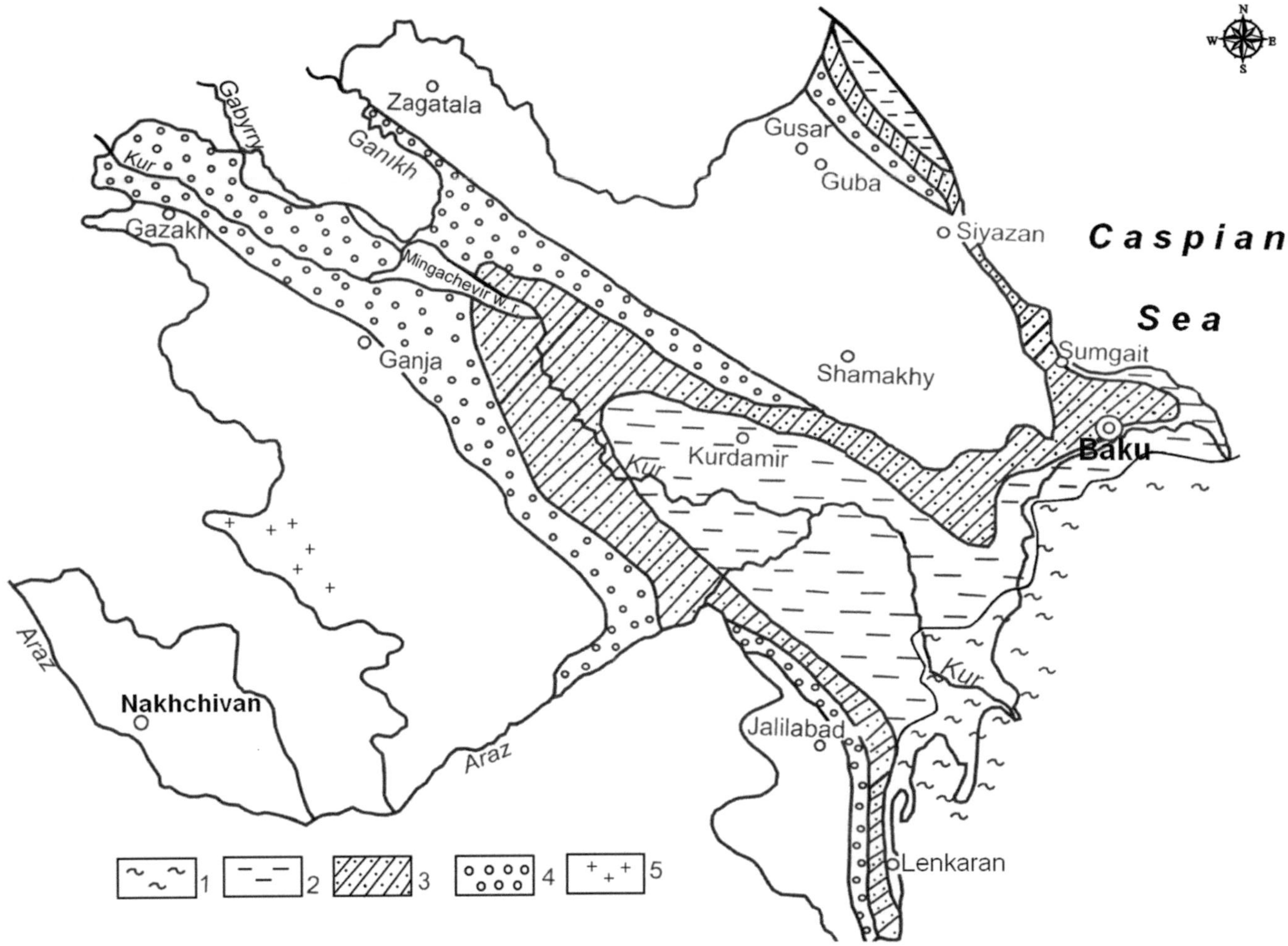

Fig. 3.2 Scheme of facies of Khazar Horizon deposits (after Alizadeh 1987): (*1*) clays, (*2*) predominantly clay-sandy sediments, (*3*) clay-sandy sediments interbedded with coarse gravel and, rarely, coquina, (*4*) continental coarse fragmental sediments, (*5*) volcanogenic rocks

south along the Alat-Langabiz ridge, Garamaryam ridge, and appear for the last time in the Mingechevir area. In this area to the Lower Khvalyn belongs an abrasion-accumulative terrace of the Mingachevir neck at the height of 30–35 m. At present, this terrace is flooded by waters of the Mingachevir reservoir.

In the eastern part of the Kur Depression, the Lower Khvalyn terraces compose separate small outcrops on the vault of Padar anticline and slopes of Mishovdagh, Kurovdagh and Bozdagh mountains, where they are presented by pebbles, coquina, sands and sandy clays (10–12 m). In the lowland of the Kur Depression, the Lower Khvalyn deposits (60–140 m) covered by thin latest generations were studied by drilling in the Kursangya, Garabaghly, Saryjalar and Daykend areas. Here they are represented by interbedding of sands and clays with rare alternating of gravel, coquina and volcanic ash.

In the sections of the Baku archipelago, the Lower Khvalyn sediments (25–60 m) are usually have subhorizontal occurrence and are characterized by sand material reduction and appearance of frequent knoll breccia layers.

The Lower Khvalyn deposits are developed in significantly different facies within the Lenkaran lowlands, where drilling operations discovered 20–25 m thickness presented in the lower part by gray laminated shales with interbedded sands and pebbles, and in the upper part mostly by sands.

In southeastern Gobustan and in the Absheron peninsula, the distribution of the Lower Khvalyn deposits is limited by the bandwidth of the modern coast. The deposits form three morphologically expressed terraces, which are traced at altitudes of 28–30, 20, and 14–15 m, and presented mostly by sands with pebbles and coquina, as well as shell limestone. The total thickness of these sediments is no more than 20–30 m.

To the northwest of the Absheron peninsula in the Caspian coast between the Yashma and Gilyazi villages, the Lower Khvalyn deposits are also represented by three levels of terraces at altitudes of 10, 30–35, and 40–50 m. The

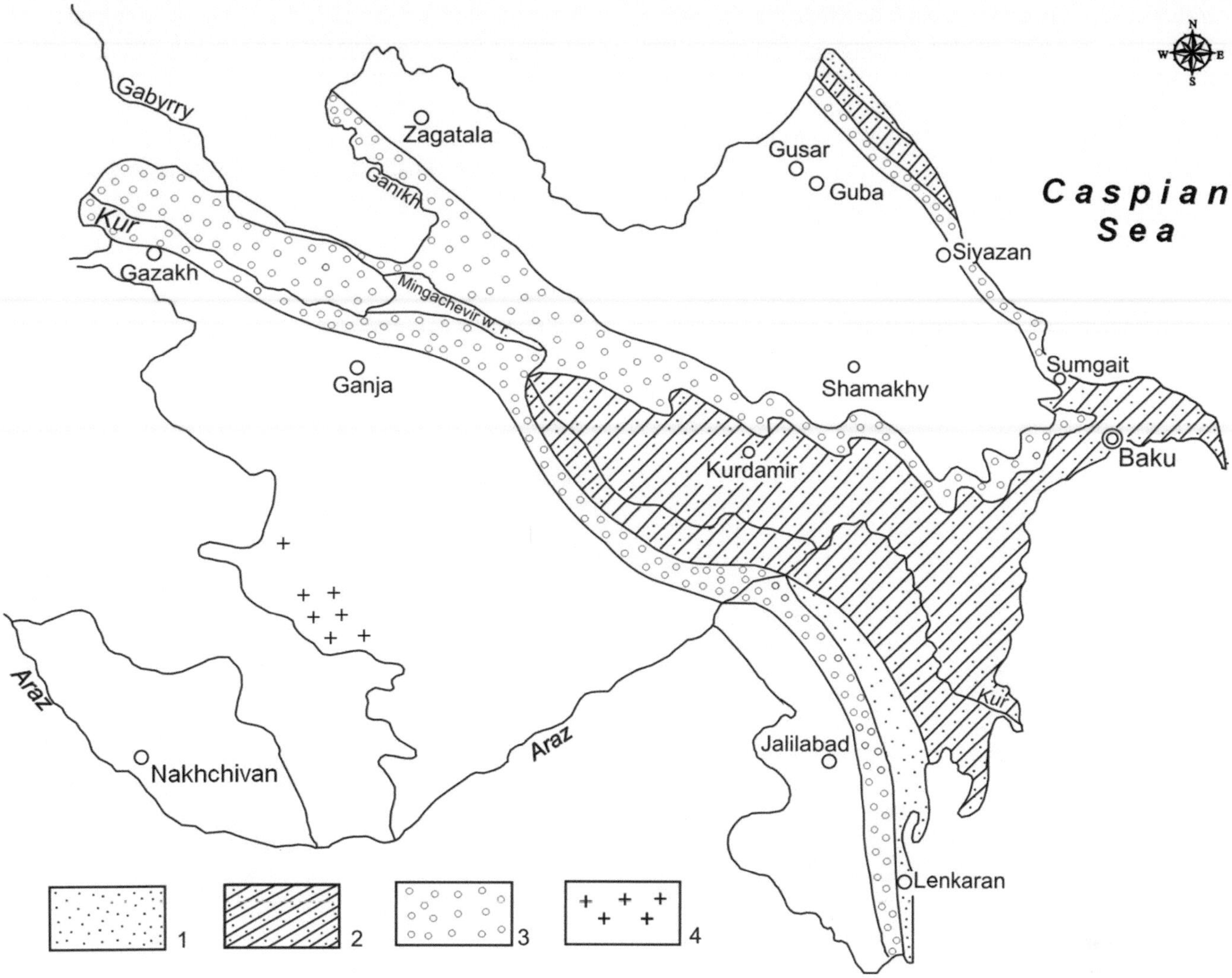

Fig. 3.3 Scheme of facies of Khvalyn Horizon deposits (after Alizadeh 1987): (*1*) sands and coquina, (*2*) clay-sandy sediments interbedded with coarse gravel and, rarely, coquina, (*3*) continental coarse fragmental sediments, (*4*) volcanogenic rocks

lower terrace is composed of sand with a typical Lower Khvalyn fauna, and two upper terraces by pebbles, gravel, sand and conglomerate. The penetrated thickness of these deposits ranges within 11–12 m.

Within the Gusar-Devechi Depression, the Lower Khvalyn marine sediments were revealed by drilling and presented by coastal facies, with sands and gravels interbedded with clay and coquina. Rudaceous rocks are often presented in the lower half of geological section.

The Upper Khvalyn deposits are distributed on a much smaller area and are commonly concentrated in the modern coastal strip of the Caspian Sea. These deposits are usually seen in the form of terraces and are characterized by high facial variability.

In the Kur Depression, these deposits are widely distributed as terraces in the eastern parts of its side areas—within the southeastern Shirvan, Mughan lowland and Lenkaran coast. They are composed primarily by sands; gravels are found only in their bottom. Coquina and clay in some places generate interlayers among the sand formations. Inside the Kur Depression, the Upper Khvalyn deposits discovered by numerous wells at the depths from 5–10 to 25–30 m. The thicknesses of these deposits in the Kur Depression vary from a few meters within the anticlinal uplifts to 60–70 m within the synclinal troughs, and in the area of the Baku archipelago reach 90–100 m.

Khvalyn deposits in the continental facies are presented by different genetic types. In the upper reaches of the Nakhchivanchay, Gilanchay and Terter rivers and in sources of rivers in the southern and northern slopes of the Greater Caucasus to this age relate the glacial deposits. They fill trough valleys and are presented by yellow-brown gravel loams with poorly rounded boulders and rubbles. Their thickness in some places reaches 35–40 m.

Alluvial deposits of this age are developed in the valleys of Kur and Araz rivers and their tributaries. They form four (and sometimes more) erosion-accumulative and accumulative terraces (see Table 3.1). Lithologically, they are presented by sand, sandy loam and pebbles with silt filler. Their total thickness reaches 70–80 m.

In the Ganykh-Ayrichay Valley, Upper Pleistocene associations compose up to 50–60 m layer of alluvial-deluvial gravel, sand and clay. With the same epoch is compared proluvial alluvial gravel accumulation in the band of the northeastern foothills of the Lesser Caucasus, at the foot of the southern slope of the Greater Caucasus (100–200 m), in the Gusar plain (60–70 m), in the western part of the Shirvan lowlands and in the Middle Araz Depression (30–40 m).

Talus-proluvial loam (20–30 m) fills a synclinal valley between the Kur and Gabyrry rivers and in the Ajinohur foothills and developed in the form of the loop at the foot of the north-eastern slope of the Talysh. Colluvial formations having form of small islands (represented in the south-eastern Gobustan by blocky talus) are composed by limestone in the Absheron stage. In the Garabagh volcanic plateau at this time were formed volcanic deposits presented mainly by basaltic andesite lavas. Deluvial and deluvial-colluvial deposits formed gravelly and gravelly-loamy entities with clay aggregates and have received extensive development in the slopes of the Greater and Lesser Caucasus and Talysh, with thickness from several to a few tens of meters.

Figure 3.4 shows the changing of lithofacies of Khvalyn deposits in the depression zones of Azerbaijan.

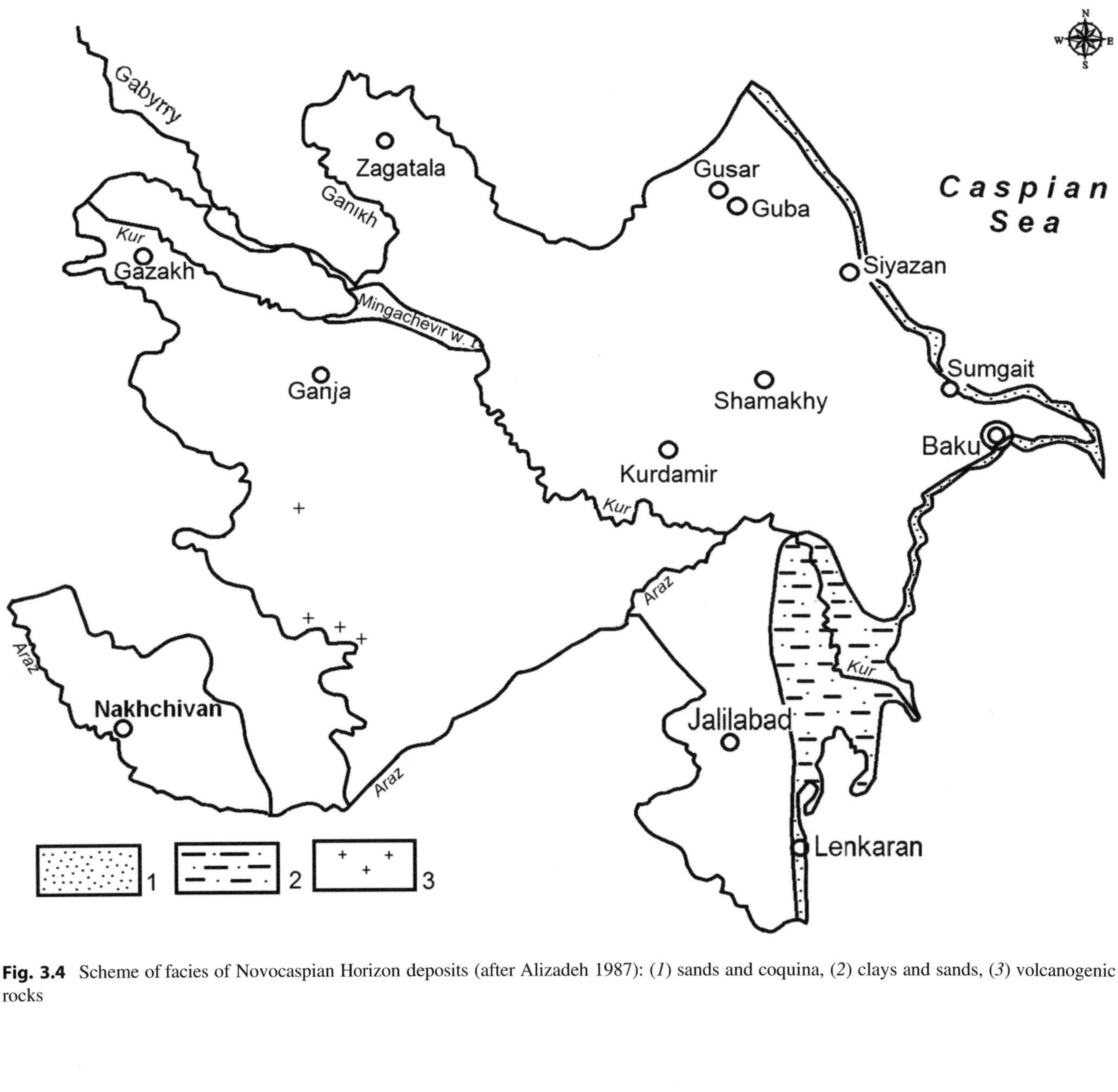

Fig. 3.4 Scheme of facies of Novocaspian Horizon deposits (after Alizadeh 1987): (*1*) sands and coquina, (*2*) clays and sands, (*3*) volcanogenic rocks

3.6.3.3 Holocene (Novocaspian Horizon)

The beginning of the Holocene period can be dated as 9–10 kyr ago. Modern deposits in Azerbaijan are widely distributed and presented by the Novocaspian horizon. These deposits are represented by several genetic types.

In modern marine facies, sediments represented the Lower and Upper Novocaspian deposits with *Cardium edule* (Linne), *Didacna trigonoides* (Pall), *D. crassa* (Eichw), D. pyramidata (Grimm), D. bacri (Grimm), etc. and occupy almost the entire territory of the Lower Kur Depression, coastal strip of the Lenkaran and Gusar-Devechi lowlands, narrow border the Absheron peninsula and compose a significant part of the Caspian Sea bottom. Almost everywhere they are separated by interruption from the underlying Khvalyn layers.

The area of Novocaspian transgression occurs predominantly the same area of the previous Caspian Sea elevation. For these deposits, lithological diversity and rapid variability of geological sections (both vertically and horizontally) are typical. This is caused by rapid and frequent changes of sedimentation conditions under frequent fluctuations of the sea level and the shoreline, ingression and transgression of the Kur ancient delta, partial erosion and re-deposition of sediments.

The Lower Novocaspian marine deposits are represented in the Lower Kur Depression mainly by bluish-gray and reddish-brown viscous clays with interbedding of sands. Near the western border of their distribution frequent interlayers alluvial (delta) of loam and sandy loam are observed; they often are predominating ones. In the coastal strip, a fairly large area of sand distribution dominates. It can be assumed that they are Kur deposits which flowed at this time in the Caspian Sea on the territory of southeastern Shirvan. These sands are overlain by a thin layer of marine clay. Strictly speaking, in this territory are developed deposits of mixed genesis (deltaic and marine); only ubiquitous distribution of shells Sardium edule (Linne) throughout the area gives the basis to interpret their origin as marine deposits. Novocaspian layer distribution within the depression is limited by isohypse of −21 m. The thickness of the deposits usually ranges within 7–10 m and sometimes reaches 40–50 m.

On the Lenkaran, Absheron and Gusar-Devechi coasts of the Caspian Sea Upper Novocaspian associations are developed, which along with the fauna discovered the lower layer of the horizon contain *Leytilaster lineatus* Gmel. and *Balanus improvisus* Darw. They form a series (up to 3) of accumulative terraces and are presented mainly by sand and coquina.

Modern bottom sediments of the Caspian Sea in the coastal strip are presented by sands and coquinas. As the distance from the coast, they are replaced by silty sand, then by sandy silt, and finally by silt that has a thickness of 30 m and more. The thickness of the modern sediments ranges from 1–2 to 8–10 m on elevations and up to 40 m and more in the depressions.

The Holocene transition facies correspond to alluvial marine deposits, with a thickness of no more than 10 m. They are presented in the southern part of the Mughan and Salyan steppes and composed by alternating alluvial and marine layers. Marine layers contain modern Caspian fauna —the most common form of which is the *Sardium edule* (Linne). Lithologically, alluvium is presented generally by lighter soils (sandy loam, loam, and less sand) than marine deposits (mainly by clay).

The continental Holocene facies are presented by lacustrine, alluvial, alluvial-proluvial, talus, colluvial, aeolian, of mud-volcanic and volcanogenic formations.

Lacustrine deposits developed as separate spots in the Kur-Araz and Lenkaran lowlands and in Absheron peninsula, where they cover areas of currently existing and almost completely dried up lakes. They are represented by alternating thin layers of clay and silt, among which are observed the lenses of small and fine-grained sand. A color of the rocks is dark and bluish-gray. A specific characteristic is an abundance of plant residues. The thickness of the deposits does not exceed 5–7 m.

Lacustrine-alluvial and lacustrine-salt-marsh deposits are widely presented at the contact between the Shirvan plain and Kur ancient riverine trees, in the Mil plain, eastern part of the Lenkaranian Mughan, in the Ajinohur Lake basin, Mughan and Salyan lowlands and in the Absheron peninsula. Lacustrine-alluvial deposits are presented by silts, heavy loams and clays, and lacustrine-salt-marsh, as well as by salt silts, sandy loam, loam with lenses and thin layers of various salts. Their thickness consists of 2–5 m.

Alluvial deposits are widespread along the Kur and Araz rivers and their tributaries. They compose rivers-beds, oxbows and low accumulative terraces with relative heights from 1.5–2.5 to 20–25 m (see Table 3.1). Their lithological composition is determined by geomorphological conditions of accumulation. Alluvium of the mountainous rivers is presented mostly by gravel with boulders, sand and silt aggregate. Alluvium of the Kur and Araz rivers is composed by sands, sandy loams, loams and clays. The thickness of the modern alluvial deposits varies from 10 to 20–30 m, increasing in the direction of river flows.

With the Novocaspian horizon, the following are compared: (1) proluvial strata of clay and loam with a thickness up to 5–6 m, which widely developed in left bank of the Araz River at the foot of the south-eastern slope of the Lesser Caucasus, (2) alluvial-proluvial gravels composing modern alluvial fans of rivers, and (3) deluvial-proluvial loam and sandy loam bordering foot of the Langabiz-Alat and Gojashen-Goychay ridges composing synclinal troughs of the Ajinohur foothills and interfluve of the Kur and Gabyrry rivers. The thickness of the modern alluvial-proluvial deposits sometimes (for instance, at the foot of the southern slope of the Greater Caucasus) reaches 50–60 m, and thickness of deluvial-proluvial deposits—20–30 m.

Colluvial formations are widely distributed in the Garabagh volcanic plateau, on the slopes of Mt. Kapaz and and the Khankendi basin in the Lesser Caucasus, and partially in the Greater Caucasus. They are composed of block debrises with gravelly-loam filler whose thicknesses reach (on the slope of the Kapaz Mt.) up to 100 m and more.

Eolian deposits are widespread in the northern part of the Absheron peninsula, in the eastern part of southeastern Shirvan, and are also found in the South Gobustan and Nearcaspian region located to the northwest of the Absheron. They are represented by sand ridges, dunes and hilly sands consisting of shelly detritus Tusa. Their thickness is typically less than 4–5 m, but it sometimes reaches 12–15 m.

Modern volcanic deposits are developed as small arrays around the young volcanic centers of the Garabagh highland and are presented by tuffs, tufolavas, slags and andesitic breccia.

The deposits of mud volcanoes developed in the Absheron peninsula, Gobustan, at the Alat high, in the Lower Kur Depression and in the Baku archipelago. This is a mud volcanic breccia, consisting of clay mass without any stratification and very homogeneous in the composition. It usually contains clastic inclusions, and sometimes very large chunks of rocks of different ages; however, the overall breccia content in these inclusions is very small. The thickness of the deposits depends on the height of the mud volcanoes and intensity of their eruption.

In the territory of Azerbaijan, besides the above-mentioned, are presented deluvial and deluvial-colluvial deposits composed respectively by gravelly loam and gravel with clay aggregates. These deposits cover the slopes of ridges of the Greater and Lesser Caucasus and Talysh and sometimes reach 40–50 m thickness.

References

Abdulkasumzadeh, M. P. (1963). *Stratigraphy and fauna of Upper Jurassic deposits of the north-eastern part of the Lesser Caucasus*. Baku (in Russian): Publ. of the Academy of Science of Azerbaijan.

Abdulkasumzadeh, M. P. (1988). *Upper Jurassic of the Lesser Caucasus within Azerbaijan Republic (Stratigraphy and Ammonite Fauna)*. Baku: Elm (in Russian).

Abdulkasumzadeh, M. P., Gasanov, T. A., et al. (1972). The Lesser Caucasus and Southern Transcaucasian (Armenia and Azerbaijan). In *Stratigraphy of the USSR. Jurassic System*. Moscow: Nedra (in Russian).

Agabekov, M. G., & Mamedov, A. V. (1961). *Geology and Oil&Gas bearing of the Western Azerbaijan and Eastern Georgia*. Baku: Azerneshr (in Russian).

Agabekov, M. G., Salayev, S. G., Averbuch, B. M., Allakhverdiyev, R. A., & Moshashvili, A. B. (1982). *Structural-facial peculiarities and perspectives of Eocene Oil&Gas bearing deposits of the depression zones of Azerbaijan*. Baku: Elm (in Russian).

Agalarova, D. A., & Mamedova, U. Ya. (1967). Upper cretaceous deposits of the Lesser Caucasus. *Transaction of the Azerbaijan Institute of Oil and Chemistry*, Vol. 19, Baku: Azerbaijan (in Russian).

Agayev, V. B. (1966). *Stratigraphy of Middle Jurassic deposits of the North-Eastern Azerbaijan and Southern Dagestan*. Baku: Azerneshr (in Russian).

Agayev, V. B. (1990). *Stratigraphy of Jurassic deposits of Azerbaijan*. Baku: Elm (in Russian).

Akhmedov, G. A. (1957). *Geology and oil-bearing of the Gobustan*. Baku: Asgosizdat (in Russian).

Aliyev, G. A., Akhmedbeyli, F. S., Ismailzade, A. D., Kangarli, T. H., & Rustamov, M. I. (2005a). In V. E. Khain & Ak. A. Alizadeh (Eds.), *Geology of Azerbaijan. Tectonics* (Vol. IV). Baku: Nafta-Press (in Russian).

Aliyev, I., Dadiyeva, T., Gasanov, F., Zokhrabova, V., Nabiyev, M., Pashaly, N., et al. (2005b). In Ak. A. Alizadeh (Ed.), *Geology of Azerbaijan. Lithology* (Vol. II). Baku: Nafta-Press (in Russian).

Aliyev, S. A., Aliyev, C. S., Askerhanova, H. Q., Balakishibeyli, Sh. A., Qasanov, A. O., Qasanov, A. B., et al. (2005c). In Ak. A. Alizadeh (Ed.), *Geology of Azerbaijan. Physics of the Earth* (Vol. V) (in Russian).

Aliyev, M. M., Aliyev, O. B., & Mamedzade, R. N. (1958). Cretaceous deposits of the south-eastern part of the Lesser Caucasus. *Transaction of the Institute of Geology, Acad. Sci. Azer. Rep.*, Vol. XIX, Baku (in Russian).

Aliyulla, H., & Babayev, R. G. (1984). On the establishment of Mesozoic reef construction in the Middle Kura Depression (with utilization of sections of boreholes OP-1 and SG-1 in the Saatly area). *Azerbaijan Oil Industry, 7* (in Russian).

Alizadeh, A. A. (1945). *Maykop suite of Azerbaijan and its oil-bearing*. Baku: Azneftizdat (in Russian).

Alizadeh, A. A. (1969). *Akchagyl of Azerbaijan*. Moscow: Nedra (in Russian).

Alizadeh, A. A. (1973). *Absheron of Azerbaijan*. Baku: Elm (in Russian).

Alizadeh, A. A. (1974). *Sarmat of Azerbaijan*. Moscow: Nedra (in Russian).

Alizadeh, Ak. A. (1966). Stratigraphic division of Upper Cretaceous deposits of Azerbaijan by belemnites. *Doklady Acad. Sci. of Azerb. Rep., Ser.: Geol, 171*(3) (in Russian).

Alizadeh, Ak. A., Khain, V. Y., & Ismailzadeh, A. D. (2000). *Saatly superdeep. Analysis of deep structure of the Kur intermontane depression by the data of saatly superdeep borehole SG-1 drilling*. Baku: Nafta-Press (in Russian).

Alizadeh, K. A. (1968). *Oligocene deposits of the eastern part of the Lesser Caucasus*. Baku: Academy of Science of Azerbaijan Republic (in Russian).

Alizadeh, K. A., Alizadeh, Ak. A., Abbasov, A., Abdulkasimzadeh, M., Agayev, V., Azizbekova, A., et al. (2007). In Ak. A. Alizadeh (Ed.), *Geology of Azerbaijan. Stratigraphy. Part 2: Mesozoic and Cenozoic* (Vol. I). Baku: Nafta-Press (in Russian).

Alizadeh, K. A., Atayeva, E. Z., et al. (1986). The south-eastern Greater Caucasus and Lesser Caucasus. Miocene. In *Stratigraphy of the USSR. Neogen*. Moscow: Nedra (in Russian).

Alizadeh, K. A., Azizbekova, A. I., & Atayeva, E. Z. (1980). *Oligocene-Miocene deposits of the Talysh Mts., the south-eastern end of the Lesser Caucasus (Jebrail Area) and Nakhchivan Depression*. Baku: Academy of Science of Azerbaijan Republic (in Russian).

Alizadeh, K. A., et al. (1989). *Regional stratigraphic scheme of Paleogene of Azerbaijan*. Baku: Elm (in Russian).

Alizadeh, S. A. (1987). *Anthropogen of Azerbaijan*. Baku: Elm (in Russian).

Alizadeh, Ak. A. (Ed.), (2012). The Modern Problems of Geology and Geophysics of the Eastern Caucasus and South Caspian Depression. Baku: Nafta-Press.

Alizadeh, S. A., Bairamov, A. A., Mamedov, A. V., & Shirinov, N Sh. (1978). *Geology of quaternary deposits of Azerbaijan*. Baku: Elm (in Russian).

Azizbekov, Sh A. (1961). *Geology of Nakhchivan Republic*. Moscow: Gosgoltekhizdat (in Russian).

Azizbekov, Sh. A., Alizadeh, K. A., Shikhalibeyli, E. Sh., & Gadjiev, T. G. (Eds.). (1972). *Geology of the USSR. Azerbaijan Republic. Part 1. Geological description* (Vol. 47). Moscow: Nedra (in Russian).

Azizbekov, Sh A, et al. (1979). *Geology and Volcanism of Talysh*. Baku: Elm (in Russian).

Babayev, Sh. A., Kangarli, T. N., & Mamedov, A. B. (2015). In Ak. A. Alizadeh (Ed.), *Stratigraphy of Nakhchivan Autonomic Republic*. Baku: Nafta-Press (in Russian).

Bagirzadeh, F. M., Kerimov, K. M., & Salayev, C. G. (1987). *Deep structure and oil&gas-bearing of the South Caspian megadepression*. Baku: Asgosizdat (in Russian).

Bagmanov, M. A. (1963). Paleogene deposits of the Talysh Mts. (Stratigraphy and Mollusk Fauna). Baku: Academy of Science of Azerbaijan Republic (in Russian).

Bagmanov, M. A. (1980). *Stratigraphic scale of the lower Palaeogene*. Baku: Elm (in Russian).

Cretaceous deposits of the Caspian Sea Framing, 1980. Moscow: Nauka (in Russian).

Cretaceous Fauna of Azerbaijan, 1988. Baku: Elm (in Russian).

Environment of Azerbaijan, Gobustan Rock Art Cultural Landscape, Climate of Azerbaijan, & Garabagh Horse. (2010). *Collection of papers*. General Books LLC.

Eppelbaum, L. V., & Khesin, B. E. (2012). *Geophysical studies in the Caucasus*. New York: Springer.

Gasanov, T. A. (1961). *Fauna and stratigraphy of lower and middle Jurassic deposits of the north-eastern part of the Lesser Caucasus (Azerb. Republic)*. Baku: Academy of Science of Azerbaijan Republic (in Russian).

Gasanov, T. A. (1967). *Lower Jurassic of Azerbaijan (The Lesser Caucasus)*. Academy of Science of Azerbaijan Report, Baku (in Russian).

Gasanov, T. A. (1973). *Middle Jurassic of Azerbaijan (The Lesser Caucasus)*. Baku: Elm (in Russian).

Gasanov, T. Ab. (1968). Lower Eocene deposits of the central part of the Lesser Caucasus. *Izvestiya, Academy of Sciences of the USSR, Series Geological*, *9*, 91–97 (in Russian).

Gasanov, T. Ab. (1983). Paleogene of the south-eastern part of the Lesser Caucasus. *Soviet Geology*, *2* (in Russian).

Gasanov, T. Ab. (1985). *Ophiolites of the Lesser Caucasus*. Moscow: Nedra (in Russian).

Grossgeim, V. A., & Khain, B. E. (1963). Stratigraphy of Cretaceous deposits of flysch zone of the Greater Caucasus. *Transaction of VNIGRI*, *220* (in Russian).

Isayev, B. M., & Grigoryantz, B. V. (1968). Stratigraphy of the Cretaceous deposits of Vandam zone (southern slope of the Greater Caucasus) in the interfluve of Agsu and Tikanlychay rivers using new materials. *Transaction of the Azerbaijan State University, Series: Geology and Geography*, *4* (in Russian).

Ismail-Zadeh, A. D., Mustafayev, G. V., & Rustamov, M. I. (2005). In Ak. A. Alizadeh (Ed.), *Geology of Azerbaijan. Magmatism* (Vol. III). Baku: Nafta-Press (in Russian).

Kangarli, T. N., Balamedov, Sh. R., Sadykhov, E. A., & Mekhtiyeva, Z. N. (2013). Sedimentational and geodynamic situations of forming Upper Jurassic complex of South-Eastern Caucasus. Paper 1—Malm of Side Ridge. *Izvestiya (Proceedings), Azerbaijan Academy of Science, Series: Earth Science*, *3*, 3–15 (in Russian).

Khain, V. E. (1947). Section and facies of Mesozoic of the south-eastern Caucasus according to the data of newest investigations. *Transaction of the Geological Institute, Academy of Science of Azerbaijan Republic*, *XIII* (in Russian).

Khain, V. E., & Akhmedbeyli, F. S. (1957). Geological structure and development of the Kusary-Devechi Sinclinorium. In *Materials on geology of north-eastern Azerbaijan* (pp. 181–385). Baku: Academy of Science of Azerbaijan Republic (in Russian).

Khain, V. E., & Shardanov, A. N. (1957). Geological structure of the northern slope of the South-Eastern Caucasus. In *Materials on geology of north-eastern Azerbaijan* (pp. 15–179). Baku: Academy of Science of Azerbaijan Republic (in Russian).

Khain, V. E., Shardanov, A. N., & Kasimova, N. M. (1955). On the stratigraphy of Middle Jurassic of the south-eastern Caucasus. *Doklady Academy of Sciences of the USSR*, *100* (in Russian).

Khalafova, R. A. (1969). *Fauna and stratigraphy of Upper Cretaceous deposits of south-eastern part of the Lesser Caucasus and Nakhchivan Republic*. Baku: Azerneshr (in Russian).

Khalilov, A. G. (1959). *The lower Cretaceous deposits of Azerbaijan part of the Lesser Caucasus*. Baku: Academy of Science of Azerbaijan Republic (in Russian).

Khalilov, A. G. (1965). *Stratigraphy of lower Cretaceous deposits of the south-eastern ending of the greater Caucasus*. Baku: Academy of Science of Azerbaijan Republic (in Russian).

Khalilov, A. G. (1978). *Stratigraphy of Azerbaijan*. Baku: Elm (in Russian).

Khalilov, A. G., & Aliyev, G. A. (1985). *Schemes of stratigraphy of Cretaceous deposits of Azerbaijan*. Baku: Elm (in Russian).

Khalilov, A. G., Aliyev, G. A., & Askerov, R. B. (1974). *The lower Cretaceous of the south-eastern ending of the Lesser Caucasus (stratigraphy and paleography)*. Baku (in Russian): Elm.

Kotlyar, G. V., & Ganelin, V. G. (1983). Main achievements and problems in studying Permian System in the USSR. *Transaction of VSEGEI*, *322*, 68–77 (in Russian).

Kovalevsky, S. A. (1936). *Continental Stratum of Adjinour (Stratigraphy and Genesis)*. Baku: Azneftegeofizrazvedka (in Russian).

Leven, E. Ya. (1975). Stratigraphy of Permian deposits of the Transcaucasian. *Soviet Geology, 1*, 96–110 (in Russian).

Mamedov, A. B. (1962). *Stratigraphy and fauna of brachiopods of the Danzik suite of Devonian in Nakhchivan Republic* (Ph.D. thesis, Institute of Geology, Baku) (in Russian).

Mamedov, A. B. (1973). *Geological structure of the middle Kura depression*. Baku: Elm (in Russian).

Mamedov, A. B. (1979). Zonal dividing of Eifel stage of Middle Devonian in Nakhchivan Republic. *Izvestiya, Azerbaijan Academy of Science, Series: Earth Science, 5*, 92–98 (in Russian).

Mamedov, A. B. (1980a). *Guide of geological excursions on typical sections of Middle Devonian in Nakhchivan Republic*. Baku: Elm (in Russian).

Mamedov, A. B. (1980b). Zonal dividing of Tournasian stage of lower Carboniferous in Nakhchivan Republic. *Doklady, Academy of Science of Azerbaijan Republic, 36*(12), 53–57 (in Russian).

Mamedov, A. B. (1980c). Zonal dividing of Middle Devonian of Transcaucasian by brachiopods. *Transaction of the session of International Stratigraphy Commission on Devonian of Azerbaijan* (Nakhchivan—Baku) (pp. 28–35). Baku: Elm (in Russian).

Mamedov, A. B. (1981). Zonal dividing of Visean stage of lower Carboniferous in Nakhchivan Republic. *Doklady, Academy of Science of Azerbaijan Republic, 37*(2), 60–64 (in Russian).

Mamedov, P. Z., Guliyev, Q. Q., Kadirov, F. A., Shikhalyev, Y. A., Guliyev, I. S., Aliyeva, E. H., et al. (2008a). In Ak. A. Alizadeh (Ed.), *Geology of Azerbaijan. Oil and Gas* (Vol. VII). Baku: Nafta-Press (in Russian).
Mamedov, A. B., Rzhonsnitskaya, M., Vuks, G., Gasanov, G., Kotlar, G., Krapacheva, G., et al. (2008b). In Ak. A. Alizadeh (Ed.), *Geology of Azerbaijan. Stratigraphy. Part 1: Precambrian and Paleozoic* (Vol. 1). Nafta-Press, Baku (in Russian).
Mamedov, A. B., Vekilov, B. G., & Pashaly, N. V. (1970). Quaternary deposits of the Middle Kura depression in connection with neotectonics problems. In *Recent tectonic movements and structures of the Alpine geosynclinal belt of south-western Eurasia*. Baku: Elm (in Russian).
Mamedzade, R. N. (1967). *Stratigraphy of Cretaceous deposits of NE part of the Lesser Caucasus (Interfluve of the Koshkarchay-Debetchay Rivers)*. Baku: Academy of Science of Azerbaijan Republic (in Russian).
Mirchink, M. F., & Shurygin, A. M. (1972). *Formation of structure of Tertiary and Cretaceous sediments of the south-eastern Caucasus*. Moscow: Nauka (in Russian).
Nevesskaya, L. A., & Trubikhin, V. M. (1984). The history of the Caspian Basin and its mollusk fauna in Later Pliocene and Early Pleistocene. In *Anthropogene of Eurasia* (pp. 19–27). Moscow: Nauka (in Russian).
Paffengoltz, K. N. (1940). *Geological Sketch of Nakhchivan Republic* (Vol. 28). Baku: Academy of Science of Azerbaijan (in Russian).
Paffengoltz, K. N. (1959). *Geological sketch of the Caucasus*. Yerevan: Academy of Science of Armenia Republic (in Russian).
Paffengoltz, K. N. (1979). *Stratigraphy of Palaeogene of the Lesser Caucasus*. Leningrad: Nedra (in Russian).
Permian System. (1966). *Stratigraphy of the USSR*. Nedra: Moscow (in Russian).
Potapov, I. I. (1954). *Absheron oil-bearing province (geological characteristic)*. Baku: Academy of Science of Azerbaijan Republic (in Russian).
Problems of the Talysh geology, 1958. Moscow: Academy of Science of the USSR (in Russian).
Rasulov, G. L., & Mamedzade, R. N. (1978). *Geological structure of Pre-Kura Region of Western Azerbaijan*. Baku: Elm (in Russian).
Rzhonsnitskaya, M. A. (1948). Devonian deposits of Transcaucasian. *Doklady, Academy of Science of the USSR, 49*(8), 1477–1480 (in Russian).
Rzhonsnitskaya, M. A., & Mamedov, A. B. (1988). Border Devonian and Carboniferous deposits of the Southern Transcaucasian. In *Boundaries of Devonian and Carboniferous in the territory of the USSR* (pp. 124–137). Moscow: Nauka i Technika (Science and Technique) (in Russian).
Salayev, S. G. (1961). *Oligocene-Miocene deposits of the southern-eastern Caucasus and their oil&gas-bearing*. Baku: Academy of Science of Azerbaijan Republic (in Russian).
Shikhalibeyli, E Sh. (1956). *Geological structure and evolution of Azerbaijan part of the southern slope of the Greater Caucasus*. Baku: Academy of Science of Azerbaijan Republic (in Russian).
Shikhalibeyli, E. Sh. (1964–1967). *Geological structure and history of tectonic evolution of the eastern part of the Lesser Caucasus (Stratigraphy, Tectonics and History of Tectonic Evolution)* (Vols. I–III). Baku: Academy of Science of Azerbaijan Republic (in Russian).
Shikhalibeyli, E Sh. (1994). *Geology and economic minerals of Nagorny (Daghliq) Garabagh of Azerbaijan*. Baku: Elm (in Russian).
Shikhalibeyli, E. Sh., Allakhverdiyev, G. A., & Babayev, Sh. A. (1976). The Paleocene deposits of the central part of the Lesser Caucasus (within Azerbaijan). *Doklady Academy of Science of Azerbaijan Republic, 32*(2) (in Russian).
Shikhlinskiy, A. Sh. (1967). *Geology and perspectives of oil&gas-bearing of Pliocene deposits of the lower Kura depression*, Baku (in Russian).
Shirinov, F. A., & Bazhenov, Yu P. (1962). *Geological structure of the foothills of the southern slope of the greater Caucasus*. Baku: Azerneshr (in Russian).
Sultanov, K. M. (1953). *Stratigraphy and Fauna of Upper Miocene of the Eastern Azerbaijan*. Baku: Academy of Science of Azerbaijan Republic (in Russian).
Sultanov, K. M. (1960). On the stratigraphy of Sarmatian deposits of the Talysh. *Doklady Academy of Science of Azerbaijan Republic, 134*(5) (in Russian).
Sultanov, K. M. (1964). *Absheron stage of Azerbaijan*. Baku: Azerneshr (in Russian).
Vekilov, B. G. (1962). *The Pontian stage of the eastern Azerbaijan*. Baku: Academy of Science of Azerbaijan Republic (in Russian).
Vekilov, B. G. (1969). *Anthropogenic Deposits of north-eastern Azerbaijan*. Baku: Elm (in Russian).

4 Magmatism

The Caucasian segment of the Mediterranean folded system is characterized by its development heterogeneity and phase heterogeneity, thus reflecting the general polycyclism and heterochronism inherent for the entire belt. In this system, Azerbaijan encompasses its eastern part, characterized by the southeastward plunging of the Greater and the Lesser Caucasus structural zones, along the western offshore of the Caspian Sea basin (Ismailzadeh et al. 2005).

The Baikal and the Hercynian age formations that preceded the Alpine epoch of the Caucasus generation are represented in metamorphic facies and are exposed fragmentarily.

Geological and geophysical data for the **Baikal period** complexes reflect their correspondence to the rigid platform base rocks that served as a basement for tectonic and magmatic cycles, manifesting themselves consequently.

In the **Hercynian period**, a deep-water basin was observed within the Greater Caucasus boundaries; at the Lesser Caucasus, the Transcaucasian massif preserved its relative elevation stability, whereas the Nakhchivan zone looked like a shallow-water basin of marginal continental type with a subplatform development mode.

According to the existing notions related to the geology of the adjacent Northern Caucasus region, the pre-Alpine basement of the Greater Caucasus is represented by amphibolites, gneisses and crystalline schists, intruded by the Hercynian granitoids. In the upper part of the pre-Alpine basement, the Triassic volcanogenic formations of andesite composition are found, which were uncovered in Azerbaijan by a prospecting borehole in the Khachmaz region (Agzybirchala village).

In the Lesser Caucasus, the pre-Alpine basement outcrops are composed within the Transcaucasian block of the Precambrian–Early Paleozoic gneisses and crystalline schists that constitute the lower structural level. The upper structural level, observed in the Nakhchivan block, is represented by the Devonian–Lower Carbon and Permian conformable beds that underwent metamorphosis in greenschist facies, chlotite-sericitic schist, quartzites and marble, which lack volcanogenic matter in their composition. The Late Paleozoic complexes are conformably overlapped by a thick series of Triassic dolomites, viewed by their metamorphic type and bedding conditions in a single Paleozoic–Triassic basement.

Noted facies features and the level of tectonic metamorphic transformations of the Caucasian pre-Jurassic complexes reflect the geodynamics of its early generation periods—from the Greater Caucasus active continental margin in the north to the Lesser Caucasus passive subplatform-type margin in the south. These geodynamical peculiarities reflect the complicated multiple-stage mode of presentation of the Alpine-Cimmerian tectonic and magmatic processes that took place during the successive periods of its generation.

The **Alpine–Cimmerian** period was a turning period for the entire Mediterranean belt, as it marked the separation from the subplatform development to the differentiated block movements. The establishment of structural and formational zones took place on the rigid heterogeneous basement of the Baikal and Hercynian bases.

During this period, the Caucasian magmatism proceeded in complex geodynamical conditions emerging on the continent lithospheric plate convergent borders during their recurrent collisions, resulting in the occurrence of different series of magmatic complexes. The following tectonic and magmatic stages are isolated in the Caucasian evolutional development:

The Cimmerian stage, with substages:

- Early Cimmerian, Early–Middle Jurassic, J_1–J_2a;
- Late Cimmerian, Middle–Late Jurassic—Early Cretaceous, J_2b–K_1nc;

A.A. Alizadeh et al., *Geosciences of Azerbaijan*, Regional Geology Reviews,
DOI: 10.1007/978-3-319-27395-2_4

The Alpine, with substages:

- Early Alpine, Cretaceous, K_1ap, K_2st, $K_2cp\text{-}m$;
- Late Alpine, Paleogene and Miocene—Quaternary, $\text{₽}_1\text{₽}_2$, $\text{₽}_3\text{N}_1$, $N_1^2–N_2$, $N_2^3–Q$.

Within the Azerbaijan Republic boundaries, the magmatic complexes of this period are found in the Greater Caucasus, the Kur Depression, and the Lesser Caucasus structures.

4.1 The Greater Caucasus

4.1.1 The Tufan Zone

In the Greater Caucasus Tufan (Tfan) zone, three formations stand out in the composition of a thick series of the Early–Middle Jurassic argillo-arenaceous deposits: sodic basalts (Early Jurassic), andesite-dacite-rhyolites (Middle Jurassic), and gabbro-diorite-plagiogranite (Late Jurassic); subalkaline gabbroids (Late Cretaceous); and in the Vandam zone, also the trachybasalt-trachyandesite (Late Cretaceous) (Fig. 4.1).

The sodic basalt formation, from the Late Pliensbachian–Early Toarcian stage (184 Ma) (Abdullayev et al. 1971) is represented by toleite basalts of pillow lavas, olivine dolerites and dolerites in sub-volcanoes and dyke-like bodies. They metamorphized from the initial stage of greenstone transformation to complete spilite formation. They are folded together with the enclosing Sinemurian–Pliensbachian rocks. In addition, they are disseminated within the western wings of the Zhikhikh-Chugar structure southern and northern steps.

According to their petrochemical characteristics, the formation rocks correspond to the sodic series by $Na_2O/K_2O = 4.4$ and 6.2, by low alumina and femicity factor of $f = 22.1–23.27$ %, with $Al' = 0.72$ and 0.66, to melanocratic basalts, and with $K_2O/TiO_2 = 0.2–0.5$, to toleite basalts and dolerites.

The average contents for iron group elements such as Ti, Cr, Ni, Co are below their percentage abundance; for V, they are increased, with a vanadium concentration factor of 1.5.

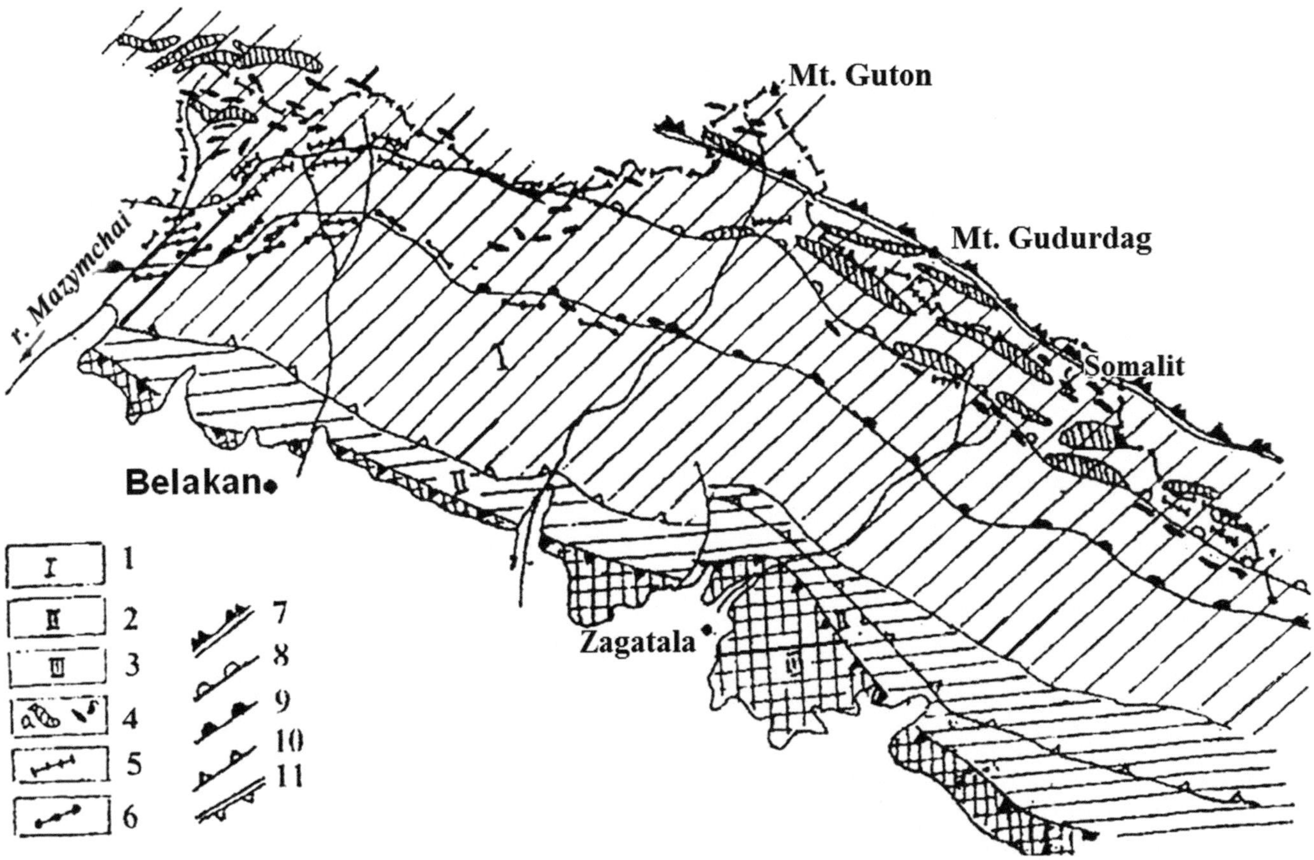

Fig. 4.1 Scheme of distribution of magmatic formations in the Tufan zone of the Greater Caucasus (based on N.K. Kurbanov, B.M. Isayev, R.B. Kerimov, etc.) (*1*) Tufan anticlinorium, (*2*) Zagatala-Govdagh sinclinorium, (*3*) Vandam anticlinorium; formations (*4–6*), (*4*) sodium basalts: (*a*) lavas, (*b*) subvolcanoes, (*5*) andesite-dacite-rhyolitic, (*6*) gabbro-diorite-plagiogranite, (*7*) Main Caucasian Thrust; faults (*8–11*), (*8*) Kehnamedan, (*9*) Jikhih, (*10*) Malkamud, (*11*) Zangin

In metallic and metallogenic elements, Zn, Sn, and F contents are higher and Cu, Pb, and Hg contents are lower than their percentage abundance for rocks with corresponding composition.

The **andesite-dacite-rhyolite formation**, Middle Jurassic (166 Ma), is represented by a subvolcanic facies such as dykes, strata, and, more rarely, boss-like bodies. It developed at the Zhikhikh-Chugar ore-bearing zone southwestern wing, within the Katsdagh field and along the Mazymchay River.

By the geochemical features of all rock types Ni, Cr and V contents are within or below the percentage abundance, with a reduction in concentration from andesites to rhyolites. The average Co content in the rocks is above the percentage abundance and amounts by the following (ppm): in andesites—24.0, in dacites—10.8, and in rhyolites—6.8. Cu, Zn, and Pb contents in the rocks are three to fourfold higher than the percentage abundance values. Content analysis of major oxides shows that evolution of chemical qualities in the formation rocks with their general correspondence to the calc-alcali series and to the 1:3:2 ratios of constituent facies (andesites, dacites, rhyolites) reflects the process of crystallization differentiation that took place in the near-surface magmatic focus.

The gabbro-diorite-plagiogranite formation, from the Late Jurassic (150 Ma, Abdullayev et al. 1971), is confined to the Tufan zone southwestern part, south of the Middle Jurassic formation, and is expressed in intrusive facies of dykes, sills, and boss-like bodies. They are characterized by immense diversity—from gabbro, diorites, quartz diorites to porphyraceous and aplitic plagiogranites. In large intrusives ($100 \times 80\ m^2$), structural zonality is observed, from the large-grain gabbro-diorites and diorites to quartz diorites of the central part. At the periphery, it is replaced by medium-grained gabbro, quartz-containing gabbro, and micro gabbro, and is accompanied by the broad hornfelsing aureole.

Petrochemical parameter analysis in the formation rocks showed the following:

- Gabbro: $Na_2O/K_2O = 7.8$ (sodic series), $Al' = 0.71$ (low-alumina), $f = 22.71$ (melanocratic differences),
- Diorites, quartz diorites: $Na_2O/K_2O = 4.06–6.5$ (sodic series), $Al' = 0.80–1.05$ (high-alumina),
- Plagiogranites: $Na_2O/K_2O = 4.67$ (sodic series), $Al' = 2.42$ (extremely high-alumina).

All of the above-mentioned facts indicate the conformity of the basic series rocks to toleite series, and of mean and acidic differences, to calc-alcali series. Generation of the gabbro-diorite-plagiogranite formation rocks is believed to be associated with the sodic series of magma differentiation.

Distribution of ore and rare elements in the formation rocks (ppm) reflects an important feature of the latter, with increased contents of Cu, Zn, Co and Sc. Along with some other data, this demonstrates its metallogenic specialization. It should be noted that the highest Cu contents are found in the near-contact parts of diorites (171.7 ppm) and quartz diorites (190.0 ppm), which that might be conditioned by post-magmatic enrichment or a re-distribution process.

The subalkaline gabbroid complex, from the Late Cretaceous, outcrops in the Main Ridge water parting site, in the Belakanchay River head. It is represented by gabbro-diabases, dolerites and large magmatic bodies composed of intermediate and basic rocks with elevated alkalinity. Such large intrusives as the Dindidagh, Somalit, Gudurdagh etc., are lenticular, 1–4 km length, with a northwest strike (290–310°) and steep upridging.

Zonation of intrusive structures is found from the edge parts to the central ones. It is expressed in substitution of biotite-containing gabbro-diabases by gabbro-syenites and monzonites. Subcentrally, syenite-diorites and leucocratic syenites are found. Gabbro-diabases with pronounced process of pyroxene uralitization and serpentine pseudomorphosis are found in the marginal parts of thin intrusives.

4.1.2 The Vandam Zone

The Late Cretaceous formations are locally distributed along the southern framing of the Greater Caucasus within the Vandam structural formational zone boundaries, considered as an independent tectonic unit of the southern slope of the Greater Caucasus. In the north, along the Zangi interzonal sutural fracture, it abuts the Main Ridge structures. The structure's southern half is concealed under a thick nappe of the Alazan-Agrichay valley Pliocene–Quaternary deposits.

The Upper Cretaceous volcanogenic-sedimentary and volcanogenic formations participate in the zone's geological structure (Fig. 4.2), outcropping in the Kakh, Sheki, and Ismailly regions. In the west, in the Belakan region, the Toarcian–Aalenian argillo-arenaceous deposits take part in the structure arrangement.

Magmatic formations of this zone (Samedova 1985; Abdullayev et al. 1991) are consolidated in the Late Cretaceous trachybasalt-trachyandesite formation, with its further subdivision into the Cenomanian subalkaline basalt-andesibasalt complex and the Late Senonian shoshonite-latite complex. Small outcrops of gabbro-syenites (the Buynuz intrusive) are dated to the Paleogene.

The trachybasalt-trachyandesite formation is from the Late Cretaceous. Volcanogenic and volcanogenic-sedimentary formations of the Cenomanian subalkaline basalt-andesibasalt complex can be traced from the Greater

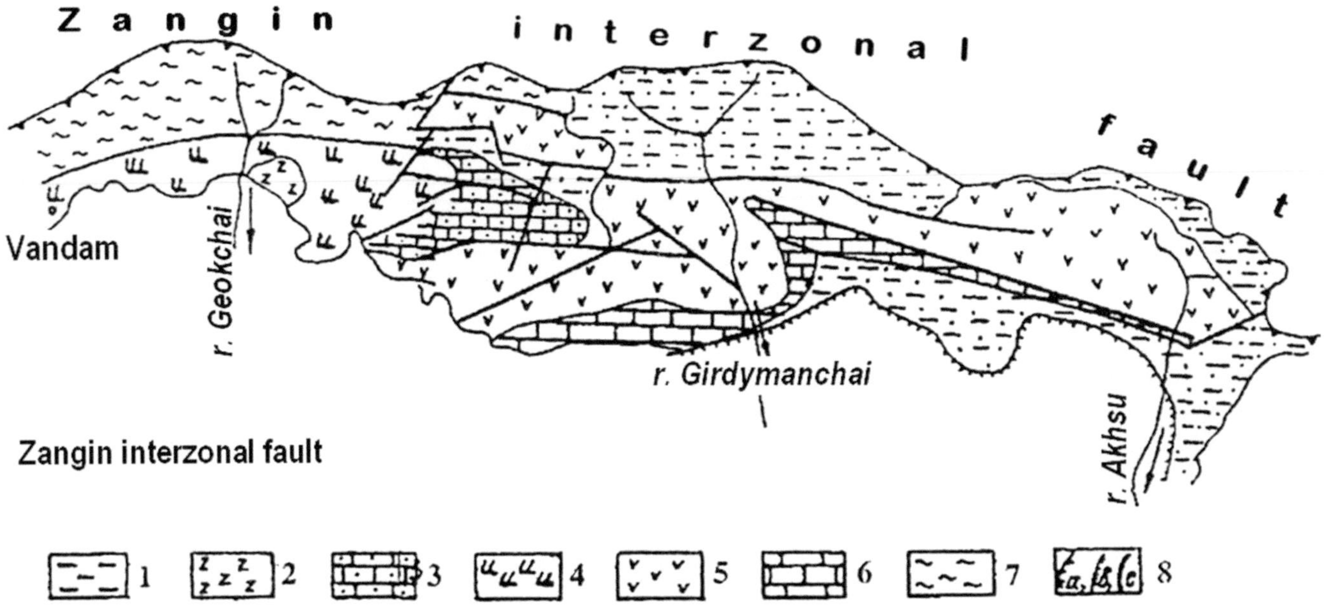

Fig. 4.2 Scheme of distribution of magmatic formations of the Vandam zone of the southern slope of the Greater Caucasus (based on M.A. Mustafayev and R.A. Samedova) (*1*) Oligocene-Miocene (Maykop series) clays, sandstone and marls, (*2*) Eocene gabbro-syenites, (*3*) Maastricht limestone, clays, and sandstone, (*4*) Late Cenonian basalts, trachybasalts and andesites, (*5*) Alb-Senomanian basalts, trachybasalts and andesites, (*6*) Neocomian limestone, (*7*) Aalenian schists and sandstones, (*8*) tectonic signatures: (*a*) interzonal Zangin fault, (*b*) fault, (*c*) shift fault

Caucasus southern slope foothills, from the Akhsuchay River in the east (500–600 m), to complete thinning-out near the Kishchay River in the west.

Volcanogenic and volcanogenic-sedimentary formations of the Late Senonian shoshonite-latite complex have local development. Their complete cross-sections are arranged along the Akhokhchay and the Geokchay Rivers.

The lava, subvolcanic, pyroclastic, and volcanogenic-sedimentary facies are separated in both formation complexes.

In the Senomanian complex, lavas are represented by basalts, dolerites, andesites; subvolcanic rocks, by stocks and sills of basalt-dolerite composition; pyroclasts, by agglomerate tuffs and tuff breccias (30–100 m); and the volcanogenic-sedimentary facies of broader development, by the alteration of tuff conglomerates, tuff gravelites and argillites.

In the Late Senonian shoshonite-latite complex (30–100 m), lavas are represented by trachybasalts and trachyandesite basalts; subvolcanic facies, by sills and stocks of essexite-diabase composition; pyroclasts (10–60 m), by tuffs and agglomeration tuffs; a volcanogenic-sedimentary facies develops only poorly.

According to their petrochemical characteristics, the zone's isolated facies are classified (Abdullayev et al. 1991) into four groups: (1) normal basalts and dolerites, (2) subalkaline basalt-trachybasalts, (3) trachyandesite basalts-latites, (4) andesites.

Unlike shoshonite-latite complex rocks, the geochemical peculiar properties of the Cenomanian basalt-andesitebasalt complex are characterized by close correspondences between lithophylic and chalcophylic elements, varying within the percentage abundance values.

Increased contents of Cu, Zn, Cr, V, Ni, and Pb in the first complex, with 5- to 20-fold exceeding percentage abundance values, are attributable to the input by hydrothermal processes. In the second complex shoshonites, the twofold percentage abundance increase of Li, Rb, Ga, and Sc results from the complex's alkalinity rise. According to some researchers (Abdullayev et al. 1991), substitution of the Cenomanian calc-alkali volcanism products by the Senonian subalkaline products corresponds to the similarity of their generation mechanisms and island arcs, and the presence of shoshonites and latites in subalkaline rocks evidences to the mature stage of the Vandam island arc.

4.1.3 The Kur Depression

In the Kur Depression, the Mesozoic volcanogenic complexes, determined from the geophysical and drilling data under the Miocene–Pliocene molasses cover, are integrated into formations, based on their petro-geochemical data (Abdullayev and Salakhov 1983; Ostroumova et al. 1995):

- Basalt-andesite-rhyolite formation: the Middle Jurassic;
- Basalt-andesidacite formation: the Late Jurassic—Early Cretaceous;
- Basalt-andesite-dacite and trachybasalt-trachyandesite formation: the Late Cretaceous.

The **basalt-andesite-rhyolite formation** of the Middle Jurassic age is represented by the following facies: lava facies—basalts, andesites, andesibasalts, dacites, rhyodacites and rhyolites; volcanoclastic facies—clastolava with cement of basalt and andesite composition and rock fragments from basalts to dacites and rhyolites; subvolcanic facies—dolerites, diabases, and gabbro-diabases.

According to their petrochemical characteristics, the Middle Jurassic volcanites of basic and intermediate composition belong to the normal alkaline-earth series. As for geochemical characteristics, at near-percentage abundance concentration of some elements in various intervals, the formation rocks display drastic increases in Cu, Zn, Ni, Co, V, and notable variations in their contents, resulting from the intensity of metasomatic alterations. It should be noted that by investigating the complex rocks from basic to acidic ones, determinate variations in petrogenic components demonstrate genetic integration. That is, they are the products of single, time-developing calc-alkali magma, comparable with those of the island arcs and active continental margins.

The **Basalt-andesite formation** of the Late Jurassic–Early Cretaceous age is represented by the following facies: lava facies—basalts, andesibasalts, rarely dacites; pyroclastic facies—tuffs, tuff breccias, tuffites of basalts, andesites, dacites; and subvolcanic facies—dolerites and diabases. By their petrochemical parameters, they correspond to the calc-alcali series of the island arc systems. According to their geochemical characteristics, the formation rocks are moderately enriched by the large-cation lithophylic elements (Li, Rb, Ba, Sr, etc.) in relation to the coherent elements (Ni, Co, Cr, Cu, V, etc.).

The **Basalt-andesidacite formation** of the Late Cretaceous is represented by the following facies: lava facies—basalts, andesibasalts, andesites, and pyroclastic facies—tuffs and tuff breccias of basalts, andesibasatls and dacites. By their petrogeochemical parameters, the formation rocks correspond to those of the island arcs and active continental margins.

The **Trachybasalt-trachyandesite formation** of the Late Cretaceous is represented by the following facies: lava facies—a broad range from trachybasalts to trachytes, shoshonites, and latites, of 15–20 m thickness, and volcanoclastic facies—clastolava, tuff breccias, and tuff, and considerably rarer subvolcanic facies—an extensive branch of intrusive granitoids and the Upper Senonian syenite-diorites.

Petrochemically, the intrusive rocks are characterized by elevated contents of alkalies and alumina. Geochemically, they have increased values of Pb, V and Hg, while the other elements have near-percentage abundance values. The petrogeochemical parameters of these formation rocks reflect their correspondence to those of the island arcs and active continental margins, with final stage of alkaline magmatism manifestation in the Late Cretaceous (Ismailzadeh et al. 2005).

4.2 The Lesser Caucasus

The Lesser Caucasus is characterized by considerable complexity caused by magmatism manifestations within structural zones, differing in their initiation time and tectonic development history. The Lesser Caucasus contemporary structures have a northeast border with the Kur Depression, along the abyssal near-Lesser Caucasian fracture, and the southeast border, with the Iranian microcontinent margin, along the Zangazur abyssal fracture. Three structural zones are found in its geological and tectonic arrangement: the Lok-Garabagh and the Gafan, filled with the Jurassic and Cretaceous volcanogenic formations; and the Goycha-Hakeri ophiolite mélange zone, separating them and looking like the ophiolite association allochthonous lamina in the middle of the Cretaceous volcanogenic-sedimentary complex (Fig. 4.3).

The Lok-Garabagh zone, situated at the Lesser Caucasus northeastern slope, is represented by the island arc formations from the Early Jurassic to the Late Cretaceous, with prevailing basic and intermediate composition of complexes. The zone is complicated by transversal fractures and is characterized by block structure, with southeastward and northeastward plunging of the latter ones. This fact conditioned magmatism zonal manifestations; for example, the thick Early Bajocian volcanites, represented by basalts and andesibasalts and developed in the southeast, are replaced in the northeast by the Late Bajocian andesites and dacites, and later on, by the Bathonian andesites, dacites, and rhyolites.

The Gafan zone is characterized by volcanogenic formations similar to those of the Lok-Garabagh one. It differs only in more developed acidic facies, probably caused by differing participation of the crust (pre-Alpine) material in magma generation and in its differentiation.

The Lesser Caucasus Mesozoic magmatic formations, according to their age relations, are subdivided into the

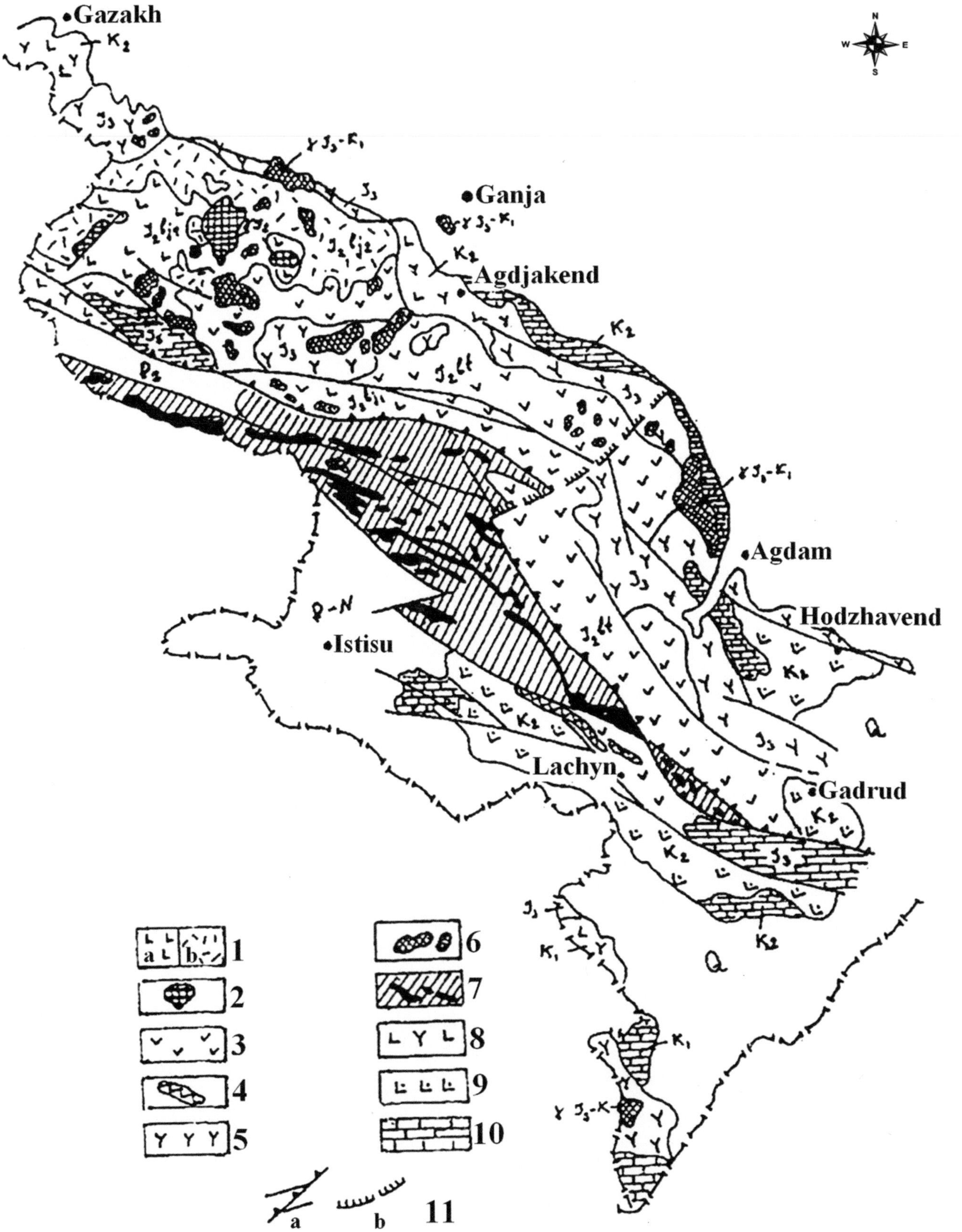
Gazakh
Ganja
Agdjakend
Agdam
Hodzhavend
Istisu
Lachyn
Gadrud
N
W
E
S
a
b
1
2
3
4
5
6
7
8
9
10
11

◀ **Fig. 4.3** Distribution of the Mesozoic magmatic formations of the Lesser Caucasus (after G.V. Mustafayev and M.A. Mustafayev) (*1*) Bajosian basalt-rhyolithic formation: (*a*) basaltic complex of the Lower Bajocian, (*b*) rhyolithic complex of Upper Bajocian, (*2*) Upper Bajocian plagiogranitic formation, (*3*) Bath basalt-andesite-rhyolithic formation, (*4*) Bath gabbro-plagiogranitic formation, (*5*) Upper Jurassic basalt-andesite-dacite formation, (*6*) Upper Jurassic—Lower Cretaceous gabbro-tonalite and gabbro-granitic formations, (*7*) ophiolithic association, (*8*) Upper Cretaceous basalt-dacite-rhyolithic formation, (*9*) Upper Cretaceous trachybasaltic formation, (*10*) Upper Jurassic and Cretaceous limestone, (*11*) (*a*) deep faults, (*b*) flexures

Middle Jurassic, Late Jurassic–Early Cretaceous, Early Cretaceous, and Late Cretaceous.

The Middle Jurassic formations

Magmatism of this period is represented within the Lok-Garabagh and the Gafan zones by the Bajocian differentiated volcanic-plutonic formations, such as basalt-rhyolites (3–3.5 km) with plagiogranite intrusives lying unconformably on the Lower Aalenian argillo-arenaceous rocks and overlapped unconformably by such Bathonian deposits as basalt-andesite-dacite-rhyolites (up to 2 km) with gabbro-plagiogranite intrusives.

The Bajocian basalt-rhyolite formation is represented by the contrasting Lower Bajocian basalt complex and by the Upper Bajocian rhyolite complex. Within the Shamkir High boundaries, basalts have 1500 m thickness and are overlapped by the nappes of the Upper Bajocian quartz plagioporphyries from rhyolite complex; in the northeast, they are overlapped by the unconformable Bathonian deposits, and further on, by the Upper Cretaceous and Middle Eocene formations. Within the Shamkir High boundaries, the complex rhyolites have peak thickness of up to 700 m. It is essentially composed of lava nappes that occupy large areas in the Dzegamchay and Tovuzchay rivers interfluve.

The formation facies composition shows predominant volcanogenic-detrital rocks (60–65 %), with a subordinate role of lava (25–35 %) and volcanogenic-sedimentary formations (5–10 %) (Mustafayev 1991). Composition of the lava facies rocks changes from basalts to rhyolites with participating andesibasalts and negligible dacites.

Petro-geochemical parameters of the basalt subformation rocks display the toleite trend with supra-percentage abundance contents of Cu, Co, Mo, Pb, Au, and Ag. The rhyolite subformation rocks demonstrate the calc-alkali differentiation trend with high Cu, Mo, Zn, and Sn concentrations. High positive correlation relationships of the latter can serve as a criterion for potential ore bearance in the rhyolite subformation metasomatites.

The Bajocian plagiogranite formation participates in the texture of the Atabek, Gilanbir, and Tovuz intrusives. Chemical and petrographic compositions of the formation rocks do not differ significantly from their effusive analogs and are the co-magmates of the Upper Bajocian plagioporphyries. The intrusives are represented by two intrusion phases: (1) plagiogranite-porphyries, granophyric plagiogranites, and (2) leucocratic granites, aplite-like plagiogranites (Mustafayev 1977).

According to Nikolsky et al. (1975), a vast negative gravity anomaly over the Atabek intrusive area testifies to the existence of a large deep-seated intrusion massif. In the intrusion aureole, the enclosing quartz plagioporphyries-rhyolites are transformed into secondary quartzites.

The contents of some ore-forming and rare elements in plagiogranite formation rocks show higher than percentage abundance values for Cu, Mo. On the other hand, Pb, Zn, Cr, and V contents are much lower than percentage abundance and Ni and Co values are near-percentage abundance ones.

The chemical activity of plagiogranite intrusives is well correlated with rhyolite (quartz plagioporphyries) chemistry, thus supporting the appropriateness of their isolation as a single volcano-plutonic association.

The Bathonian basalt-andesite-dacite-rhyolite formation is the headmost developed in the Lachin, Garabagh, Mrovdagh structures, in their volcanogenic and volcanogenic-sedimentary facies. It represents a continuously differentiated series consisting of lavas and lava breccias, volcanic breccias, agglomerate tuffs, tuff breccias of extrusives and subvolcanoes. Flows of lava and basalt lava breccias of 2–5 to 60–80 m thicknesses that comprise up to 70 % of section thicknesses are characterized by significant continuity and length, while those of acidic volcanites (e.g., rhyolites and dacites) undergo fast thinning out. Extrusives are present in all structures and correspond to andesibasalts, andesites and rhyolites. Subvolcanic rocks are represented by stocks, dykes, and sills composed of dacites and rhyolites, less frequently, by basalts.

Petrochemical characteristics of the formation basalts correspond to the toleite series, while other differentiates correspond to the calc-alkali series. The formation volcanites are related to the sodic and potassic-sodic series. The average chemical composition of rocks in the Lachin anticlinal formation is plotted on the AFM graph in the field of alkaline olivine-basalt series, and for the Garabagh anticline, in the field of high-alumina basalts.

Among the Bathonian volcanites (Ostroumova and Tzenter 1986), high-magnesia andesites are found. They are the analogs of the marianite-boninite series, showing up as pipe-like body of 300 m diameter and over 100 m depth. The authors believe that it is an eroded volcanic neck.

Geochemical parameters for basic and intermediate rocks of the Garabagh and Lachin zones establish pair correlation of significant positive relations between Ni, Cr, Mn, Co and Cu. Co peak concentrations are noted in basic and intermediate rocks of the zones, and minimal concentrations in the acidic ones.

Petrochemical data of the formation volcanites correspond to those of the island arc systems. The primitive liquid was developed first by the toleite differentiation way, and then, by the calc-alkali way.

The Bathonian **gabbro-plagiogranite formation** was found in Lachin district as the Byulyuldyuz massif that has shaped during two phases: the first one (about 2 %) created gabbro (gabbro, gabbrodiorites, less frequently, quartz diorites), and the second one (about 98 %) created plagiogranites (plagiogranites, quartz diorites, granodiorites, granites).

With silica acidity growth, alkalinity increase, and iron content drop, variations in rock chemical compositions correspond to the trend of calc-alcali series.

Geochemical peculiarities of the formation plagiogranites are nearly identical to those of the Bajocian age, except for anomalous contents of Cu and Mo, inherent in the Atabek intrusive plagiogranites. In the majority of ore elements—Cu, Pb, and Zn—the contents are near-percentage abundant. In Cr, Ni, W, and Mo, they are below the percentage abundance.

The Late Jurassic–Early Cretaceous formations

The basalt-andesite-dacite formation relates to the Late Jurassic–Early Cretaceous age. The formation volcanic rock assemblages are extensively developed in the Lesser Caucasus and are among the essential ore-containing rocks of endogenous mineral deposits. According to Mustafayev (1977), the formation is subdivided into basalt-andesibasalts, developed in the Dashkesan and Ajakend troughs, and andesite-dacites found within the Shamkir High.

The formation of Early Cretaceous volcanites are inseparable from those of the Late Jurassic age and create the integrated Oxfordian–Early Cretaceous complex of 2500 m thickness, composed of the lava and volcano-detrital rocks of andesite and andesitebasalt composition, alternating with reef limestone benches (Ismailzadeh et al. 2005). Hypabyssal intrusive massifs are composed of gabbro, diorites and quartz diorites; together with enclosing effusives, they constitute a volcanic-plutonic association. Intermediate types (andesibasalts, andesites) are most commonly encountered among the formation rocks, basalts are less frequent, and the least widespread are acidic rocks (dacites, rhyodacites, rhyolites).

Pertochemical parameters of the formation rocks occupy two fields: the Shamkir High basalts occupy the field of subalkaline basalts and trachybasalts, and the Dashkesan and Agjakend trough basalts occupy the intermediate position between the subalkaline and normal basalts. Intermediate rock types are related by their acidity to normal types. Acidic members, developed in the positive structures, are more alkaline. All members of this formation create an integrated development trend towards enrichment with alkali and alumina ("the Bowen" trend), which is typical for the differentiation of calc-alkali magma.

Geochemical characteristics of the rocks (Mustafayev 1977) describe them as having low percentage abundance values for Ni, Y, Yb, Cs, Cr, Zn, Pb, and Mo. Co content is twofold higher than the percentage abundance and its distribution coincides with the normal model; Cu content is high in pyroclastic facies and low in lavas and subvolcanic rocks. Higher Cu contents are noted in the stem facies that are also enriched by sulphide ores. High Cs content is also found in the formation rocks.

The gabbro-tonalite formation of the Late Jurassic–Early Cretaceous age integrates the intrusives of the Garabagh, Shamkir, Mekhmany, and Mrovdagh groups. They are relatively small (up to 60 km^2) mono- and polyphasic bodies with early phases represented by gabbroids, and second phases by granitoids.

According to (Nikolsky et al. 1975; Eppelbaum and Khesin 2012), vast negative gravity anomalies in mountainous conditions may designate an occurrence of a large deep-seated intrusive massifs. The intrusives transect the Middle and Upper Jurassic volcanogenic and sedimentary rocks and produce contact-metasomatic effect on them. In the Gedabey intrusive aureole, major effusives are hornfelsed and carbonates are skarned, forming vesuvianite fields.

Polyphase intrusives (Gedabey, Chanakhchi) are represented by the major differentiates such as gabbro and gabbro-norites. Their marginal parts are composed of gabbro-diorites, with transformation into quartz diorites and diorites. The monophase bodies are composed of tonalities, quartz diorites (the Mekhmany and Mrovdagh groups), and less frequently, granodiorites (the Barum massif).

Gabbroid area does not exceed 20 %. Dyke formation shows predominant development of basic rocks, from lamprophyres to diorite-porphyres; pegmatites and aplites are infrequent and of low thickness.

According to their petrochemical characteristics, the gabbro-tonalite formation rocks occupy the field of normal alkalinity rocks on (Na_2O + K_2O)–SiO_2 graph. The AFM graph shows weak differentiation of magma. In the Ab-Q-An graph, a normative Ab and Q increase with the normative anortite decrease.

Geochemical peculiarities show high Cu content in all types of complex rocks with excessive 1.5- to 3.0-fold percentage abundance, as well as high Zn, with low Pb. If Cu and Zn high contents can be interpreted as the result of their ore specialization, then quite low Pb content is indicative of basaltoid genesis.

The gabbro-granite formation of the Late Jurassic–Early Cretaceous is typical for the Dashkesan and Uchtapa-Gyzylgai intrusives. The Dashkesan intrusive of 40 km^2 area is confined to the same-named trough of the

Lok-Garabagh zone. Three stages are differentiated in its generation (Kashkay 1965):

(1) gabbroids (20 %)—normal, hornblende, uralite, olivine gabbro, gabbro-norites, norites, quartz gabbro; gabbro are frequently facially substituted by gabbro-diorites and diorites;
(2) granites, granitodiorites (70 %), and in the marginal parts, their hybrid differences are developed such as diorites and quartz diorites;
(3) granite-aplites, alaskites (10 %), less frequently, leucocratic granitoids and granosyenites.

The Uchtapa-Gyzylgai intrusive is located in the pre-Lesser Caucasus fracture zone with its exposed area of about 2 km^2. The polyphase stock is composed of granodiorites and granites.

According to their petrochemical peculiarities, the rock compositions correspond to the field of normal alkalinity formations on ($Na_2O + K_2O$)–SiO_2 classification graph; on the AFM graph, they correspond to the differentiation trend of calc-alcali series; the FeO^x/MgO ratio is dual.

Their geochemical features show that granites and granodiorites are dramatically depleted of alkaline elements: Li and Rb are two to threefold in relation to percentage abundance. Co, V, Mo, Sn, and Fe are major ore-generating elements in the Dashkesan intrusive; they also have low relative abundance values. They show stable positive bond to F and Br volatile components, and this bond can be used as one of granitoid ore-bearing criteria that is related to iron-ore mineralization. Normalized La/Yb relations testify to the correspondence of the magma-melting source not to the continental crust but to the melt with the composition close to the oceanic crust basalts (Mustafayev et al. 1983).

Generally, Fe, Mg, Co, Ni, V distributions are characterized by the left symmetric type, thus reflecting the low-content stability in these elements. Slightly elevated Co contents and decreased Cr and Ni also reflect their geodynamical formation conditions, supported by rather low La/Yb = 2.8–3.4 relationship in the intrusive complex basic and acidic rocks, whereas in similar formations of mature island arcs this relationship never drops below 6.5.

Petrochemical and geochemical data of gabbro-tonalite and gabbro-granite formation rocks reflect complex geodynamical conditions of intrusive formations that correspond both to the mature island arcs and to the active continental margins.

The trachyandesite-trachydacite formation of the Early Cretaceous age is found in the Lok-Garabagh zone and is represented in lava facies, unlike the Late Jurassic formations exposed in volcanogenic-sedimentary facies. Trachyandesites and trachydacites (80–85 %) and latites in small numbers are indicative rocks of the Early Cretaceous formation. On ($Na_2O + K_2O$)–SiO_2 classification graphs, the formation rocks occupy the central part of subalkaline difference field; they are related to the potassium-sodic type, with high alumina and low FeO/Fe_2O_3 ratio. The subalkaline content of the Early Cretaceous volcanites dramatically differentiates them from the Late Jurassic volcanites that correspond to the calc-alcali series, and it is a crucial criterion of their independent manifestation.

The Late Cretaceous formations

The Lesser Caucasus northeastern slope displays a series of transversely superimposed depressions deepening and opening towards the Kur Depression. From northwest to southeast orientation, these depressions are represented by the following sequence with distinguishing volcanic series: the Gazakh and the Agjakend with calc-alkali basalt-andesite-dacite-rhyolite rock series, and the Khojavend, Hadrud (Aghoghlan), and Goycha, with predominant alkaline and subalkaline basaltoids.

The **basalt-andesite-dacite-rhyolite formation** of the Late Cretaceous age displays itself in the Gazakh depression by the Coniacin–Lower Santonian basalts, andesites and andesibasalts and by the Upper Santonian rhyolites; in the Agjakend depression, dacite-rhyolites are predominant, later replaced by andesibasalts-basalts. In both troughs, the formation bottoms are represented by volcanogenic and volcanogenic-sedimentary rocks such as tuffites, tuff sandstones with limestone intermediate layers, and their tops, by lava flows.

From their petrochemical parameters, basalts (and dolerites) of the Gazakh and Agjakend depressions are grouped into 4 series with regard to K_2O/SiO_2: (1) toleite, (2) calc-alkali, (3) high-potassium calc-alkali, and (4) subalkaline (Mustafayev 1983). The overall magmatism development trend is oriented towards alkalinity rise.

As for geochemical features, relatively high Cu, Pb and Ag contents in basalts and dolerites deserve attention, while the iron group elements are characterized either by near percentage abundance values or by significantly lower values. High Cu contents are essential due to the presence of numerous Cu ore occurrences in the Lok-Garabagh zone and modalilies of potential ore bearance in the Cretaceous magmatism.

The trachybasalt-trachyandesite formation of Late Cretaceous age is common in the Khojavend, Gadrut, and Gocha depressions situated at the Lesser Caucasus extreme southeast. In the Khojavend depression, vulcanite thickness varies in wide range, from 4 m at the periphery to 600 m in the volcanic center areolas. Rock compositions are various basalts (Akhundov 1986; Mamedov 1990). In the Gadrut depression, volcanogenic formation is represented by tuff breccias, agglomerate and crystalloclastic tuffs, alternating with lavas and basalt lava breccias, of 235 m thickness. In the Goycha depression, subalkaline rocks are extensively developed; andesites also occur. The thickness facial lateral variability is noted; the depressions northwest part manifests shallow-water conditions for limestone accumulation and central-type volcanism.

The formation petro-stock shows the presence of subalkaline picrobasalts in the Khojavend depression, as well as of potassic alkaline basalts such as olivine tuffites and leucitic tuffites; the presence of andesites and dacites in the Gadrut depression, and the presence of andesites in the Goycha depression.

The rock petrochemical specificities such as the predominance of basalts of subalkaline and alkaline series, high K_2O and Na_2O content, as well as TiO_2 and P_2O_5, helped to determine indicative types of rocks, the similarity of their parameters, and those of mature island arcs and continental collision zones. Shikhalibeyli (1994) tended to ascribe the Khojavend depression to rift-related structures.

The Goycha depression volcanites, represented by trachybasalt and andesite complexes, with all their existing differences, are characterized by the succession of certain chemical activity features and petrochemical parameters that permit to interpret them as single magma derivatives. In the set of geological-petro-geochemical parameters, the major mass of effusives is represented by alkaline olivine-basalt series that has defined the generated image of the depression Late Cretaceous rocks (Grigoryev and Sokolov 1978). In general, these volcanites are comparable with subalkaline effusives of rift-related structures.

A complex of subalkaline gabbroids shaped as minor intrusives has been developed in the Khojavend depression, where trachybasalts are represented extensively. According to some researchers (Ostroumova and Tzenter 1986), they might be the cumulative residue of the initial magma melt.

Petrochemically, the subalkaline gabbro are regarded as the starting link of the tephrites-trachybasalts-trachyndesites series and can be related to the ultrabasic rocks of increased alkalinity. However, basing on their modal mineralogical composition with distinct basic plagioclase prevalence (Ostroumova and Tzenter 1986), they belong to gabbroids. Their correspondence to ultrabasic accumulative rocks of the intrusive body bottom part is assumed.

4.2.1 The Nakhchivan Zone

The basalt formation of the Early Jurassic age is found on the Greater Caucasus southern slope and in the Nakhchivan zone. Within both structural zones, formation of volcanogenic complexes is related to rifting, as rift-related trough in the Greater Caucasus and as continental rifting in the Nakhchivan zone. Peak thickness of the Nakhchivan zone Early Jurassic volcanites is 205–235 m, with predominant subvolcanic facies of diabases and dolerites in the cross-section.

Petrochemical characteristics of basalts relate them to the subalkaline series: moderately aluminous, with iron sum of 9–14 %, they are depleted of MgO, with high $TiO_2 = 3$ % concentration. By their TiO_2/K_2O ratio, basalts occupy the intermediate position between island arc trends and continental rifts, with pronounced bent to the latter.

According to their geochemical characteristics, with silica content increase, basalts display reduced Cr, Ni, Co content and increased Y, Ba and V. According to Karyakin (1989), Ti, Zn, Y and Nb contents in the Lower Jurassic basalts are comparable with the basalts of intra-oceanic islands and continental rifts. As long as the setting in intra-oceanic islands does not correspond to geological position of the Lower Jurassic vulcanites, the latter are set equal to the continental rift basalts.

The basalt-rhyolite formation of the Late Cretaceous age is developed in the same areas as the Early Jurassic magmatites of the Nakhchivan zone. They are represented by tuffs of rhyolites, dacites, andesibasalts and basalts of various dimensions, with 300 m thickness. According to Azizbekov (1961), the deposits belong to the Albian age, whereas Rustamov (1979) attributes them to the Coniacian age.

The Negram cross-section volcanogenic formation consists of basalt lavas and lava breccias, alternating with thin flows of olivine basalts or with rare flows of andesite lavas and lava breccias (100 m). Contrasts in petrographic rock types as basalts and andesibasalts, on the one hand, and rhyolites and dacites, on the other hand, predetermined their petrochemical differences. According to petrochemical parameter data, the basalt series rocks are similar to the Early Jurassic basalts in this zone, and it permitted the researchers to state the similarity of geodynamic conditions of their generation in the continental rifting mode. Inherited petrochemical peculiarities and geological situation of the formation rhyolites put them in paragenetic kinship with basalts, but they are separated by time differentiation and by contamination of later doses of the primary toleite magma. The researchers made the same conclusion with regard to geochemical parameters of the formation basalts.

4.2.2 The Goycha-Hakeri Zone

Ophiolite association

The Lesser Caucasus ophiolite association shaped an ophiolite mélange belt is the eastern branch of the Pontian-Transcaucasian ophiolite belt in the Mediterranean folded system. It is developed in three structural zones: the Goycha-Hakeri (Sevan-Akera) zone that is of the Mesozoic age at the Transcaucasian microplate northern margin; the Yerevan-Ordubad (Vedi-Nakhchivan subzone) zone that is of the Mesozoic age at the Iranian microplate northern framing; the Zangazur zone that is the Early Alpine suture zone in the connecting strip of these two above mentioned paleomargins.

The researchers' consecutive survey shows evolution of the views concerning the ophiolite formation conditions: from purely intrusive (magmatism stages) to the complexes typical for oceanic crust found at present in allochthonous bedding.

Various aspects of the "ophiolite problem" in the Lesser Caucasus have been studied for many years by a few generations of geologists: Z.B. Abdullayev, Sh.I. Allakhverdiyev, Sh.A. Azizbekov, V.M. Babazadeh, G.Kh. Efendiyev, R.K. Gasanov, T.Ab. Gasanov, M.A. Kashkay, V.E. Khain, L.N. Leontiev, R.S. Malyutin, V.F. Morkovkina, K.N. Paffenholtz, S.M. Suleimanov, E.Sh. Shikhalibeyli. Special mention should go to the new cycle in the ophiolite problem covering related to the development of plate tectonics concepts that found their reflection in the publications by T.Ab. Gasanov, A.D. Ismailzade, A.L. Knipper, S.A. Sokolov, and G.A. Zakariadze.

The Lesser Caucasus ophiolite association under discussion is a classically expressed belt of spatially and structurally integrated complexes of ultrabasites, gabbroids and effusive radiolarites. Its most extensive and broad development is seen in the Goycha-Hakeri zone, looking like a narrow trough controlled by the Mrovdagh and Garabagh thrusts in the north, and by the Lachin-Bashlybel fracture in the south. The zone consists of two echelon arranged structural subzones—the Tourzachay and the Sarubaby subzones that are linearly stretched along the Shakhdagh and the Garabagh ridges from Goycha Lake northeastern coast to the Lysogorski Pass in the southeast and have the length of over 200 km. Maximal ophilolite width within the zone is 25–30 km, with narrowing to 1.5–2.0 km near the Lysogorski Pass. At the Goycha Lake northeast coast they plunge under the Paleogene deposits of the Armenian Pambak-Shirak zone, again outcropping in the northeast in the Amasia area.

According to the "plate tectonics" concept, the association-constituting complexes correspond to the oceanic type crust and are characterized by allochthone occurence. Two nappe laminas are isolated in its structure (Knipper and Sokolov 1972):

1. The lower Ipyak, gabbro-serpentinite lamina: serpentinized dunites, harzburgites, lherzolites, troktolites, gabbro (1000 m); gabbroid age, according to Sm/Nd method, is related to the Later Triassic (the Carnian, the Norian ages) (Sokolov et al. 1996);
2. The upper Nagdali, volcanogenic-sedimentary lamina: toleites, diabases, aphiric basaltoids, carbonaceous-siliceous and siliceous rocks, radiolarites, jaspers and micritic limestones (1500–2000 m) of the Later Jurassic–Early Cretaceous age.

The allochthone lamina basement displays the Senomanian–Early Senonian olistrostrome thickness, lying conformly with gradual transfer to the autochthon Albian–Senomanian deposits. The olistostrome thickness is represented by chaotic hummock of olistolites, ultrabasites, gabbroids, rocks of the effusive-radiolarite series, enclosing frame, and the ancient basement metamorphic rocks.

Allochthone lamina are unconformly overlapped by the Late Cretaceous neoautochthone series of the Coniacian-Santonian volcanogenic-terrigenic thickness that can be most completely traced at the belt southeast, in the Sarybaba subzone, and fragmentarily, at the northwest, in the Touragachay subzone.

The Vedi ophiolite belt (the mélange zone) is located within the Yerevan-Ordubad zone of Nakhchivan microcontinent. The belt can be traced from the Turkish border southeastward, to the Vedi River basin, and has not been determined farther. The belt length is about 50 km, with 2–7 km width. In its structure are distinguished the autochthon, allochthone, and neoautochtone complexes.

The autochthon complex is composed of the Paleozoic deposits (the Permian and Triassic limestones) transgressively overlapped by the Late Cretaceous rocks (the Senomanian–Turonian limestone-terrigenic flyschoid thickness) replaced by the Early Coniacian olistostrome formations.

The allochthone complex consists of two parts: the lower Late Jurassic–Early Cretaceous effusive-radiolarite (spilites, diabases, albitophyres, radiolarites) part (800–1000 m), and the upper gabbro-serpentinite part represented by serpeninized peridotites, gabbro, gabbro-amphibolites (500–700 m) (Abdullayev 2007).

The neuautochtone complex, the Late Coniacian calcareous-terrigenic thickness (300 m), is characterized by sharply transgressive manner of occurrence with basal conglomerates at the bottom (1–25 m). It is stated that the ophiolite generation age is the Early Coniacian, and the effusive-radiolarite series generation age (based on the determination of foraminifera and corals therein) is the Late Jurassic–Early Cretaceous epoch.

The Zangazur zone of the Araz belt is characterized in its allochthone occurrence only by the manifestations of effusive-radiolarite series rocks as well as by the separate ultrabasite fragments. Ultrabasites found at the Kafan zone border are in tectonic contacts with the Devonian limestones. The ultrabasite lamina length is less than 500 m with 5–20 m thickness. Along with them, the boreholes showed spilites with radiolarite lenses, probably of the Late Jurassic–Early Cretaceous age. Ultrabasites and spilite-radiolarites age down to the pre-Early Senonian period (Ismailzadeh et al. 2005).

Ultrabasite outcrops in the Goycha-Hakeri zone are exposed to metamorphism and tectonic processing in various degrees. Metamorphism is expressed in substitution of pro-anhydrous ferro-magnesian silicates by magnesium hydrous silicates, the serpentinite group minerals, such as lizardite, chrysotile and antigorite. Early serpentinization with the participation of reduced gases (H_2, CH_4, CO_2), brought by ascending mantle fluids, and late serpentinisation of superimposed different-depth hydrothermal-metasomatic type associated with the impact of endogenous processes accompanied by stress deformations, are isolated (Allakhverdiyev 1985).

The gabbro complex represented by gabbro, gabbro-norites, and norites forms in its contact ribbon with untrabasites some kind of a "transient ribbon" represented by feldspar peridotites, olivine gabbro and troctolites. Such formations are associated with the complex multistage process of ultrabasite metasomatic processing under the effect of gabbro intrusions (Fig. 4.4) (Ismailzadeh 1984).

The effusive-radiolarite complex overlaps various ultrabasite-basite nappe horizons, and in rare cases, lies directly on olistostrome thickness. Two unconformly lying nappes are differentiated in its composition, based on their structural and facial peculiarities and the pateontological data:

1. alternating lava flows with cherts containing the Upper Jurassic corals, and less frequently, sandstones (the Late Jurassic–Early Cretaceous age);
2. cherts with horizons and lenses of gravelites, sandstones and fine detrital breccias, represented by ultrabasic rocks, various effusions, plagiogranites and gabbro (the Late Cretaceous and the Early Senonian).

The manifestation sequence in isolated different-aged and facially differing types of effusive-radiolarite complex cannot be traced because of their participation in thrust sheets and congestion in cross-sections. Rustamov (1979) found the buried 105-km long Nakhchivan ophiolite belt in the Ordubad zone, with its northwestern extension being the Vedi ophiolite zone.

The Goycha-Hakeri and the Vedi mélange ophiolite zones of the Lesser Caucasus display parallel features in their strike, composition similarity of their constituents, and participation in closely spaced tectonic processes. Geodynamically, they are viewed in a unified genetic aspect, as the derivatives of a single oceanic crust, obducted during the Cretaceous period onto continental margins of the Nakhchivan and Transcaucasian microplates. According to the researchers, the nappe plate root (rift-related) zone was located in the following zones: the Goycha-Hakeri (Knipper 1971; Gasanov 1985; Rustamov 1995), or was noticed north of the contemporary ophiolite outcrops, and is at present overlapped by the Lok-Garabagh (Somkhit-Agdam) zone structures (Sokolov 1977).

The latest version concerning the interrelations of the Lesser Caucasus ophiolite belts is the proposition that in the early Late Cretaceous age an ophiolite lamina was forced out of the Zangazur zone (now, the Lesser Caucasus Mesotethys suture zone). It was obducted to the northern (Goycha-Hakeri) and the southern (Vedi) zones of continental margins of the Mesotethys closing basin (Knipper 1975; Lomize et al. 1980; Gasanov 1985; Zonenshain et al. 1987; Ismailzadeh 1990; Khain 1995).

According to the data of certain researchers, the olistostrome thickness generation period and its movement to the allochthone lamina Goycha-Hakeri zone corresponds to the interval between the Senomanian and the Early Senonian periods.

Manifestation regularities for the Mesozoic magmatism

Presented factual materials on geology and petrogeochemistry of the Jurassic and Cretaceous volcanogenic and intrusive formations in the Lesser and Greater Caucasus

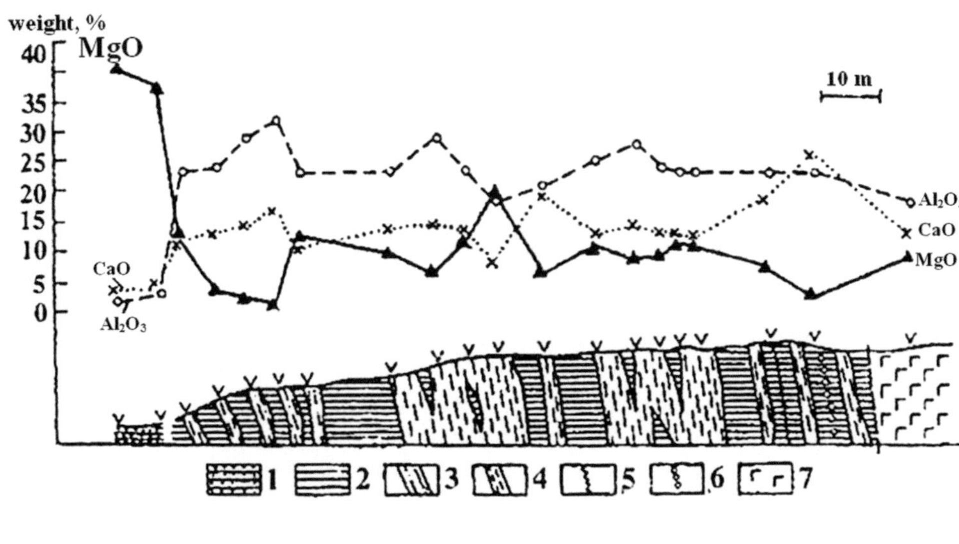

Fig. 4.4 Schematic geological profile across troctolites of the Lev massif (composer: A.D. Ismailzade) (*1*) harburgites, (*2*) plagioclase peridotites, (*3*) anorthosites, (*4*) troctolites, (*5*) gabbro-pegmatites, (*6*) prehnites, (*7*) olivinic gabbro

(Azerbaijan) allow us to report basic manifestation regularities of Mesozoic magmatism in the region's different structural zones.

The Early Jurassic magmatism revealed itself simultaneously at two continental margins that rim the Transcaucasian Middle massif in the extension zones: along the Eurasian platform southern frame on the Scythian microcontinent (the Greater Caucasus water dividing part) in the north, and the Iranian platform northern frame on the Nakhchivan microcontinent (the Sharur and the Julfa horst-anticlines, the Daralagez zone) in the south. It showed itself in basalt and diabase facies, corresponding to the poorly differentiated basalts of toleite series, similar to those of continental rifts.

Subsequent magmatic formations of the Greater Caucasus are represented (Fig. 4.5):

- the Middle Jurassic age—by andesite-dacite-rhyolites of calc-alkaline series;
- the Late Jurassic age—by gabbro-diorite-plagiogranites of subalkaline series.

The Middle Jurassic and the Late Jurassic–Early Cretaceous magmatism encompassed the entire Lesser Caucasus and the Kur Depression, while the Late Cretaceous magmatism is developed in superimposed transversal depressions, opening towards the Kur Depression. The Late Jurassic magmatism is also found limitedly in the Nakhchivan zone and, partly, on the Greater Caucasus southern

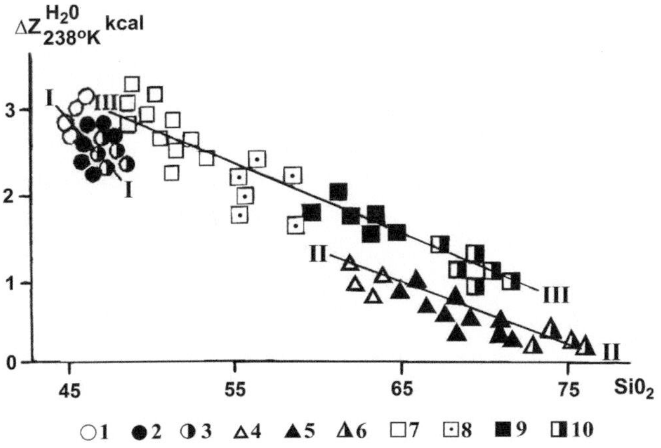

Fig. 4.5 SiO_2—$\Delta Z^{H_2O}_{328\,°K}$ kcal for the rocks of magmatic formations of the southern slope of the Greater Caucasus (after G.I. Kerimov; Ismailzadeh et al. 2005). Trends of differentiations for formations (*I–III*), (*I*) sodium basalts, (*II*) andesite-dacite-rhyolite, (*III*) gabbro-diorite-plagiogranite (*1*) olivine basalts, (*2*) dolerites, (*3*) basalts, (*4*) andesites, (*5*) dacites, (*6*) rhyolites, (*7*) gabbro, (*8*) diorites, (*9*) quartz diorites, (*10*) plagiogranites

slope. The researchers attribute it to magmatism intensification in the Lesser Caucasus and the Kur Depression.

Petrophysical composition and petrochemical specific features of the Lesser Caucasus Early Bajosian and Bathonian basalts correlate with the toleite series with the Fenner differentiation trend, and the Late Bajosian rhyolites and the Bathonian andesites, dacites, rhyolites correspond to the calk-alkali series with the Bowen differentiation trend. The Bajocian age complex on the Mrovdagh and Shamkir Ridges contains high-alumina andesibasalts and andesites, whose petrochemical properties are close to toleites, while Turshsu village region of the Garabagh Ridge is noted for high-magnesia andesites attributed to the rock equivalents of boninite-marianite series (Ostroumova et al. 1991).

The Kafan block, with similar to the Lok-Garabagh block structure, is characterized by the manifestation of the Middle Jurassic formation of basalt-dacite-rhyolite suite in low- and normal potassium calk-alkali series.

Magmatic melt of the Jurassic, Jurassic–Early Cretaceous volcanism in the Lesser Caucasus has medium acid, calk-alkali composition, reflecting its subductive generation conditions; the same melt of the Late Cretaceous period belongs to medium basic, subalkaline series, corresponding to the mature formation age of island arcs.

Major petrochemical characteristics present the Lesser Caucasus Middle Jurassic magmatic rocks (the Bajocian, the Bathonian ages) as the derivatives of hybrid magma with basalt constituent correspondence to the toleite series, and middle and acid constituents, to the calc-alkali series.

The Jurassic volcanogenic belt of the Lesser Caucasian paleo-island arc shows pronounced northern polarity with K_2O content increase from south to north. Low- or normal potassium calk-alkali rocks predominate at the zone southern periphery, and high-potassium calk-alkali series, at its northern edge. Growth of rock acid facies is observed in the same direction.

The post-Bathonian folding period witnesses marine regression accompanied by the accumulation of subcontinental facies that have been developed in the zone western parts, i.e. the destruction and tension conditions were replaced by compression and uplift.

By the manifestation antidromic manner, the Middle–Late Jurassic volcanism in the Kur Depression corresponds to the calk-alkali series.

The Kur Depression Late Cretaceous volcanism is subdivided into three formations: normal calk-alkali basalt-andesite-dacite; subalkaline trachybasalt-trachyandesite, and alkaline shoshonite-latite one. These two latter are typical only for the Vandam zone Late Cretaceous formations at the Greater Caucasus southern slope as well as for the Gadrut, Khijavend, and Goycha troughs of the Lesser Caucasus.

The Late Jurassic–Early Cretaceous formation is widely developed within the Lesser Caucasus and is the chief ore-hosting entity for the region main endogenous fields that are genetically or paragenetically related to various facies of volcanogenic or intrusive constituents. This formation is represented by differentiated volcanogenic suite of basalt-andesite-dacites with polyphase intrusives of gabbro-tonalite (Gedabey, Mekhmana) and gabbro-granite Dashkesan) compositions so that in basalt-andesibasalt facies they are confined to troughs, and in andesite-dacite facies to elevations.

Generally, in the Lesser Caucasus, the Early Cretaceous volcanism is closely similar, both by its composition and by its occurrence manner, to the Late Jurassic one and forms a unitary basalt-andesite-dacite complex of calk-alkali series with potassium-sodic alkalinity and homodromic development sequence. Spatiotemporal contingency, affinity of geological generation conditions, petro-geochemical peculiarities of the Late Jurassic and Early Cretaceous volcanogenic and intrusive formations reflect their co-magmaticity and conformity to the unitary volcano-plutonic association.

In such a manner, the Lesser Caucasus and the Kur Depression volcanism evolves in the Mesozoic age from basic to acidic and even to subalkaline and alkaline series, decreasing eruption volume in this direction.

Petrochemical studies of the Lesser Caucasus ophiolite complexes indicate that the peridotite-gabbro trend is not uniform, due to absent succession in differentiation and correlation of ultrabasite and basite compositions, while gabbroid and diabase fields (the effusive complex toleites) are overlapped thus reflecting their petrochemical uniformity.

The Cenozoic magmatic formations

The Cenozoic magmatism is most actively manifested within the Lesser Caucasus. After the Late Cretaceous calm, expressed in the generation of thick Campanian–Maastrichtian carbonate complexes that sealed all intracontinental and marginal basins, a new magmatism outburst is related to the Laramian phase of tectogenesis, when during its progress some extended tensile zones serving as the Paleogene magmatism channels emerged in the Mesotethys enclosed basin. Geological and petrochemical peculiarities of its manifestation at Azerbaijan territory were reflected in publications of G.I. Allakhverdiev, Sh.A. Azizbekov, A.S. Bairamov, T.G. Gadjiev, A.D. Ismailzade, M.A. Kashkay, A.I. Mamedov, M.N. Mamedov, Sh.F. Mekhtiyev, M.I. Rustamov, E.Sh. Shikhalibeyli, M.M. Veliyev, etc.

The Cenozoic volcanogenic formations stretching as a nearly noninterruptible tract through Turkey, Lesser Caucasus and Iran were for a long period considered as the formations of a unitary "andesite" belt. Evolution of collisional geodynamics starts between the interior arc-plate type plates, rarely, between the arc-arc type plates, is replaced by the continent-continent type, and continues at an increasing pace in the Late Alpine tectogenesis late orogenic phases. Studies of occurrence peculiar features in various structural-formation zones revealed ambiguities of their areal distribution and correspondence to the integrity of belts based on various structural-nature crusts of continental types.

During the Paleogene cycle, volcanism occurred solely in marine and subaerial conditions, and continental conditions of volcano-plutonic belt development took place only at the final stage of granitoid massif generations and at the collisional invasion substage of near-surface small intrusives.

According to recent research, three volcanic-plutonic belts were isolated in the Lesser Caucasus Paleogene "andesite" belt that are represented:

(1) the southern one, in the Ordubad zone: the successively differentiated basalt-andesite-dacite-rhyolite formation (Lower Eocene) with gabbro-granite formation (Middle Eocene); trachybasalt-latite-trachyandesite formation (Middle Eocene) with plutonic gabbro-monzonite formation (Late Eocene); subporphyritic granitoid formation (Late Oligocene–Early Miocene age); andesite formation (Lower—Middle Oligocene);
(2) the central one, in the Kyalbajar trough, superimposed on the Goycha-Hakeri and Kafan zones: basalt-andesite-dacite-rhyolite formation (Middle Eocene) with granosyenite-granite (Oligocene—Miocene age) and gabbro-diorite-granodiorite (Middle–Late Eocene age) intrusive formations (Kyalbajar trough); trachybasalt-trachyandesite-trachyrhyolite formation with gabbro-monzonite (Middle Eocene) formation (Shakhdagh trough); trachyandesibasalt (Middle–Upper Eocene) formation (the Gazakh trough); the Early and Middle Eocene rocks of calk-alkali series are the most abundant in the compositions of these two volcanic-plutonic belts; volcanism manifestation area is reduced in the Late Eocene and, in parallel, general alkalinity increases, up to the occurrence of subalkaline and alkaline differentiates;
(3) the northern one, in the Talysh zone, as alkaline olivine-basalt series with trachybasalt-trachyandesite (Early–Middle Eocene age), trachybasalt-phonolite (Late Eocene), and gabbro-pyroxenite—gabbro-syenite (Late Eocene–Oligocene) formations.

The southern and central volcanic-plutonic belts, due to petrological similarity of the formations, correspond to the unitary geodynamic generation environment conforming to the island arc complexes of continental margins at the regional collisional development stage. Differentiation degree and petro-geochemical features of vulcanite composition in the northern belt conform to the passive rift-induced tension in the Elburs island arc system backarc zone.

The southern and central volcanogenic plutonic belts of the Cenozoic magmatism mate along the Zangazur suture zone, which is the Mesotethys suture in the Lesser Caucasus.

4.2.3 The Central Volcano-Plutonic Belt

Basalt-andesite-dacite-rhyolite formation. The Cenozoic volcanism in the Lesser Caucasus central portion is confined to the Kyalbajar, Kalaboiny, and less frequently, to the Mykhtokyan troughs, with transversal arrangements with regard to the ancient structures. In these troughs, volcanism begins in the Middle Eocene period, whereas the Lower Eocene shows the development of sedimentary facies.

In the troughs under study, the most developed among the Cenozoic complexes is the Paleogene one, characterized by the successively differentiated series: basalt-andesite-dacite series in the Kyalbajar trough, and trachybasalt-trachyandesite-trachydacite-trachyrhyolite series in the Shakhdagh trough.

The Late Eocene volcanism is poorly exhibited only in the Kyalbajar trough and is manifested by volcanogenic and volcanogenic-sedimentary thicknesses. Petrochemical characteristics of the investigated formation rocks correspond to the calk-alkali (less frequently, toleite) and subalkaline (less frequently, alkaline) series.

Geochemically, basic and middle members of the normal and subalkaline series are characterized by rather wide distribution variation of incoherent elements such as K, Na, Li, Rb, Sr, U, Th, and by low bulk earth values of Cr, Co, Ni, Ti. Low Ni/Co < 15 ratio reflects the correspondence to the island arc formations. Low K/Rb and Ba/Sr values of the relationships provide the examples of the crust active participation in magma generation.

Granosyenite-granite formation. The Dalidagh granitoid batholith (30–35 Ma, the Oligocene—Lower Miocene) is the volcanogenic formation plutonic phase and it is the largest intrusive (90 km^2) in the Lesser Caucasus central part, stretching northwestward with 20 km length and 4–6 km width. Tectonically, it is bordered by the Terter and Lachin abyssal fractures and is located on the Dalidagh-Mekhmany high (Shikhalibeyli 1966). The Dalidagh intrusive is double-phase and polyfacial, composed of syenite-diorite and granite-granodiorite series, but M.A. Kashkay and A.I. Mamedov believe it to be a single-phase intrusive composed of the above-mentioned series.

The first phase (series) rocks are of poor occurrence. They compose the intrusive hypabyssal part and are hybrid rocks represented by syenites, granosyenites, quartz syenites, quartz syenite-diorites, and biotite and hornblend differences. The second phase (series) rocks, which constitute the massif's major part, correspond to biotite granites, granodiorites, granite-aplites, and quartz syenites. Their petrochemical characteristics correspond to the calk-alcali series rocks of normal and subalkaline type. The presence of smooth petrochemical parameters in the differentiated facial series reflects their petrological similarity and correspondence to the unitary parental acid magma that underwent differentiation.

Geochemically, the Dalidagh intrusive rocks show increased content of radioactive elements, from the early syenite-diorite phases to the subsequent granites, granosyenites (Geidarov 1971) that appears to be related to potassium metasomatism in the intrusive activity late stage. Co, Ni, and Ti contents have two to threefold excess of percentage abundance for acid rocks, while Mn and Cr are lower.

Gabbro-diorite-granodiorite formation is a constituent of volcanic-plutonic association within the Lesser Caucasus central part. It is represented by the Middle and Late Eocene intrusives among the Cretaceous sedimentary and the Eocene volcanogenic complexes (Mamedov 1974, 1981).

The Middle Eocene intrusives are developed in the Levchay, Jamilli, and Galaboinu River basins as stock-like elongated intrusive bodies. They were generated during the same magmatism phase and are represented by gabbro-diorites, diorites (in the center) and porphyry differences (margins).

The Late Eocene intrusives are found in the Tutkhun River basin and consist of the following massifs: Gazykhanly—2.5 km^2, Gyzylitan-Zarguly—4.5 km^2, Garasu-Agjakend—3.2 km^2, Mozchay-Asrik—2.0 km^2. Morphologically, they are shaped as stocks, discordant bodies, and laccoliths (Kashkay et al. 1964). They were generated during three phases: (1) the gabbroid phase, facial series: gabbro-gabbro-diorite-monzodiorite-syenite-diorite-diorite, (2) the granitoid phase, facial series: quartz diorites-granodiorites-adamellites-granites, (3) the porphyry-like granites, facial series: diorite-porphyries—granodiorite porphyries—granite-porphyries. The first phase rocks of the Tutkhun group intrusives are subject to contamination and correspond to hybrid formations. Calculation of petrochemical parameters in the Levchay-Galaboiny group Middle Eocene intrusive rocks permitted to establish their correspondence to the high-aluminium, potassium and potassium sodic calk-alkali series.

Petrochemical parameters of the Tutkhun formation Late Eocene intrusives are similar to those of the Middle Eocene but with higher crystallization medium oxidation level.

By their geochemical parameters (Geidarov 1971), the Middle Eocene intrusive rocks are characterized by lower bulk earth values for Be and Ba content, as well as for Co, Ni, V, Mn and rare-earth elements. The Upper Eocene intrusive complex rocks contain in their gabbroids Be value higher than bulk earth values, and Ba value lower than bulk earth values; Cr, Co, Ni, and V are higher than bulk earth values. The complex in general is characterized by low radioactivity.

The **Trachybasalt-trachyandesite-trachyrhyolite formation** is the northwestern part of the central volcano-plutonic belt. It is distributed throughout the Shakhdagh syncline in volcanogenic-sedimentary facies, intruded by the Darvadagh (~ 20 km^2) and Shakhdagh (~ 5 km^2) granitoid intrusives. Petrogenous stock of the formation rocks includes trachybasalts, shoshonites, trachyandesibasalts, latites, trachyandesites, quartz latites, trachyrhyolites, as well as pantellerites and comendites.

Petrochemical parameters of the Middle Eocene rocks correspond to three trends of calk-alkali series: (1) moderate-K, andesite-trachydacite—trachyrhyodacite, (2) high-K, trachyandesite-trachydacite, (3) subalkaline, trachybasalt-quartz latite-trachyte. The Late Eocene rocks correspond to two trends of calk-alkali series: (a) moderate-K, trachydacite-trachyrhyolite, and (b) high-K, trachydacite-pantellerite-comendite.

The set of petrochemical data makes the Middle and Late Eocene rocks the derivatives of primary alkaline olivine-basalt magma that underwent complete differentiation. Based on a number of petrochemical features of the Middle and Late Eocene rocks, the Shakhdagh structure can be thought as corresponding, on the one side, to modern island arcs, and on the other side (due to presence of pantellerites and comendites) to continental rifts (Abdullayev et al. 1991).

The above described volcanogenic formation is associated with the plutonic **Gabbro-Monzonite formation** represented by the Darvagh intrusive (≈ 1 km^2), formed by two intrusive phases: (1) subalkaline gabbroids, and (2) monzonites and quartz monzonites (major volume of the massif rocks).

Petrochemical characteristics of monzonites and quartz monzonites correspond to subalkaline diorite-monzonites and quartz monzodiorites. Content of Co, Ni, Cr, Pb ore elements as well as of Li, Rb, Sr is reduced in gabbro and monzonites are reduced; in quartz monzonites, it is sharply increased.

Trachyandesite-basalt formation, from the Middle Eocene age, is developed within the Lock-Garabagh zone Gazakh trough. In volcanogenic facies (down to 200 m), it is represented by subvolcanic intrusives of Didivan Gyamishalchay, Demirchilyar (45–50 Ma, the Upper Eocene age).

Facial composition of these formations: lavas and volcanoclasts as plagioclase trachyandesibasalts and trachyandesites; subvolcanic rocks, by trachydolerites. Rock groundmass contains numerous rounded and elongated amygdales filled with hydrous ferric oxides, chlorite, ceolite, reflecting the intensity of hydrothermal processes According to the aggregate of petro-geochemical parameters, the volcanites under study correspond to subalkaline series of island arc systems.

4.2.4 The Southern Volcano-Plutonic Belt

Commencement of the magmatism vigorous Paleogene cycle in the Lesser Caucasus is marked with the formation of independent volcano-plutonic southward, central, and northward belts that are exposed in superimposed rift-related depressions.

In the southern belt (Rustamov 1983), after the Late Senonian tectonic stabilization, rift-related tension and intensive plunging of the Ordubad, Arpi, Vedi superimposed depressions takes place along the abyssal fracture Zangazur zone.

The **Basalt-andesite-dacite-rhyolite formation**, from the Lower Eocene, is developed at the Nakhchivan Autonomous Republic (AR) area and at the Zangazur Ridge slopes, and is also noted as separate fragments on the Armenian Megri-Gyunei and Bargushat ridges. The Sakkarsu, Alyagi, Uchurdagh, Gorukhlar, and Garanguch formations stand out within is scope and, according to their volcanism and tectonic confinedness type, characterize the volcanogenic thickness suites confined to the axial graben and frontal flank blocks along the Zangazur suture zone.

The formation petrostock is characterized by basalts—42 %, andesibasalts—17 %, andesites—22 %, andesite-dacites and dacites—12 %, rhyodacites, and rhyolites—7 %. Their petrochemical parameters correspond to the toleite (39 %), subalkaline and high-potassium calk-alkali (35 %) and calk-alkali series.

Basaltoid alkaline varieties are of limited development and have been reduced at the volcanism early stage. Cosubordination of melt differentiations integrates significant composition variations for volcanogenic formations with tectonic positions of structural zones: a toleite trend is confined to the axial trough, while the calk-alkali trend is observed in adjoined blocks with elevation tendency. For this reason, basic petrochemical parameters of the basalt-andesite-dacite-rhyolite formation rocks are related to the toleite (39 %) and calk-alkali (26 %) series.

The toleite series volcanites are met in all formations, except the Gorukhlar one. They are represented by basalts and andesibasalts in the Sakkarsu and Alyugi formations, by basalt-andesidacite family, in the Uchurdagh formation, and by andesite-rhyolite family, in the Garanguch formation. The calk-alkali series is limitedly developed among the Sakkarsu and Alyugi intrusive rocks but is extensively manifested in the Uchurdagh and Garanguch intrusives, respectively, in their basalt-andesite and andesite-rhyolite associations. The subalkaline series of rocks, from trachybasalts to trachyrhyodacites in differing associations, is developed in each one of the complexes.

Geochemical parameters of the formation basalts characterize them as having reduced content of incoherent alkaline elements such as K, Rb, Li and alkaline-earth Ba, with the peak Sr content in the Sakkarsu intrusive basalts. Ni and Co contents, similar in the toleite and calk-alkali series (Ni = 9.5–10.0 ppm, Co = 16–20 ppm), increase towards the subalkaline values (Ni = 14 ppm, Co = 26 ppm). The value of Ni/Co relation, similar for all complexes (0.4–1.4), testifies to their correspondence to the magmatic differentiation products of the mantle-crust genesis. Cr content in the basalts of all complexes slightly increases from toleite series (4.9–18 ppm) to calk-alkali (18.2 ppm) and subalkaline (25.6 ppm) series (Fig. 4.6).

Correlation analysis of petro- and geochemical parameters for all series of basalt-andesite-dacite-rhyolite formation shows similarity to the peer toleite and subalkaline series rocks of the Afar depression and Rio-Grande rift and to the calk-alkali series of the Andine continental margin in various geodynamic environments.

Petrogenesis of the formation rocks is explained by the evolution of primary toleite-basalt magma generation focus during three volcanism stages in the respective geodynamic conditions, and by the essential role of the magma generation focus depth. In this process, melt differentiation in the

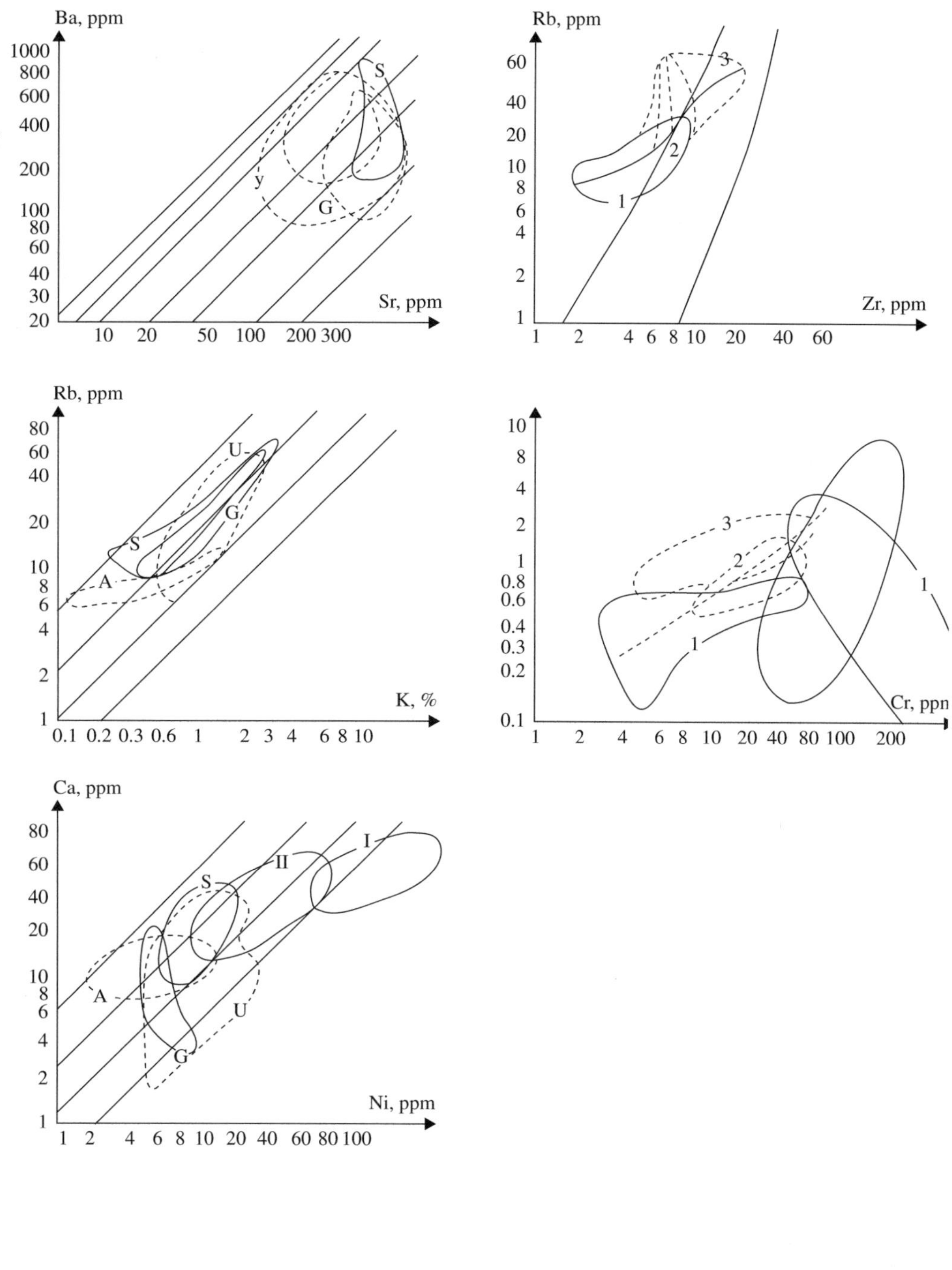

Fig. 4.6 Geochemistry of rocks of basalt-andesite-dacite-rhyolite formation. Fields of rocks of complexes: *S* Sakkarsuin, *A* Alyagin, *U* Uchurdag, *G* Garanguch. Fields of basalts of basalt-dacite-andesite formation (*1*) tholeiite, (*2*) limestone-alkaline, (*3*) subalkaline and trends. Fields of basalts: (*I*) oceanic, (*II*) island arcs

axial graben appeared to occur at all its substages according to the toleite trend, and in the adjoined flanks, according to the calk-alkali trend, due to enrichment with water fluids and increase in fo_2 melts in shallow lying peripheral foci.

Gabbro-granite formation, the Middle Eocene, extends within the Ordubad syncline southeastern and eastern parts at the Nakhchivan AR and Armenian areas.

Intrusive complexes make up the heterogenous polyformational Megri-Ordubad batholith as well as a series of separate double-phase massifs such as Kovshut, Gekh, etc. in the Armenian Bargushat group of intrusives.

According to the majority of researchers, the Megri-Ordubad batholith was formed during the Upper Eocene–Oligocene period as a result of three intrusive phases—the monzonite, the granosyenite, and the granite ones (Gukasyan and Meliksetyan 1965). In the batholith Ordubad section, the adamellite and the granosyenite phases of intrusive activity during the Oligocene age have been differentiated (Azizbekov et al. 1979).

The gabbro-granite formation within the Megri-Ordubad batholith is represented by the Gyandzi (178 km^2) and the Megri-Gyunei (62 km^2) complexes in the batholith western and eastern parts that resulted from the intrusion of gabbroid and granitoid phases. The phase I gabbroids are limitedly developed universally, at times even within phase II (Soyutdagh Mountain piedmont), and occupy up to 5–7 % of the formation volume.

The formation rock stock features diversified compositions of rocks generated by the intrusion of successive phase with their subsequent assimilation and hybridism, and extensively displayed in the Ganja (the Zangazur High) and the Agsaggal (the Okhchi High) complex granitoids (Fig. 4.7).

Gabbroids are composed of leucocratic and pyroxene-amphibolic gabbro, gabbro-diorites and diorites, while granites consist of granodiorites, adamellites, granites and their hybrid petrotypes, less frequently, of apical facies leucogranite outliers, and marginal facies contacting with the Upper Cretaceous carbonate thicknesses include melanocratic petrotypes of monzonites and quartz monzonites, leucocratic quartz monzonites and, occasionally, granosyenites.

According to their petrochemical characteristics, the formation rocks are related to normal, less frequently, to subalkaline series, and are represented by the toleite, calk-alkaline, high-potassium calk-alkaline, and rarely, by

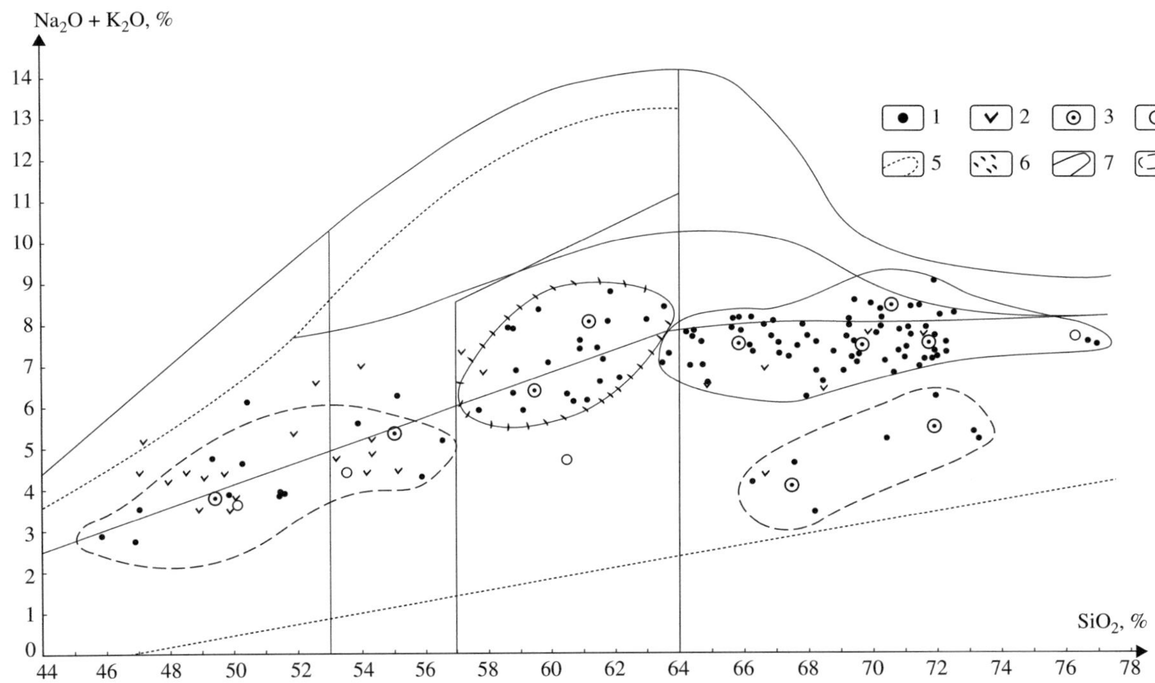

Fig. 4.7 Systematics of rocks of gabbro-granite formation in coordinates of SiO_2–(N_2O + K_2O) (*1*) typical rocks, (*2*) dykes, (*3*) average composition of typical rocks of gabbro-granite formation, (*4*) average composition of typical rocks of central volcanic-plutonic belt of the Lesser Caucasus, fields of distribution: (*5–8*), (*5*) gabbroids, (*6*) monzonitoids of margin facies, (*7*) granitoids, (*8*) plagiogranites

shoshonite-latite series. Plagiogranites and plagiogranodiorites are distinguished by their decreased general alkalinity and are segregated in the toleite series field.

Contents and relationships of trace elements in the gabbro-granite formation rocks, in rock-forming and accessory minerals characterize the granite generation manner at the Paleogene magmatism early development stage in the Ordubad syncline. Thus, according to Tauson (1982), Rb and Li alkaline elements, alkaline-earth Ba and Sr bring the gabbro-granite formation closer to andesite and latite series of granitoids, originating in basaltoid magmas of the mature continental crust areas. The fact is also confirmed by the Ni/Co ratio in them—that is, in correlation with the collision areas of mature island arcs and active margins with powerful continental crust.

In the formation rocks, primary isotope ratio of $^{87}Sr/^{86}Sr = 0.70319$–$0.70755$, along with geological-petrographical and geochemical data, witnesses to the origin of the gabbro-granite formation rocks that took place during the interaction of mantle and crust melts by the generation of magma foci that ended in rock melting at the crust level.

The **Trachybasalt-latite-trachyandesite formation**, the Middle Eocene, first found by Rustamov (1976), is represented in pyroclastic thickness volume within the Ordubad syncline, and its age corresponds to the Middle Eocene Lutetian Stage (40–43 Ma). The formation composition is characterized by high quantitative relations of pyroclasts such as volcanic breccias (scree breccia, etc.), tuffs with subordinate role of lava flows and tuffaceous-sedimentary (tuff conglomerates, tuff gravelites, tuff sandstones) and sedimentary rocks. Volcanic breccias and scree tuffs are composed of block, agglomerate, and bomb-like varieties; lapilli and vitric tuffs are noted, frequently with spheroidal jointings.

Petrographic composition of this formation is represented by subalkaline (85–90 %) and normal (10–15 %) rocks, constituting the association of three evolutionary series: (a) shoshonite-trachyte series, (b) trachybasalt-trachydacite series, and (c) andesite-andesibasalt series.

Petrochemically, the association rocks differ only in their alkalinity level, while the majority of alternative parameters do not show essential differences.

Analysis of petro- and geochemical characteristics of all association rocks in the formation under study reveals their correlation with the complexes, being generated under continental margin collision conditions. Confinement of calk-alkali series volcanites to the formation cross-section tops is indicative replacement of vigorous tension by compression in transversal strike-slip faults of rift-related trough.

Lava flows are occasionally met in volcanic field cross-sections and are extensively developed in intercaldera facies. For the first time in the development of southern volcano-plutonic belt this formation (Rustamov 1976) is isolated as the volcanism final stage that comprises three cross-section types: the facies of volcanic edifices and of intracaldera type; the volcanic field facies, and the remote volcanic field and shelf facies

Gabbro-monzonite formation (37–40 Ma) that completes the Late Lutetian stage of subalkaline olivine-basalt volcanism and is developed along the abyssal fracture of Zangazur zone. Intrusive formations of this entity have wide occurrence in the Armenian and Nakhchivan AR areas, composing the Megri-Ordubad batholith and constituting individual large massifs of the Arpy and Vedy troughs.

The Misdag-Gapychyg, Kajaran, Megri, and Lyakyatach complexes of this formation are its principal intrusive complexes. Three phases are isolated in the Misdag-Gapychyg complex structure: (1) the subalkaline gabbro, gabbro-monzonites, (2) the quartz and quartz-free monzonites and monzodiorites, (3) monzogranodiorites with quartz diorite apical facies.

The Lyukyatakh complex of monzoitoid diorites is exposed in the Alinjachay River head and is represented by three confocal massifs such as Damirli, Dashyurt, and Nallypir, and by numerous dykes confined to the Middle Eocene volcano-tectonic high with subsidence caldera. Petrographically, subalkaline gabbroids (phase I) are represented by gabbro, gabbro-monzonites, gabbro-diorites, while quartz and quartz-free monzonites and quartz monzodiorites (phase II), the formation indicative rocks, are frequent in less eroded environments and in hanging endocontact; in their final stage, they are accompanied by pyroxene monzonites replaced by quartz monzonites at the depth.

Petrochemical parameters of the formation rocks correspond to the alkaline series, and their geochemical indices, especially, for the iron group elements (Co, Ni, Cr, V), to geodynamic compression condition in the regions with thick continental crust.

The **Pseudoporphyritic-granitoid formation**, from the Late Oligocene–Early Miocene age, is composed of pseudoporphyritic granitoids of increased alkalinity and completes the evolution of multistage Paleogene magmatism in the Ordubad syncline. Its key feature is the vigorously manifested processes of potassium and sodium metasomatism, confined to the late magma stage of intrusive generation (Azizbekov et al. 1964; Azizbekov and Rustamov 1972). Also of note is the presence of thick and extended granodiorite-porphyry dyke derivatives of pseudoporphyritic granitoids.

The formation intrusives are confined to the most mobile structures of the Zangazur abyssal fault. They are represented by the Gekhi-Sakkarsu, Alangez, Lyakyatakh, etc. Lower and Middle Eocene volcanic-tectonic highs, which according to the set of their geological and geomorphological parameters, are the central type intrusions.

Petrographical composition of pseudoporphyritic granitoid formation shows such typical rocks as quartz syenite-diorites, granosyenites, granodiorites, granites, and aplogranitoids that have similar lithofacial composition but differ in their structural-compositional image and qualitative relationship of minerals.

Petrochemically, the formation rocks correspond to the subalkaline and normal series and are comparable with crust calk-alkaline series palinogenic granites (Tauson 1982) (Fig. 4.8).

The formation rock peculiar features are high contents of incoherent and amphoteric group coherent (Zr, Nb, Ta) elements that reflect geochemical profile and high productivity of potential ore-bearance in pseudoporphyritic granitoid formation. Cr, Ni, Co, and V contents vary within a broad range, and, except for Cr, have higher than bulk earth values. All formation rocks are characterized by sharply elevated contents of light rare-earth elements. In general, the rare-earth element relative and absolute values are indicative of the magma generation crust nature, and distinct Eu peak indicates the correlation between melted magmas and the levels of plagioclase parageneses.

The metasomatosis process plays significant role in the formation rock petrogenesis. It emerged during the introduction of the second and third phases and serves as the determining factor of intensive hydrothermal activity. Apparently, acid crust magmatic focus was the K source during rock kalifeldsparization in the intrusive chamber. Potassium enrichment and magma transformation into subalkaline composition are related to the assimilation of upper substrate carbonate thicknesses rather than to the abyssal supply of K. At the same time, acid magma generation took place at the expense of crust amphibolite substrate, under the impact of abyssal fluids and heat from decompressed mantle waveguide and deep-seated basaltoid focus that resulted from this latter. Lamprophyre dykes are the derivatives of basaltoid deep-seated focus rather than of the crust acid one.

The **Andesite formation**, from the Lower–Middle Oligocene, is abundant at the eastern and northeastern parts of the Nehram plateau, in the Lakinjachay River lower course basin, at Darridagh, Shakh-Karash, and Dagusti mountain area.

The formation petrochemical stock displays tuff breccias and tuff conglomerates, with relatively poor manifestations of effusives such as autoclastic lavas and lava breccias of andesites, andesite-dacites, and, less frequently, dacites. It is volcanoclastic by its facial composition, with low-thickness nappe of autoclastic lavas at the top. The formation maximal

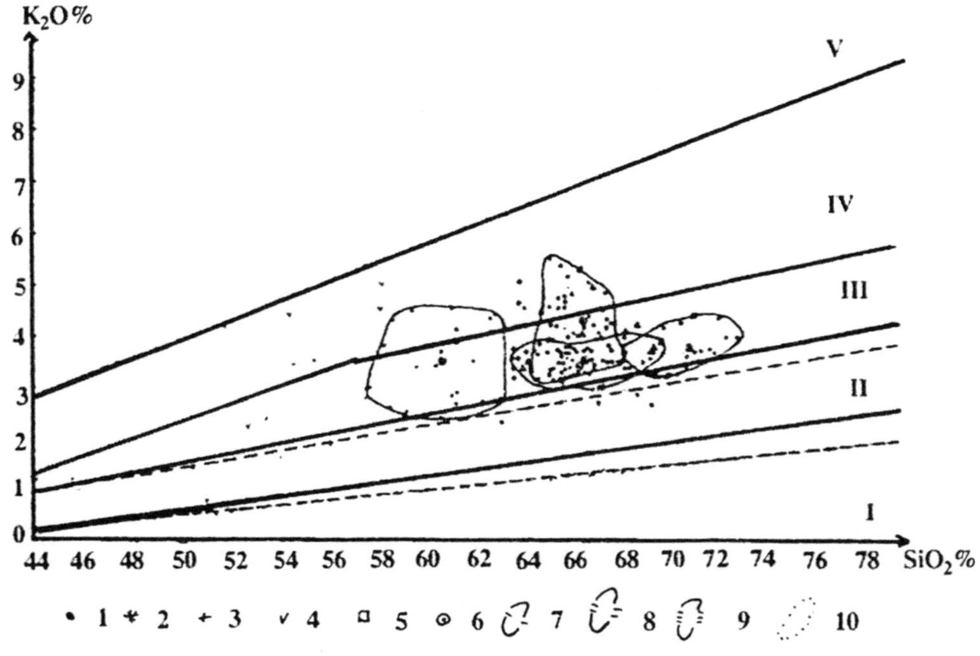

Fig. 4.8 Systematics of rocks of porphyric granitoids in coordinates SiO_2–K_2O (Ismailzadeh et al. 2005) (*1*) typical rocks of Alangez complex of the Megri-Ordubad batholith (ACMOB) and Sakkarsuin intrusive complex, (*2*) aplogranitoids, dyke complexes (*3–4*): (*3*) granodiorite-porhyrites etc., (*4*) lamprophyres and diabases, (*5*) average composition of rocks of other massifs of southern and central volcanic-plutonic belts, (*6*) average composition of rocks of porphyric granitoid formation, fields of distribution of different phases and complexes of porphyric granitoid formation (*7–10*): (*7*) quartz syenite-diorites of the first phase of the ACMOB, (*8*) porphyric grano-syenites and granites of the second phase of the ACMOB, (*9*) porphyric granodiorites of IIIrd phase of the ACMOB, (*10*) porphyric granodiorites of the Sakkarsuin intrusive complex

thickness of 883 m is observed between Cheshmabasr—Gyzylja villages and Darrydagh Mountain, with gradual northwestward displacement of the trough axis.

Petrochemical parameters of autoclastic lava and clast andesites and andesite-dacites correspond to the normal alkalinity rocks, i.e. to the calk-alkali series. Low content of ferrous iron, with considerable predominance of ferric iron, reflects regeneration in oxidative conditions that were beneficial for the isolation of punctulated hematite aggregates, hornblende and biotite opacitization, and ferriginization of the principal rock masses. It indicates their formation in shallow basins, in conditions similar to the continental ones.

As for geochemical parameters, the near-bulk earth values for iron group element contents such as Cr, V, Co, and Ni; the low bulk earth values for Mo and Zn; and the high bulk earth values for Cu and Pb, which in certain cases show 30-fold excess of percentage abundance, reflect the correspondence between the formation rocks and the basic magma differentiates.

4.3 The Talysh Zone

4.3.1 The Northern Volcano-Plutonic Belt

This belt is being developed in the Talysh structural and formational zone confined to the Lesser Caucasus–Elburs folded system northeastern edge and looks like an isolated horst northwestward plunging under the Neogene–Quaternary formations of the Lower Araz depression, and northeastward plunging under the Kur Depression Quaternary formations.

According to geophysical data, continental crust of 35–38 km thickness experiences westward to eastward plunging with surface cover thickness from 5–6 to 10 km in the direction of Astara town and corresponding rise of mantle surface from 37 to 31 km. All these features are indicative of the Talysh continental crust thinning towards the South Caspian depression.

The Talysh volcano-plutonic belt structure is comprised of volcanogenic alkaline-basaltoid series with trachybasalt-trachyandesite-trachyte (the Early–Middle Eocene), trachybasalt-phonolite (the late Eocene), and the intrusive gabbro-pyroxenite—gabbro-syenite (the Late Eocene–Oligocene) formations.

Trachybasalt-trachyandesite-trachyte formation is most extensively (3000 m thickness) developed in the Kosmalyan trough. It is represented by the thicknesses of: picrite-trachybasalt and leucite-tephrite (450 m) lavas and pyroclasts, andesibasalt volcanic conglomerate breccias (440 m), sedimentary tuffaceous thickness (200–600 m), trachyandesite and trachyandesibasalt autoclastic lavas and volcanic breccias (450 m), flyschoid sedimentary-tuffaceous thickness (600–800 m), plagioporphyric trachyandesibasalts and trachyandesites (1000–1200 m). The formation volcanic rocks are typically manifested in the lava-pyroclastic facies (pillow and autoclastic lavas, tuffs and volcanic breccias) and in subvolcanic facies (nekks, sole injections, stocks and dykes). The lava to pyroclastic facies relation is 2:3. The area of its trough manifestation is >2000 km^2. The relation between basic, middle, and proper alkaline rocks is 25:60:15 %.

The formation petrostock is constituted by trachybasalts, olivine-leucite tephrites, trachydolerites (essexites), trachyandesibasalts, plagioporphyric trachyandesites, leucitic trachyandesites, and trachytes.

Trachybasalt-phonolite formation is developed in the Dyman trough, located northwestward of the Kosmalyan one. It is represented by the thicknesses of: tuffaceous-sedimentary (800 m), lavas and pyroclasts of subalkaline vitrobasalts, trachybasalts ($\sim$500 m), and leucitic phonolites ($\sim$200 m).

Its volcanism is characterized by the central-fissure eruption type with manifestations in the lava (pillow-block lavas) and, less frequently, in pyroclastic (tuffs and bombs) facies. The subvolcanic facies is represented by small specular bodies, domes, and low-thickness sole intrusions. The lava to pyroclastic facies relation is 3:1. The area of its Dyman trough manifestation is >1500 km^2.

The formation petrostock is constituted by subalkaline vitrobasalts (42 $\pm$ 2 Ma), trachydolerites (essexites), and leucitic phonolites.

The **Gabbro-pyroxenite—gabbro-syenite formation**, from the Late Eocene–Oligocene, is confined to the areas of intersection of the northwest positive structures and northeast elevations. The intrusives are represented by stocks, sole injections, and complex concentric bodies situated in the Danish-Paleocene tuffaceous-sedimentary complex (Nyudis-Galasy, Palikesh), in the Middle Eocene flyschoid thickness (Ganarat, Aliabad), and in the Late Eocene sedimentary-tuffaceous thickness (Alashar-Yelagach).

The intrusives structures are prominent with the facies corresponding to the central parts such as gabbro-pyroxenites, and to the marginal parts such as alkaline gabbro, gabbro-syenites, intersected by hydrothermal metasomatic veins represented by xenolith, pectolite, amphibole, zeolites, and calcite. Said differential series is the most completely exposed in the Gamarat-Aliabad intrusive. In the area of intrusive outcropping, bedding bodies (0.5–3 m) and dykes (0.7–1.8 m) of leucocratic and melanocratic lamprophyres that have northwestward striking, are noted in the sedimentary thickness.

For all formation rocks, including gabbro-pyroxenites, strong amphybolization and biotitization are typical.

Geochemically, high values of the iron group elements, Cr/V and Ni/Co relationships reflect the correlation between

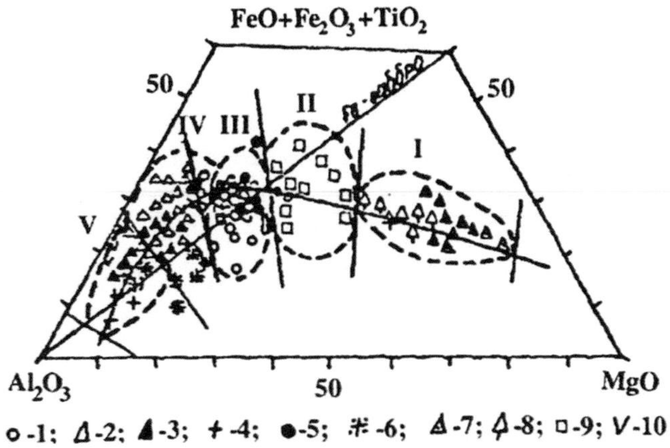

Fig. 4.9 Diagram of Al_2O_3–($FeO + Fe_2O_3 + TiO_2$)–MgO for rocks of volcanogenic and intrusive formations of Talysh. ***Volcanogenic formation***. Early–Middle Eocene (*1*) trachybasalts (trachydolerites, shoshonites, olivine-leucitic tephrites), (*2*) trachyandesito-basalts, trachyandesites, (*3*) plagioporphyric trachyandesito-basalts, trachyandesites, (*4*) alkaline trachytes. Late Eocene: (*5*) trachybasalts (vitrobasalts), (*6*) leucitic phonolites. ***Intrusive formation***. Late Eocene–Early Oligocene: (*7*) picrites, (*8*) gabbro-pyroxenites, (*9*) alkaline gabbro, (*10*) gabbro-syenites. ***Rock associations***: *I* alkaline-ultrabasic, *II* alkaline-gabbroid, *III* trachybasaltic, *IV* trachyandesite-basaltic, *V* trachyte-phonolithic

volcanogenic formation basic rocks and abyssal melts with relatively high content of coherent elements and low content of non-coherent ones. Increase in these latter during magma differentiation corresponds to the "crust" concentration type, i.e. to the differentiation type at the Earth crust level.

Analysis of petrochemical data brings to the conclusion that volcanogenic and intrusive formation rocks are the derivatives of unitary alkaline basaltoid magma, where its volcanogenic part experiences differentiation and assimilation in intermediate foci, and its intrusive part has skipped these processes (Fig. 4.9).

Structural disunity of volcanogenic formations, along with petrological and petrochemical parameters, permits to suggest their formation in the Early–Middle Eocene, in the independent foci above asthenospheric roof with less deep bedding that promoted the formation of differentiated series with high-magnesium, low-titanium, and low-silicon major elements such as trachybasalts, as compared to the deeper lying Late Eocene titanian, silicious, and ferrian basalts.

Petrochemical parameters of the intrusive formation ultrabasic differentiates correspond to the cumulative formations of alkaline olivine-basalt magma. Manifestation of volcanogenic formations took place at the early-collision stage of regional development, while the intrusive formation occurred during the orogenic stage, when the uplift of heavy ultrabasic cumulate matter resulted from the activation along deep permeable zones.

This association has the properties that characterize the mantle melt with the participation of continental crust matter corresponding to the alkaline-basalt melts displayed in the backarc passive rift genesis conditions confined to the transient zone from the Iranian plate continental margin to the Caspian basin suboceanic plate.

4.4 Comparative Analysis of the Paleogene Magmatic Formation Series

The material cited on the Paleogene volcanic formations and the comparison results of its series within the Lesser Caucasus various structural zones allow to establish certain general regularities in their manifestations.

1. Paleogene volcanism is confined to three volcano-platonic belts, characterized by differing abyssal structures and geodynamic conditions: the southern belt—the Nakhchivan microcontinent northern margin (the Ordubad zone), the central belt—the Transcaucasian median massif southern margin (the Goycha-Hakeri zone), and the northern belt—within the Elburs zone northern margin.
2. Thick (>50 km) continental crust with powerful "granite" layer served as the foundation for the southern and central volcano-plutonic belts. The northern belt emerged on the crust (of up to 40 km) of subcontinental (transient) type in the juncture band with the South Caspian "granite-free" crust projection.
3. Definite regularity in temporal and lateral manifestations of volcanic formations is observed within the belts.

In the southern belt, formation of the Early–Middle Eocene volcanic formation occurred in the strip from the Zangazur High to the Paradasht depression, with the movement of the Middle and Late Eocene volcanogenic-sedimentary and sedimentary complexes toward the central of this latter.

In the central development belt of the Middle–Late Eocene volcanic formations, volcanic processes are characterized by the general movement tendency from the Kyalbajar and Shakhdagh Ridges northward, to the Gazakh depression.

In the northern belt (southeastern segment), lateral regularity is determined quite distinctly within the Early-Middle and Late Eocene, by the northward movement of volcanic activity.

4. The regularity of volcanism manifestation within volcano-plutonic belts was also accompanied by the changes in volcanic formation compositions. Thus, during the Paleogene within the southern belt, the volcanism that originated in the Zangazur Ridge with toleite and calk-alkali low-titanium, high-alumina, sodium

volcanites transforms into subalkaline K-Na volcanites, also of low-titanium, high-alumina content and with increased difference alkalinities. The central belt is characterized by the development of calk-alkaline series of K-Na type, also of low-titanium, high-alumina content but with slightly increased alkalinity. Altogether, within both belts, volcanic formations, rather similar in their petrochemical parameters, are observed that culminate in the formation of large granitoid batholiths. In the northern volcanic belt, by progressing northward from the early phases of the Eocene volcanism manifestations to the late, the lateral compositional changes are displayed in the occurrence of more alkaline facies with considerable K predominance over Na and with higher correspondence to the shoshonite series differentiates.

5. Volcanic and plutonic formations are represented in the southern and central belts by a similar set of basic and acidic differentiates of normal and elevated alkalinity, and in the northern belt, by alkaline basic and ultrabasic differentiates with pronounced differences in primitive magma liquids and their manifestation conditions.

Analysis of rare element distribution manner in major volcanites of the Lesser Caucasus Cenozoic volcanic formations permitted to reveal the following regularities:

- the selected Paleogene volcanic formations such as the basalt-andesite-dacite calk-alkali formation and the trachybasalt-trachyandesite-trachyte subalkaline formation are characterized by differing composition relationships for incoherent and coherent elements;
- within the southern belt, major differentiates of the basalt-andesite-dacite formation have very low values of alkaline, alkaline-earth, radioactive, light rare-earth elements and iron group elements (except V) as compared to the other Cenozoic formation volcanites, and correspond to toleite and calk-alkali basalts of island arcs, typical for the developed island arcs of the Kurily-Kamchatka arc type;
- within the central belt, basalts show high contents of alkaline and radioactive elements, low values of the iron group elements, and correspond to the calk-alkali series matching to the mature island arcs of the Japanese arc type;
- in the northern belt, subalkaline basic differentiates of the trachybasalt-trachyandesite-trachyte formation are characterized by the considerably higher content of alkaline, radioactive, alkaline-earth, and rare-earth elements, and according to the iron group elements, by Cr and Ni, matching to olivine basalts of rear or marginal troughs of the Japanese sea type.

4.4.1 The Lesser Caucasus Neogene Formations

The Neogene volcanic formations are spatially confined to the Paleogene volcanism areas and are independently developed within the Nakhchivan, Miskhan-Kafan, and Goycha-Hakeri zones. The selected formations correspond to the region development collisional stage.

The Neogene stage has the following typical magmatic formations (Ismailzadeh et al. 2005):

- quartz-diorite one, the Early Miocene;
- trachyandesite-teshenite one (the Kashirdagh and Saradak complexes), the Middle Miocene;
- andesite one (the Bichenag thickness), the Early Pliocene;
- andesite-dacite-rhyolite one (the Basarkechar suite), the Mio-Pliocene;
- rhyolite one, the Pliocene;
- trachybasalt-trachyandesite one (the Kyalbajar depression), the Late Pliocene-Quaternary period.

The **Quartz-diorite formation** is from the Early Miocene. During the Lesser Caucasus Miocene age, magmatic processes had limited development and were associated with intrusive formations corresponding to finalizing complexes of the Paleogene granitoid formations. The following massifs are related to this Neogene intrusive formation: Ilandagh, Nagajir, Alinja, Yashylkend, Berdyk, Kazanchi, Khanaga, Khoshkeshin, Paradasht, Gyal, Ortakend, Bashkend, and Geidagh. They are confined to the Nakhchivan AR eastern part within the Miocene Paradasht trough. The intrusives are characterized by diversified morphology of the bodies such as bysmaliths, laccoliths, sphenoliths, chonoliths, stocks, and sills with small (100–200 m) erosion truncation.

The intrusive tectonic position shows evident confinement to the intersection nodes of transversal and longitudinal fractures that define their bysmalith-like morphology. The absolute age of these intrusives in the Ilandagh, Berdyk, Nagajir, Alinja and Yashylkend massifs is 16.3–18 Ma, thus corresponding to the Lower Miocene.

Petrographic composition of intrusive rocks is rather heterogeneous—quartz-diorite-porphyry—and reflects the generation single-phase feature. The intrusion second phase was found as biotite-hornblende granodiorite-porphyries only in the extremely eastern intrusive of Geidar that is also represented by quartz diorite-porphyries differing from the first-phase rocks by their appearance and appreciably inferior to the first-phase rocks in volume (Rustamov 1987). The formation rocks are classified as the normal line of calk-alkali, and less frequently, of high-potassium calk-alkali series, cf. the Nagajir, Ortakend, Bashkend, Kazanchi, Gyal, and Geidagh massifs.

Chemical and mineralogical uniformity of intrusive rocks defines their unified geochemical development. Distribution and composition manner of coherent and incoherent elements, namely, increased Li contents and decreased Rb contents, with Sr bulk earth values, as well as variations in the iron group elements such as Ni, Co, Cr, V, are conditioned by magma fractionation by non-ferrous minerals—pyroxene, hornblende, and biotite,—and by the sequence of mineral crystallization at various levels, up to the intrusive chamber.

Petrological and geochemical analysis of Nakhchivan intrusives, "laccoliths," permits us to establish their correlation with the intrusive magmatism independent stage in the region geotectonic development. Generation of intrusives occurred in the near-surface conditions with serial-porphyry rock structures. They are the derivatives of crust magmatic melt resulting from palingenesis in orogenous (collisional) environment—that is, in geodynamic compression conditions, in the region of thick continental crust, without the participation of primitive mantle liquid. Crust origin of the formation rocks is also confirmed by the isotope ratios of $^{87}Sr/^{86}Sr = 0.7065$.

The **Trachyandesite-teshenite formation** is from the Middle Miocene. The late collisional period of the region development witnessed considerable reduction in sedimentation basin and appearance of the Miocene evaporate facies in the Nakhchivan trough cross sections. The Abrakunis thickness (300–500 m) was formed during the Oligocene development stage of this intermontane trough as lagoon-continental reddish brown volcanomictous rough molassic deposits, subrounded blocks (up to 0.8–1.1 m) of granitoids of the Megri-Ordubad batholiths and the Lyaky-atakh intrusive. This thickness is everywhere conformly overlapped by the Upper Oligocene–Lower Miocene red beds (100–160 m), which makes its Oligocene age origin certain.

Unilateral graben-like plunging and stretching are accompanied by the volcanic activity manifestation at two Nakhchivan trough sites: in the eastern Gashirdagh-Ashabi-Kahf-Gyrkhlardagh area, and at the western site between the Arpachay River and Sadarak village.

The Gashirdagh complex is composed of volcanic outbursts, alternating with the flows of trachyandesite block breccias and agglomerate tuffs. On the Gyrkhlardagh Mt., the only lava flow of trachyandesibasalts participates in the complex structure. The complex petrostock also contains small intrusive teshenite bodies of Arazin, Kerimkulidiz, as well as Babekgalasy teshenite dolerits, etc.

The Saradak complex has sedimentary tuffaceous lithofacies and is represented by the alternation of copious flows (up to 55 m) of trachyandesite tuffs and their agglomerate differences and the Tarkhan-Chokrak argillo-arenaceous deposits of 160–265 m total thickness.

The trachyandesite-teshenite formation petrostock is the most abundant in pyroclastic rocks, represented by non-sorted block-agglomerate volcanic breccias and ash crystalloclastic tuffs. In the Gashirdagh complex, pyroclastic facies are occasionally interrupted by effusive facies as short flows of lavas, lava breccias or autoclastites, representing, the trachyandesite series rocks with deviation toward andesi-dacites, on the one hand, and trachyandesibasalts, on the other hand.

Small intrusives of this formation are composed by alkaline and subalkaline rocks, with marked presence of amphibole teshenites of Arazin, Gashirdag, and Kerimkulidiz intrusives. The trachyandesite-teshenite formation rocks fall under subalkaline, and rarely, alkaline (teshenites) types as high-alumina and very high-alumina differences of potassium-sodium series.

The formation volcanic rocks are characterized by decreased contents of incoherent Li and Rb and by increased contents of alkali-earth Ba and Sr. Low values of the iron group elements (Ni, Co, Cr, V, Cu) appear to be associated with peripheral focus depletion, resulting from the contamination by magma differentiation.

Petrogeochemical peculiar features of the formation rocks reflect its link to peripheral focus, where contamination of primitive alkaline-basalt magma by crust material and its differentiation took place. Various types of borates as well as copper, radioactive, and rare metal mineralization are related to this formation.

The **Andesite formation** is from the the Early Pliocene. The volcanism Pliocene—Quaternary stage was manifested in surface conditions due to the Meotian stage of tectogenesis in post-collisional conditions. Volcanogenic formations of this period are represented by the "Bichenag 900–1000 m thickness" that comprises the Zangazur and the Daralagez Ridges (Fig. 4.10).

The thickness structure demonstrates multiple alterations of lavas and pyroclasts, intruded by extrusive domes of Araji, Jindagh, Zernel, by dykes and subvolcanic intrusives. The formation petrostock composition and texture are represented by various volcanic breccias and mudflows (lahars). Andesite and andesibasalt coarse and fine-medium-clastic and block volcanic breccias are abundant. In lava facies, breccia-like andesites and andesidacites are prominent by their textured image, and occasionally, by their lava breccias with rock fragments from previous eruptions. At the eruption final stages, pyroclast outbursts and, to a lesser degree, lava outflows, differenting their composition from olivine basalts to andesites, play major role.

The rocks of vent or extrusive and subvolcanic-intrusive facies occupy insignificant volume in the formation. Among these latter, the Kyuki intrusive is distinguished composed of

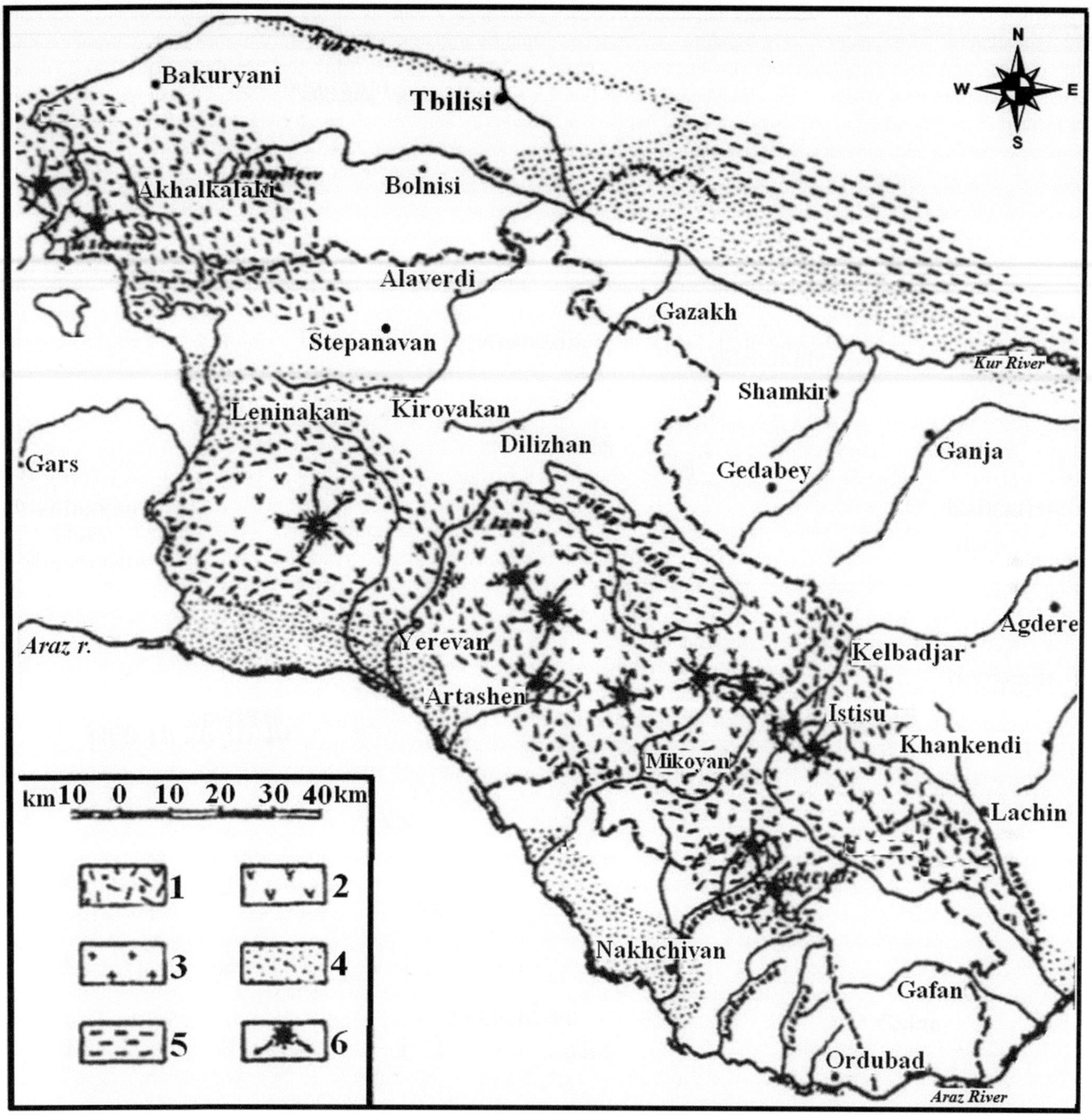

Fig. 4.10 Map of the initial distribution of Early Pliocene volcanic associations (composer: M.I. Rustamov) (*1*) piroclastics rocks (volcanic breccia, tuffs, and some others) mainly of andesitic, sometimes of dacitic, liparitic, andesite-basaltic composition with separate flows of andesites, andesite-basalts and basalts, (*2*) lavas of predominantly andesitic and andesite-basaltic composition with interlayers of piroclastolites, (*3*) lavas of liparitic, dacitic and andesite-dacitic composition, (*4*) tufogenic material in alluvial and proluvial deposits of depressions, (*5*) the same, in lacustrine sediments, (*6*) centers of volcanic explosions

biotite-hornblende andesites, as well as some small peak-like protrusions, the Gualingai, Kechili, Dibakly, Kyzylkai intrusions that serve as roots of one-time lava outflows of andesites and andesite-dacites at the large stratovolcano field. At the Turkish trough in various denudation levels, the Araji stratovolcanoes composed of andesites are prominent; in Jindagh, they are composed of andesites, Zernel volcano with extrusive andesite-dacite dome; and the Batabagh trough hosts stratovolcano that erupted basalts and andesibasalts.

Petrochemically, the formation rocks fall under normal and subalkaline differences corresponding to toleite, calk-alkaline and high-potassium calk-alkaline series. The formation predominant rocks of basalt-andesite-dacite line are located at the boundary of calk-alkaline series fields. The rocks of extrusive and subvolcanic facies such as dolerites, andesites, and andesite-dacites also belong to these latter.

The formation rock geochemical features are characterized by the increased content of Ba, V, Cu, Pb, Mo, Ga, Ta, and by the decreased content of Li, Rb, Sr. Ni/Co ratio indicates the relationship between peripheral foci and primitive basalt magma. The rare earth elements (REE) spectrum displays regular reduction of element concentrations from light lantanoids to heavy ones with noted Eu anomaly ($Eu/Eu^x = 0.42$) that testifies to the role of the crust sialic matter by generation of peripheral volcanic foci; it results in continuing differentiation of basalt and andesite melts and to the replacement of toleite trend of magma crystallization by the calk-alkali trend.

The **Andesite-dacite-rhyolite formation** of the Mio-Pliocene was isolated by Kashkay et al. (1950) in the Kyalbajar trough as the "Basarkechar" suite. Later studies by Kashkay and Mamedov (1961), Kashkay et al. (1964) and Allakhverdiyev (1967) isolated the Agjagyz and the Basarkechar thicknesses in the Mio-Pliocene formation composition. The Agjagyz thickness (the Late Sarmatian) consists of the volcanites of middle and acidic composition and of sedimentary layers (350 m), while the Basarkechar thickness is composed of megaplagioporphyric andesites, andesite-dacites, quartz latites and their pyroclasts. Subvolcanic facies are represented by laccoliths and stock-like bodies of andesite-dacites and rhyodacites of Nardivan, Keti and Gyzylkay. The formation rocks are characterized by hydrothermal metasomatic processes expressed in acid rock propylitization and silification. Secondary quarzites are the most developed rocks. According to Mamedov (1974), andesites, trachyandesites, quartz latites, dacites, and rhyolites are isolated in the composition of andesite-dacite-rhyolite formation.

Following from Imamverdiyev's (1988) studies, the Neogene volcanic formations of the Kyalbajar trough can be subdivided into the Mio-Pliocene trachyandesite-trachyrhyolite series and the Quaternary trachyandesibasalt series. Most abundant in the formation petrostock are andesites, trachyandesites, and quartz latites. The rocks' petrochemical parameters permit is to consider them as corresponding to normal and subalkaline series of calk-alkali line, close to the subalkaline latite series.

Basing on geochemical specialization of the formation rocks, the alkaline such as Rb, Li and alkaline-earth elements such as Ba, Sr play the subalkaline role. With silica increase the contents of the former increases and of the latter decreases. Contents of the iron group elements in acid and middle differentiates is slightly higher than reference values for andesite. Petrochemical analysis of the Neogene andesite-dacite-rhyolite formation rocks shows that the rock diversity is related to basalt magma fractional crystallization in the intermediate foci.

The **Rhyolite formation**, from the Late Pliocene, is developed in the Lesser Caucasus central part, in the Kyalbajar trough, along the Terter abyssal fracture. Following from the research by Paffengoltz (1934), Kashkay et al. (1950), Mamedov (1974), Makhmudov (1974), Ismailzadeh and Geidarov (1996) et al., the rhiolite formation is the completing in the Oligocene-Miocene volcanism of andesites, andesibasalts, and andesite-dacites. It is represented by volcanic domes, extrusive formations, as well as by lavas, nappes and, less frequently, pyroclasts.

The most significant Kyalbajar trough volcanic domes are Kechaldagh and Small Deve-gezu, composed of rhyolites that pass up into obsidians, perlites and pumices. Rhyolites with spherolite structure are the intermediate differences between obsidians, crystalline rhyolites, and rhyodacites.

Based on water contents and types as well as on the structural and textural features, rock glasses can be subdivided into four varieties: obsidians (up to 1 % of fossil water), obsidian-perlites, perlites (more than 1 % of water, with perlite structure), and pitchstones. Pumices, the naturally expanded perlite light differences, are found as fragments and blocks in the Kechaldagh and Large Deve-gezu volcanic domes. Petrochemically, according to Marakushev (1979) and Mamedov (1981), acid rock glasses are not the trend continuation of basalt-andesite-rhyolite series but represent potassium and sodium types.

Rhyolite rock glasses are subdivided into the potassium-sodium, sodium, and potassium groups with genetic base. This formation corresponds to the normal series, and according to FeO/MgO ratio, to the calk-alkali type; with regard to Na_2O/K_2O, it corresponds to the K–Na series. Based on geochemical characteristics, alkaline and radioactive elements are contained in the formation rocks at the bulk earth value level for acid rocks; however, low Li, Zr, Mo, Pb, Sn, and Cu values are peculiar for the formation and also do not exceed bulk earth values for acid rocks; Cr value is higher than the bulk earth value, Ni and Mg values are close to it. Among the REE, Y and Yb are also close to the bulk earth values.

Researchers (Mamedov et al. 1974, 1981) have found that generation of rhyolite formation rock glasses is associated with acid melt polymerization. Analysis of the Neogene formations allowed us to reveal that, according to the manner of geochemical element distribution, they are close to the Paleogene formations. Thus, increased values of radioactive elements typical for the formation of calk-alkali series, combined with close values of alkaline and alkaline earth values, reflect their relation to the residual Paleogene magmatic foci.

4.4.2 The Late Pliocene–Quaternary Formation

The **Trachybasalt-trachyandesite formation**, from the Late Pliocene–Quaternary period, developed in Azerbaijan in the Kyalbajar and Gochasy troughs. Quaternary nappes are found within the first one, and Late Pliocene–Quaternary nappes within the second one.

A large team of geologists—A.T. Akhverdiyev, G.I. Allakhverdiyev, G.L. Aslanov, N.A. Imamverdiyev, A.D. Ismailzade, M.A. Kashkay, S.A. Makhmudov, A.I. Mamedov, M.N. Mamedov, E.Sh. Shikhalibeyli—dealt with the issues of regional geology, tectonics, and volcanism. Previous researchers have separated calk-alkali and subalkaline series, represented by andesite-trachiandesibasalt formation at the Kyalbajar plateau. Later researchers (Imamverdiev 1988; Ismailzadeh et al. 2005) also separated the trachybasalt-trachyandesite formation.

Topography of volcanic edifices shows heights of 200–350 m, with crater size of 200–250 m. It is composed of lavas and pyroclasts. The volcanoes are polygenetic (less frequently, monogenetic) and are practically not exposed to secondary transformations.

Trachyandesites, quartz latites, trachyandesibasalts, latites, trachydolerites and subalkaline basalts are distinguished in the formation petrochemical content (Imamverdiyev 1988).

The formation rocks show the inclusions of deep-seated rocks and overpressured megachrists, represented by clinopyroxenites, amphybolites, gabbroids, quartz-feldspar rocks, quartzites, as well as by megacrysts of sanidine, pyroxene, amphibole and quartz that are the Earth crust xenoliths and xenochrists. Unlike enclosing rocks, deep-seated formations have higher TiO_2, FeO, MnO, MgO, CaO values reflected in their femicity, $f = 14–36$.

Petrochemical characteristics of the formation rocks correspond to poorly differentiated K-Na subalkaline series, differing from typical calk-alkali series by its high alkalinity and ferruginous properties, and from K- shoshonite-latites, by low K_2O/Na_2O ratio. The series peculiar feature is the transfer from olivine-norm composition of basic differences to hypersthene-norm composition of the middle ones, and ferruginous property increase with SiO_2 rise.

Geochemically, the rocks are characterized by high contents of light REE as well as of Sr, Ba, Zr, F, P and U. It should be noted that such specificity of subalkaline basaltoid volcanites is inherent in the entire Lesser Caucasus eastern part.

Petrochemical studies of molten inclusions determined the formation rock generation during crystallizational differentiation of primitive subalkaline olivine-basalt magma (Imamverdiyev 1988).

4.4.3 Comparative Analysis of Series of the Lesser Caucasus Quaternary Volcanic Formations

Manifestations of anthropogenic volcanism in the Lesser Caucasus show certain regularity that reflects structural features of a more ancient structural level, complicated by transversal structures, the most essential of them being the Transcaucasian transversal uplift. These regularities, known from the literature (e.g., Tolstov et al. 1980) and from our studies (Ismailzadeh 1986), are summarized as follows.

1. Development of the Late Pliocene–Quaternary volcanism took place on the block foundation that had been broken up already at the epi-Baikal period of tectogenesis and that experienced variable-sign differentiated movements during the Alpine time.
2. Distribution of volcanic formations demonstrates the predominance of non-differentiated olivine basalts and dolerites in the Transcaucasian Uplift north, of andesite-dacite-rhyolites, in the south, of trachybasalts and trachyandesibasalts-andesites and, locally, dacites of normal and increased alkalinity, in the central part of the Ankavan-Ishykhly zone, and of vasanites, in the southeast.
3. Distinct tendency of large-ion lithophil increase in the northwest to southeast direction is observed. They are alkaline, and, particularly, K, alkaline-earth, radioactive, as well as increase in light lanthanides, and a corresponding decrease in the iron group elements and in heavy lanthanides.
4. Petrogeochemical peculiarities of volcanic formation basalts reflect their correspondence to magmatic melts of varying melting levels under different geodynamic conditions.
5. The trend of geochemical element distributions in the rocks finds its correspondence in magmatic chamber, under the conditions of its elevating location level in consolidated crust and under the conditions of its matter involvement in magma generation. The chamber is physically and chemically tight and promotes concentration of abyssal fluids, primarily alkali.
6. Volcanism proceeded in the conditions of various Earth crust stress states, which resulted in the emergence of mantle transcrust basalt melts, such as dolerite basalts (Lori-Kechut), in the region western part (the

Transcaucasian Uplift) under pulling stress and in the emergence of andesite melts (the Gegam, Vardenissi, Kyalbajar, Ishykhly uplands) in the eastern part. They were formed in the earth crust under compressing efforts. The Kafan block basanites seem to be generated as a result of brief mantle melt interaction with crust matter in the intermediate level, which brought to the formation of mineralogical differences such as olivine and hornblende basanites

7. Complicated geodynamic conditions of volcanism manifestation at a sufficiently limited site and in narrow time interval reflect the inhomogeneity of crust-mantle processes by generation of folded-blocked areas at the final stages of collisional development period.

4.4.4 Rock and Megacryst Inclusions in the Lesser Caucasus and Talysh Volcanogenic Complexes

Study of plutonic inclusions (plutonic xenoliths) of rocks and megacrysts has fundamental importance for the clarification of structures and compositions in lower strata and upper mantle that encompass the magma generation regions at continental margins and under island arcs.

In the Lesser Caucasus, inclusions and megacrysts were found in the Paleogene subalkaline complexes of Ajaro-Trialtia and Talysh and in the Late Pliocene–Quaternary calk-alkali and alkali complexes of the Lesser Caucasus volcanic highland (Kyalbajar trough of Azerbaijan).

In the Talysh zone, plutonic inclusions are noted in the manifestation areas of the Eocene volcanism early phases (the Kosmalyan trough) such as dykes, subvolcanic formations, lava breccias of trachyandesites and trachyandesibasalts (Ismailzadeh et al. 1982). Their composition is represented by amphibolites, feldspar pyroxenites, pyroxenites and alkaline basalts (from 2–3 to 20 cm). Along with rock inclusions, the inclusions of biotite and amphibole megacrysts were noted (1–3 cm).

Interactions of deep-seated formations and enclosing rocks are expressed by the temperature action border rims as thin (2–3 mm) interrupted dark cryptocrystalline mass. Plutonic inclusions are characterized by holocrystalline middle-large-grain structure. Their structures are allotriomorphic, pseudoporphyrytic, occasionally hypidiomorphic-granular (pyroxenites); their textures are taxite.

Petrographic and chemical investigations permitted to establish the predominance of dark-colored components over light-colored ones and the predominance of ferro-magnesian constituent over calcium one. Under saturation of inclusion rocks by silica was also revealed. Trace element distribution in amphibolites, pyroxenites and feldspar pyroxenites shows that certain regularity is observed with relatively similar content of lithophylic (Ba, Sr, Mn, Ti), chalcophylic (Ga, Cu), and siderophylic (Co) elements that reflects the xenolith composition: increased contents of V and Ni are noticed in pyroxenites and feldspar pyroxenites, and high contents of nearly all above mentioned elements are found in alkaline basalts.

Plutonic inclusions and megacrysts in the Quaternary volcano lavas of the Kyalbajar depression are emphasized in the publications of Solovkin (1939), Mamedov et al. (1986), Ismailzadeh et al. (1982). Inclusions and megacrysts out of the Quaternary andesite and andesibasalt volcanic lavas and cones of the Terter River left bank are represented by gabbroids, less frequently, by pyroxenites and amphybolites, while megacrysts are represented by feldspars, pyroxenes, amphiboles, and quartz. The composition of inclusions in the Talysh Paleogene alkaline-basaltoid formation is comparatively uniform, and more diversified, with large quantity of leuco- and melanocratic types of inclusions in the Lesser Caucasus Quaternary subalkaline and calk-alkaline formation. The presence of similar petrographic differences of inclusions in volcanites of differing facies and compositions as well as the presence of thermally differently transformed set of inclusions in the same enclosing rocks are indicative of their xenogenic nature.

The Kyalbajar depression in the Late Pliocene–Quaternary formation also demonstrates cumulative or restitic inclusions resulting from melting of intracrust andesite magma. Unlike xenoliths, they are equilibric with melt (part of gabbroids and pyroxenites).

The Earth crust formation in the regions under study can be represented as matter basification under the amphibole facies progressive metamorphism, when melting foci emerge with ascending thermal flux. In this process, parental magma for the Talysh alkaline-basaltoid formation corresponded to melting from sub-crust level with subsequent fractional crystallization in the Earth crust consolidated part (inclusions of amphibolites and pyroxenites), while for the Kyalbajar depression calk-alkaline and subalkaline formation, it was direct melting from the Earth crust consolidated part (inclusions of cumulative or restitic gabbroids, less frequently, amphybolites and pyroxenites), but in both cases, with the participation of mantle alkaline fluids.

4.5 Regularities of the Cenozoic Magmatism Manifestations

Investigation of the Cenozoic volcanic formations permitted to find extensive areal (belt) distribution and, simultaneously, decrease in manifestation scope and differentiation level throughout the Paleogene–Neogene–Quaternary periods. The results of petrographic and multi-dimensional statistical

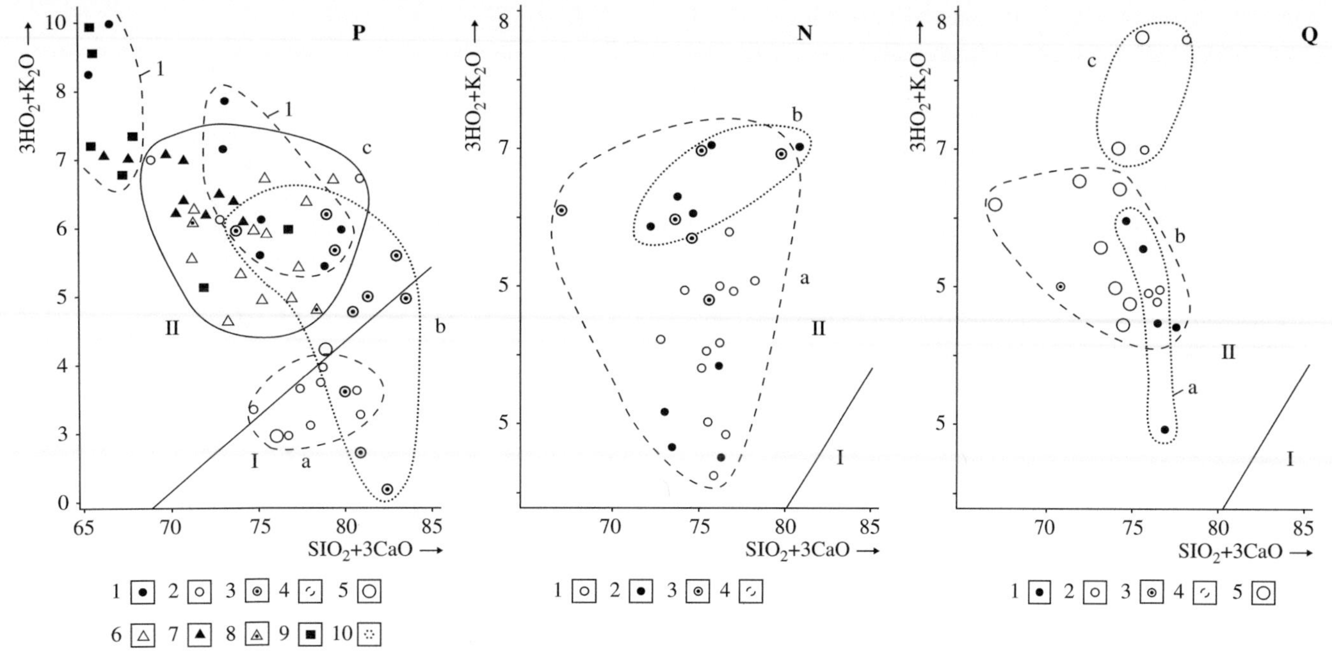

Fig. 4.11 Separation of Cenozoic rocks of volcanogenic formations to the island arc and continental-margin types on the basis of petrochemical parameters (Piskunov 1987). ***Paleogene*** (***P***): (*a*) tholeiitic basalts of: (*1*) Yerevan-Ordubad zone, (*2*) Adzharo-Trialet zone; (*b*) subalkaline basalts: (*3*) Paradasht depression, (*4*) Kelbadjar depression, (*5*) Shakhdagh depression, (*6*) Goycha depression; (*7*) Gazakh depression, (*8*) Talysh depression, (*9*) Goycha-Shirak depression, (*10*) field of distribution, (*c*) trachybasalts of Talysh zone. ***Neogene*** (*N*): (*a*) andesites and trachyandesites: (*1*) Nakhchivan Depression, (*2*) Bichenag plateau; (*b*) trachyandesit-dacites of: (*3*) Kelbadjar depression, (*4*) Daralagez uplift. ***Quaternary*** (***Q***): (*a*) subalkaline dolerites of Lori-Kechut subzone, (*b*) trachyandesites and trachyandesito-basalts of: (*2*) Gegam subzone, (*3*) Kelbadjar subzone, (*4*) Ishikhly subzone; (*c*) basanites of (*5*) Kafan subzone. Fields of: *I* island arc associations, *II* continental margin associations

analysis of the Cenozoic formation rock compositions revealed their correspondence to various-alkalinity series and to isolation of groups with regionally pronounced peculiarities that are developmentally transformed during collisional and post-collisional stages of the region's development (Fig. 4.11).

Petrogeochemical compositions of the formation basite rocks allowed us to differentiate typomorphic peculiar features that reflect the difference in primitive liquids and in formation conditions. Distribution analysis of petrogenic and rare elements in the Cenozoic volcanites revealed their inherent features of petrogeochemical evolution associated with the mantle melt transformation under crust conditions:

- in the Paleogene basalts, from topeite and calk-alkali series to subalkaline series, the increasing values of lithophylic elements with regard to siderophylic reflect the enhanced role of assimilation and contamination in primitive mantle liquid;
- in the Neogene calk-alkali and subalkaline differentiates, the manner of magmatism evolution inherited from the Paleogene period with increased crust process intensity is observed;
- in the Quaternary formations, the relationships of rare elements reflect the mantle-crust nature of primitive liquid, but without significant differentiation in crust foci.

4.6 Geodynamics of Azerbaijan Alpine Magmatism

4.6.1 Geodynamics of the Mesozoic Magmatism

The publications of numerous researchers, such as Sh.A. Adamia, O.A. Bogatikov, A.D. Belov, T.Ab. Gasanov, A.D. Ismailzade, Yu.V. Karyakin, A.L. Knipper, V.E. Khain, G. V. Mustafayev, M.A. Mustafayev, A.S. Ostroumova, M.I. Rustamov, S.D. Sokolov and others, are dedicated to various geodynamical aspects of Alpine magmatism in the Caucasian Azerbaijan sector and are based on modern concepts of lithospheric plate tectonics.

The Alpine cycle within Azerbaijan commences from the active Lower Jurassic magmatism in the Greater Caucasus southern slope, in the Main Caucasus (Tufan) structural zone, and in limited scope, in the Nakhchivan zone–the Lesser Caucasus extreme south.

The Mesozoic Alpine plunging and accumulation of kilometer-deep thicknesses of sedimentary terrigenic matter at the southern slope is inherited from the Paleozoic age. However, manifestations of the Early Jurassic basic magmatism as sodium basalts (spilites) along the main Caucasian abyssal fracture and along its feathering faults predetermine the establishment of the abyssal trough at the Greater

Caucasus marine basin. It is witnessed by huge thicknesses of the southern slope Jurassic terrigenic-sedimentary deposits that reach up to 10 km (Kurbanov 1982). It should be emphasized that no magmatism was found to be present by the region Late Permian plunging and accumulation of 2.4-km thickness of sedimentary rocks (Khain and Seslavinsky 1991). In the Lower Jurassic period, active submarine basalt volcanism with pillow lavas was displayed. By their petrochemical parameters they correspond to low-potassium toleites of ensimatic island arcs that generally constitute the incipient base of island arc systems, their young stage. The subsequent Middle Jurassic basalt-andesite-rhyolite and Upper Jurassic gabbro-plagiogranite formations also match the island arc toleite series, basing on their $SiO_2/(Na_2O + K_2O)$, $SiO_2/(FeO/MgO)$ ratios and other petrochemical parameters.

In the Nakhchivan zone, the Lower Jurassic magmatism manifested itself locally, outcropping along the abyssal fracture in a number of points. Its composition corresponds to magmatic derivatives of subalkaline olivine toleites with high titanium content (TiO_2 = 1.2–3.4 %). According to some of their petrochemical parameters (TiO_2/K_2O, FeO/MgO, etc.), basalt compositions occupy intermediate positions between the island arc trends and the continental rifts, with higher tendency to these latter, and it is well correlated with vulcanite geological positions that actually correspond to the continental rifts (Fig. 4.12).

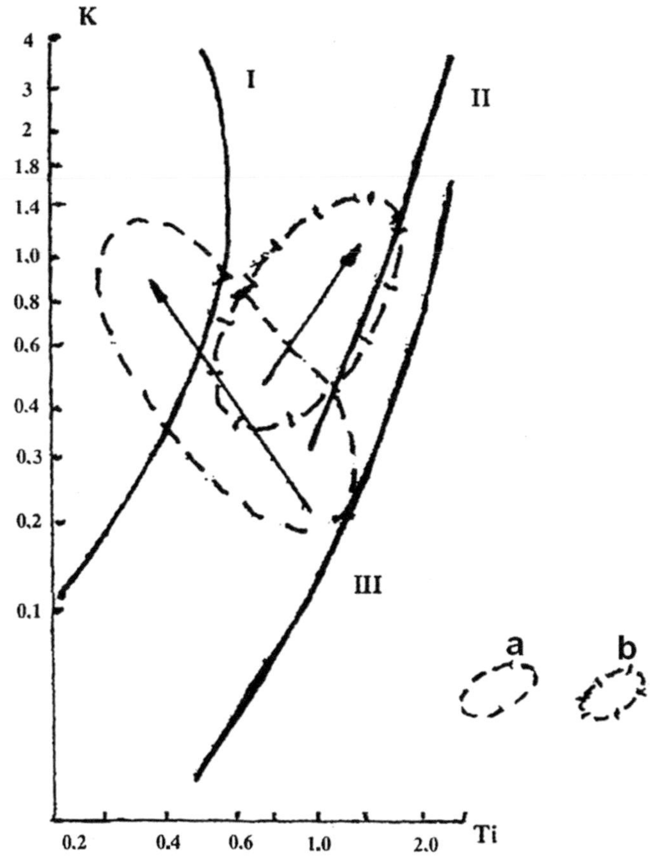

Fig. 4.12 Fields of associations of Jurassic basalts of the southern slope of the Greater Caucasus (*a*) and Nakhchivan zone (*b*). Composition trends (according to Lutz 1980): *I* island arcs, *II* continental rifts, *III* middle oceanic ridges

It should be noted that inversion of a certain structure of both the Greater and the Lesser Caucasus took place during the Middle Jurassic (Bajocian) period. As a result, in the southern slope Tufan zone, sodium basalts are replaced by the medium-acidic volcanites of andesite-dacite-rhyolite formation, and magma composition corresponds to the calk-alkali and toleite series. At the Lesser Caucasus, the inversion also results to the replacement of toleite basalts by calk-alkali magma products similar to the island arc ones. In both cases, magmatism products form a homodromic series, and the differentiation trend is directed from the basic to acidic rock types, which is confirmed by their petrographic composition and petrochemical parameters.

The duality of petrochemical parameters that correlate either with toleite or with calk-alkali series appears to evidently reflect the complex geodynamic environment for the manifestation of the Middle Jurassic magmatism in the Lok-Gafan structural and formational zone that generated in the transition area from oceanic to island arc environment.

Numerous discovered intrusives of the Upper Jurassic–Lower Cretaceous age are grouped (Mustafayev 1991) into two formations: the gabbro-tonalite (Gedabey, Mekhmana, Barum, etc.), and the gabbro-granite (Dashkesan, Uchtapa-Gyzylkay). The intrusive early gabbroid phases reflect magmatic toleite composition, while later granitoid phases reflect its calk-alkali composition.

By the beginning of the Upper Cretaceous period, continental crust formation was at its termination. The Upper Cretaceous period magmatism that occurred in certain isolated structures, such as the Lesser Caucasus transversal depressions, was already imposed upon the Jurassic foundation. It is characteristic that despite the manifestation, independent magmatism of these structures (Gazakh, Agjakend, and Khojavend, Gadrud, Gochass) possesses common petrographic rock types and petrochemical parameters. They are also united by the fact that, during the Upper Cretaceous age, magmatism of these structures essentially showed itself in the effusive facies. In the northwest structures, in the Gazakh and Agjakend trough, the basalt-andesite-rhyolite formation was generated during the Upper Cretaceous age (with antidromic differentiation trend in the Agjakend trough). In the southeast structures (the Khojavend, Gadrud, Gachassy troughs), along with calk-alkali eruption products, subalkaline and, less frequently, alkaline rocks found extensive occurrence and formed the trachybasalt formation.

According to Ostroumova et al. (1991), the products of two petrochemical series are noted in the Gochass trough:

highly sodium—alkali-olivine-basalt, and subalkaline—basalt-andesite-trachyte series. According to Abdullayev and Mustafayev (1995), the entire integrity of rocks is the product of single alkaline olivine-basalt magma. Unlike the Gazakh and Agjakend troughs, the geodynamic environment of the Gochass, Khojavend, and Hadrud (Aghoghlan) troughs is characterized by vigorous stretching, which might be associated with tectonic movements. According to Karyakin (1989), this dissects the Lok-Gafan paleoisland arc into two subzones. For this reason, magmatic evolution of these troughs is comparable to those of oceanic islands and continental rifts. Geological arrangement of these structures makes plausible their comparison with continental rift generation specifically. It should be marked that the Goycha-Hakeri zone that admittedly originated during the Upper Jurassic–Lower Cretaceous age is also characterized by the association of the Upper Cretaceous alkaline and low-potassium basalts, and its petrochemical parameters are similar to those of intraoceanic islands and volcanic series of primitive island arcs (Zakariadze et al. 1982).

During the Mesozoic stage, geodynamic environment of the intramontane Kur Depression, another Greater Caucasian structure, corresponded to those of island arcs and active continental margins (Ostroumova et al. 1995) (Fig. 4.13).

Comparison of the Kur trough's Mesozoic magmatism with those of the Greater and Lesser Caucasus complementary structures, conducted by Abdullayev and Salakhov (1983), showed that geological and petrological peculiarities of the Kur Depression's Upper Jurassic–Lower Cretaceous and Upper Cretaceous magmatic formations are quite comparable with the same-age vulcanites of the Lesser Caucasus northeast slope. These regions might have been a single geotectonic element during the Mesozoic era (Fig. 4.14).

Kremenetsky et al. (1990) compared rare earth element (REE) distribution in SD-1 volcanites as contrasted to the reference one, corresponding to various geotectonic conditions. They noted that early basalts (basalt thickness) are similar to the island arc toleites, whereas late basalts (basalt thickness I) and their comagmatic intrusive formations are similar to continental toleites. The youngest andesibasalt sills are also close to continental basalts. Based on the cited data, Kremenetsky et al. (1990) argued that evolution of geotectonic conditions in the investigated crust block occurred in the direction from island arcs to continental one. Affinity of vulcanite series (separated by him) to the island arc magmatic formations is proved by SiO_2–TiO_2/P_2O_5 diagram and by the values of isotope primary relationship: $^{87}Sr/^{86}Sr$—0.7032, for the I basalt rocks, and 0.7039, for andesibasalt thickness. However, it is further maintained that the Kur Depression experienced rift-related conditions. Kremenetsky et al. (1990) differentiates three large stages: the pre-rift, the early rift, and

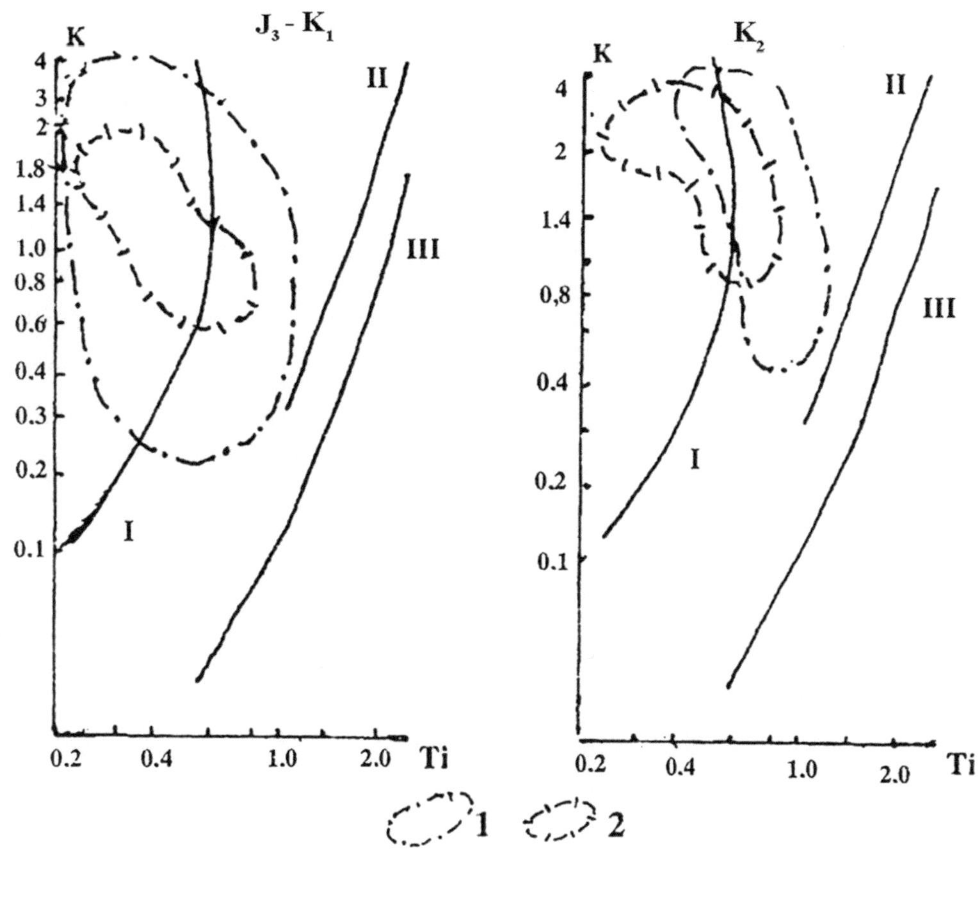

Fig. 4.13 Composition of basalts on the Dobretsov's diagram (after Abdullayev and Salakhov 1983). Magmatic formations: (*1*) Kur Depression, (*2*) Lesser Caucasus. Composition trends (according to Lutz 1980): *I* island arcs, *II* continental rifts, *III* middle oceanic ridges

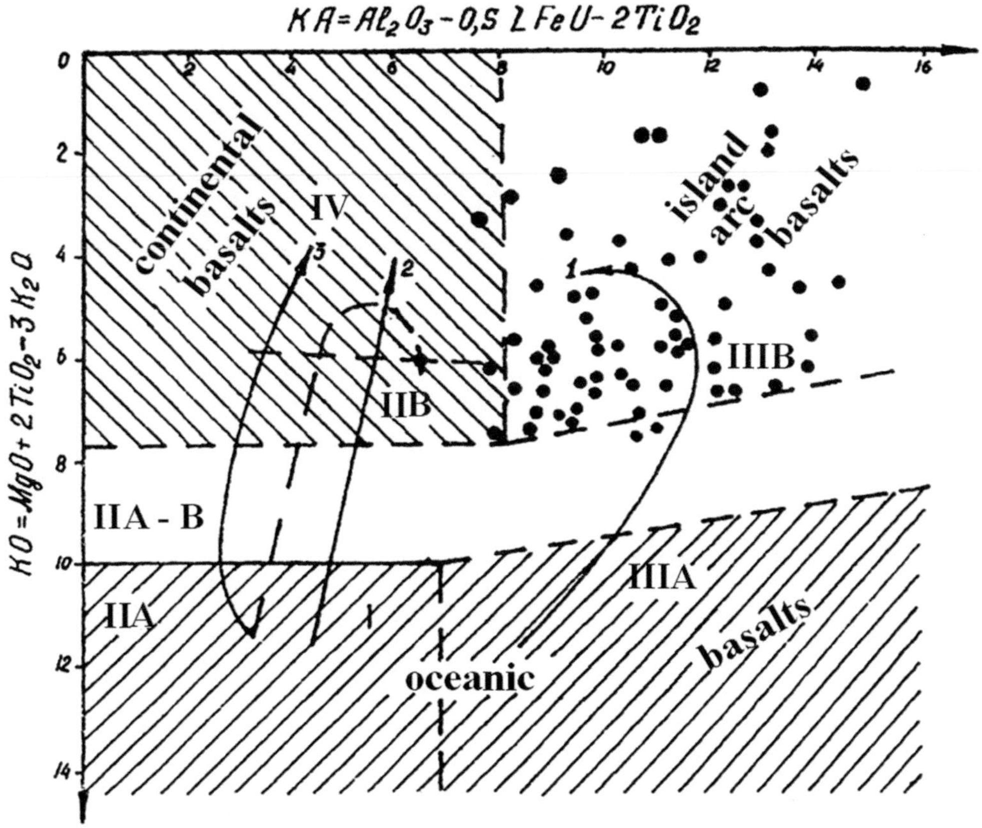

Fig. 4.14 Fields of vulcanite associations of the Lesser Caucasus (*1*) and Kur Depression (*2*) of Jurassic and Late Cretaceous in the diagram K–Ti. Composition trends (according to Lutz 1980): *I* island arcs, *II* continental rifts, *III* middle oceanic ridges

the late rift one, in the Kur Depression development history that was established on sialite foundation of deep-seated middle massif (Ismailzadeh 2009).

The pre-rift stage (Pz-T) was manifested in differentiated movement of large foundation blocks, specifically in uplift and denudation of the Middle Kur Depression's Saatly-Geokchay block. At the early rift stage (J) during the Jurassic period, rift-related structures filled with dacite-andesite-basalt series of mantle volcanites were being established. The antidromic trend of vulcanite development is explained by magma contamination by the crust sialite matter. As a result, the thickness geochemical data become similar to the island arc series. The late rift stage (K) that encompasses the Cretaceous period is characterized by cardinal reconstruction of the Kur Depression's structural geometry. During this historic period of geological development of the Kur Depression and of its framing Greater and Lesser Caucasus mobile belts, similarly composed vulcanites of calc-alkali and subalkaline series of main and middle compositions are being generated. The late rift stage culminates in active folding in the Greater and Lesser Caucasus mobile belts, and by poor folding in the Kur Depression paleorift. The post-rift stage (*P*) witnesses further plunging of the Saatly-Geokchay block, development on depression thereupon, filling with terrigenic deposits. Tectonic activity of this period is recorded as dyke and sill intrusion into the Lower Cretaceous vulcanite thicknesses and into overlapping limestones.

Summing up the research results for the Mesozoic effusive and intrusive formations in Azerbaijan territory, which actually encompasses the entire magmatism development region of the Eastern Caucasus, the following conclusions can be made.

Magmatic rocks, by their predominant lithofacial composition—and, mainly, by their age—can be subdivided into the Lower Jurassic, Middle Jurassic, Upper Jurassic–Lower Cretaceous, and Upper Cretaceous formations. The integrity of their petrographic mineralogical and petrological geochemical characteristics allows us to isolate the toleite, calk-alkali, subalkaline and alkaline series of magmatites. Geodynamical environments of their manifestations in various structural and formational zones were dramatically different.

The Lower Jurassic magmatic complexes of toleite and basic subalkaline series were exposed at the Greater Caucasus southern slope under island arc conditions of ensimatic type, and in the Nakhchivan zone under the continental rift-related conditions.

Establishment of the Middle Jurassic complexes of toleite, calk-alkali and boninite series of the Lesser Caucasus and Kur Depression took place in the conditions of elementary geodynamic environments, with generation of island arcs of young and developed series on the suboceanic and continental crusts.

The Late Jurassic–Early Cretaceous complexes are characterized by inhomogeneity of both serial and paleogeodynamic environments, which is enhanced in the Upper Cretaceous era. It is characteristic that the Upper Jurassic–Lower Cretaceous intrusive formations bear similar features both with island arcs and with active continental margin, while the Upper Cretaceous formations are similar to island arcs, rift-related structures and intraoceanic islands. These similarities might be caused by various reasons. In one case, under magmatism antidromic development, it might result from the sialic crust assimilation, and in another case it may be explained by island arc attachment to the continents and their build-up. At least, duality of indications in the Upper Jurassic and Cretaceous magmatic complexes, their belonging to different geodynamic conditions testify to the complicated geodynamic environments of their generation or their change, when new hallmarks are added up to the inherited signs of early generation stage. All of these find their reflection in chemical activity and petrogeochemical parameters of magmatic complexes that assist in reconstructing the manifestation paleogeodynamic environments.

4.6.2 The Cenozoic Magmatism Geodynamics

In the Late Cretaceous period, the central Caucasian-Iranian segment of the Mediterranean orogenic belt has entered the collisional stage of tectonic-magmatic development. It was conditioned by the convergence of Afro-Arabian and Eurasian plates and by final closure of the Mesotethys oceanic basin.

Paleogene magmatism is manifested as volcanic-plutonic (VP) belts in the Lesser Caucasus, Garadagh-Talysh, in pre-Elburs region, in Central and Eastern Iran and around the Lut Ring (Fig. 4.15). The VP belts are abundant in the interior zones of the Mediterranean belt's central segment and are completely lacking in its exterior tectonic zones (the Greater Caucasus, Kopetdagh, Zagros, Urmiya-Khamadan, etc.). Paleogene VP belts have lengths exceeding 2400 km, being traced continuously from the Lesser Caucasus via Iranian Azerbaijan and Central Iran high mountains towards the Gulf of Oman. Thereby a large planetesimal lineament of northwest-southeast striking is emerging.

The Mesotethys basin with oceanic crust in the central Caucasian-Iranian segment was completely closed in the Late Cretaceous as a result of the Austrian phase of tectogenesis. The Late Senonian thickness (400–1000 m) limestones that span the entire segment area between these continents serve as an indicator of stabilization of the Early Alpine movements in tectogenesis Austrian phase. Because of that, following the Austrian phase tectogenesis, this segment, including the Lesser Caucasus, lacked zones or troughs with oceanic crust, were exposed to subduction, and thus terminated the initial stage of mature continental crust generation. Due to this process, the segment tectonic zones entered throughout the early collisional development mode—the magmatism of new Paleogene cycle.

At the Lesser Caucasus, contrary to the notions on andesite belt separated by ophiolite suture, researchers (Knipper 1975; Sokolov 1977; Rustamov 1979, 1983; Lordkipanidze 1980; Ismailzadeh 1990) believe that Paleogene VP belts are controlled by two ophiolite sutures as the abyssal fracture areas representing parallel sutures of the Mesozoic paleosubduction between three continental plates. The plates are the Transcaucasian Plate with Baikal-Hercynian foundation in the north, the Eastern Anatolian–Lesser Caucasian plate in the center, and the Nakhchivan block of the Iranian Plate Hercynian consolidation in the south. The southern VP belt has traced the Zangazur suture and the plate-arc collision type—namely, the passive continental slope of the Nakhchivan block and the island-arc slope of the Miskhan-Gafan block. The central VP belt has traced the Goycha-Hakeri suture and the arc-arc collision type; the Miskhan-Gafan and the Lok-Garabagh island-arc blocks of Mesozoic consolidation are conjugated here.

The southern VP belt passes into Iranian Azerbaijan, preserving a similar geotectonic position and collision type, and is controlled by the Garadagh ophiolite suture zone of the east-southeast strike. The Eocene magmatism of various compositions, predominantly of alkaline and shoshonite-latite series in Talysh and of calk-alkali and subalkaline series in Garadagh, manifested itself along the suture zone's divergent abyssal fractures of the Talysh and Garadagh flank slopes of a single rift-related trough. Talysh is the northwestern termination of the Elburs Mesozoic system.

Paleogene magmatism development sets opposite development tendencies along belt strikes. It has northwestern ranges in the southern VP belt and southeastern ranges in the central VP belt. The intensity of volcanism and plutonism, characterized by longitudinal migration of processes accompanied by tension and plunging reduction, decreased in the same directions. Paleogene cross-section of up to 4–4.5 km thickness is composed of volcanogenic, volcanogenic-sedimentary, tuffaceous-flyschoid thicknesses, and less frequently, low-thickness carbonate suites generated under shallow-water and offshore conditions.

Volcanic and intrusive formations of each VP belt constitute several independent regional formations consisting of separate same-age intrusive complexes and different-age volcanic complexes. The formation composition is subdivided into toleite (T), calk-alkali (CA), high-potassium

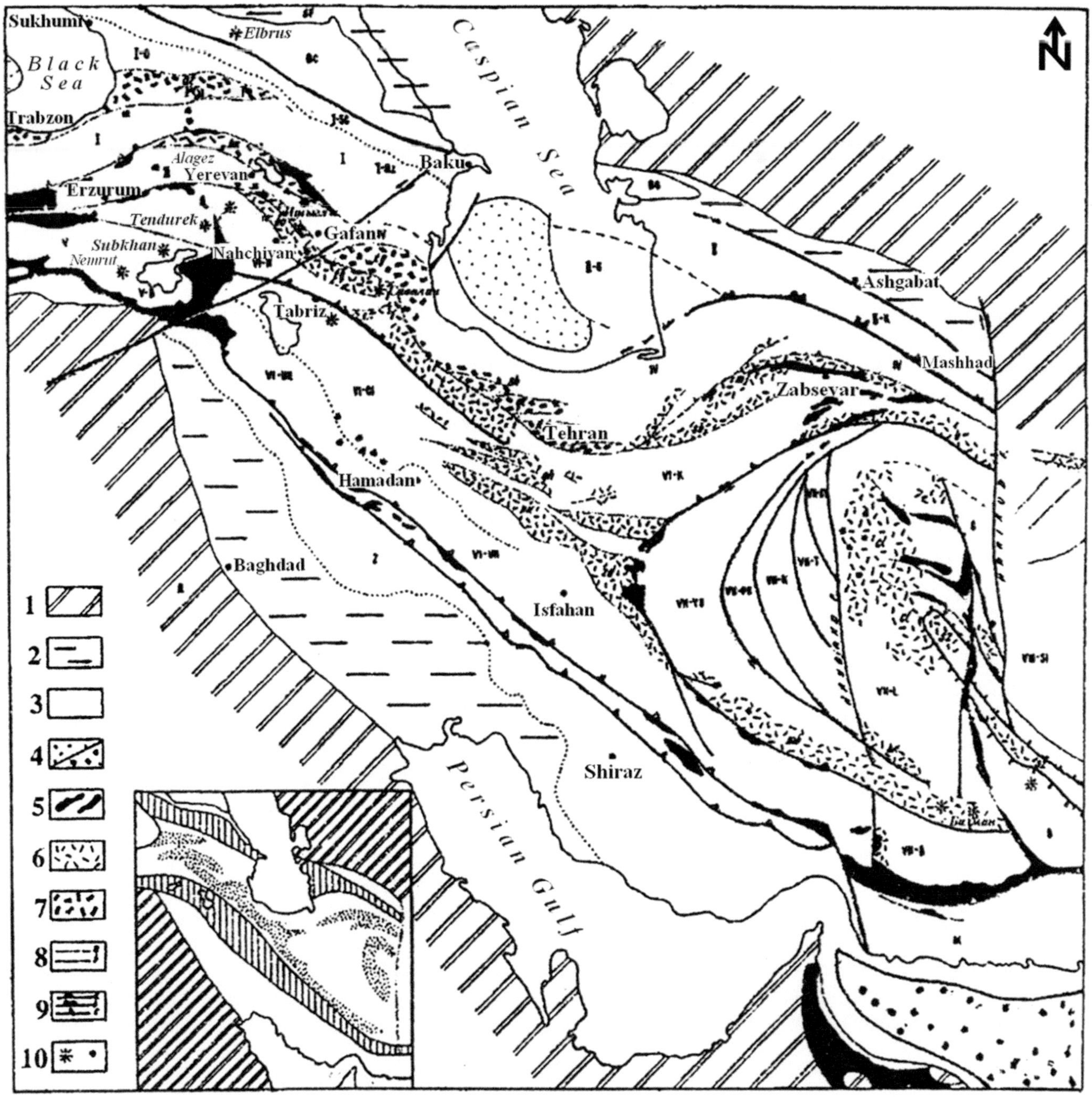

Fig. 4.15 Tectonic position of Paleogene volcanic-plutonic belts of the central segment of the Mediterranean tectonic belt (after M.I. Rustamov)

calk-alkali (subalkaline (SA), shoshonite-latite (SL), and alkaline (A) series, which, unlike the island-arc magmatism and the Andean type of active margins, are characterized by the diversity of same-age series, their spatial overlapping, and are devoid of lateral petrochemical zonation, having a unified structural tectonic control of eruptive centers. Petrochemical zonation is distinctly expressed in the course of time, and it is typical for the evolution of all magmatism geodynamic types.

Three types of the Paleogene VP belts are differentiated at the Lesser Caucasus and Talysh within the Caucasian segment:

(1) *Concordant belts*, where troughs are aligned parallel along the suture zones (Ordubad trough, Bazum-Kyalbajar trough). The following magmatism sequence is found for the Ordubad trough: the Lower

Eocene basalt-andesite-dacite-rhyolite formation with T, CA, SA, and rather rarely, A series. Volcanism culminates in the gabbro-granite constituent. The Middle Eocene trachybasalt-latite-trachyandesite formation is represented by the associations of shoshonite-latite-trachyte and trachybasalt-trachyandesite-trachydacite series of rocks, with local development of the CA series intermediate rocks. The Upper Eocene–Oligocene gabbro-monzonite formation is co-magmatic to it. During the trough formation inversion stage, the Paleogene magmatism ended in the intrusion of the Oligocene nephelinic syenites and the late Oligocene–Early Miocene pseudoporphyric granitoids of increased alkalinity and proper plutonic type that do not possess volcanic co-magmates.

(2) *Discordant belts*, where troughs are situated at acute angles to the sutures. In this case, despite the presence of plate-arc collision type, rifting is superimposed on the elevated island arcs with the Early Alpine consolidation (Talysh trough, Kazvin-Talegan, Bakharan, etc.). The Talysh rift-related depression is characterized by intensive manifestation of alkaline-basalt multistage volcanism, starting from the late Paleogene and to the Late Eocene, inclusively. A thick mass of volcanites alternating with flyschoid deposits (4 km) is represented by the alkaline series rocks (analcitites, phonolites, tephrite, basanites, etc.), in close spatial and suture relationship to volcanites of SL, SA, and CA series, sometimes with toleite basalts. As this takes place, the fractionation level in the association of all series rocks does not exceed the content of andesite level silica. During the inversion stage, magmatism ends in the generation of the Late Eocene–Oligocene gabbro-pyroxenite–gabbro-syenite formation (Fig. 4.16).

(3) *Concordant belts* with uncomforble flank troughs are typical for the regions, where arc-arc collision took place and where rifting lateral migration was noted (the Pambag trough, Shakhdagh, Mosakhet-Basman, etc.). The Shakhdagh rift-related trough at the Goycha-Hakeri belt northern wing is characterized by rifting migration towards the Early Alpine consolidation island arc. The Eocene volcanites of SA, SL, A, CA, T series with multiple subvolcanic bodies combined into the trachybasalt-trachyrhyolite formation with consecutively differentiated and bimodal complexes end up in gabbro-monzonite formation. Southern wing of the Pambag rift-related trough is characterized by rifting migration towards the Miskhan massif Baikalides of the Meso-Cenozoic activization. Volcanites (600 m) belong to the basalt-trachyte-phonolite formation, and their intrusive co-magmates, to pseudoleicite-nephelinesyenite formation of the Upper Eocene–Oligocene age.

Therefore, complex geodynamic environment for the development of collisional folding areas manifests itself in vulcanite polyserial features and in consistency of petro-geochemical indicators corresponding to active continental margins and to less oceanic rocks. The Paleogene VP belts of the Lesser Caucasus and Talysh show high similarity to continental rifts in the peculiarities of rock composition changes in time and space for each formation and formation family, in iniquitousness of subalkaline and alkaline

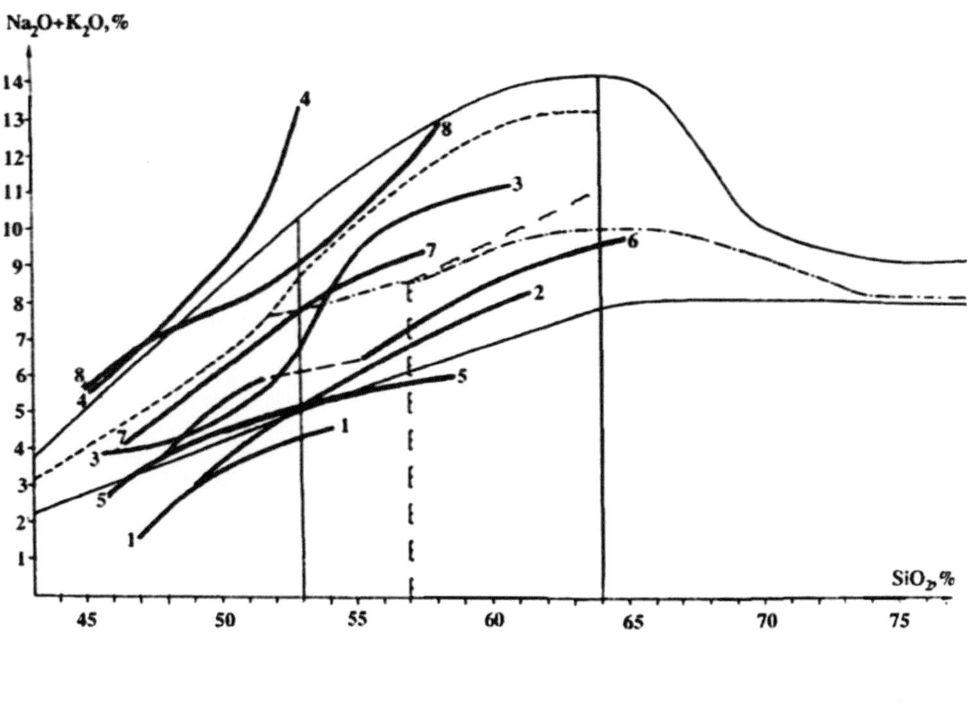

Fig. 4.16 Evolution of collision-riftogenic magmatism of discomfort volcanic-plutonic belts (Ismailzadeh et al. 2005). ***Volcanic series of the Talysh type***: (*1–1–2*) calc-alkaline basalts and trachyandesites, (*3–3*) shoshonites-latites-trachytes, (*4–4*) alkaline basalts-analcimites-phonolites. ***Volcanic series of Elburs type***: (*5–5–6*) calc-alkaline basalts-andesites and subalkaline trachybasalts-trachyandesites-dellenites, (*7–7*) shoshonites-latites-trachytes, (*8–8*) alkaline basalts-tephrites-phonolites

volcanites, in eruption of toleite series volcanites in axial zone during rifting early stage, in the increased role of intermediate and acidic rocks in the flank zones, as well as in the period of tension decelerations, decrease in potassium contents in the axial zone rocks and some other features.

The continuity of some petrochemical peculiarities of active continental margins in the Paleogene VP belt rocks is logical because the establishment and development of rift-related troughs is confined to the juncture boundary of two plates, and their geodynamics is controlled by elevations of decompacted mantle substrate.

Rare earth elements reflect geodynamic environment of the Paleogene volcanism manifestation most obviously. The graph of light and heavy REE correlation with potassium alkalinity shows that the rocks of each Lesser Caucasus formation under study were generated in complex geodynamic environments, in the conditions of continental rifting and active continental margins. In this process, the toleite and subalkaline series rocks are in all formations characterized by the tension environment, and the calk-alkali series rocks by the compression environment. It is of interest to remark that trend directions in differentiated associations from basalts to rhyolites show the change in geodynamic environment, when rifting tension is replaced by compression towards the period of generation of final acid members. It is typical for the association of toleite and subalkaline series rocks, while the differentiated association of calk-alkali series features permanent compression environment. It seems that each volcanism stage witnessed the change in geodynamic conditions, which resulted in the formation petrostock diversity. It is worth mentioning that increase in potassium alkalinity in toleite and subalkaline series rocks is accompanied by minor drop in La/Yb relation; in calk-alkali series rocks, the La/Yb relation increases critically at the expense of increase in light REE.

The above data confirm that Paleogene volcanism belongs to collisional rifting manifested in complex geodynamic environment that, according to the concepts of Bogatikov et al. (1984), is defined by governing combinations of magmatism products, indicative of various elementary geodynamic environments, in geological structures, for instance, the magmatism of active continental margins, of continental rifts. The investigations show that petrochemical polyseriality feature for each formation petrostock that represents separate magmatism stages in a specific geological structure is one of determining peculiarities of complex geodynamic environment during the development of mobile regions and evolutional direction of mature continental crust.

The complex of developed subalkaline and alkaline differentiates, serving as the alkaline-basalt magma derivatives in the Talysh folded zone is characterized by magmatism manifestation in the backarc passive rifting conditions, in spatiotemporal association with island-arc volcanism of the Elburs folded system.

Summing up all above mentioned, it can be noted that during the Paleogene period active continental margins of Andean type existed in the Lesser Caucasus central part, whose orogeny (collision) during the Eocene–Early Oligocene promoted final closure of the Mesotethys basin. It should be pointed out that volcanite differentiation (observed during the Paleogene age within both belts—the southern and the central) towards intermediate and acidic compositions and generation of large granitoid batholiths (the Dalidagh and the Megri-Ordubad ones) reflect areal granitization, associated with continuing crust continentalization (Fig. 4.17).

The Oligocene formations are developed in the Yerevan-Ordubad zone, in sedimentary (normal marine) and volcanogenic facies, and are confined to the Nakhchivan depression edge part.

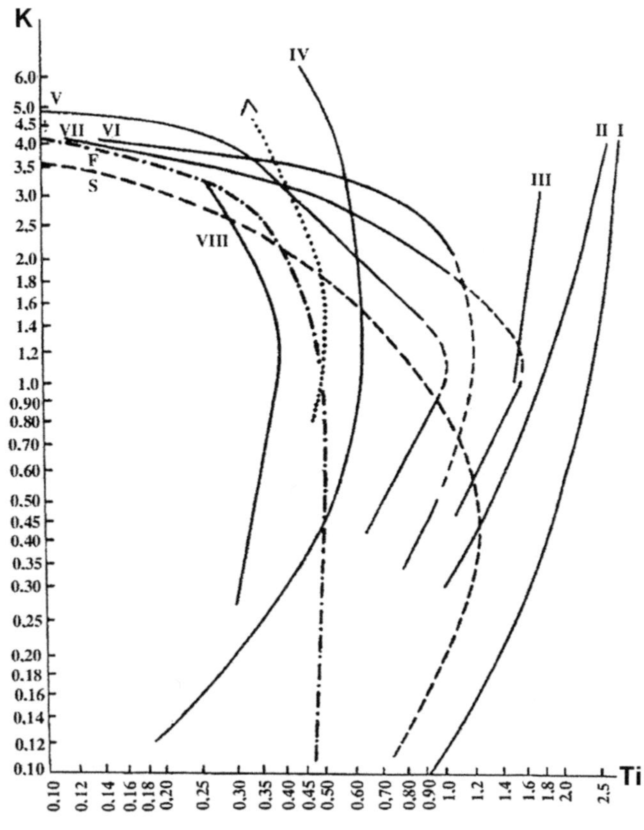

Fig. 4.17 Correlation diagram of Paleogene volcanic formations of the Lesser Caucasus. Volcanic associations: *I* COX and abyssal plains, *II* continental rifts, *III* Baikal rift, *IV* island arcs, *V* province of basins and ridges (USA), *VI* Mesozoic rifts of the Mongol-Okhot belt, *VII* basalt-comendite series of the East African rift, *VIII* Adjar-Trialet rift. *F* and Λ are the formations of the South belt: *F* basalt-andesite-dacite-rhyolite (Ꝑ $_2^1$), Λ trachybasalt-latite-trachyandesite (Ꝑ $_2^2$), *S* trachybasalt-trachyryolite formation (Ꝑ) of the Central belt

The Neogene stage. During the entire Miocene–Early Pliocene period, compression mode was observed in both volcanic-plutonic belts in the Zangazur block junction zone, accompanied by brief tension periods. Neogene volcanism within the southern and central volcanic-plutonic belts is confined to active continental margins, but basing on their petrogeochemical peculiarities and factor analysis data (Ismailzadeh 1990), they do not constitute a unified differentiation trend with the Eocene deposits but rather have a paragenetic bond only.

The belt formations are represented by andesite-dacite-rhyolite series (andesite-dacite in the Yerevan-Ordubad zone, and andesite-dacite-rhyolite, in the Goycha-Hakeri zone).

The Miocene–Early Pliocene formations find occurrence in the Kur, Lower Arazy and Hakeri depressions, where they are represented by molasses deposits of considerable thickness. They are considerably less abundant in volcanogenic facies within the Yerevan-Ordubad and Goycha-Hakeri zones (the Goycha-Shirak and the Kyalbajar troughs).

The Neogene volcanism of the southern and central volcanic belts shows the development of intermediate and acidic differentiates of calk-alkali and subalkaline series. However, subalkaline poorly differentiated formations of trachyandesite-trachytes (P_3^3–N_1^1) are also associated with the Lesser Caucasus Paleogene–Neogene period: as the layer of Elpin (subalkaline basalts, trachyandesites, trachyrhyolites, trachytes), and the formations of trachyandesite-phonolite (N N_1^3–N_2^1), as the Daralagez thickness (trachyandesites, trachytes, leucitic phonoliths) that are confined to the Ankavan-Zangazur zone abyssal faults and correspond in their chemical activity to alkaline olivine basalts (Ostroumova 1967). These subalkaline series, in contrast to the Paleogene ones, are characterized by predominant subalkaline intermediate and acidic differences and subordinate basic ones.

Thus, active continental margins and backarc basins are replaced by compression and stretching of continental blocks during the Paleogene era and by generation of internal and external molasses seas during the Neogene.

The Quaternary stage. New volcanism outburst took place in the Late Pliocene–Quaternary period and covered a sizable strip along the Zangazur zone and the Transcaucasian Uplift.

In the north, in the Transcaucasian zone (the Lori and Kechut subzone), the abyssal melt derivatives are observed such as the subalkaline basalts of non-differentiated trap rock type; more southward, the derivatives of calk-alkali magmas of the crust generation levels. Along the Ankavan-Ishikhly fracture (the Gegam and Kyalbajar highlands), volcanism is represented by basic and intermediate differentiates of normal and increased alkalinity, corresponding to the crust-mantle smelts (Tolstov et al. 1980), and more southeastward by alkaline basanites of intraplate rift-related formation types. Widely developed mass overthrust and thrusting correspond to the manifestations of plastic displacement in the lithosphere, thus reflecting the cessation of subductive processes as the major generators of magmatic melts (Koronovsky and Korzhinsky 1979).

References

Abdullayev, R. H., & Mustafayev, M. A. (1995). Types of the Mesozoic volcanic series of the Lok-Garabagh zone of the Lesser Caucasus. *Transactions of the Geological Instituites of the Azerbaijan Acadamy of Science 25*, 14–25 (in Russian).

Abdullayev, R. N., Ismet, A. R., Tuayev, A. C., Abdullayev, I. A., & Bagirbekova, O. D. (1971). Absolute age of small intrusions of gabbro-diabasic formation and sulfur mineralization of the Greater Caucasus. *Transactions of the XVII Session of Commission on Determination of the Absolute Age of Geological Form*, 112–113 (in Russian).

Abdullayev, R. N., Mustafayev, M. A., Samedova, R. A., Shafiyev, H. I., & Mamedov, M. N. (1991). *Petrology of magmatic complexes of the southern slope of the Greater Caucasus (Vandam Zone)*. Baku: Elm (in Russian).

Abdullayev, R. N., & Salakhov, A. C. (1983). Comparative characteristics of Jurassic volcanites of the Kurdamir-Saatly uplift and NE part of the Lesser Caucasus. *Izvestiya, Azerbaijan Academy of Sciences Series Earth Sciences, 4*, 14–23 (in Russian).

Azizbekov, Sh A. (1961). *Geology of nakhchivan autonomous republic*. Moscow: Gosgoltekhizdat (in Russian).

Allakhverdiyev, Sh. I. (1967) *Mineralogy of the weathering crust of ultrabasic rocks of the Lesser Caucasus*. Azerneshr: Baku (in Azerbaijanian).

Allakhverdiyev, Sh. I., (1985). Hyperbasites of the Lesser Caucasus, peculiarities of their secondary transformations and ore-bearing (Azerbaijan Rep.). *D.Sci. Thesis*. Leningrad (in Russian).

Azizbekov, Sh. A. (1961). *Geology of Nakhchivan Autonomous Republic*. Gosgoltekhizdat, Moscow (in Russian).

Azizbekov, Sh. A., Bagirov, A. E., Veliyev, M. M., Ismailzadeh, A. D., Nizheradze, N. Sh., Emelyanova, E. N., & Mamedov, M. N., (1979). *Geology and volcanism of Talysh*. Baku: Elm (in Russian).

Azizbekov, Sh. A., Gadjiev, T. G., Emelyanova, E. N., & Rustamov, M. I. (1964). *Petrology of intrusives of araz tectonic zone of the Lesser Caucasus*. Baku: Elm (in Russian).

Azizbekov, Sh. A., & Rustamov, M. I. (1972). Tectonic conditions for forming granitoid intrusions (on example of Megri-Ordubad batholith). *Geotektonika, 6*, 106–117 (in Russian).

Akhundov, F. A. (1986). Upper cretaceous volcanic formations of the Lesser Caucasus and associated with them economic deposits. *D. Sci. Thesis*, Tbilisi: Georgian Academy of Sciences (in Russian).

Bogatikov, O. A., Zonenshain, L. P., & Kovalenko, V. I. (1984). Magmatism and geodynamics. Petrology. *Transactions of the 27th International Geological Congress* (Vol. 9, pp. 3–14). Moscow: Nauka (in Russian).

Eppelbaum, L. V. & Khesin, B. E. (2012). *Geophysical Studies in the Caucasus*. Heidelberg - N.Y.: Springer.

Gasanov, T. Ab. (1985). *Ophiolites of the Lesser Caucasus*. Moscow: Nedra (in Russian).

Geidarov, A. C. (1971). Uranium and thorium in granitoids of the Lesser Caucasus (Azerbaijan Republic). In *Investigations in oil and physical chemistry* (pp. 338–355). Baku: Elm (in Russian).

Grigoryev, V. N., & Sokolov, S. D. (1978). Later Senonian volcanism of the Gochas sinclinorium (Lesser Caucasus). *Izvestiya, Acadamy of Science of the USSR, Series: Geological, 10*, 43–53 (in Russian).

Gukasyan, R. H., & Meliksetyan, B. M. (1965). Absolute age and peculiarities of generation of complex Megri pluton. *Izvestiya Academy of Sciences Arm, XYIII*(3–4, 8–26; 5), 20–38 (in Russian).

Imamverdiyev, N. A. (1988). Petrological-geochemical peculiarities of the late orogenic volcanism of the Kelbadjar covered trough (Lesser Caucasus). Ph.D. Thesis (in Russian).

Ismailzadeh, A. D. (1984). Gabbro- troctolite-anorthosite association of the hyperbasitic complex of the Lesser Caucasus. *Izvestiya, Academy of Science of the USSR, Series: Geological, 5*, 51–59. (in Russian).

Ismailzadeh, A. D. (1986). Petrological-geochemical peculiarities and geodynamics of the Late Pliocene—Quaternary volcanism of the Lesser Caucasus. *Izvestiya, Azerbaijan Acadamy of Science, Series Earth Sciences*, 4 (in Russian).

Ismailzadeh, A. D. (1990). Evolution of the Cenozoic basite volcanism of the Lesser Caucasus. *D.Sci. Thesis*, Baku: Geological Institute of the Azerbaijan Academy of Science (in Russian).

Ismailzadeh, A. D. (2009). Bipolar conjugation of volcanic-plutonic and ophiolite belts in the Caucasus. *Izvestiya, Azerbaijan Acadamy of Science Series Earth Sciences, 1*, 40–53 (in Russian).

Ismailzadeh, A. D., & Geidarov, A. C. (1996). Behaviour of alkaline and radioactive elements in evolution of the Cenozoic volcanism of the Kelbadjar depression. In *Transactions of the Institute of Geography of the USSR Academy of Science* 120–128.

Ismailzadeh, A. D., Genshaft, Yu. S., Emelyanova, E. N., & Mamedov, M. N. (1982). Facial composition and peculiarities of distribution of abyssal inclusions in Paleogenic volcanites of Talysh. In *Physical-Chemical Investigation of Products of Abyssal Magmatism* (pp. 85–107). Moscow: Institute of Physics of the Earth, USSR Academy of Science (in Russian).

Ismailzadeh, A. D., Mustafayev, G. V., & Rustamov, M. I. (Ak. A. Alizadeh, Ed.). (2005). *Geology of Azerbaijan, Magmatism* (Vol. III). Nafta-Press, Baku (in Russian).

Karyakin, Yu. V. (1989). *Geodynamics of forming volcanic complexes of the Lesser Caucasus*. Moscow: Nedra (in Russian).

Kashkay, M. A. (1965). *Petrology and metallogeny of Dashkesan*. Moscow: Nedra (in Russian).

Kashkay, M. A., Aliyev, V. H., Mamedov, A. I., Makhmudov, S. A., & Aliyev, A. A. (1964). *Petrology and metallogeny of petrographic formations of the Tutkhun River Basin*. Baku: Academy of Science of Azerbaijan (in Russian).

Kashkay, M. A., Khain, V. E., & Shikhalibeyli, E. Sh. (1950). On the stratigraphy of Paleogene of the Akera and Terter head rivers and adjacent part of the Lake Sevan basin. *Izvestiya, Academy of Science Azerbaijan, 3*, 54–67 (in Russian).

Kashkay, M. A., & Mamedov, A. I. (1961). *Pearlites and Obsidians*. Baku: Academy of Science Azerbaijan (in Russian).

Khain, B. E., & Seslavinsky, K. B. (1991). *Historical geotectonics*. Paleozoic. Moscow: Nedra (in Russian).

Khain, V. E. (1995). On the relationship between processes taking place on the boundary of tectonic plates. *Geotectonics (Geotektonika), 2*, 99–102 (in Russian).

Knipper, A. L. (1971). *Internal structure and age of serpentine mélange of the Lesser Caucasus*. Geotectonics (Geotektonika), No. 5, 11–26 (in Russian).

Knipper, A. L. (1975). *Oceanic crust in the structure of Alpine folded region*. Moscow: Nauka (in Russian).

Knipper, A. L., & Sokolov, S. D. (1972). Pre-Upper Senonian tectonic nappes of the Lesser Caucasus. *Geotectonics (Geotektonika), 6*, 74–80 (in Russian).

Koronovsky, N. V., & Korzhinsky, D. S. (1979). Palaeogene volcanism in the geological history of the Near and Middle East. *Bulletin of the Moscow State University, Series Geological*, 2, 3–16; 3, 30–42 (in Russian).

Kremenetsky, A. A., Lapidus, A. V., & Skryabin, V. Yu. (1990). *Geological-geochemical methods of deep prognosis of economic minerals*. Moscow: Nauka (in Russian).

Kurbanov, N. K. (1982). Main stages of forming complex copper-polymetallic deposits and their interrelation with stages of evolution of the Alpine terrigenous geosyncline of the Greater Caucasus. In *Peculiarities of location of sulfur deposits in flyschoid stratums. Transactions of TzNIGRI, 168*, Moscow, 3–18 (in Russian).

Lomize, M. G., Sukhanov, M. K., & Zvetkov, L. A. (1980). Caucasian margin of the Tethys in the beginning of Alpine stage. In: *Geotectonics of the Mediterranean Belt* (pp. 127–179). Moscow: Nauka (in Russian).

Lordkipanidze, M. B. (1980). *Alpine volcanism and geodynamics of the central segment of the Mediterranean folded belt*. Mezniereba: Tbilisi (in Russian).

Lutz, 1980. *Geochemistry of oceanic and continental magmatism*. Moscow: Nedra (in Russian).

Makhmudov, H. I. (1974). *Volcanic glasses of Azerbaijan*. Baku: Azerneshr (in Russian).

Mamedov, A. I. (1974). Cenozoic intrusives of the Sevan-Garabagh zone. In *Mineralogy and ore deposits*. Baku: Elm (in Russian).

Mamedov, A. I. (1981). Conditions of generation of Cenozoic magmatic complexes of the central part of the Lesser Caucasus. In *Problems of Mineralogy, Geochemistry and Petrology of Azerbaijan* (pp. 51–61). Baku: Elm (in Russian).

Mamedov, M. N. (1990). *Petrology and geochemistry of late cretaceous and eocene magmatic formations of the Lesser Caucasus and Talysh*. Baku: Elm (in Russian).

Mamedov, M. N., Khalafov, A. A., Akhundov, F. A., & Makhmudov, C. A. (1986). Fe–Ti-oxide minerals of Late Cretaceous volcanites of the Lesser Caucasus. *Izvestiya, Academy of Science Azerbaijan Series: Earth Science, 4*, 45–52 (in Russian).

Marakushev, A. A. (1979). *Petrogenesis and ore generation*. Moscow: Nauka (in Russian).

Mustafayev, G. V. (1977). *Mesozoic granitoid formations of Azerbaijan and peculiarities of their metallogeny*. Baku: Elm (in Russian).

Mustafayev, G. V. (1991). On the ore-bearing of the Dashkesan granitoid intrusive. *Soviet Geology, 2*, 70–78 (in Russian).

Mustafayev, M. A,. et al. (1983). Facial analysis of the Upper Jurassic volcanogenic associations of the Somkhit-Agdam zone of the Lesser Caucasus. *Izvestiya, Academy of Science Azerbaijan Reports Series. Earth Sciences, 2* (in Russian).

Nikolsky, Yu. N, Milai, T. A., & Kogan, L. Z. (1975). *Geological-Geophysical investigations of tectonics, magmatism and metallogeny of the Caucasus*. Leningrad: Nedra (in Russian).

Ostroumova, A. S. (1967). Basalt-trachyte formation of the Lesser Caucasus. In *Alkaline Volcanic Formations of Folded Regions* (pp. 6–129). Leningrad: Nedra (in Russian).

Ostroumova, A. S., Abdullayev, R. N., Tzenter, I. Ya., Mudrenko, S. V., & Shatkova, L. N. (1995). Saatly super-deep borehole. Model of deep structure of the Meso-Cenozoic Earth's crust. In *Superdeep Boreholes of Russia and Adjacent Regions* (pp. 155–190). VSEGEI: Sankt-Petersburg (in Russian).

Ostroumova, A. S., & Tzenter, I. Ya. (1986). Analogues of rocks of the marian-boninite series in Jurassic volcanites of the Garabagh Ridge (Lesser Caucasus). *Doklady of the Academy of Science of the USSR, 290*(2), 441–445 (in Russian).

Ostroumova, A. S., Tzenter, I. Ya., Avdeyev, A. G., & Bugrova, E. M. (1991). *Late Cretaceous volcanic series of the Gochas depression (Lesser Caucasus)*. Soviet Geology, No. 7, 65–72 (in Russian).

Paffengoltz, K. N. (1934). On the results of geological investigation in the Lake Gokcha basin. *Problems of Soviet Geology*, *5* (in Russian).

Piskunov, B. N. (1987). *Geological-petrological specifics of island arcs volcanism*. Moscow: Nauka (in Russian).

Rustamov, M. I. (1976). Calderas with confocal granitoids - new morphogenetic type of magmatism development in the Lesser Caucasus. Doklady Acad. Sci. USSR, Vol. 228, No. 6, 1403–1406.

Rustamov, M. I. (1979). Olistostrome generations and problems of ophiolites of Nakhchivan zone. *Izvestiya, Academy of Science Azerbaijan Reports Series. Earth Science, 5*, 84–91 (in Russian).

Rustamov, M. I. (1983). New data on the evolution of Palaeogene magmatism of the Lesser Caucasus. *Transactions of All-Union Conference "Evolution of Magmatism in the Main Earth's Structures"* (pp. 65–66). Moscow: Nauka (in Russian).

Rustamov, M.I. (1987). Magmatic formations of Nakhchivan zone and problems of further investigations. In: Problems of Geological Petrology of Azerbaijan. Elm, Baku, 44–64 (in Russian).

Rustamov, M. I. (1995). Collision-riftogenic magmatism of Palaeogene of the central segment of the Mediterranean belt. In *Transactions of Russian Petrographic Symposium "Magmatism and Geodynamics"*(pp. 179–181). Russia: Ufa (in Russian).

Samedova, R. A. (1985). Cretaceous volcanism of the Vandam zone of the southern slope of the Greater Caucasus. Ph.D. Thesis. Geological Institute of the Azerbaijan Academy of Science (in Russian).

Shikhalibeyli, E. Sh. (1994). History of tectonic evolution. In *Geology and Economic Minerals of the Nagorny (Daghliq) Garabagh of Azerbaijan* (pp. 171–195). Baku: Elm (in Russian).

Shikhalibeyli, E. Sh. (1966). *Geological structures and history of tectonic evolution of the eastern part of the Lesser Caucasus* (Vol. II). (*Tectonics and Magmatism*). *Academy of Science Azerbaijan Report,* Baku (in Russian).

Sokolov, S. D. (1977). Olistosrrome strata and ophiolitic nappes of the Lesser Caucasus. In *Proceeding of the Geological Institute Academy of Science USSR* (Vol. 296, pp. 1–94). Moscow: Nauka (in Russian).

Sokolov, S. D., Silant'ev, S. A., Zakaridze, G. S., & Konenkova, N. N. (1996). Metamorphic rocks in ophiolites of the Sevan-Akera zone of the Lesser Caucasus. *Petrologiya* (*Petrology*), *4*(2), 171–182 (in Russian).

Solovkin, A. N. (1939). *Intrusions and intrusive cycles of Azerbaijan Republic*. Baku (in Russian).

Tauson, L. V. (1982). Geochemistry and metallogeny of latite series. *Geologiya Rudnykh Mestorozhdenii (Geology of Ore Deposits), 3*, 3–14 (in Russian).

Tolstov, M. I., Ostafaichuk, I. M. et al. (1980). Composition, physical properties and problems of petrogenesis of the newest volcanic generations in Armenia. *Academy of Sciences of Armenia*, Erevan, 206–222 (in Russian).

Zakariadze, G. S., Knipper, A. L., & Lordkipanidze, M. B. (1982). Experience in correlation between the Mesozoic volcanism of the ophiolite belt of the Lesser Caucasus and zones of its flanking. In *Problems of the Geodynamics of the Caucasus* (pp. 117–121). Moscow: Nauka (in Russian).

Zonenshain, L. P., Derkur, Z., Kazmin, V. G., et al. (1987). Tethys evolution. In History of the Tethys Ocean. Acad. of Sci. USSR, Moscow, 104–115 (in Russian).

Tectonics 5

5.1 Basic Structural Elements

The territory of the Azerbaijan Republic spans over the eastern part of the Caucasian segment of the Alpine-Himalayan (Mediterranean) folded belt, which was generated in the area of collision junction (collision) of the Eurasian and Afro-Arabian lithospheric plates. The fold-mountain structures that constitute this belt form a number of branches along its striking, of which, according to Khain (1984, 2001), the Dobruja-Crimean-Caucasian-Kopetdagh branch, the Pyrenean-Alborzian (Pyrenean-Lesser Caucasian) branch, and the Dinar-Zond branch as well as the Rioni and Kur intermountain troughs participate in the Caucasian Isthmus structuring. The characteristic feature of the region's crust is its tectonic inhomogeneity, expressed in the complex relationship of its constituent structural and formational units, having differing lithological and stratigraphic sequences, deformation natures, and geological development history. Within Azerbaijan's borders, the geological sequence on the daylight surface is essentially represented by sedimentary, volcanogenic, volcanogenic–sedimentary and intrusive formations of the Cenozoic, Mesozoic, and partially Paleozoic (Phanerozoic) periods (Fig. 5.1).

The country's tectonic structure is subdivided into the Greater and Lesser Caucasus fold-mountain systems, the Kur megadepression that separates them, and the Talysh folded region; these include a number of subordinate structural elements, such as tectonic megazones, zones, and subzones (Khain and Alizadeh 2005). The meridionally oriented Caspian Megadepression is unconformably superimposed on the land's heterogeneous tectonic elements, which plunge under recent marine sediments in the east (Fig. 5.2).

Along the northeastern side of the Greater Caucasus Azerbaijan, the Gusar-Devechi superimposed depression can be traced; under its Pliocene–Quaternary formations, the Jurassic, Cretaceous, and Paleocene–Miocene deposits of Dagestan are submerged from the Samur River's left bank. The Lesser Caucasus fold-mountain system is limited by the Nakhchivan superimposed depression from the southwest, whereas in the southeastern direction it is periclinally immersed under Upper Pliocene–Anthropogenic sediments of the Lower (Ashaghy) Araz superimposed depression. The latter separates the Lesser Caucasus structural zones from the Talysh folds, which are located at the northeastern side of the Alborz fold-mountain system that rims the Southern Caspian (Janubi Khazar) abyssal trough from the southwest and south. The complex joint assembly of the Greater Caucasus southeastern submersion, the Ashaghy (Lower) Kur and the Southern Caspian troughs are characterized by vigorous manifestations of mud volcanism, as it is the surface expression of the oil-and-gas potential of the earth's crust deep horizons.

The Caucasus contemporary structure was formed at the Alpine stage of tectogenesis within the spatial borders that encompass the southern edge of the Eurasian continent and the northern edge of the South-Azerbaijan segment of the Central Iranian microcontinent (microplates, quasi-platforms). The following megastructures play determining roles in the structure of eastern Azerbaijan in the Caucasus (from north to south):

1. *The Northern Caucasus continental microplate* is part of the Scythian-Turanian epi-Hercynian platform of the Eurasian continent's southern edge, which is involved in peri-cratonic subsidences related to the establishment and development of the Mesotethys Alpine marginal marine basin. In the superficial geological structure, it corresponds with the Side Range (Yan Sirt) elevation (megazone) of the Greater Caucasus, represented within Azerbaijan's borders by the Tahirjal, Sudur, Shakhdagh-Khizi and Guton-Gonagkend tectonic zones, which are made of Jurassic and Cretaceous rocks. The elevation's northern side, by its flexural subsidence, is traced under the Pliocene–Holocene molasses of the Gusar-Devechi superimposed depression towards the Caspian Sea.
2. *The Southern Slope (Janub Yamaj) trough* is a large, linearly outstretched tectonic unit (megazone) of the Greater Caucasus that corresponds to the axial section of

A.A. Alizadeh et al., *Geosciences of Azerbaijan*, Regional Geology Reviews,
DOI: 10.1007/978-3-319-27395-2_5

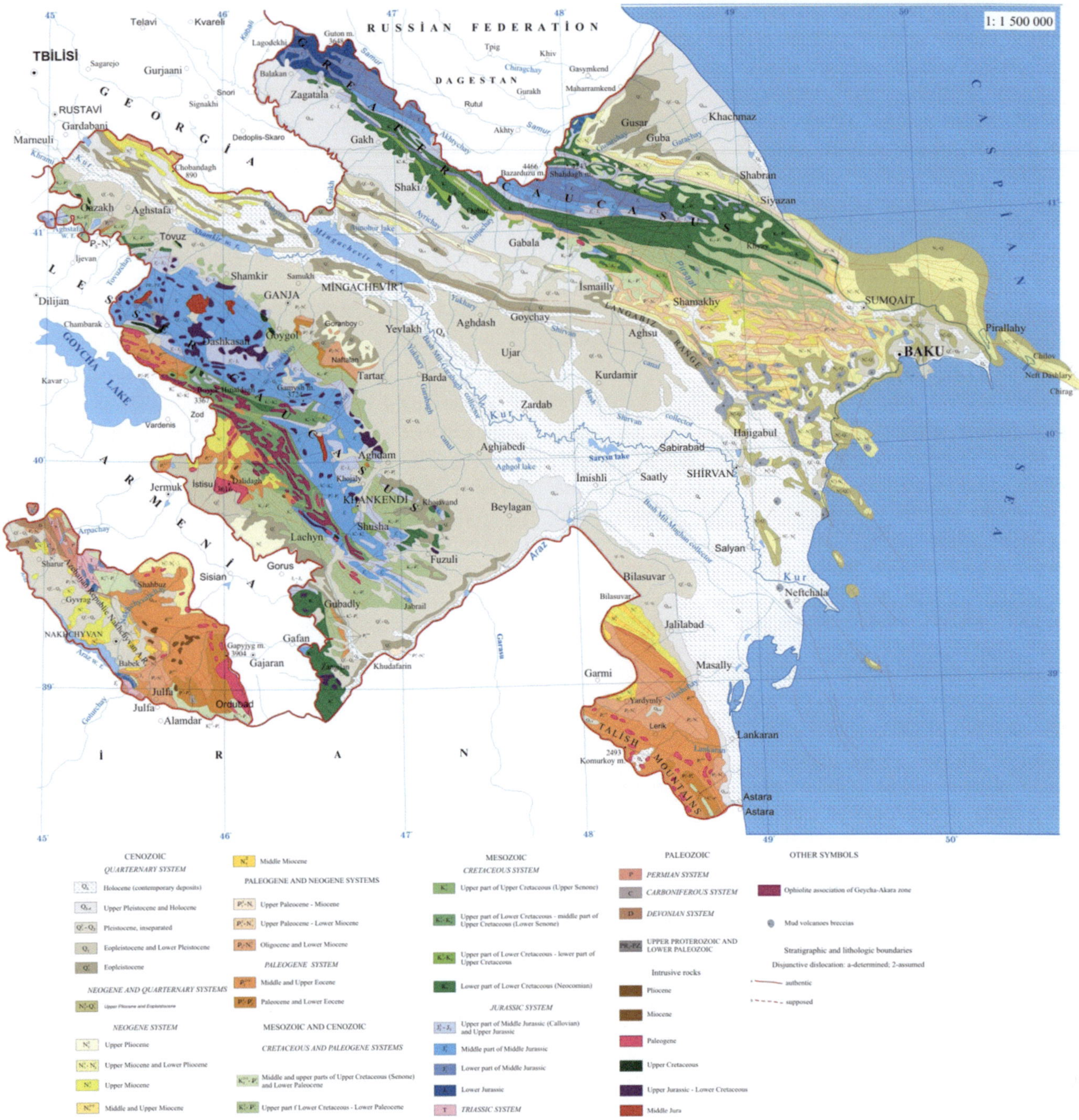

Fig. 5.1 Geologic map of the territory of the Azerbaijan Republic (National Atlas 2014; T.N. Kangarli)

the Alpine marginal marine basin, whose consolidated crust is subjected to destruction and thinning. In the contemporary structure, the megazone composed of the Jurassic rock laid the Speroza-Tufan (in the north) zone and is represented by the Cretaceous–Paleogene rock Zagatala-Govdagh (in the south) zone. It is characterized by intensive folding and tectonic layering, corresponding to the accretion prism, pressed between the Northern Caucasus and the Southern Caucasus continental microplates and is composed of a series of different-scale tectonic laminae. The southern side's structural and compositional complexes allochthonously superpose the Southern Caucasus microplate northern edge as a system of low-angle overthrusts that serve as the branches of the regional Kbaad (Krasnopolyan)—Zangi deep-seated fracture.

3. *The Southern Caucasus continental microplate* is a fragment of the Gondwana passive margin that was torn from the mainland during the Paleotethys opening and attached to Eurasia during movements of the Hercynian cycle of tectogenesis. In the Alpine cycle of tectogenesis, the Southern Caucasus microplate looked like an island arc system that separated the Greater Caucasus marginal sea from the Lesser Caucasus arm of the Mesotethys. In the contemporary structure, the plate central section corresponds to the Kur megadepression (intermountain trough). Its lateral sections, composed of volcanogenic and sedimentary complexes of the Middle Jurassic, Cretaceous, and Cenozoic rocks, participate in the generation of fold-mountain structures of the Greater (Vandam-Gobustan megazone) and the Lesser (Artvin-Garabagh megazone) Caucasus. These megazones, in their turn, consist of smaller tectonic areas: Dashaghyl-Lahyj, Gulluk-Dadagunash, Ayrichay, Jangichay, Keyvandi-Shahgaya and Langabiz-Alat in the north; and Lok-Garabagh, Goycha-Hakeri (its structure includes the allochthonous ophiolite complex), Gafan and Kalbajar zones in the south. The Kur megadepression is subdivided into Middle (Orta) and Lower (Ashaghy) Kur megazones that consist of the second-order tectonic areas (including the buried ones).
   The extreme southeastern segment of the Southern Caucasus continental microplate corresponds, within Azerbaijan's borders, to the Talysh folded region (megazone) that is represented by the Astara, Lerik-Yardymly, Buravar, Jalilabad and Southern Mughan (Janubi Mughan) tectonic zones composed of volcanogenic sedimentary Paleogene and sedimentary Miocene rocks.
4. *The Southern Azerbaijan segment* is a fragment of the Central Iranian microcontinent represented within the Azerbaijan Republic's borders by the Araz megazone. The latter is composed of a massive Paleozoic-Triassic subplatform complex (the Sharur-Julfa tectonic area), whose northeastern edge underwent destruction and rifting that was completed in the Oligocene period (the Ordubad segment of the Lesser Caucasus back-arc rift). In its central section, it is superposed by the Oligocene–Miocene formations of the Nakhchivan superimposed depression.
5. *The Caspian (Khazar) megadepression* is a submeridionally imposed structure that was formed 10 Ma ago during the late Miocene period. It consists of the Middle Caspian (Orta Khazar) and Southern Caspian (Janubi Khazar) sedimentary basins subdivided by the Absheron latitudinal threshold in the sea bottom relief. The sedimentary mantle age includes the Meso-Cenozoic period in the Middle Caspian and Oligocene–Holocene period in the Southern Caspian. The Permian-Triassic formations in the Middle Caspian, the Jurassic, Cretaceous and Paleocene–Eocene formations are partially included in the basin's folded basement and partially in their mantle. Respectively, the Middle and Southern Caspian basins correspond to those megazones, under whose contemporary marine molasses, arriving from the west, the Meso-Cenozoic period complexes of the Greater Caucasus in the northern part of the area's Azerbaijan sector and the Lower Kur Depression and Talysh in the southern parts of the area's Azerbaijan sector are submersed.

5.2 The Greater Caucasus Fold-Mountain System

According to contemporary views (Adamia et al. 1981; Khain 1984; Dotduyev 1986, 1989; Baranov et al. 1990; Kangarli 1999, 2011; Khain 2001; Akhmedbeyli et al. 2002), the Greater Caucasus—as one of the segments of the Dobruja-Crimea-Caucasus-Kopetdagh branch of the Alpine-Himalayan folded belt—is represented as a complex folded cover structure resulting from Alpine (Late Cimmerian and proper Alpine) tectonic movements out of three major Mesozoic structures: the Northern Caucasus continental microplate (the southern side of the Scythian-Turanian epi-Hercynian platform or plate), the Southern Caucasus continental microplate, and the vast Greater Caucasus marginal marine basin of Tethys that separates them (Figs. 5.3 and 5.4).

In the contemporary structure, the Northern Caucasus continental microplate edge that has experienced Alpine deformations is thrust along the Main Caucasus fault over the Southern Slope area folds. The latter are torn off their basements and are shifted along the Kbaad-Zangi deep upthrust to the structural facial complexes of the Gagra-Java (in the west) and Kakheti-Vandam-Gobustan (in the west) zones due to the counter motion of the Southern and Northern Caucasus microplates. The Ganykh-Ayrichay-Alat fault is the southern border of the Kakheti-Vandam-Gobustan zone. Along this fault, the upthrust overlapping of the Middle and Lower Kur thrust structures by the zone formations takes place. Along its strike, the Greater Caucasus structure is not monolithic. It is subdivided into four transversely separated tectonic segments separated by the Pshekh-Adler, Kazbek, and Aghdash-Darband (Samur) flexural-fault areas, which are defined in geological literature as the North-Western, the Central, the Eastern, and the South-Eastern Caucasus. Within Azerbaijan's borders, the eastern segment (southern slope) is situated partially and the south-eastern segments are situated completely; their structures, in turn, are dissected into relatively elevated and downdip blocks by means of transversally oriented (northwestern and northeastern) subjacent fault and wrench-fault dislocations—often of significant horizontal

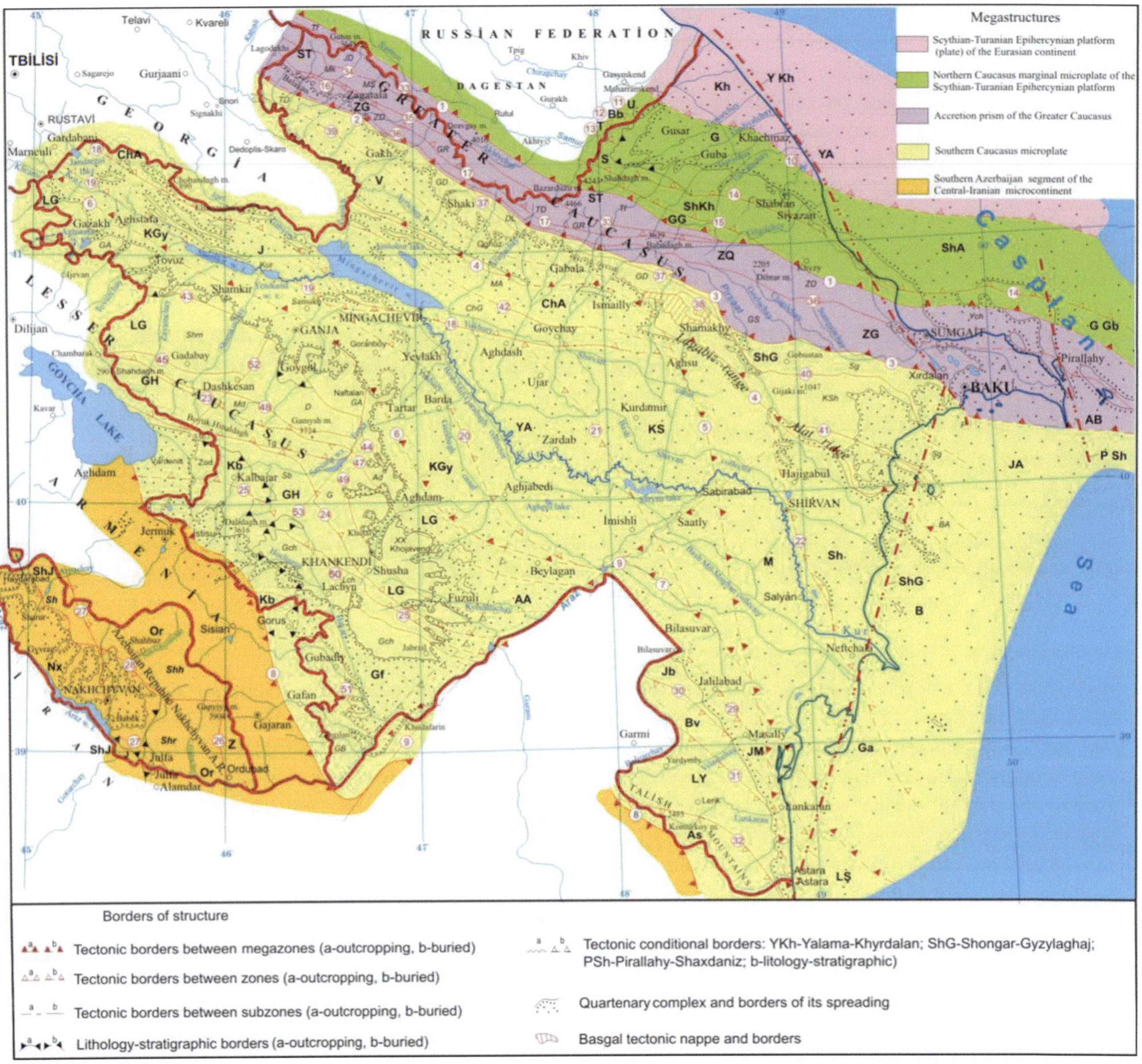

offset. The largest of them are (from north-west to south-east) the Tinovroso, the Zagatala, the Akhvay, the Kish, the Girdymanchay-Valvalachay, the Western-Caspian, and the Pirsaat-Salyan-Neftchala.

The structure of the Azerbaijan sector of the fold-mountain structure reflects the movements of the Late Cimmerian (or Early Alpine) periods of tectogenesis and of the proper Alpine (Late Alpine) periods of tectogenesis, which were accompanied by cover and fold-fault deformations (Khain 1950, 1975; Azizbekov et al. 1972; Akhmedbeyli et al. 2002; Kangarli 2005, 2009; Khain and Alizadeh 2005).

Late Cimmerian deformations are diagnosed as stratigraphic and structural unconformities in the upper layers of Middle Jurassic and in the base and cross-section of Upper Jurassic periods. These deformations play a decisive role in the contemporary structure formation of the tectonic zones composed of the Lower–Middle Jurassic complex and are represented, specifically, by the wedge-like protrusions and high-amplitude inclined plates of rootless slices in the Main Range area. Among other things, the entire Lower–Middle Jurassic complex (and, probably, the pre-Jurassic basement as well) of the Side Range elevation is brought to an allochtonic condition in the Eastern and South-Eastern Caucasus under the effect of the crystalline substrate underthrust of the Greater Caucasus marginal basin under the Northern Caucasus continental microplate. To the south of the Main Caucasus thrust line, the Goytkh-Tufan cover complex of the Early–Middle Jurassic deposits was formed.

◀ **Fig. 5.2** Tectonic scheme of the territory of the Azerbaijan Republic (National Atlas 2014; T.N. Kangarli). **Hercynic structures: *Araz megazone*: ShJ**—Sharur-Julfa zone: segments: *Sh* Sharur; *J* Julfa. **Alpine structures: Mountain-folding system of the Greater Caucasus: *Gusar-Devechi megazone (Miocene-Pliocene superimposed depression)***: zones: *Kh* Khachmaz; *G* Guba. ***Side Range (Yan Sirt) megazone***: zones: *U* Ulluchay; *Bb* Beybulagh; *T* Tahirjal; *S* Sudur; *ShKh* Shakhdagh-Khizi; *GG* Guton-Gonagkend. ***Southern Slope (Janub Yamaj) megazone***: zones: *ST* Speroza-Tufan; *ZG* Zagatala-Govdagh; subzones: *Tf* Tufan; *JD* Jikhikh-Dindidagh; *MS* Mazym-Sarybash; *Mk* Megikan; *GR* Galal-Rustambaz; *ZD* Zagatala-Dibrar; *TD* Talachay-Duruja; *GS* Govdagh-Sumgayit. ***Kakheti-Vandam-Gobustan megazone***: zones: *V* Vandam; *ShG* Shamakhy-Gobustan; subzones: *DL* Dashaghyl-Lahyj; *GD* Gulluk-Dadagunash; *A* Ayrichay; *Jg* Jangichay; *KSh* Keyvandi-Shahgaya; *LA* Langabiz-Alat. **Kur intermountain depression (Miocene-Pliocene superimposed depression): *Middle Kur (Orta Kur) megazone***: zones: *ChA* Chatma-Ajinohur; *J* Jeyranchol; *KGy* Near-Lesser Caucasus (Kichikgafgazyany); *YA* Yevlakh-Agjabedi; *KS* Kurdamir-Saatly; subzones: *MA* Mirzaani-Arash; *ChG* Chatma-Goychay. ***Lower Kur (Ashaghy Kur) megazone***: zones: *Sh* Shirvan; *M* Mughan. **Mountain-folding system of the Lesser Caucasus: *Artvin-Garabagh megazone***: zones: *LG* Lok-Garabagh; *GH* Goycha-Hakeri; *Qf* Gafan; *Kb* Kalbajar (Eocene–Pliocenic superimposed depression); *AA* Lower Araz (Ashaghy Araz—Eopleistocene–Holocene superimposed depression); subzones: *GA* Gazakh-Aghburun; *Shm* Shamkir; *Ad* Aghdam; *D* Dashkasan; *XX* Khachynchay-Khojavand; *Md* Murovdagh; *Gb* Garabagh; *Tg* Toragaychay; *Sb* Sarybaba; *Lch* Lachyn; *Hch* Hochaz; *GB* Gafan-Basitchay. ***Araz megazone***: zones: *Z* Zangazur; *Or* Ordubad (segments: *Shb* Shahbuz; *Shr* Shurud); *Nx* Nakhchivan (Miocenic superimposed depression). ***Talysh megazone***: zones: *As* Astara; *LY* Lerik-Yardymly; *Bv* Buravar; *Cb* Jalilabad; *JM* Southern Mughan (Janubi Mughan). **Caspian megadepression (Miocene-Pliocene superimposed depression): *Middle Caspian (Orta Khazar) megazone***: zones: *YA* Yalama-Aghzybirchala; *ShA* Northern Absheron (Shimali Absheron); *GGb* Gilazi-Neargubadagh (Gilazi-Gubadaghyany); *AB* Absheron-Nearbalkhan (Absheron-Balkhanyany); subzones: *YCh* Yashma-Chyrag; *AG* Absheron-Gurgan. ***Southern Caspian (Janubi Khazar) megazone***: zones: *JA* Southern Absheron (Janubi Absheron); *B* Bandovan; *Ga* Gyzylaghaj; *LSh* Lenkaran-Shahaghaj; subzones: *P* Puta; *Sg* Sangachal; *BA* Baku archipelago. **Faults: on the boundaries of megazones:** *1* Main Caucasus fault; *2* Mamrukh-Galajig; *3* Zangi-Garajuzlu; *4* Ganykh-Ayrichay-Alat; *5* Western Caspian (Garbi Khazar); *6* Pre-Lesser Caucasus; *7* Bilasuvar; *8* Girratagh; *9* Lower Araz (Ashaghy Araz) fault (Palmira-Absheron lineament); **on the boundaries of zones:** *10* Imamgulukend-Khachmaz-Aghzybirchala; *11* Ashaghy Maku; *12* Puya-Tindi-Keydi; *13* Tahirjal; *14* Uruj-Khnov-Siyazan; *15* Shakhdagh-Gonagkend; *16* Megikan; *17* Gamarvan; *18* Arikdar-Bozdagh-Padar; *19* Kur; *20* Southern Kur (Janubi Kur); *21* Goychay-Imishly; *22* Pirsagat-Salyan-Neftchala; *23* Murovdagh; *24* Garabagh; *25* Lachyn-Bashlybel; *26* Eastern Ordubad (Shargi Ordubad); *27* Havush-Nehram (Anabadgadik); *28* Nakhchivan-Goygel; *29* Pre-Talysh (Talyshonu); *30* Buravar; *31* Haftoni; *32* Astara; **on the boundaries of subzones:** *33* Khuray-Malkamud; *34* Kehnameydan; *35* Suvagil; *36* Gaynar-Gozluchay-Aladash; *37* Dashaghyl-Mudrisa; *38* Muju; *39* Shambul-Ismailly; *40* Galabughur-Jangichay; *41* Sangachal; *42* Ajinohur-Surkhaykhan; *43* Yasamal; *44* Aghdara; *45* Aryghdam; *46* Arysu-Zivlan-Ballygaya; *47* Aghdam; *48* Sarsang; *49* Meydanchay-Shusha; *50* Deyhan; *51* Hakeri; *52* Goshgarchay; *53* Terter-Injachay

Alpine deformations are also associated with this fault, which to the north of the Main Caucasus thrust resulted in the generation of Shakhdagh-Beshbarmag overthrust of the Malmian-Neocomian limestone in the Siyazan thrust and to the retro mass overthrust of the Eastern Caucasus Northern frontal thrust, which is transformed into the Dagestan wedge cover (i.e., superimposition of the Paleozoic–Mesozoic deposits onto the Late Mesozoic–Cenozoic sediments). In the southern slope area, the Novorossiysk-Dibrar fold-nappe complex was formed, composed of flysch formation of the Jurassic, Cretaceous, and Early Paleogene periods and confined by the Bekishey-Gamarvan disjunctive in the north and by the Kbaad (Krasnopolyanskiy)-Zangi disjunctive in the south. In the west, the nappe complex is mass overthrust by para-autochthonous substrate of the Chvejipsa-Govdagh area and, overlapping it in the extreme northwest, is superimposed onto the Akhtsu-Dzykhra barrier reef structures and the Southern Caucasus lagoon. In the east, the nappe complex (with the Diz Series of Middle–Upper Paleozoic period in the cross-section basement) appears from under the Main Caucasus thrust in the Inguri River head and can be traced to the Caspian Sea as the eastward opening and the southward oversteepened tectonically dislensed flysch depression. From the south, the nappe complex is constrained by the subthrusted side of the Kakheti-Vandam bordering uplift, a homolog of the Gagra-Java area in the east. Along the contact, the Aalenian argillites of the Talachay-Duruja tectonic plate, pressed out of the nappe complex Jurassic basement, are protruded in the narrow suture zone.

The scaly-folded structures of the flysch depression southern side in the eastern and southeastern Caucasus are represented by the roots of south-overthrust cover plates of the Chvejipsa-Govdagh cover complex eastern segment. The latter is composed of the Cretaceous, Paleogene and Miocene complexes mass overthrust onto heterochronous deposits, from the Jurassic to Miocene periods and occasionally down to the Lower Pliocene, developed at the Southern Caucasus continental microplate northern edge. In Eastern Georgia, the complex is superimposed on autochtonic formations of the Gagra-Java and Georgian blocks. In Azerbaijan, it overlaps the Shamakhy-Gobustan depression's northern edge and can be traced within the borders of the Absheron peninsula, being levelled by the neoautochtonic Pliocene.

The principal structural elements of the fold-mountain structure within Azerbaijan's borders are the megazones of the Side Range (Yan Sirt), the Gusar-Devechi, the Southern Slope (Janub Yamaj), and the Kakheti-Vandam-Gobustan megazones (Fig. 5.5).

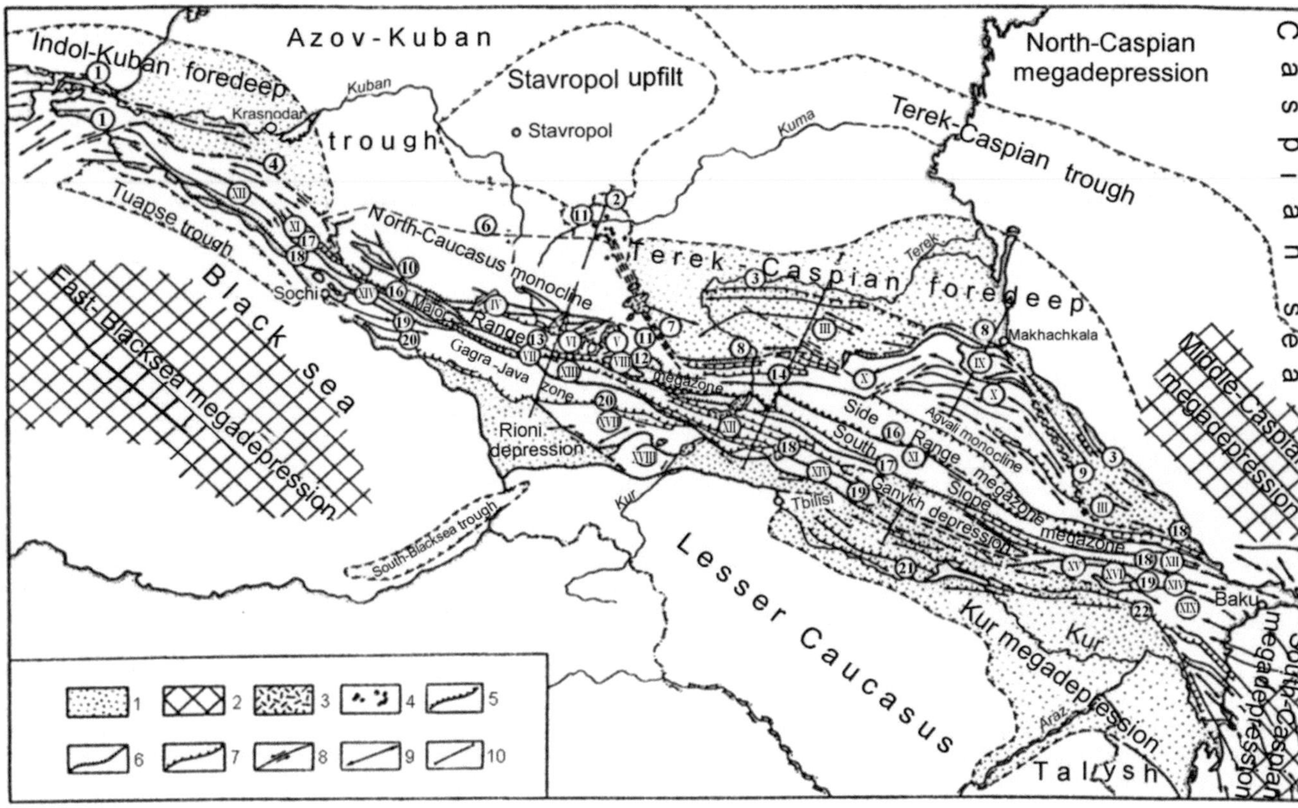

Fig. 5.3 The main elements of the Alpine structure of the Greater Caucasus (Dotduyev 1986; Baranov et al. 1990): (*1*) margin and intermontane depressions; (*2*) parts of marine depressions with suboceanic type of the Earth's crust; (*3*) modern effusions; (*4*) Neogene intrusions; (*5*) Main Thrust of the Greater Caucasus; (*6*) thrusts limiting nappe complexes; (*7*) other thrusts and overthrusts; (*8*) displacements; (*9*) axes of anticlines in periphery of mountainous constructions; (*10*) lines of profiles. *I* Kerch-Taman folded zone; *II* Mineralovodskiy nose; *III* folded zone of the Terek-Caspian margin depression; *IV* uplift of Fore Range; *V* Balkar-Digor ledge; *VI* Teberda ledge; *VII* Sofia ledge; *VIII* Shkhar ledge; *IX* Shatskiy nappe package; *X* Limestone Dagestan cover; *XI* Goytkh-Tufan nappe complex; *XII* Novorossiysk-Dibrar nappe complex; *XIII* Svaneti zone; *XIV* Chvejipsa-Govdagh nappe complex; *XV* Vandam uplift; *XVI* Basgal nappe; *XVII* Okrib-Sachheri uplift; *XVIII* Dzirul ledge; *XIX* Absheron-Gobustan zone. Numbers in *circles*: (*1*) Temruk ledge; (*2*) Nagut fault; (*3*) Terek-Darband thrust; (*4*) Akhtyr ledge; (*5*) Pshekh displacements zone; (*6*) buried Circas fault; (*7*) Kabarda fault zone; (*8*) Northern frontal thrust of the Eastern Caucasus; (*9*) Western Caspian fault zone; (*10*) Pshekish-Tyrnyauz fault; (*11*) Chegejar thrust; (*12*) Ullukam-Shtula fault; (*13*) Sofia-Kitlod fault; (*14*) Buron-Lar fault; (*15*) Uruj-Khnov-Siyazan thrust; (*16*) Main Caucasus thrust; (*17*) Bekishey-Gamarvan thrust; (*18*) Kbaad (Krasnopolyanskiy)-Zangi thrust; (*19*) Vorontsov-Orekhevi thrust; (*20*) Abkhaziya-Lechkumi fault; (*21*) Erikdar-Bozdagh-Padar fault; (*22*) Ganykh (Alazani)-Ayrichay-Alat thrust

5.2.1 The Side Range Megazone

The Side Range (Yan Sirt) megazone is a natural continuation of the Greater Caucasus Central uplift in the east and is limited by the Puya-Tindi-Keydi zone of rupture dislocations (in the north) and the Main Caucasus thrust (in the south). In the east, the uplift experiences plunge and is subdivided into four tectonic areas (from north to south): the Tahirjal upraised, Sudur transitional, Shakhdagh-Khizi dropped, and Guton-Gonagkend upraised zones (Khain and Alizadeh 2005). The extreme southern and northern areas are composed of Middle Jurassic rocks, while two central ones are composed of Upper Jurassic, Cretaceous, and occasionally, preserved Paleogene–Neogene rocks. On the northern side (Tahirjal and Sudur areas), the megazone is flexurally immersed under the Miocene–Pleistocene sediments of the Gusar-Devechi depression. On the southern side (the Guton-Gonagkend zone), it closes periclinally east of Girdymanchay-Valvalachay flexure. Only in its central section (Shakhdagh-Khizi zone) can it be traced into the Caspian Sea bounds, thus participating in the formation of the Absheron-Balkhanyany (Absheron-Nearbalkhan) elevation zone's northern side.

The interpretation of geophysical materials leads to the conclusion of a root-free location of the Liassic-Lower Bajocian compositional complex that outcrops in the Guton-Gonagkend and Shakhdagh-Khizi structural zones, north of the Main Caucasus thrust (Kangarli 2009, 2011). The location of the latter at the pre-Jurassic basement level, according to the same data, is routed 25–30 km to the north,

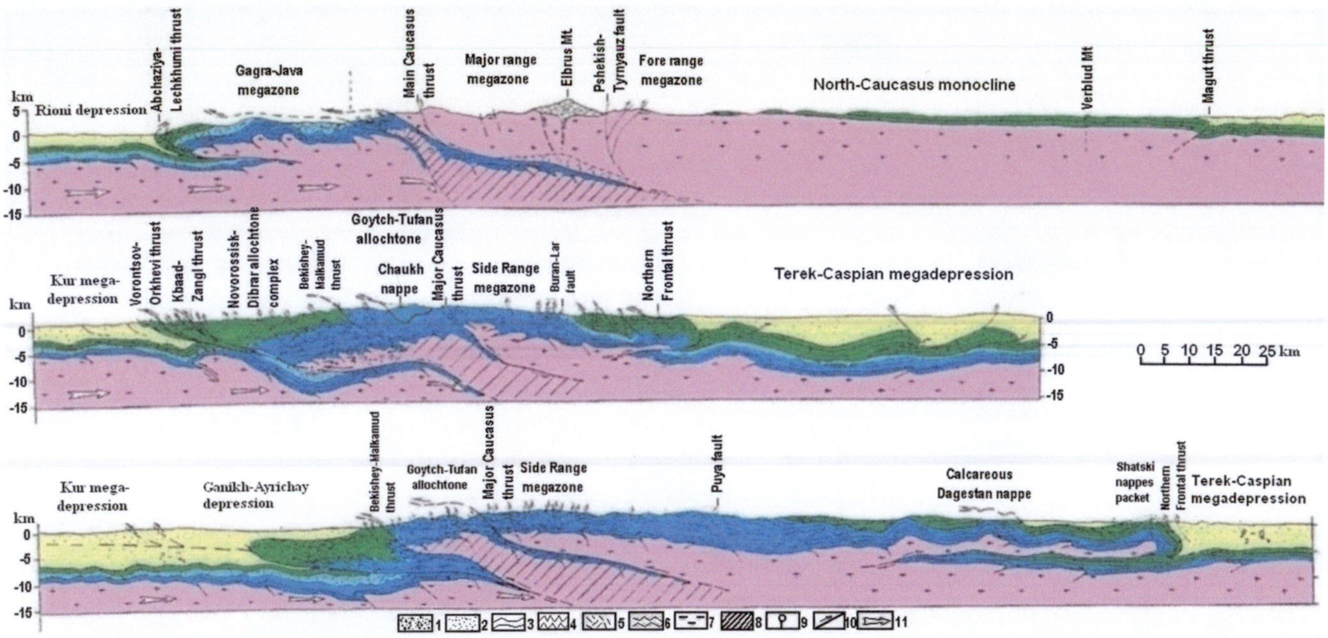

Fig. 5.4 Profiles across the Greater Caucasus with assumed deep structures by Elbrus (I–I), Chaukh (II–II) and Dagestan (III–III) intersections (Dotduyev 1986; Baranov et al. 1990): (*1*) Late Pliocene–Holocene effusions of Elbrus volcano; (*2*) molasses; (*3*) Malmian-Eocene shallow deposits; (*4*) the same, subflysh and flysh; (*5*) Bajos-Bathonian deposits of the Transcaucasian (Southern Caucasus) plate (including porphyry series); (*6*) Lower and Middle Jurassic deposits; (*7*) Pre-Jurassic basement of the Transcaucasian (Southern Caucasus) and Scythian plates; (*8*) subtraction of Goytkh-Tufan and Novorossiysk-Dibrar nappe complexes; (*9*) Alpine intrusions; (*10*) faults and base of nappes; (*11*) direction of predominant direction of crust mass displacement. *P–T* Psekish-Tyrnyauz fault

under the Mesozoic and Cenozoic sea basin of the Gusar-Devechi depression, at the Akhty-Nugadi-Gilazi field traverse (Fig. 5.6). It allows isolation of the *Shahnabad-Jimi nappe complex*, with its autochthonic (parautochthonic) basement being the terrigenic-flysch formation of the Liassic-Dogger period, constituent of the Bejeta and Tufan lamination of the Goytkh-Tufan fold-nappe complex that corresponds to the Southern Slope (Janub Yamaj) trough's northern side. The neoautochtonic complex is formed by the Late Bajocian–Early Bajocian terrigenic complex (mainly sandstone). The Late Bathonian is not represented in this section because it was cut by the ancient pre-Callovian erosion or did not sediment at all (Khalifazadeh 1967). The erosion on the allochthone northern side resulted in complete disintegration of the late Bajocian-Bathonian complex. The terrigenic-flysch compositional complexes of the Callovian, Malmian, and Neocomian periods serve essentially as neoautochtone on the continental slope and piedmont facies, which form the major section of the Shakhdagh-Khizi zone outcropping section; at the same time, they serve as an autochthone for the *Shakhdagh-Beshbarmag overthrust sheet*.

5.2.1.1 The Tahirjal Zone

The Tahirjal zone corresponds to the extreme northern area (within Azerbaijan) of the Middle Jurassic (Aaleinan-Bathonian) basin. It is situated in the Agdash-Darband fault strip (left transtensional fault) of the northern-eastern course, which serves as the border between the Greater Caucasus's eastern, relatively upraised segment and the southeastern lowered segment. In the Tahirjalchay River basin, flexural plunge of the Greater Caucasus's northern side structures (from north to south) under Tertiary formations of the Gusar-Devechi superimposed depression can be observed in the Ulluchay (Jufidag) anticlinal and the Beibulag synclinal zones of the Mountain Dagestan exterior section, as well as the Tahirjal (in Dagestan of the Samur Ridge area) anticlinal rising zone of the Side Range (Yan Sirt) uplift. The first two structures have relatively smooth structures in Dagestan and are simplified by approaching the Samur River, degenerating into single folds. The Tahirjal zone as a whole is mapped as a 5-km wide band of intensive linear folding, confined by the large-amplitude Puya-Tindi-Keydi fracture from the north, and by the Tahirjal fracture from the south. The zone's internal structure is represented by three large arc-shaped anticlines: the Zukhul, the Garakh, and the Maku. The folds are separated by narrower synclines and are complicated in their huckles and sides by small-size folds and low-amplitude ruptured dislocations of heavy southern and northern pitches.

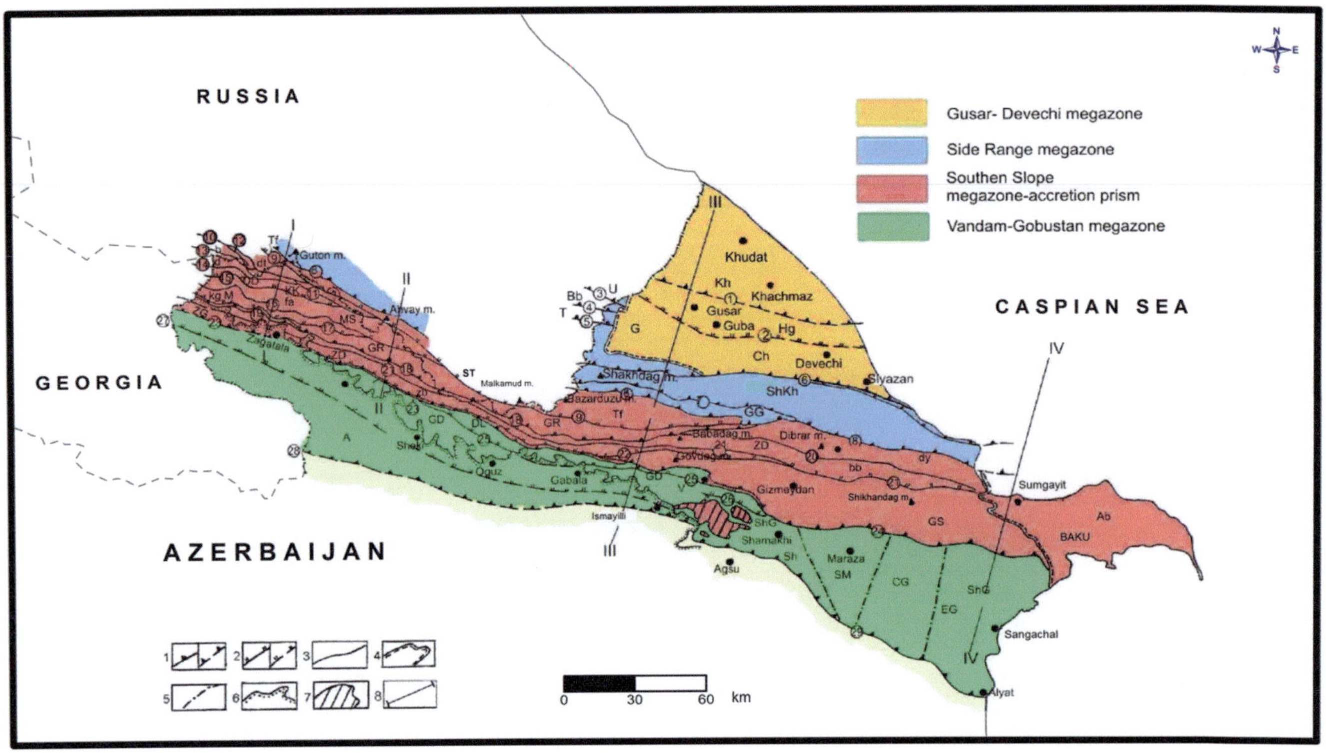

Fig. 5.5 Tectonic scheme of Azerbaijanian in the Greater Caucasus (Khain and Alizadeh 2005; T.N. Kangarli). Boundaries of structures: (*1*) interzone tectonic boundaries (*a* traced on surface; *b* buried); (*2*) tectonic boundaries between subzones (*a* traced on surface; *b* buried); (*3*) boundaries of tectonic slices; (*4*) stratigraphic boundaries; (*5*) conventional boundaries; (*6*) distribution boundary of modern sediments in the Ganykh-Ayrichay superposed depression; (*7*) Basgal nappe; (*8*) lines of synthesized geological-geophysical sections (Figs. 5.6, 5.12, 5.13 and 5.18). **Structures: *Gusar-Devechi megazone***: zones: *Kh* Khachmaz; *G* Guba. ***Side Range (Yan Sirt) megazone***: zones: *U* Ulluchay; *Bb* Beybulag; *T* Tahirjal; *S* Sudur; *ShKh* Shakhdagh-Khizi; *GG* Guton-Gonagkend. ***Southern Slope (Janub Yamaj) megazone (accretionary prism)***: zones: *ST* Speroza-Tufan; *ZG* Zagatala-Govdagh; *Ab* Absheron (western segment of the Absheron-Nearbalkhan zone); subzones: *Tf* Tufan; *JD* Jikhikh-Dindidagh; *MS* Mazym-Sarybash; *M* Megikan; *GR* Galal-Rustambaz; *ZD* Zagatala-Dibrar; *TD* Talachay-Duruja; *GS* Govdagh-Sumqayit; tectonic slices: *jt* Jurmut-Tunsaribor; *kh* Khalakhol; *r* Rokhnor; *b* Boskal; *j* Jikhikh; *kk* Kasdagh-Kasmala; *fa* Filizchay-Attagay; *kg* Katekh-Gumbulchay; *dy* Dibrar-Yashma; *bb* Balakan-Babadagh; *zb* Zagatala-Burovdal. ***Kakheti-Vandam-Gobustan megazone***: zones: *V* Vandam; *ShG* Shamakhy-Gobustan; subzones: *DL* Dashaghyl-Lahyj; *GD* Gulluk-Dadagunash; *A* Ayrichay; *J* Jangichay; *KSh* Keyvandi-Shahgaya; *LA* Langabiz-Alat. **Fractures:** (*1*) Imamgulukend-Khachmaz; (*2*) Khazra-Guba-Kuchay; (*3*) Ashagy Maku; (*4*) Puya-Tindi-Keydi; (*5*) Tahirjal; (*6*) Uruj-Khnov-Siyazan; (*7*) Shakhdagh-Gonagkend; (*8*) Main Caucasus (Bash Qafqaz); (*9*) Khuray-Malkamud; (*10*) Joakhor-Gudurdagh; (*11*) Khalakhol; (*12*) Kasmaldagh; (*13*) Machkhalor; (*14*) Jikhikh-Chugak; (*15*) Kohnamadan; (*16*) Hamzagor-Sarybash; (*17*) Suvagil; (*18*) Gamarvan; (*19*) Megikan; (*20*) Altyaghaj; (*21*) Ilisu-Aladash; (*22*) Gaynar-Gozluchay; (*23*) Mamrukh-Galajyg; (*24*) Zangi-Garajuzlu; (*25*) Dashaghyl-Mudrisa; (*26*) Muju; (*27*) Shambul-İsmailly; (*28*) Galabughur-Jangichay; (*29*) Sangachal; (*30*) Ganykh-Ayrichay; (*31*) Ajichay-Alat

5.2.1.2 The Sudur Zone

The Sudur zone is characterized by the development of essentially box-shaped folds that involve the sediments from the Lower–Middle Jurassic to the Lower Cretaceous ones. The zone segregation takes place south of the Tahirjal fracture. In Azerbaijan, the zone has maximal width (16 km) in the intersection along the Tahirjalchay River, where all structural elements participating in its generation can be traced. In this process, the zone's central section and its northern side gently dip in the southeastern direction under the Miocene–Pliocene formations of the Gusar-Devechi superimposed depression's northwestern edge. For this reason, the length of this section of the Sudur zone within Azerbaijan's borders does not exceed 7 km. At the same time, the zone's southern side structures can be traced at a distance of up to 50 km, having an average width of about 5 km and an oblique cut by the superimposed depression's southern edge, up to the complete pinching out in the Gudiyalchay-Garachay interfluve. The zone's southern border is the Siyazan overthrust (the rear branch of the Main Caucasus overthrust), which is known in the west, within Dagestan's borders, as the Uruj-Khnov thrust.

At its erosional section, the Sudur zone is composed of Middle Jurassic and Upper Jurassic-Neocomian sediments as well as of partially preserved Upper Cretaceous and Paleocene–Miocene sediments. Argillo-arenaceous Middle Jurassic sediments (Aalenian) exposed by the Tahirjalchay River's deep incision are dislocated into two pairs of full

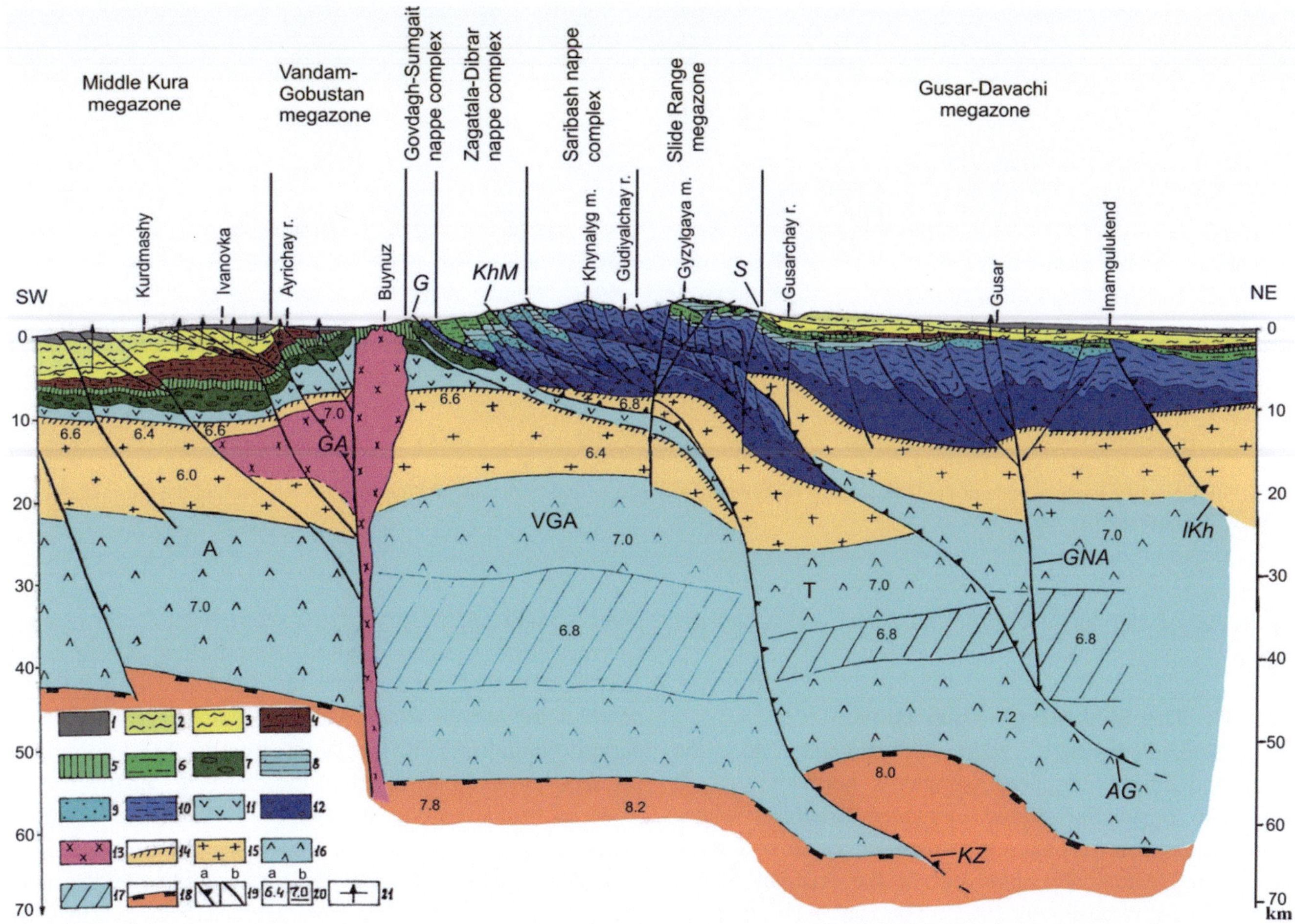

Fig. 5.6 Synthesized geology-geophysical section (III–III in Fig. 5.5) constructed along the earth crust's Kurdmashy-Imamgulukend traverse (Kangarli et al. 1994b; Kangarli 2007, 2012). (*1*) Pleistocene–Holocene; (*2*) Upper Pliocene–Eopleistocene (Akchagyl and Absheron regiostages); (*3*) Middle Miocene–Lower Pliocene; (*4*) Paleocene–Lower Miocene; (*5*) Upper Cretaceous; (*6*) Lower Cretaceous; (*7*) Upper Jurassic and Lower Cretaceous; (*8*) Upper Jurassic; (*9*) Upper Jurassic (Bajocian and Bathonian stages); (*10*) Middle Jurassic (Aalenian stage); (*11*) Lower–Middle Jurassic of Vandam-Gobustan megazone; (*12*) Lower Jurassic; (*13*) syenite-diorites (Buynuz intrusion); (*14*) surface of Prejurassic basement; (*15*) "granite" layer; (*16*) "basaltic" layer; (*17*) "wave conductor"; (*18*) Moho boundary; (*19*) fractures: *a* on the tectonic step boundaries of the consolidated Earth's crust (*IKh* Imamgulukend-Khachmaz; *AG* Axty-Nugadi-Gilazi (Main Caucasus); *KZ* Kbaad-Zangi; *GA* Ganykh-Ayrichay-Alat); *b* other dislocations, to include overthrusts: *S* Uruj-Khnov-Siyazan; *KhM* Khuray-Malkamud; *G* Gamarvan; (*20*) speed of the seismic waves: *a* boundary; *b* layer; *21* wells. Tectonic steps of the consolidated earth crust (longitudinal blocks): *Kh* Khachmaz; *GNA* Gusar-North Absheron; *T* Tufan; *VGA* Vandam-Gobustan-Absheron; *A* Ajinohur

folds (syncline + anticline). The Middle Jurassic section is heightened by the transgressive shallow dipping complex of the Malmian-Neocomian deposits; its overall width tends to increase from north to south and from west to east (from 300 to 1000–1100 m). The complex consists of two formations: the lower sulphate-terrigenous lagoon formation ($J_2c + J_3o$) and the upper carbonate shelf one ($J_3km - K_1n$).

These sediments armor the Middle Jurassic complex and altogether form a megasyncline with a gently dipping northern side and recumbently steep southern side. Its northern monocline side (5.5–6.0 km in width), plunging southward, is collapsing into five linear conformal flat-topped folds (Gukhur, Sudur, Kusnet, Laza, Douk ones) that are occasionally complicated by small-size folds and are separated by narrow synclines (Fig. 5.7). In the south-eastern direction, the folds successively plunge under the Gusar-Devechi superimposed depression with Miocene–Pliocene sediments; only the Laza anticline can be traced along the entire length of the Sudur zone, up to the Garachay River valley. At some points, the syncline bowls preserve separate outliers of Campanian-Maastrichtian (caving conglomerates of the Upper Jurassic-Neocomian chalkstone and carbonate flysch), Sarmatian and Pontian (clays and chalkstones) sediments that superpose the Neocomian ones with

Fig. 5.7 Folds of the southern side of the Sudur zone—compliance between tectonic structure and relief on the left bank of the Gusarchay River (*photo* by I.T. Kangarli)

deposition of the most Upper Cretaceous and Cenozoic stages.

5.2.1.3 The Shakhdagh-Khizi Zone

The Shakhdagh-Khizi zone corresponds to the Shahnabad-Jimi overlapping complex's northern side. It looks like a syncline-shaped elongated structure, whose western extension (in Dagestan) hosts the Samur or Gardib graben that tectonically thins out in the Avarian Koysu valley (Guschin and Panov 1992). The zone's length up to the Caspian shore is about 320 km (within Azerbaijan's borders, above 150 km), with its width being up to 5 km in the west and 12–15 km in the east. The zone is constrained by the Uruj-Khnov-Siyazan thrust-overthrust from the north; in the south, it borders the Guton-Gonagkend elevation along the large-amplitude Tlyagda-Shakhdagh-Gonagkend fault (from 2–3 km in the west to 0.8–1.0 km at the Mt. Shakhdagh meridian). After tectonic thinning-out of this elevation in the east (the Istisuchay-Gyzylgaya interfluve), the Main Caucasus overthrust becomes the zone's southern border, as the thrust horizontal offset at the Caspian coast, according to the drilling data, equals to 1800 m. The zone has a fold-overlapping structure and is represented by autochthonous, allochthonous, and neoautochthonous complexes at its daylight surface (Isayev et al. 1981; Shekinsky et al. 1985; Kangarli et al. 1994b; Kangarli 2005, 2007, 2009, 2011, 2012).

The allochthonous complex is traced at a distance of up to 120 km west of Shakhdagh and Gyzylchay massifs and east of Chyrag-Gala castle and Mt. Beshbarmag. It is composed of Malm-Neocomian reef carbonates. The most complete cross-section is preserved at Mt. Shakhdagh, with up to 1400 m thickness. In the contemporary structure, primary allochthone is disintegrated in different-scaled gently bent plates (Shakhdagh, Gyzylgaya-Zukhur, Agh-Gadik, Sohub, Chyrag, Dashly-Jalgan, Sagholjan) and clips (Beshbarmag, Jek) that are embossed among the dislocated Lower Cretaceous autochthonic strata. The amplitude of visible duplication overlap is 7–10 km in the west to 1–3 km and the primary hundreds of meters in the east (Fig. 5.8).

In its frontal section and foothill as well as at a certain distance, in front of the allochthonous plates, the tails of chaotic complex of allochthonous and autochthonous rocks can frequently be observed (sedimentary mélange); they are unconformably located at the autochthone corrugated surface and are the primary mass overthrust constituent. The allochthone roots are confined to the Siyazan fracture zone, where the steeply set Upper Aalenian argillites and Malmian-Neocomian chalkstones protrude in the narrow band and can be traced from the Usukhchay and Tahirjalchay Rivers' divide to Tanga sluice pipes. Furthermore, the gash zone punches out, and a tectonic suture between compositional complexes of autochthone and the Gusar-Devechi depression can be observed up to the structures' common plunge at the Caspian coast. In the west, allochthon (Gyzylgaya-Zuhur massifs) is overthrusted to northwest, overlapping (along the Gyzylgazma upthrust) both the scar zone and Lower Cretaceous limestone which overlie by Paleogene, Sarmatian and Pontian deposits of the Sudur zone.

The autochthonous complex forms a synclinally shaped elongated structure, which is essentially composed of Jurassic and Lower Cretaceous (including Lower Aptian) terrigenic formations with facies that are dramatically different from the allochthone coeval deposits. They are represented by three compositional complexes: Aalenian-Lower Bajocian argillo-arenaceous flysch, coarse Callovian-Tithonian terrigenous flysch, and Lower

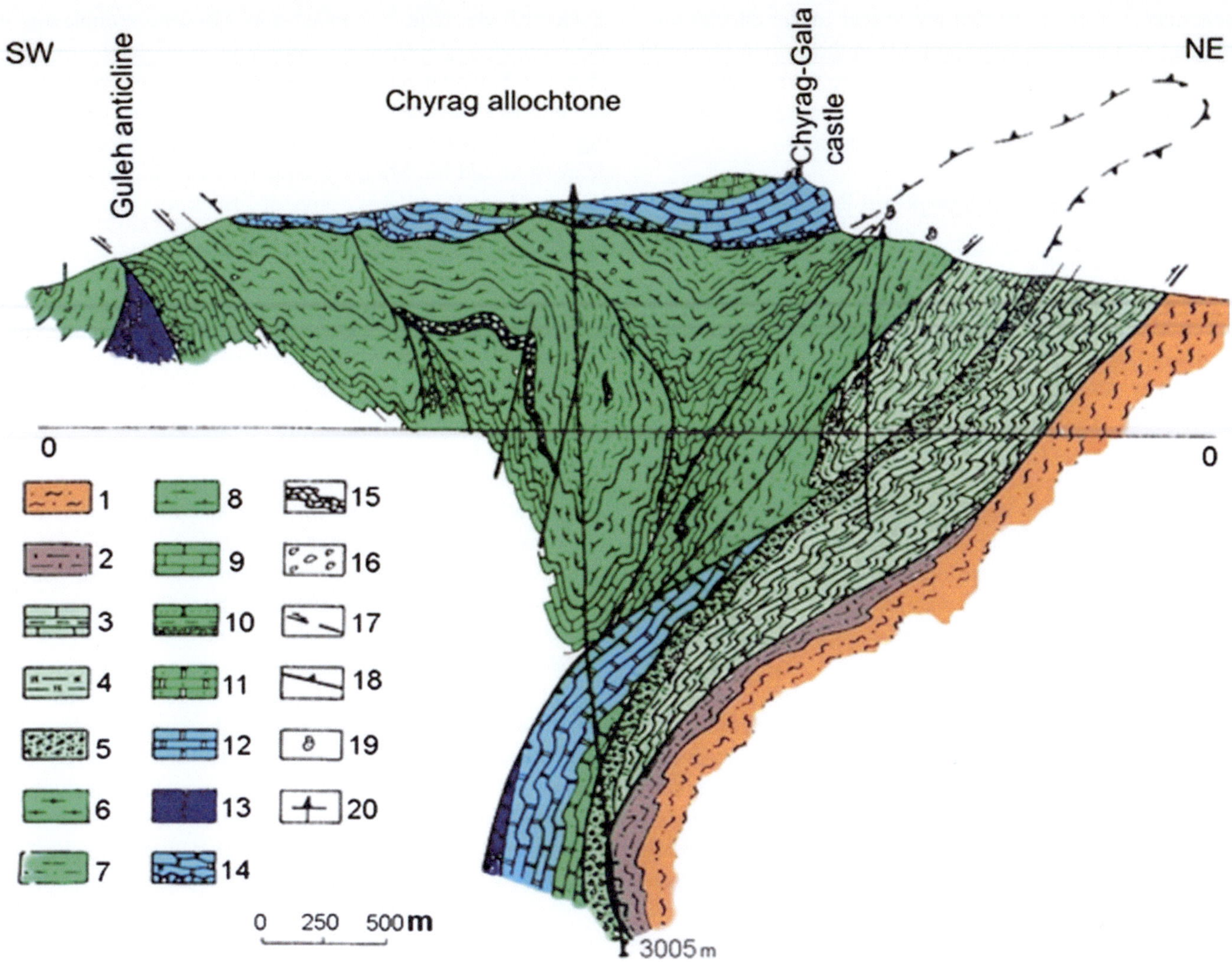

Fig. 5.8 Geological section crossing the Chyrag nappe sheet of the Shakhdagh-Beshbarmag allochthonous complex (Khain and Alizadeh 2005; Kangarli 2009, 2012). (*1*) Oligocene–Lower Miocene: clays, sands, aleurolites; (*2*) Paleocene–Eocene: clays, sands, marls; (*3*) Upper Campanian–Danian: marls, sandstone, limestone, gritstones; (*4*) Santonian–Lower Campanian: clays, sandstone, limestone, marls, gritstones; (*5*) Upper Turonian–Coniacian: conglomerates; (*6*) Upper Aptian—Albian: clays, marls, sandstone, limestone, gritstones; (*7*) Barremian—Lower Aptian (lower part): argillites with marls and limestone parting; (*8*) Hauterivian: argillites with concretions and marls and limestone parting; (*9*) Valanginian: carbonaceous-terrigene flysch; (*10*) Berriasian: basal conglomerates, argillites, sandstone, limestone; (*11*) Neocomian: carbonaceous flyschoid; (*12*) Upper Jurassic: limestone ("Shakhdagh" facies); (*13*) Lower Jurassic: argillites, sandstone; (*14*) tectonic crushing zone; (*15*) olistostromes; (*16*) olistolites; (*17*) thrusts and upthrust-overthrusts; (*18*) sole of nappe; (*19*) stations for faunistic justification of the rocks' age; (*20*) wells

Cretaceous "wild flysch" that contains Shakhdagh chalkstone olistholiths, olistostromes, and olistoplaques (Vassoyevich 1938; Khain 1950; Isayev et al. 1981; Shekinsky et al. 1985). Jurassic formations are exposed in the eroded domes of comb-shaped anticlines, obliquely located towards the zone's general strike. The lower Cretaceous wild flysch serves as an immediate underlayer for the allochthonous plate; its spots protrude at the Shakhdagh lamina front in the west and are completely exposed from under the primary mass overthrust preserved fragments east of the Gudiyalchay River. The Upper Jurassic–Lower Cretaceous autochthone interval as well as its superimposed Shakhdagh allochthone plate steeply drops along the Agdash-Darband transtensional fault in the Shahdagh massif western escarpments. Only the Middle Jurassic section of the Shahdagh-Khizi zone autochthonous complex, without any allochthone participation, protrudes within the limits of elevated fault ring in Dagestan.

The neoautochthonous complex in the shape of low-angled Upper Jurassic troughs with basal conglomerates in their basement ("the Khizi conglomerates") levels concertedly the autochthonous and allochthonous complexes; they are located as separate fragments either on the allochthone plate or on the eroded and corrugated autochthone surface. In places, some thin Upper Aptian, Albian, and

Senomanian formations that are absent in the allochthone-situated fold northern sides protrude in the section basement in the trough's southern sides. They are also absent in the allochthonous plates bottoms, which, finally, show their participation in neoautochthone composition.

5.2.1.4 The Guton-Gonagkend Zone

The Guton-Gonagkend zone generates the Shahnabad-Jimi overlapping complex's southern side and is situated on the natural extension of anticline benches in the frontal root-free slices of the Main Caucasus's thrust zone in the Central Caucasus, which includes the Kion and Chaukh nappes of 10–15 km amplitude. According to Russian geologists (Shempelev 1978; Romanov 1983; Makarovskiy 1982; Guschin and Panov 1992; Panov 2002), this fracture has no roots in the mountainous zone; it remains in a "suspended" condition. At a depth of about 8 km, it either breaks away along the Jurassic sediment bottom or turns into a large mass overthrust in a pre-Jurassic basement, by which the entire Side Range (Yan Sirt) crust plate is shifted southward to the Southern Slope (Janub Yamaj) zone, an axial trough of the Greater Caucasus's Early Jurassic basin.

By and large, the zone looks like a large Shahnabad-Maza para-fractured anticline intruded by thrust-fault into a series of digitations in the Main Caucasus's thrust suspended side, to the north confined by the Tlyagda-Shahhdagh-Gonagkend fault. It is composed of Lower–Middle Jurassic terrigenic flysch and traced (completely or without fault-cut core and southern side) from the outcrops of the Central Caucasus Daryal granitoids to the Jimichay River at a distance of above 610 km and a width from 3.5–4.0 to 6–7 km. Coming east of the Shahnabadchay River within Azerbaijan's bounds, the anticline is complicated by additional higher-order folds as a result of fold plunge and virgation. At the Jimichay River's right bank in the Girdymanchay-Valvalachay flexural zone, the Middle Jurassic submerges under the Malmian-Neocomian coarse flysch (the neoautochthonous complex) and is further exposed from under the complex to the surface in "anticline" windows at the deep river valley shut-ins, up to and including the Istisuchay River (the Gilgilchay River right tributary); immediately after that (20 km of the Jimichay River), it plunges conclusively.

5.2.2 The Gusar-Devechi Megazone

This Gusar-Devechi megazone is the superimposed depression of the Agdash-Darband transversal fracture downthrow side that was established on a heterogenous basement in the Late Miocene (Sarmatian age) and was finally formed in the Early Pliocene as the western closing of the Middle Caspian North-Absheron Depression (Khain and Alizadeh 2005). The megazone structure is determined by the levelling of dislocated Mesozoic and Paleocene–Miocene complexes of the Greater Caucasus's northern side and of the Scythian-Turanian platform's southern edge by continental and marine molasses of the Pliocene–Quaternary period (see Fig. 5.6). A conditional border between the superimposed depression and the Middle Caspian area is drawn along the Yalama-Zarat-Khyrdalan abyssal fracture that extends southeast along the Caspian Sea coast.

Geological and geophysical sources demonstrate the existence of a deep pre-Sarmatian washing-away, which resulted in complete destruction of the Middle Miocene, Oligocene-Lower Miocene, Paleocene-Eocene, Cretaceous, and Upper Jurassic sediments within the megazone's large areas (Fig. 5.9). Subsequent relative elevation that took place in the Late Miocene–Early Pliocene was replaced by intensive descending movements and powerful sediment accumulation of the Early Pliocene–Quaternary period with active rising of the Side Range internal zones. From here, the detrital material, up to the coarse one, was ablated, generating continental molasses that were laterally eastward and southeastward substituted by the marine ones.

Seismic survey results demonstrate the intensive fracturing of buried Mesozoic–Cenozoic formations made by folded-ruptured dislocations, with subsequent subdivision into different-scaled fault blocks. In this process, the mountainous zone and Seacoast Dagestan dislocations can be traced in tectonics of the buried Mesozoic complex, which, with rare exception, do not find direct correlation with the Mesozoic surface's rugged topography. From a hypsometric viewpoint, higher interface surfaces repeat the Mesozoic surface topography in large part. The structures of the Side Range uprising's northern side (the Sudur and Tahirjal zones) and the Mountainous Dagestan (the Ulluchay and Beybulag zones), which plunge into the lightly-pitching molasses of the superimposed depression's southwestern edge and its central section, find their natural extension in the buried Guba tectonic zone that is delineated on the basis of combined geophysical data analysis. Based on the same data, the depression's northeastern edge corresponds to the Khachmaz tectonic zone at the molassa-buried extension of the seacoast Dagestan Cretaceous and Tertiary folds. The latter structurally belong to the Samur-Peschanomys (Samur-Kumdi Muis) elevation's southern side of the Scythian-Turanian epi-Hercynian platform, which in the south is confined by the Imamgulukend-Khachmaz fracture.

5.2.2.1 The Khachmaz Zone

The Khachmaz zone is the southeastern extension of the Dagestan seacoast's anticlinal zone. Within the zone in Azerbaijan, seismic survey and deep-hole prospecting results displayed local brachyo-elevations such as Yalama, Khudat, etc.; in their arches, the Mesozoic surface is stripped under various Cenozoic horizons (from Paleocene to Lower

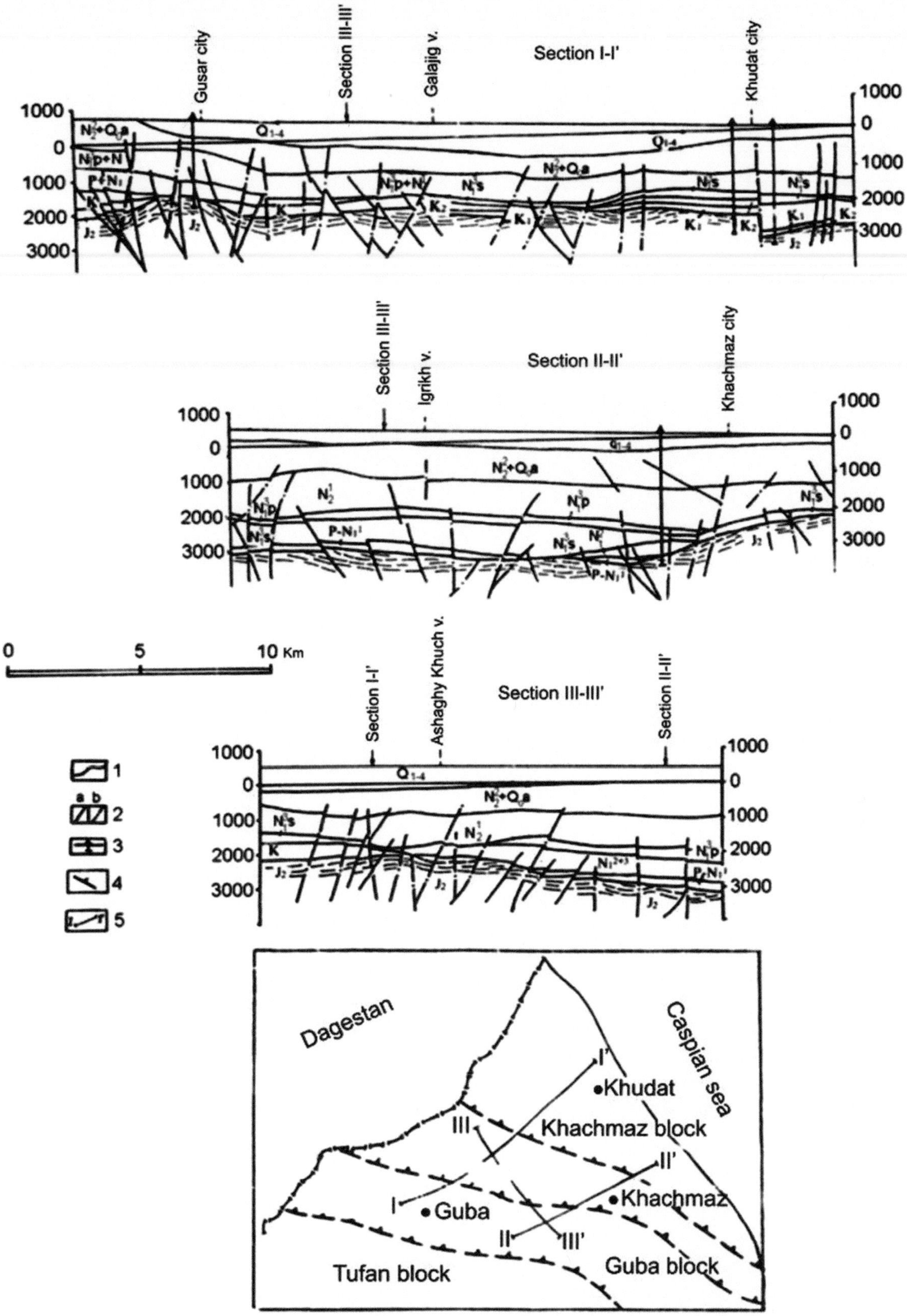

Fig. 5.9 Geologic-geophysical sections of the Gusar-Devechi megazone by Urva-Khudat (I–I′), Nugadi-Khachmaz (II–II′) and Galajig-Pirvahid (III–III′) traverses (composer: Kangarli et al. 1994a; Khain and Alizadeh 2005): (*1*) stratigraphic contacts; (*2*) faults of Common-Caucasian (**a**) and Anti-Caucasian (**b**) strike; (*3*) boreholes; (*4*) faults at the boundaries of structural zones; (*5*) lines of geological-geophysical sections

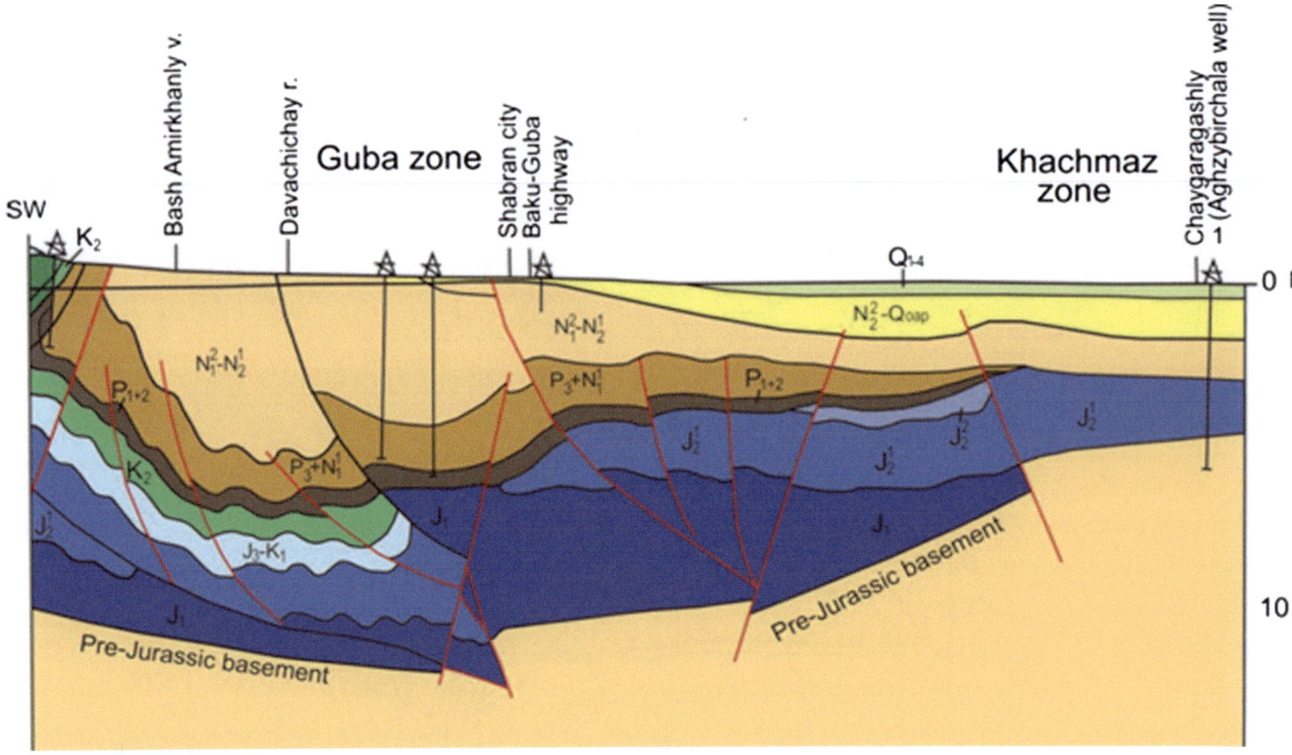

Fig. 5.10 Geologic-geophysical section of the Gusar-Devechi megazone along the Bash Amirkhanly-Chaygaragashly traverse (Kangarli 2011)

Pliocene) at depths from 1800 to 3000 m, with the immersion tendency from northwest to southeast and from northeast to southwest. At the same time, along the elevation course, the Mesozoic bed thickness reduces: at the extreme southeast, near the Devechi coast (Agzybirchala area) and within the borders of a buried local elevation, a well downhole at 4700 m entered and esites (Early Jurassic); at 4830 m, it exposed the Permian-Triassic metamorphic rocks. In the Middle Caspian water area, the Yalama-Agzybirchala zone of the Samur-Kumdi Muis elevation's southern periphery serves as its natural extension. In the southwest, the zone, along the Imamgulukend-Khachmaz buried fracture, borders the lowered Guba tectonic zone; the buried fracture amplitude (based on seismic survey results) at the pre-Jurassic base level reaches 3.5–4.0 km in the northwest (see Fig. 5.6), reducing to 1.0–1.5 km in the southeast (Fig. 5.10). The Yalama-Agzybirchala serves as the zone's natural extension in the Middle Caspian water area.

5.2.2.2 The Guba Zone

The Guba zone on the southwestern side borders the Shakhdagh-Khizi zone of the Side Range elevation along the Uruj-Khnov-Siyazan dislocation; on its northeastern side, it is confined by the Imamgulukend-Khachmaz fracture (see Fig. 5.10). The downthrown block of Earth's crust located within these borders generally corresponds to the narrowed edge of the Northern Caucasus's continental microplate, which can be traced to the Middle Caspian borders as a consolidated basement of the Northern Absheron depression. In the strip adjacent to the Siyazan thrust, the Pliocene formations east of the Gudiyalchay River are fold-collapsed together with underlying Miocene and Paleogene sediments. North of this strip, the structure of the Alpine complex hidden under Quaternary molasses is characterized by its complex composition, which generates second-order structures (from the south to the north):

1. ***The Guba edge depression***, situated at the Sudur zone's southeastern extension of the Side Range elevation, is clearly expressed in Pontian and older sediments, down to Cretaceous sediments. It has a maximum width of 18–20 km at the town of the Guba meridian. The Cretaceous–Paleogenic sediments that compose the depression's southern side (the Siyazan monocline) generally correspond to tectonic slices with steep and northeastern oversteepened beds. The Rustov-Gilvar syncline is located north of this strip. At its core, Paleocene–Eocene sediments are exposed by drilling at a depth of more than 4000 m.
2. ***The Talabi-Gyzylburun anticline belt*** stretches to a distance of up to 80 km, from the Agchay River in the northwest to the sea coast near the Atachay River's estuary. It consists of imbricately substitutions along the

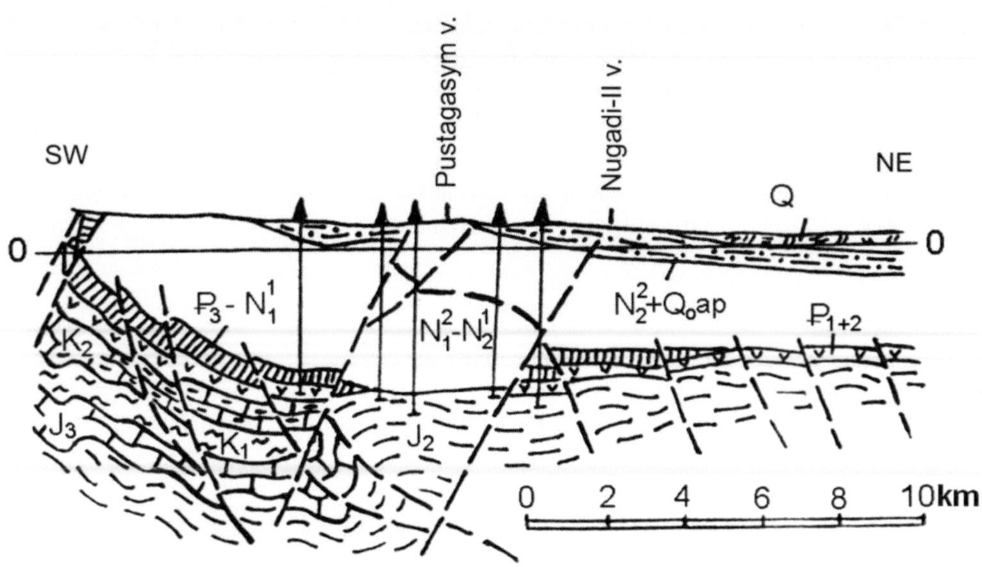

Fig. 5.11 Geological section across the Talabi plateau (Khain and Alizadeh 2005; T.N. Kangerli)

strike of the Nugadi, Talabi, West-Gaynarja (Shabranchay), Gaynarja and Gyzylburun land folds, as well as the Gyzylburun-deniz and Jorat-deniz elevations that extend to the sea. According to drilling data on the Talabi elevation dome, the Lower Sarmatian sediments lie at a depth of over 4280 m on the Middle Jurassic argillites (Fig. 5.11). Fold cores and sides are complicated by ruptured faults of general- and anti-Caucasus directions.

3. ***The Gusar buried high*** is situated on the buried extension of the Mountainous Dagestan Ulluchay anticlinal zone. It is expressed by a local maximum in the Bouguer gravity: maximum axis extends from Ashagy Galankhur village southeastward toward the town of Guba. The maximum fails to reach the Valvalachay River, 7–8 km northeast of the Talabi fold. The wells drilled in the town of Gusar area exposed argillites and Middle Jurassic shales that are transgressively superimposed by the Sarmatian stage sediments at a depth exceeding 2400 m.
4. ***The Zeykhur-Gilgilchay depression*** is expressed in the Bouguer gravity by a local maximum. It can be traced from the Samur River right bank to the Caspian coast in the Gilgilchay River estuary. It extends to more than 85 km and has a maximum width of 18–20 km. The depression is rather clearly expressed in the Miocene–Lower Pliocene sediments and is considerably weaker in the Akchagyl and Absheron age layers. A well drilled in the depression's central section (in the Imamgululend village area) has exposed the descending cross-section of the Holocene, Pleistocene, Eopleostocene, Pliocene, Miocene (the Sarmatian and the Chokrakian stage), and Paleocene (the Danian stage) under modern detrital deposits. A well point at a depth of 2250 m entered the Upper Cretaceous layer, whose thickness in the Maastrichtian-Turonian interval exceeded 750 m, according to drilling data.

5.2.3 The Southern Slope Megazone

The Southern Slope (Janub Yamaj) megazone (trough) in its first approximation corresponds to the southern oversteepened megasynclinal with an upturned northern edge. It is represented by the southeastern segments of the Goytkh-Tufan and Novorossiysk-Dibrar fold-nappe complexes that stretch to the Caspian Sea and are further traced in the structure of the Absheron-Nearbalkhan (Absheron-Balkhanyani) elevation zone's southern side (Khain and Alizadeh 2005; Khain et al. 2007). In its contemporary structure, the megazone corresponds to the accretion prism confined between the Main Caucasian Ridge (in the north) and the Kbaad-Zangi (in the south) dislocations and the Southern and Northern Caucasus (Janubi and Shimali Gafqaz) continental microplates (when the first one underthrusts the second one), which formed during a pseudo-subduction interaction that took place at various Alpine tectogenesis phases and subphases (from Late Cimmerian to Walachian ones). This was the reason for tectonic stratification of the accretion prism's Alpine formations into different-scaled lamina of southern vergence, which are united into overlapping complex (Kangarli 2007, 2009, 2011, 2012). The prism-underlying autochtonic substrate is represented by the Mesozoic–Cenozoic complex of the northern Vandam-Gobustan margin (megazone) of the Southern Caucasus continental microplate. The complex, in its turn, is collapsed and dispensed into southern-detached tectonic plates that gently overlap the Kur megadepression's northern edge along the Ganykh-Ayrichay-Alat thrust. Thus, the

North-Caucasian margin of the Scythian-Turanian platform is the allochthone cover, and the Alpine complex of the Southern Caucasus continental microplate's northern edge is the autochthone (parautochtone) basement of the accretion prism.

At the Cimmerian development stage, the megazone was the most immersed section of the paleobasin, with accumulations of thick clays, aleurolites and sandstone in combination with manifestations of tholeiitic basalt magmatism (diabase-porphyrite-spilite and albitophyre formations). In continuation, up to the period of total uplift to the orogene development stage, the megazone developed under the conditions of inherited subsidence, as evidenced by its maximal thicknesses and the most complete cross-section of the Upper Jurassic, Cretaceous, and Paleogene flysch formations. In the southeastern depression strip, the section is raised by argillo-arenaceous (carbonate); in the upper part, it is Miocene rocks. Under regional southeastern plunge conditions of the structures, the Main Caucasus thrust cut the depression's northern edge at an acute angle, with complete tectonic overlapping of its Jurassic structural and compositional complex west of the Girdymanchay-Valvalachay flexure. From the south, the Southern Slope's trough is limited by the Kbaad-Zangi abyssal fracture (underthrust), which is expressed at the surface by the Gaynar-Zangi group of elementary thrusts; along them, the Jurassic and Cretaceous formations torn off their basement are shifted to the Alpine structures of the northern edge of the Southern Caucasus continental microplate.

The megazone's composition is subdivided into the following:

(1) The Bejeta zone, which is tectonically pinched along the Main Caucasus thrust at the approaches to Azerbaijan territory (in the Jurmut River head) and represents a graben-like structure composed of the non-metamorphosed argillo-arenaceous sediments of the Upper Toarcian and Lower Aalenian age with the north-thrusted Side Range Domerian-Lower Toarcian shales;
(2) The Speroza-Tufan zone (eastern segment of the Goytkh-Tufan fold-nappe complex), which has a smooth structure on the northern side and a southern side that is collapsed into south-overtilted and thrust-cut folds;
(3) The Chiauri-Dibrar zone (the eastern segment of the Novorossiysk-Dibrar fold-nappe complex), which consists of two echelon segments:
(a) The Georgian segment (the Chiauri depression that gives rise to the Mountainous Kakhetian overthrust sheets), which is submersed under contemporary drifts of the Ganykh-Ayrichay depression
(b) The Azerbaijan segment (the Zagatala-Govdagh zone), which emerges from under this drift at the Mazymchay River left bank; it looks like a narrow zone separated on the Bekishey-Gamarvan dislocation lowered side, which, when broadening, spans to the general Caucasian direction. The principal structural element of this zone is the Zagatala-Dibrar depression, which is composed of the Upper Jurassic, Cretaceous, and Paleogene sediments. Along its southern edge, a narrow Talachay-Duruja plate of Jurassic rocks is exposed in the west and, after its closing, the Govdagh-Sumgayit overthrust-folding structure (a homolog of the Mountainous Kakheti Alisisgori-Chinchveli overthrust sheets and of the North-Western Caucasus Chvezhipsa allochthonous complex) in the east. Thus, the extreme southeastern complexes of the Novorossiysk-Dibrar and Chvezhipsa-Govdagh regional complexes are represented as structural elements of a unified Zagatala-Govdagh zone and participate in its composition as subzones.

Within the megazone borders, the Alpine cover is allochthonously located on the substrate made of the Middle Jurassic igneous-sedimentary rocks in Vandam facies; its presence is defined by a set of geophysical techniques. It is represented by the north-to-south overriding heterochronous fold-nappe complexes of structural zones such as Tufan and Sarybash in the Speroza-Tufan structure, as well as Talachay-Duruja, Zagatala-Dibrar and Govdagh-Sumgayit in the Zagatala-Govdagh structure. In turn, these consist of allochthonous slice (subzone) packages and separate higher-order tectonic plates. The Speroza-Tufan zone's allochthonous plates owe their origin to the Late Cimmerian deformations; those of the Zagatala-Govdagh zone are from Late Alpine deformations.

5.2.3.1 The Speroza-Tufan Zone

Within Azerbaijan's borders, the Speroza-Tufan zone is represented by the Tufan, Jikhikh-Dindidagh, Mazym-Sarybash, Megikan, and Galal-Rustambaz subzones. The first corresponds to the independent nappe plate; the others, as allochthonous schuppens, participate in the structure of the Sarybash nappe complex (Figs. 5.5, 5.12 and 5.13).

The Tufan nappe complex spreads from the Jurmut riverhead in Dagestan in the southeastern direction, at a distance of more than 195 km, through to the Jimichay River meridian, where it pinches out tectonically at the Girdymanchay-Valvalachay flexure. *Allochthone* is confined to the north by the Main Caucasus thrust and to the south by the Khurai-Malkamud thrust. It is represented by the tectonic plate of the Lower–Middle Jurassic argillo-arenaceous flysch sediments (from the Toarcian to the Lower Bajocian) in the first dislocation front, superimposed on the accretion prism central section structures; despite their allochtonous condition in the contemporary structure, they bear the function of

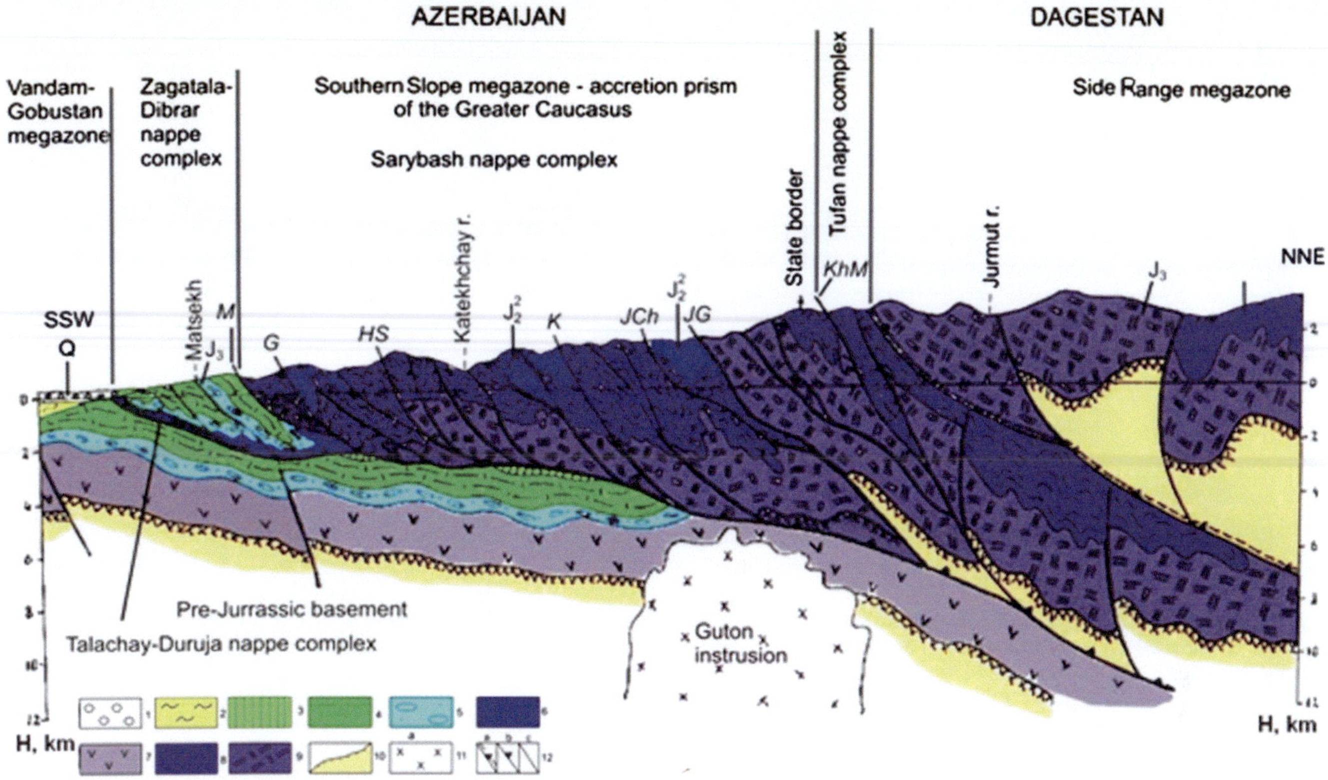

Fig. 5.12 Synthesized geological-geophysical section (I–I in Fig. 5.5) constructed along the Matsekh-Jurmut River's traverse passing through the nappe complexes of Northwestern Azerbaijan (Kangarli 2007, 2012). (*1*) Pleistocene-Holocene; (*2*) Eopleistocene (Absheron regio-stage); (*3*) Upper Cretaceous; (*4*) Lower Cretaceous; (*5*) Upper Jurassic; (*6*) Middle Jurassic (Aalenian stage); (*7*) Lower–Middle Jurassic (volcanogenic–sedimentary formation of the Vandam zone); Lower Jurassic (*8–9*): (*8*) Toarcian stage; (*9*) Sinemurian and Pliensbachian stages; (*10*) surface of prejurassic basement; (*11*) Guton intermediate-basic intrusion (according to geophysical field interpretation—see Chapter V of Volume II); (*12*) fractures: *a* Main Caucasus thrust; *b* Kbaad-Zangi (Mamrukh-Galajig) thrust; *c* other faults, to include overthrusts: *M* Megikan; *G* Gamarvan; *HS* Hamzagor-Sarybash; *K* Kokhnamadan; *JCh* Jikhikh-Chugak; *JG* Joakhor-Gudurdagh; *KhM* Khuray-Malkamud

autochthone (parautochtone) for the cover described. The nappe roots are absorbed in the consolidated crust pseudo-subduction zone in the Akhty-Nugadi-Gilazi fracture strip, which reflects the Main Caucasus's thrust position at the basement level. In an *autochthone* cross-section, two sub-complexes are separated: the Liasic-Lower-Bajocian, which served as an autochtone substrate for the early subphase; and the Upper-Bajocian-Titonian, which served as a substrate for the late subphase of the Late Cimmerian tangential stresses; this is the reason for the two-act scenario of the Tufan nappe generation. The *neoautochthonous complex* is represented by the lower section preserved after the drainage; ir comprises the stratigraphic interval from the Late Bajocian to the Callovian ages.

An advanced analysis of the magnetic field over the Guton intrusion is presented in Figs. 4.37 & 4.38 in Volume II.

The Sarybash nappe complex, composed of Lower–Middle Jurassic terrigenous flysch (in combination with the manifestations of tholeiitic-basalt magmatism) is the principal structural element of the Southern Slope megazone. Coming from the borders of Georgia and Dagestan to the Azerbaijan territory, it can be traced as a narrow side wedge at a distance of more than 220 km up to the Jimichay River. East of the river, it immerses flexurally and is tectonically superposed by the Main Caucasus thrust. From the south, the allochthone is limited by the Gamarvan thrust. In its interior structure, it is subdivided into different-scaled tectonic schuppens of southern vergence, such as the Jikhikh-Dindidag, Mazym-Sarybash, Megikan, and Galal-Rustambaz zones. In the riverheads of the Samur River's right tributaries (the Kurdul and the Akhtychay), a transversal narroside of the Speroza-Tufan zone takes place, with longitudinal south-eastern immersion and cross-section rejuvenation.

In this process, the Main Caucasus thrust and the Khuray-Malkamud thrust situated in its recumbent wing within Dagestan (after the Akhvay Mountain meridian) experience deviation from the General Caucasian to a more southern direction, forming a broad, south-bent arch. This results in tectonic thinning-out (allochtonous superposition) of the nappe complex's northern plates along these

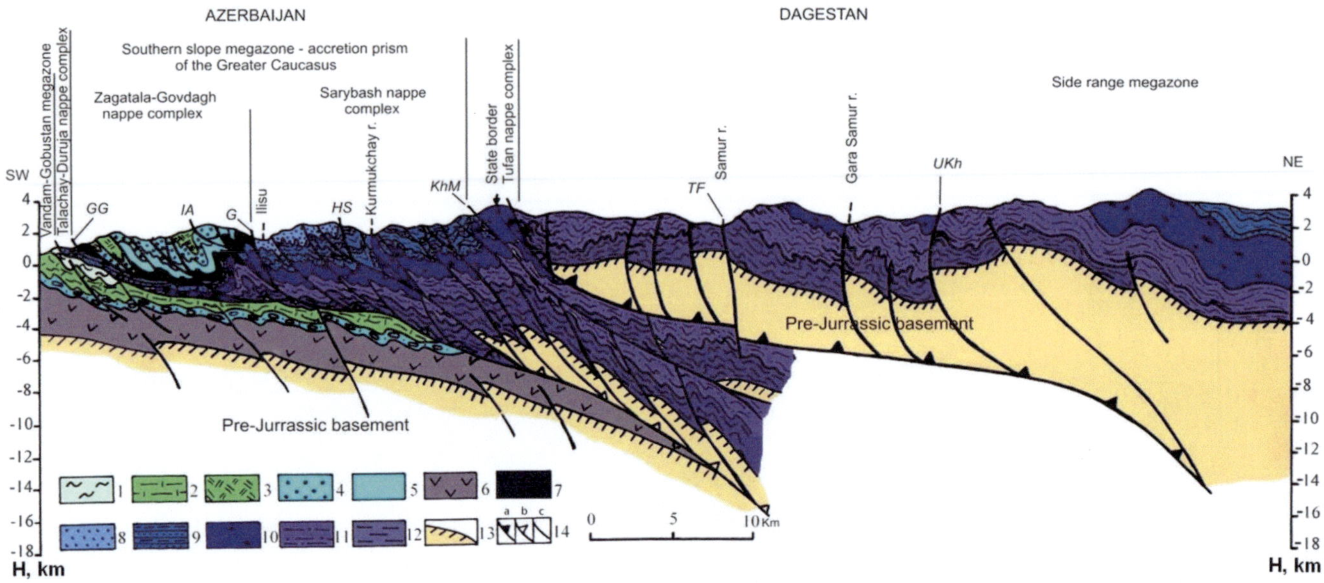

Fig. 5.13 A synthesized geological-geophysical section (II–II in Fig. 5.5) constructed along the Ilisu-Gara Samur River's transverse passing through the cover complexes of Northwestern Azerbaijan and Mountainous Dagestan (Kangarli 2009, 2012). (*1*) Upper Cretaceous (Vandam zone); (*2*) Lower Cretaceous (Vandam zone); (*3*) Lower Cretaceous (Berriasian and Valanginian stages); Upper Jurassic (*4–5*): (*4*) upper substage of the Oksfordian stage, Kimmeridgian and Tithonian stages, (*5*) lower substage of the Oksfordian stage; (*6*) Lower–Middle Jurassic (volcanogenic–sedimentary formation of the Vandam zone); Middle Jurassic (*7–10*): (*7*) Callovian stage; (*8*) upper substage of the Bajocian stage; (*9*) Upper Aalenian stage and lower substage of the Bajocian stage; (*10*) Lover Aalenian stage; Lower Jurassic (*11–12*): (*11*) Toarcian stage; (*12*) Sinemurian and Pliensbakhian stages; (*13*) surface of Prejurassic basement; (*14*) fractures: *a* Main Caucasus thrust; *b* Kbaad-Zangi (Mamrukh-Galajig) thrust; *c* other faults; overthrusts: *GG* Gaynar-Gozluchay; *IA* İlisu-Aladash; *G* Gamarvan; *HS* Hamzagor-Sarybash; *KhM* Khuray-Malkamud; *TF* Tlyagda-Falfan; *UKh* Uruj-Khnov

dislocations. Only the extreme southern Galal-Rustambaz wedge scales are traced up to the Jimichay River; west of it is cut by the thrust bottom of the Main Caucasus. The *autochthonous complex* is represented by the Lower–Middle Jurassic formations of the Kakheti-Vandam-Gobustan megazone. Within Azerbaijan's territory, the complex is not exposed; however, along its western trend within Georgian borders—due to deep disintegration of Mesozoic rocks in the central part of the Southern Slope megazone and southern side—it emerges onto the surface, thus forming the lower part of the Gagra-Java zone cross section. The formations, from the Sinemurian to Pliensbachian age, essentially have terrigenic-flysch composition and are superimposed by the thick volcaniclastic Bajocian "porphyritic suite," represented by calci-alcalic, weakly differentiated, basalt-andesite-rhyolite series.

The *neoautochtonous complex* is represented by terrigenic-flysch formations stratified in the Callovian-Lower Oxfordian age interval, which have been preserved in the contemporary structure only at the Galal-Rustambaz plate cross-section. The complex was formed within the period between two subphases of the Late Cimmerian deformations: the Middle pre-Cellovian and the Late pre-Cretaceous, which were accompanied by large tectonic movements and regional breaks in sediment accumulation. The Upper Jurassic and Lower Cretaceous horizons are absent in the neoautochtone cross-section because of later disintegration. In the contemporary structure, under the impact of the pre-Cretaceous and later Alpine movements, the complex is folded together with allochtone.

5.2.3.2 The Zagatala-Govdagh Zone

This zone is the most extended and structurally pronounced fabric element of the Greater Caucasus Azerbaijan part. By its typical features, it is subdivided into three subzones: the Zagatala-Dibrar, the Talachay-Duruja, and the Govdagh-Sumgayit. These subzones differ in the stratigraphic section, type, and behavior of their constituent structural shapes (see Fig. 5.5).

The Talachay-Duruja nappe complex was formed during the pre-Callovian subphase of the Late Cimmerian deformations. At the boundary between the Cretaceous and the Paleogen ages, it was repeatedly south overthrust (together with neoautochtone cover of Calovian-Titon rocks) to the Vandam high borders (see Figs. 5.12 and 5.13). The *allochtonous complex* is exposed by two fragments (wedge scales):

1. The northern Talachay massif of the Toarcian terrigenous-flyschoid rocks combined as minor folds and exposed within the Balakan-Kakh transverse uplift at the

bottom of the Upper Jurassic–Lower Cretaceous series of the Zagatala-Govdagh northern side

2. The southern Duruja fragment, which can be traced along the southern edge of the same zone for 125 km as a narrow (up to 2.5–3.0 km) course of outcrops of the Aalenian argillaceous slates, aleurolites and sandstone with tuff lava and tuff breccia flows, having andesite-dacite composition in the section lower parts.

The plate is limited by frontal slices of the Zangi abyssal fracture: the Gaynar-Gozluchay north thrust and the Mamrukh-Galajyg south thrust. In general, it corresponds to the south overthrust anticline with cut southern side. The *autochtonous complex* is represented by volcanogenic–sedimentary Cretaceous rocks of the Vandam uplift's northern edge. It protrudes along the entire Duruja plate front and is stripped (Upper Cretaceous age) at 300 m depth under its bottom near Bash Lai village (the Shinchay River valley). The part of *neoautochtone* is assigned to the formations of the Zagatala-Dibrar nappe complex of a synclinal structure that occurs transgressively on the Talachay schuppen eroded surface and is in tectonic contact with the Duruja wedge scales of Middle Jurassic rocks. The neoautochtone cross-section is composed of various rocks from Callovian-Tithonian, which mostly protrude in the west and along the autochtone's northern periphery, to the Upper Cretaceous–Lower Paleocene that compose steep troughs in the east.

The Zagatala-Dibrar nappe complex protrudes in the west from under contemporary alluvium of the Ganykh-Ayrichay superimposed depression and spans from the Balakanchay River meridian to the Caspian Sea coast at a distance of about 300 km. Its width is 4–6 km in the northwest to 15–18 km in the southeast. The allochthonous complex forms the Zagatala-Govdagh northern side and the central, most downwarped part. It is expressed by the Upper Jurassic and the Cretaceous–Lower Paleogene terrigenic-carbonate flysch (see Figs. 5.6, 5.12 and 5.13). The trough's northern edge is lowered and overlapped by the Gamarvan thrust, and east of the Jimichay meridian by the Main Caucasus thrust. Its southern borders are the Gaynar and the Gozluchay thrust, which serve as its continuation at the Southeastern Caucasus; both of them are branches of the Kbaad-Zangi dislocation, which helps the structure constituting the Southern Slope's megazone to be wholly shifted southward and allochtonously overlap the Kakheti-Vandam-Gobustan megazone's northern edge. By its course, from the northwest to the southeast, the cover's complex internal structure experiences changes related to the regional sinking of structures and their transversal block boundaries.

The eastern fragment finds its correspondence in the south-sloping depression, whose northern side is raised stepwise and south shifted, which results in the separation of two tectonic plate (or wedge scales): the northern Balakan-Babadagh and the southern Zagatala-Burovdal, which are separated by the Ilisu-Aladash thrust. At the Southeastern Caucasus in the Garachay River head in the Gamarvan thrust lying side, the extensive Dibrar-Yashma trough is generated and opens southeast-ward; its southern side is thrust onto the Balakan-Babadagh plate's northern edge along the Altyaghaj dislocation. West of the Girdymanchay-Valvalachay flexure, the Main Caucasus thrust cuts the Gamarvan dislocation and allochtonously overlaps the Dibrar-Yashma plate's northern side. Within the same segment, tectonic thinning-out of the southern Zagatala-Burovdal plate takes place along the Ilisu-Aladash thrust and is accompanied by the separation of rootless schuppens in its hanging side east of the Sarydashchay River, such as the Aladash, Kemchi, and Kemishdagh wedge scales. Microfauna ascribed to the Cretaceous, Paleogene, and Miocene (not younger than Meotian) ages, which are found in the products of mud volcanos located in the rear of these slices, indicate the allochtone nature of the Balakan-Babadagh plate. The Talachay-Duruja allochtone serves as *parautochthonous substrate* of the Zagatala-Dibrar nappe complex, and the Upper Paleocene–Pliocene formations participating in the cross-section of the complex's southeastern plunge at the fringes of the Caspian Sea coast serve as *neoautochthone*. The latter has been preserved only in the extreme southeastern part of the Zagatala-Dibrar sag (approaching the coast plane), where it occurs at the Maastrichtian age flysch with deep-seated washout of the Lower Paleocene rocks.

The Govdagh-Sumgayit nappe complex is the extreme southern structural element of the Greater Caucasus accretion prism within Azerbaijan's territory. It is superimposed on the Paleocene–Miocene series of the Kakheti-Vandam-Gobustan megazone northern side at the South-Eastern Caucasus (see Figs. 5.5 and 5.6). The *allochthone* is separated at western approaches to the Vandamchay River valley, in the Gaynar-Gozluchay thrust lying wall, and imbricately exposes the Duruja plate's Middle Jurassic formation from the north (Fig. 5.14). After tectonic thinning-out of the latter, the cover complex expands its development field, appearing as a package of differently scaled tectonic plates (wedge scales) that wedge-like open to the southeast. The plates are predominantly composed of the Upper Cretaceous flysch formations, which experience regional plunge with the appearance of Paleocene and Miocene rocks in their synclinal depressions. The Lower Cretaceous age is represented by layers not older than the Barremian ones and crops out only at the joints and sides of a number of western and northwestern anticlines.

According to the Upper Cretaceous outcrops, the nappe complex can be traced in the general Caucasian direction for

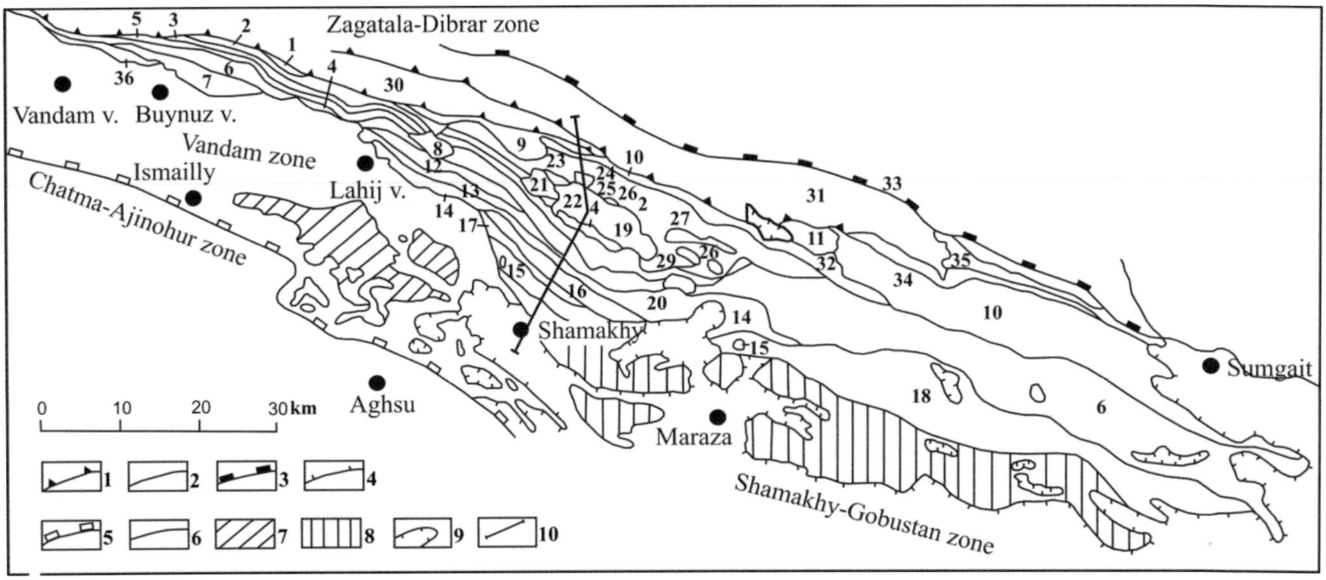

Fig. 5.14 Teconic scheme of the development belt of overthrust sheets within the Zagatala-Govdagh zone's boundaries (Khain and Alizadeh 2005; Kangarli 2006) Fractures (*1–6*): (*1*) Gaynar-Gozluchay thrust; (*2*) İlisu-Aladash thrust; (*3*) Altyaghaj thrust; (*4*) Mamrux-Galajig thrust; (*5*) Ganykh-Ayrichay-Alat thrust; (*6*) boundary of allochthonous schuppens (nappe plates); (*7*) Basgal nappe and its outliers; (*8*) parautochthonous shuppens development strip within the Shamakhy-Gobustan zone; (*9*) neoautochthonous complex; (*10*) line of the geological section (Fig. 5.15). **Allochthonous plates or schuppens (numbers in circles):** Govdagh-Sumgayit subzone (nappe complex) (*1–29*): (*1*) Bughur; (*2*) Chandakhar; (*3*) Govdagh; (*4*) Varna; (*5*) Valasin; (*6*) Shakamdagh; (*7*) Churukchay; (*8*) Garagushdagh-Beimli; (*9*) Matrak-Sardakhan-Kaykha; (*10*) Shikhandagh; (*11*) Tudar; (*12*) Khimran; (*13*) Ahan; (*14*) Jafalachay; (*15*) Nabur; (*16*) Chukhuryurd; (*17*) Galaybugurd; (*18*) Angakharan; thrust slices of Gyzmeydan nappe (*19–29*): (*19*) Astrakhan; (*20*) Nam Gobu (Mokroy Balki); (*21*) Archiman; (*22*) Khokhlas; (*23*) Gurd Yaylag; (*24*) Shiralidagh (Sharalyov); (*25*) Gumdagh (Peschany m.); (*26*) Lalay (Lalkhi); (*27*) Shikhlar; (*28*) Turfi; (*29*) Gurudagh (Suxoy m.); Zagatala-Govdagh subzone (nappe complex) (*30–35*): (*30*) Zagatala-Burovdal; (*31*) Balakan-Babadagh; (*32*) Aladash; (*33*) Dibrar-Yashma; (*34*) Kemishdagh; (*35*) Kurkachidagh; (*36*) Duruja subzone (nappe slice)

145 km, encompassing the northern Gobustan area, up to the Sumgayitchay River meridional bend at its inflow into the Caspian Sea. Further to the east, flexural plunge of the Cretaceous–Paleocene rocks takes place, but separate structural uplifts at the Eocene–Miocene formations can be traced into western and even central Absheron. From west to east, the allochthone transversal dimensions increase from 3 to 7 km in the Goychay-Girdymanchay interfluve to 20–22 km at the central section, and decrease to 18–20 km at the Sumgayitchay River. Tectonically, the cover complex is expressed on the surface as a strip of differently scaled tectonic plates overthrust from north to south. The complex separate plates have been delineated and described already in the 1930s to the 1950s (Khain 1937, 1952; Vassoyevich and Khain 1940; Solovkin 1945; Shardanov 1953; Shardanov and Molchanov 1954; Voskresenskiy 1958; Voskresenskiy et al. 1963). The idea of the allochthone nature of the entire Govdagh-Sumgayit strip of the Cretaceous–Miocene complex was conjectured in the late 1970s (Isayev et al. 1981).

In general, the nappe complex is represented by two packages of allochthone wedge scales; the lower one can be traced along its entire course, whereas the upper one corresponds to the large Gyzmeydan plate and a series of surrounding minor plates and schuppens (Kangarli 2006, 2009). Overthrust faults that south-confine each lower package plate merge at the allochthonous complex bottom into a single basic stripping surface that conducted the southward mass displacement. The surface that screens the entire system of described covers from below protrudes at the erosion cross-section as the Zangi-Garajyuzlu frontal thrust of the Govdagh-Sumgayit mass overthrust (Fig. 5.15). The Basgal nappe, located in the Girdymanchay-Aghsuchay interfluve and composed of the rocks not older than the Turonian ones and not younger than the Lower Miocene ones, is also related to the lower allochthone package. It represents a large fragment of the Govdagh-Sumgayit mass overthrust separated from ground allochthone masses by the Nialdagh ridge, which is composed of autochthonous volcanogenic–sedimentary Cretaceous rocks. The Gaynar-Gozluchay thrust zone corresponds to the cover suture (i.e. to the zone of allochthonous complex gashes or roots). The observed surface structure of the Gaynar-Gozluchay thrust's lowered side gives rise to virtually all wedge scale lamina of the northern allochthone strip. In this process, a right imbricate suite of schuppens that drifts away from the Gaynar-Gozluchay thrust is observed

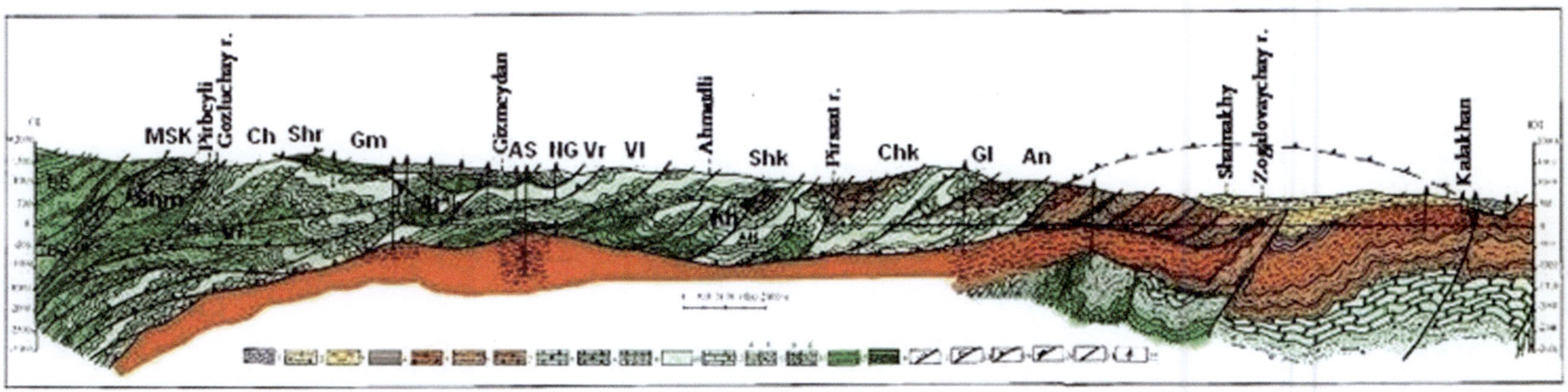

Fig. 5.15 Geological section along Pirbeyli-Kalakhan traverse passing through Govdagh-Sumgait nappe complex (Kangarli 2006, 2009, 2012). *1* mud volcano breccias; *2* Upper Pliocene (Akchagyl regiostage); *3* Upper Miocene (Pontian regiostage); *4* Middle–Upper Miocene (Tarkhanian, Chokrakian stages and Diatomian suit); *5* Oligocene–Lower Miocene (Maykop suite); *b* Eocene (Govun suit); *7–8* Paleocene: *7* Sumgait suit; *8* Ilkhidagh suit; *9* Upper Cretaceous (Vandam zone); *10* Upper Campanian–Maastrichtian (Agburun suit); *11* Santonian–LowerCampanian (Yunusdagh suit); *12* Upper Turonian–Coniacian (Kemchi suit); *13* Cenomanian–Lower Turonian: *a* Kemishdagh and Zarat suits; *б* Yelgadik and Namazgakh suits (volcanogenic–sedimentary formation of the Vandam zone); *14* Albian: *a* Altyagaj andKululu suits, aucellin level; *б* Lahyj and Julyan suits (volcanogenic–sedimentary formation of the Vandam zone); *15* Upper Aptian (Khanagah suit); *16* Barremian–Lower Aptian (Khalchay suit və Septarium level); *17–21* fractures: *17* İlisu-Aladash thrust; *18* Gaynar-Gozluchay thrust; *19* bottom of the Govdagh-Sumqaitnappe complex; *20* bottom of the parautochthonous shuppens; *21* other fractures; 22 wells. Nappe plates (schuppens): *An* Angakharan; *Gl* Galaybugurd; *Chk* Chukhuryurd; *D* Jafalachay; *Ah* Ahan; *Kh* Khimran; *Sh* Shakamdagh; *Vl* Valasin; *Vr* Varna; *NG* Nam Gobu; *As* Astrakhan; *Ar* Archiman; *Gm* Gumdagh; *Shr* Shiralidagh; *Ch* Chandakhar; *Shn* Shikhandagh; *MSK* Matrak-Sardakhan-Kaykha; *ZB* Zagatala-Burovdal; *BB* Balakan-Babadagh

from east to west; these schuppens raise over one another on the northern sides and form a package of nappe plates. The plate roots are not exposed on the surface. In the contemporary structure, they are buried under the Gaynar-Gozluchay thrust bottom. It is as if they were "dragged down" under its plane during the process of the Zagatala-Burovdal plate's southward "thrust."

The *autochthonous complex* is represented by the essentially argillous Paleocene–Miocene (except Pontian) rocks that transgressively (with sharp angular unconformity) lie on the Mesozoic sedimentary volcanogenic formations of the Kakheti-Vandam-Gobustan megazone's northern side. In the west, up to the Aghsuchay River head, the autochthone's southern side protrudes at the cover's complex front, experiencing swellings and constrictions (depending on the allochthone front's disintegration level). Further on, it finally plunges under the allochthone's Cretaceous rocks and parautochthonous schuppens of the Paleocene–Miocene rocks that protrude at its front; they are the south-displaced erratic blocks of the autochthone's buried central and northern parts. The duplication overlap of the Tertiary thicknesses was confirmed by geophysical data analysis and by stratigraphic drilling at the Angakharan, Malham, Chukhuryurd, and Gyzmeydan (Astrakhanka) areas, where typical Oligocene–Lower Miocene formations (see Fig. 5.15) were stripped at the bottom of a structurally complicated nappe complex composed at various depths (800–2500 m) of the formations from Barremian to Danian ages, inclusively. Participation of the Paleogene and Middle–Upper Miocene rocks in the autochthone's north structure is witnessed by microfaunal determinations from the mud volcanic breccia of the Gyzmeydan (Astrakhanka) group, as well as from the Gasymkend, Kehnagadik, and Kurkachidagh mud volcanos. The Basgal nappe occupies a special position because it is superimposed on the later (post-Pontian) planation surface.

The neoautochthonous complex corresponds to the Pontian formations that rest erosively on various Miocene and older rock horizons, thus participating in syncline structure generation by the middle and upper substages. In the contemporary structure, the Pontian cap rock has been preserved at small sites—within the Sundu-Maraza superimposed trough, as well as in centroclinal folds located on the cover complex surface. At the eastern plunge of the latter, within the western Absheron borders, the Pontian sediments unconformably lie on the Meotian argillaceous slates and are overlapped (with notable angular unconformity) by the productive strata. For the Basgal nappe, which is genetically associated with the Govdagh-Sumgayit allochthone, the neoautochthone function is performed by the Akchagyl non-fractured formations that "seal" its frontal contact with the autochthonic Paleogene age. The substrate underlying the nappe plate is composed of the deeply eroded Paleogene and Miocene rocks (including the Pontian one), which indicates a more prolonged process for its mass overthrust, with its termination being fixed by a younger neoautochthone cover.

5.2.4 The Absheron Zone

The Absheron zone is represented as an area of periclinal overlapping of the Zagatala-Govdagh zone's allochthonous and para-utochthonous structure for the Greater Caucasus accretion prisms that strike eastward through the Absheron peninsula and south- and north-adjacent sections of the Caspian Sea area. In the contemporary structure, the Absheron zone, as a constituent of the intensive sinking area of the meridionally oriented Caspian megabasin established at the Miocene–Pliocene border, looks like a trough along the Upper Miocene (Pontian)–Pleistocene neoautochtpone formations. Under azimuthal and angular unconformity, it contrastively overlaps the Zagatala-Govdagh zone structures that plunge from west to east and are composed by Cretaceous and Paleocene–Miocene rocks at the daylight surface. According to its Paleocene–Miocene and Cretaceous complexes that underlie the Pliocene formations, the zone corresponds to the eastern segment of the Absheron-Balkhanyan (Absheron-Nearbalkhan) elevation zone, which serves as the connective link between the Greater Caucasus fold-mountain systems and the Kopetdagh.

Being a transitional structure between the ground and marine buried parts of the accretion prism and representing a morphologically pronounced macrorelief positive form, the zone is considered as the Greater Caucasus geological structural element. However, in the tectonic zonation map (see Fig. 5.2), it is naturally included in the Caspian megatrough composition. As a segment of the Absheron-Balkhanyan elevation corresponding to the southern edge structure of the Scythian-Turanian platform's Middle Caspian block, the zone is separated within the borders confined by the Yalama-Khyrdalan submeridional dislocation to the west and the Pirallakhy submeridional dislocation to the east. The buried extension of the Gaynar-Gozluchay-Aladash thrust is subdivided into the northern (marine) Yashma-Chyrag and the southern Absheron-Gurgan subzones, represented by their western (Yashma and Absheron) segments.

The Yashma-Chyrag subzone is represented within the Absheron zone borders by its western Yashma segment and is situated in the eastern buried extension of the Zagatala-Dibrar allochthone package of the Greater Caucasus accretion prism. In its central part (within sea area borders), the subzone manifests itself as a chain of echelon-like locations, expressed in the Miocene–Pliocene complex asymmetrical brachy-elevations: Kurkachidah-deniz, Novkhany-deniz, Kurdakhany-deniz, Nardaran-deniz and Mardakan-deniz, which are complicated by transversal and longitudinal

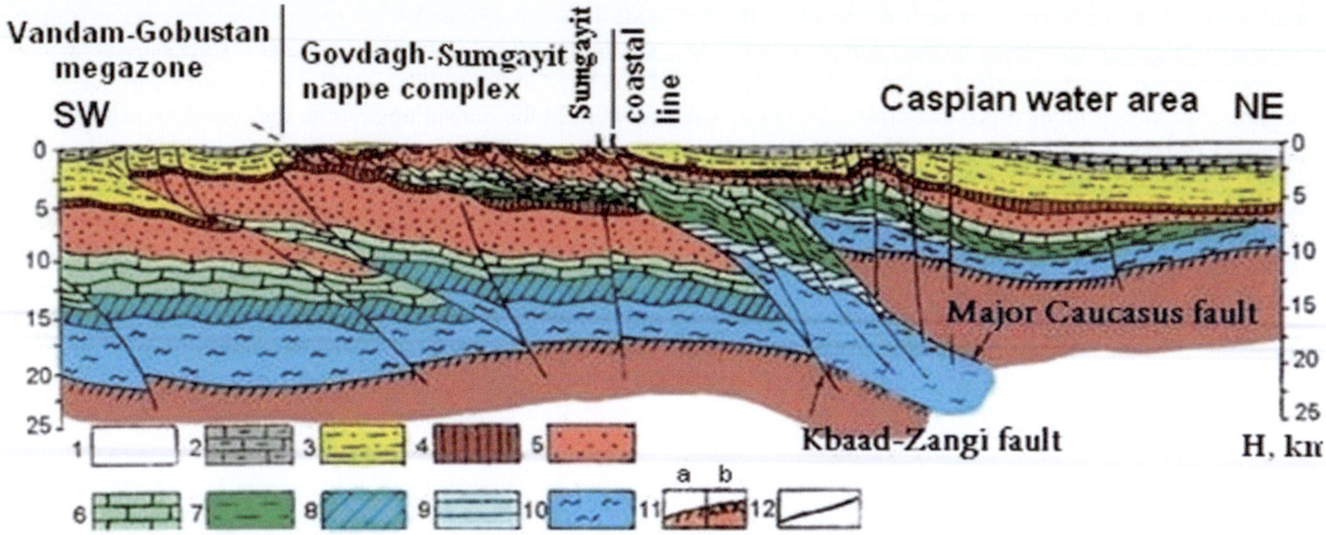

Fig. 5.16 Synthesized geological-geophysical section (*V–V* in Fig. 5.5) constructed along the Shahgaya-Sumgait profile (Kangarli and Aliyev 2005; Khain et al. 2007; Kangarli 2012)

fractures. In the subzone's southern side (inland) that stretches from the Absheron peninsula's northern coast, a strip of Paleocene–Miocene elevations is mapped in the hanging wall of the Gaynar-Gozluchay-Aladash thrust. In the west, the dome parts of the Sarygayabashy, Jorat-Novkhany and Northern Fatmai brachy-anticlinal, composed of Oligocene–Miocene rocks, crop out. At their eastern extension, the Kurdakhany-Mashtagha-Buzovna is buried high, concealed under a relatively thin (1500–2000 m) mass of the Pontian–Absheron formations. Farther to the east, the high is traced via the marine structures of Pirallakhy, Chilov, Neft Dashlary, Gunashli, etc., where the Miocene is exposed or buried under modern bottom sediments.

Within the borders of the described zone, ***the Absheron-Gurgan subzone*** is indicated by its Absheron segment that encloses the Peninsula central and southern sections, as well as the south-adjacent offshore, thus corresponding to the buried eastern extension of the Govdagh-Sumgayit cover complex in the southeastern Caucasus. Analysis of numerous geological and geophysical data (Kangarli and Aliyev 2005; Khain et al. 2007) leads to the conclusion that the thickness of the buried allochthone package of the Cretaceous–Miocene rocks of the Zagatala-Govdagh zone does not exceed 4.5–5.0 km within the Absheron borders, and the cover sole beds the parent Paleogene–Miocene formations of the Vandam elevation's downfold northern side (Fig. 5.16). In this process, roof hypsometry of autochthone Cretaceous rock at various sites of the Peninsula's central and southern sections is characterized by depths of up to 8–11 km; at the pre-Jurassic basement, depths are 17–22 km.

Buried Mesozoic and Paleogene–Miocene rocks generally preserve their sublatitudinal (general Caucasus) orientation and are generally represented by two anticlinal highs that can be traced eastward to the offshore area boundaries. As this takes place, young sediments that level the buried structural geometry experience dramatic changes in their strike—from near latitudinal in the west to meridional in the Peninsula central and eastern parts. This latter fact is explained by the change of stress vector and movements that are responsible for the establishment of the Caspian meridional megatrough during the continental development stage, starting from the late Miocene.

In the superficial structure, the Pliocene–Pleistocene deposits are predominant, and their tectonics differ considerably from the tectonics of the Paleocene–Miocene formations that are exposed at separate sites of the zone's western part or are exposed by drilling (Azizbekov et al. 1972; Khain and Alizadeh 2005). In doing this, the difference can be found not only in fold shapes and sizes and in the degree of their burdening by rupture dislocations but also in the dislocation course in these structural and compositional complexes—sublatutudinal in the lower of them and submeridional in the higher one.

Regionally, the fold zones have common Caucasian and latitudinal courses in the west and northwest, with a southern course in the east. The axes of anticlinal folds, which are relatively large and widely spaced in comparison to the adjacent Gobustan dislocations (and generally adhere to these two prevailing directions), frequently experience some diversified—and sometimes rather sharp—bends by their strike, up to semiannular twists around bowl-like troughs. The merging of folds that differ in their course is not uncommon and is accompanied by a sudden fading by one of them or by the appearance of some additional folds at a certain angle to the basic ones, as well as the echelon-like replacement of certain elevations by the alternative ones. At the same time, for most synclinal troughs due to fanciful arrangement of their framing highs and drastically expressed crenulation of bends, their irregular outlines totally depend on their anticlinal limitation: they mostly look like different-scaled, predominantly large, poorly concave structures with upraised edges—box-like and cup-like.

Distribution analysis of lithofacies and thicknesses of the Pliocene–Miocene sedimentary complex shows that the dislocations of submeridional (NW–SE, NNW–SSE) and sublatitudinal (W–E, WNW–ESE) directions had the determining significance for the formation of contemporary structures of the area under review, as they are related to the most contrasting vertical and horizontal displacements, as well as the northeastern strike fractures (SW–NE, WSW–ENE). These dislocations are mostly buried under the Late Pleistocene–Holocene formation caps but are exposed in more ancient rock outcrops.

As a whole, thetectonics of the Upper Miocene (Pontian)–Pleistocene structural and compositional complex that determines the Absheron zone surface structure is displayed as an alternation of submeridionally oriented highs and troughs, with subdivisions into the folloside structural belts (from west to east): (1) the Sumgayitchay-Baku synclinal belt; (2) the Fatmai-Zykh-Shakh-deniz anticlinal belt; (3) the Bina-Govsan synclinal belt; (4) the Gala-Zira anticlinal belt; (5) the Shuvalan-Dubandi synclinal belt; and (6) the Pirallakhy anticlinal belt. The latter is a part of the large Absheron-Cheleken zone of anticlinal highs that confine the largest (of the explored) marine oil deposits in Azerbaijan.

5.2.5 The Kakheti-Vandam-Gobustan Megazone

The Kakheti-Vandam-Gobustan megazone corresponds to the northern edge of the continental microplate in the Southern Caucasus. It consists of the Mesozoic–Cenozoic sedimentary and volcanogenic complexes, which have complicated interrelations. Within Azerbaijan's borders, the megazone is represented by two first-order tectonic elements: the western Vandam structural zone and the eastern Shamakhy-Gobustan zone. The former is situated at the southeastern continuation of the Georgian Dzau-Kakheti Mesozoic elevation, which is overlapped by the accretion prism allochthone complex on its northern side; on the southern side, along the Ganykh-Ayrichay fracture, it borders the Middle Kur Depression structures (Gamkrelidze 1977a).

Approaching the Azerbaijan border, the zone plunges under the Eopleistocene–Holocene continental molasses of the Ganykh-Ayrichay superimposed depression; its northern side is again exposed west of the Garachay River (on the southern slope of the Greater Caucasus). In the Goychay-Girdymanchay interfluve, its dome part is exposed as well. In the remaining section, the elevation is levelled by the same superimposed depression molasses complexes. The western plunge of the Vandam zone in the Mesozoic complex hosts the Shamakhy-Gobustan zone, corresponding to the depression composed mainly of the Paleogene–Neogene argillaceous formations. It is characterized by a complex fold-nappe structure. Its southern side, along the Ayrichay-Alat thrust (which is the southeastern extension of the Ganykh-Ayrichay dislocation) tectonically overlaps the Lower Kur Depression's northern edge. The depression's northern side is plunged under the Govdagh-Sumgayit mass thrust's allochthone lamina. Its central section and southern side, approaching the Caspian Sea area plunge successively, are filled by the shallow-water Pliocene sediments. Together with Absheron structures, they create a single Absheron-Gobustan pericline depression, facing the Southern Caspian basin.

5.2.5.1 The Vandam Zone

The zone generally corresponds to the Mesozoic elevation. The Cretaceous sedimentary and volcanogenic formations, upraised on the elevation's northern side east of the Aghlygchay River (the Khalkhalchay River) by the transgressive Paleogene rocks (Khain and Alizadeh 2005), are at most superficially exposed within its borders. At the same time, the southern side and most of the elevation dome are deeply eroded and overlapped by the Eopleistocene–Holocene continental formations of the Ganykh-Ayrichay superimposed depression, with their thickness reaching up to 1500–2000 m. The presence of Mesozoic volcanogenic sedimentary and intrusive formations in the continental molasses bottom is confirmed by single geophysical method analysis and by the interpretation of geophysical fields based on the integration of determining parameters (Babayev et al. 1978; Kangarli and Akhundov 1988; Kangarli et al. 1994a). The elevation length from the Georgian border to the Aghsuchay River (the Gyrkhbulag-Selibasan water parting), where the final plunge of sedimentary volcanogenic Cretaceous rock under Oligocene rocks takes place, is 230 km. The zone's width at the erosion cross-section is 20–25 km, reducing to 8–10 km at the extreme northeast and to 5–6 km at the southeastern fragment located east of the Girdymanchay River, up to a complete merging near Chukhuryurd village. On the northern side, the elevation is confined by the Gaynar-Gozluchay thrust; in the south, it is confined by the Ganykh-Ayrichay thrust. In the first approximation, it looks like a south-sloped and complex structured mega-anticline.

A typical feature of the elevation's lithological stratigraphic cross-section is extensive participation of the andesite-basalt and trachy-basalt magmatism derivatives that are represented in the Bajocian and Albian-Cenomanian volcanic-clastic, lava and subvolcanic facies ("porphyritic Jurassic rock"). A subsequent Upper Cretaceous cross-section is developed at the Vandam elevation's southeastern depression, where it is represented by carbonate terrigenic flysch with considerable tuff input. In the Vandamchay-Akhokhchay interfluve, the Upper Senoninan abnormal facies (the Maastrichtian one) is exposed, represented by a volcanic clastic complex of trachybasalt, trachyandesite, and andesite-basalt composition and intruded by hypabyssal syenite-diorite block (the Byunuz intrusion) whose age, based on the radiological data, corresponds to the Paleogene. The cross-section terminates in the carbonate-terrigenous flyschoid (Paleocene–Middle Eocene) and argillo-arenaceous (Upper Eocene–Oligocene) Paleogene facies that are exposed in both Vandam elevation sides, with partly preserved Upper Miocene (the Sarmatian) outliers.

In the contemporary structure, the Vandam zone is subdivided, from north to south, into longitudinal high-order tectonic elements: the Dashaghyl-Lahyj, the Gulluk-Dadagunash, and the Ayrichay subzones.

The Dashaghyl-Lahyj subzone has a graben-synclinal structure and corresponds to the Vandam elevation's northern side (stepped-down descending along the Dashaghyl-Mudrisa high-angle fault), stretching from a site at the Shinchay-Kishchay water parting in the west up to the left bank of the Gyrkhbulag (Aghsuchay) River in the east. In the superficial structure, the subzone's length reaches 135 km, with a maximum width of 6.5–7.0 km in the Dashaghylchay-Galachay interfluve. From the north, the subzone in confined by Gaynar-Gozluchay thrust branches, such as the Mamrukh-Galajyg (in the west) and the Zangi-Garajuzlu (in the east) dislocations, which assist in allochthone overlapping of its northern side by the Duruja allochthonous plate and by the Govdagh-Sumgayit mass overthrust, respectively.

In the west, the subzone looks like a synclinal depression, made by a sedimentary volcanogenic Albian-Senomanian stage with preserved (in the most depressed sections) Turonian, Coniacian, and Santonian sediments. Here, volcanogenic Cretaceous rocks are collected in a number of relatively large flat-topped and crest-like anticlinal folds, which are separated by narrow synclines and additionally complicated by small isocline folds. Starting from the Aghlygchay River (Khalkhalchay River) meridian and, farther to the east, in the subzone's northern side, the Upper Cretaceous rocks are transgressively overlapped by the Oligocene siltstone and shale mass, folded as small, relatively low-angle folds. The Oligocene field expands successively. At the Vandamchay River, it finally overlaps the Cretaceous formations, which are only once again exposed in the subzone's southern near-edge strip (i.e. in the Inakchi area at the Akhokhchay-Mudrichay interfluve), where broken rocks, from the Albian to the Maastrichtian, strip from under the Oligocene rocks in an erosive window and generally form the synclinal depression.

Submersion of the Dashaghyl-Lahyj trough bend north- and southeastward with simultaneous lateral accretion of the Govdagh-Sumgayit allochthone complex southern side, due to the emergence of new tectonic plates, results in a dramatic reduction of the subzone transversal dimensions in the observed superficial structure located east of Lahyj village. Allochthonous Cretaceous plates totally overlap the subzone's northern side and most of its southern one, which ultimately disappears from the surface at the Gurkhbulag-Selibasan river interfluve. Drilling documentation confirms the participation of tectonically overlapped Oligocene–Lower Miocene sediments in the geological cross-section, thus indicating the subzone's course to the far southeast in its buried location (see Fig. 5.15). A similar picture can be observed at the subzone's northwestern merging site (in the superficial structure) on the right bank slopes of the Gaynar River valley (the Kishchay River right tributary), where the Senonian volcanic clastic formations are cut at an acute angle by the Duruja plate bottom.

The Gulluk-Dadagunash subzone transversal cross-section looks like a double-humped anticlinal elevation separated by a median syncline that is north-confined by the Dashaghyl-Mudrisa disjunctive and south-confined (according to the geophysical data) by the buried Shambul-Ismailly (Vandam) disjunctive. Along the entire subzone length, its southern side is buried under the plume of alluvial-proluvial formations of numerous river debris cones that supply detrital material to the Ganykh-Ayrichay depression from the north. At the same time, a general low-angle structural-geometry plunge is observed, from the most elevated Gulluk-Oghuz segment in the Shinchay-Aghlygchay river interfluve (where the geological cross-section is represented by "porphyritic Jurassic rock," the Neocomian carbonate-terrigenic flysch, and the Senomanian volcanic-detrimental rocks) in both directions. In the northwest in the Garachay river valley, the zone is plunged under the Eopleistocene–Holocene gravels and can be traced in this position to Georgia. In the southeast, the bend undulation first results in its complete overlapping by modern detrital deposits in the Aghlygchay-Damiraparanchay river interfluve (the Khalkhal-Tikanly segment), which is further replaced by successive uprising in the Damiraparanchay-Aghsuchay river interfluves (Gabala-Dadagunash segment), and, finally, to the final Mesozoic complex plunge under the Shamakhy-Gobustan depression's Oligocene rocks at the Gyrkhbulag-Selibasan water parting. In general, the subzone length is comparable with the length of the Vandam elevation itself, with a maximal width (in the east) of 12–14 km.

The elevation's Gabala-Dadagunash segment is characterized by rejuvenated cross-section with the participation of volcanogenic and carbonate terrigenic flysch complexes in the

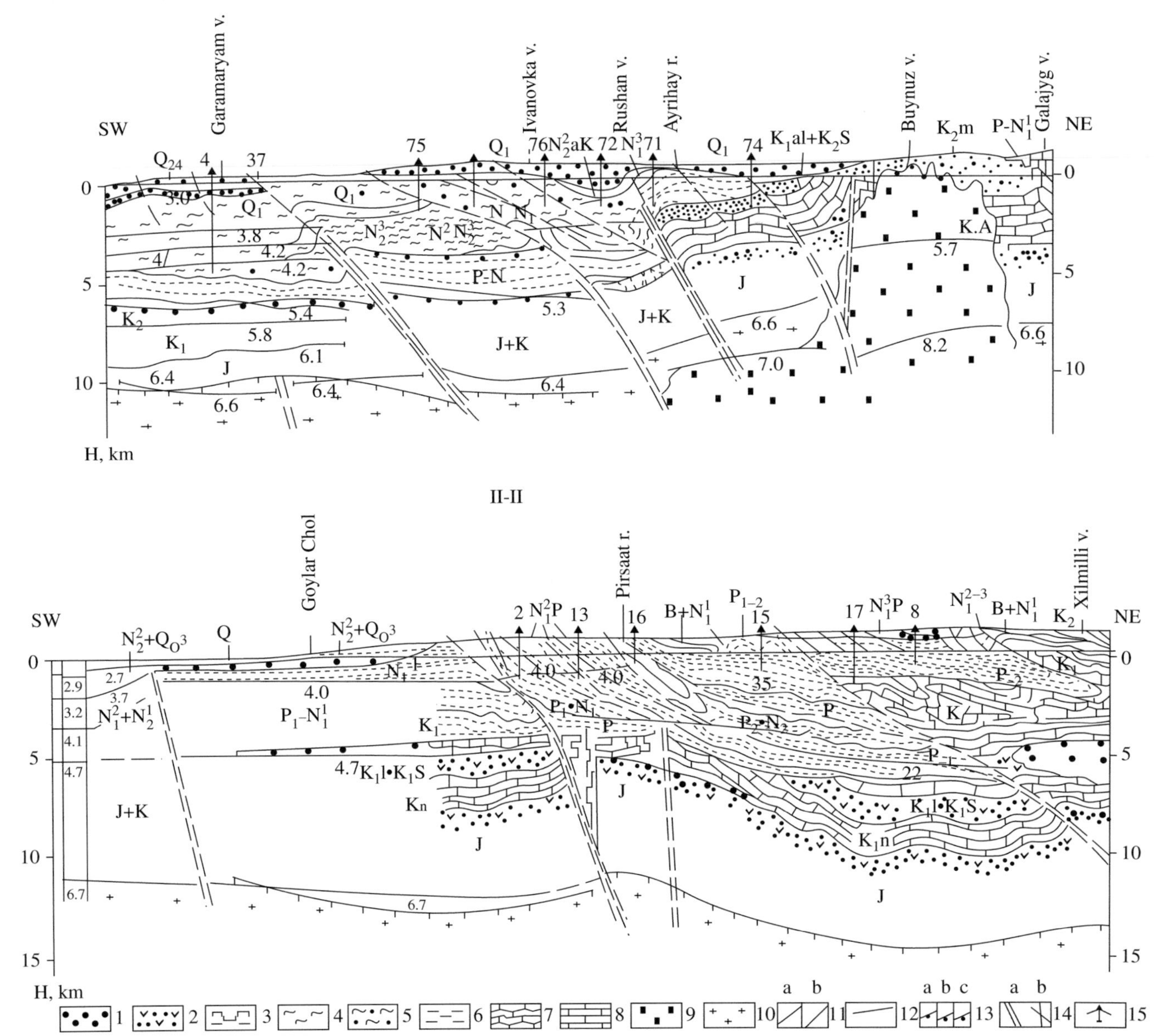

Fig. 5.17 Geology-geophysical sections across Vandam elevation along the profiles of Garamaryam—Galajyg (*I–I′*) and Goylar Chol—Khilmilli (*II–II′*) (Kangarli and Akhundov 1988; Khain and Alizadeh 2005). (*1*) gravel-pebble deposits; (*2*) volcanogenic–sedimentary deposits; (*3*) sandstone and clays; (*4*) clays, sands and sandstone; (*5*) clays, limestone and sandstone; (*6*) clays, mudstone, and sandstone; (*7*) terrigenic-carbonaceous flysh; (*8*) limestone; (*9*) intrusive bodies; (*10*) metamorphic rocks of Prejurassic basement; (*11*) boundaries in sedimentary cover: stratigraphic (*a*) and tectonic (*b*); (*12*) seismic boundaries and boundary velocities; (*13*) geoelectric boundaries: *a* in Cenozoic complex, *b* in Mesozoic complex, *c* on the surface of Prejurassic basement; (*14*) deep faults: *a* penetrating to the consoled crust, *b* reaching mantle; (*15*) boreholes. The numbers indicate the average P-wave velocity (in km/s)

observed superficial structure. They are stratified within the Upper Oxfordian–Paleogene time interval. In broad terms, the site structure is comparable with that of the western Gulluk-Oghuz segment, but it differs in general structural-geometry, with an eastward plunge and virgation of basic structures along the course. A sharp plunge of the elevation bend east of the Girdymanchay River, associated with movements in the Girdymanchay-Valvalachay flexure zone, results in complete overlapping of its eastern side by the superficial-filling Maykop series of the Shamakhy-Gobustan depression. At the same time, its northern side can be traced farther eastward, plunging under the transgressive Paleogene between the Dadagunash and Chukhuryurd villages at the Gyrkhbulag-Selibasan water parting.

In terms of location, the ***Ayrichay subzone*** can be compared to the Ganykh-Ayrichay superimposed depression, thus corresponding to the Vandam elevation's southern side, which is stepwise dropped along the Shambul-Ismailly dislocation (Fig. 5.17, section I-I′). The subzone is established from the results of combined interpretation of geophysical research materials, claiming that some sedimentary volcanogenic Mesozoic formations are buried under the Quaternary cover, and their geophysical parameters are comparable with those exposed within more northern structures of the Vandam elevation. Within Azerbaijan's borders, the subzone length (from the Georgian border up to the Girdymanchay River) exceeds 200 km, having an average width of 10–12 km. The Ganykh-Ayrichay fracture that south-confines the subzone as well as the entire Vandam uplift can be traced on the surface as separate fragments, being preferentially concealed by contemporary detritus; however, it is clearly revealed in potential geophysical data analysis, land seismic surveys, and data of parametric drilling. It is also deciphered in space photos as a single sub-east-west strike lineament. In the west, in the Ganykh-Alijanchay River interfluve, the fracture is superficially expressed by low-angle thrust of the Absheron and Akchagyl sediments onto the Lower Quaternary strata (the Dashuz-Amirvan elevation zone). The thrust plane falls northeast-ward at low-pitched angles (25–40°) at an amplitude of up to 1500–2000 m. Based on the set of geological and geophysical data, the Mesozoic and Paleogene sediments folded as the system of south-overturned folds lie under the Pliocene–Quaternary complex of the Ganykh-Ayrichay superimposed depression. It is also confirmed by the drilling data made near Akhmeta and Telavi villages in Georgia, where the south-overturned and thrust-complicated folds along the Paleozoic and Mesozoic sediments were found under nearly horizontally lying Pliocene–Quaternary sediments of the Ganykh-Ayrichay depression's western periphery (Gamkrelidze 1977b).

5.2.5.2 The Shamakhy-Gobustan Zone

The Shamakhy-Gobustan zone is considered to be a trough made by the Cenozoic sedimentary complex, isolated in the Girdymanchay-Valvalachay flexure downthrow side and corresponding to the tertiary basin that has absorbed the Vandam Mesozoic elevation's eastern part (see Figs. 5.15 and 5.17—profile II–II′). A folded structure of the trough's superficial filling stems is found on one side, from the overlapped nature of the Cenozoic sediment stripping bedding on the sedimentary volcanogenic Cretaceous facies of Vandam type; on the other side, it is from the development of intraformational strippings in the plastic Paleocene–Miocene mass caused by tangential compression.

The trough's northern edge is overlapped by a package of allochthone slices of the Govdagh-Sumgayit mass thrust, whereas its southern side (the Langabiz-Alat anticlinal zone) is thrust over the southwestward vergent Pliocene and Eopleistocene formations of the Lower Kur thrust along the Ayrichay-Alat dislocation (Fig. 5.18). The latter is the southeastern continuation of the Ganykh-Ayrichay thrust,

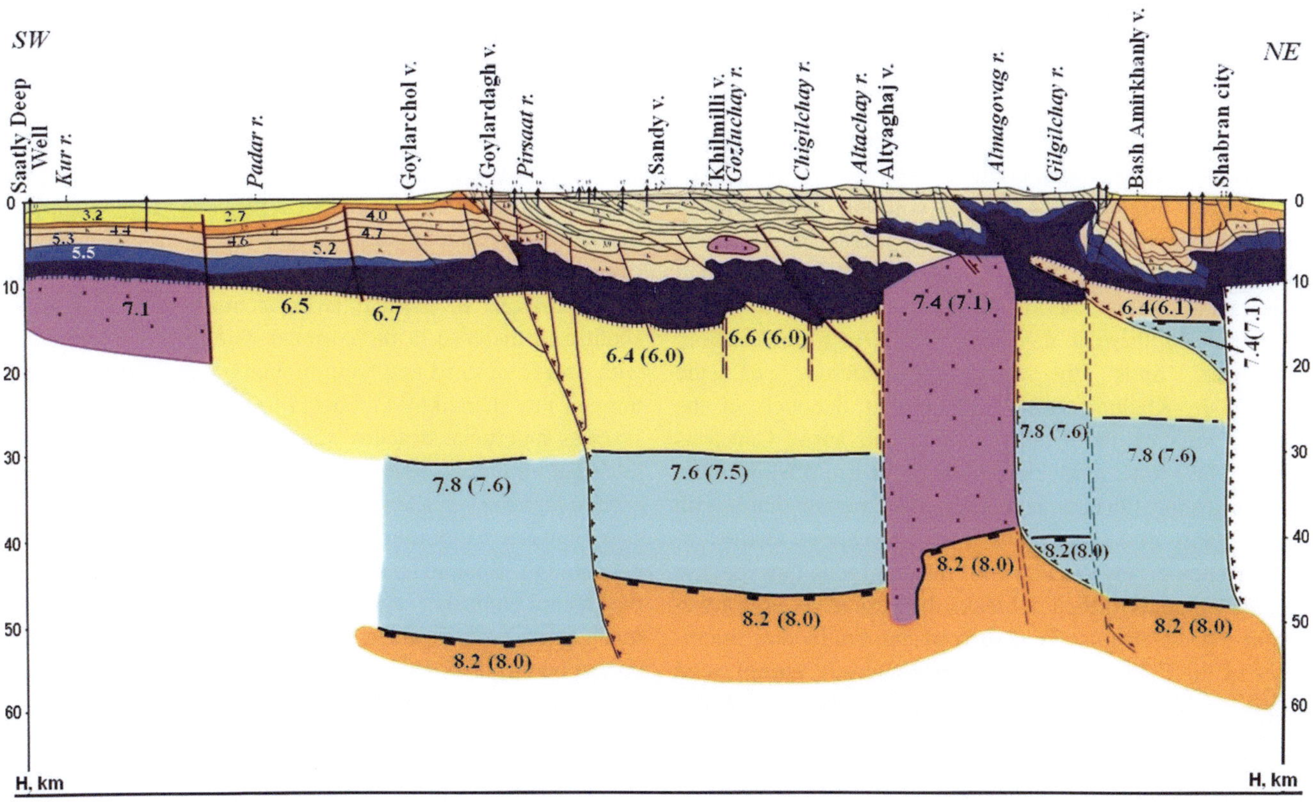

Fig. 5.18 Synthesized geological-geophysical section (*IV–IV* in Fig. 5.5) of the Earth's crust along the Padar-Shabran traverse (Kangarli 2011, 2012). The numbers indicate the average P-wave velocity (in km/s)

which is right displaced at a distance of up to 5 km along the recent fracture; this complicates the Girdymanchay-Valvalachay flexure-rupture zone's merging side.

Geological and geophysical data (Kangarli and Akhundov 1988; Khain and Alizadeh 2005) indicate that all Mesozoic structures of the Vandam elevation find their buried extensions within the limits of the Shamakhy-Gobustan trough. In the Western Caspian (Garbi Khazar) fracture strip, these structures experience a dramatic flexural twist at the pre-Alpine basement level, with a maximal dextral horizontal offset of 40–42 km. At the daylight surface, this phenomenon is reflected as the Pirsaat-Salyan-Neftchala submeridional right displacement zone—the largest in the Southeastern Caucasus (Kopp and Kurdin 1980; Kurdin 1987).

The Bouguer gravity reflect different anomalies that characterize the structures of the Alpine cover's lower part and bottom. The buried extension of the Vandam zone's Gulluk-Dadagunash elevation finds its expression in the Yavandagh-Sangachal zone of the gravity positive field; moving farther east, in a chain of local maxima that stretches from Gijaki to Kaftaran; and after that, via Garadagh, to the Caspian Sea area, where a large maximum of the Baku archipelago is prominent. At the same time, along the extension buried under the Govdagh-Sumgayit allochthone, the Dashaghyl-Lahyj subzone can be traced from the Jangichay to the Absheron peninsula bounds, where it is marked by the Gala gravity maximum, traced through Pirallakhy Island into the sea. North of said band of maxima, the Shikhlar elevated field zone and the Tudar reduced field zone can be distinguished; this serves as physical manifestation of the northern complication by additional structures for the Vandam elevation's buried part (Kangarli and Akhundov 1988; Kangarli et al. 1994b).

Of special interest is the crust section that corresponds to the Dibrar gravity maximum (see Fig. 5.18). It is located between the Babachay River and the Caspian Sea coast and is characterized by abnormally high layer velocity (7.0–7.2 km/s) and thermal flow (70–90 mW/m^2). In addition, here one finds the ascending sources of the Jimi-Khashy-Khaltan thermal group that debouch in the merging area of the Gamarvan and the Main Caucasus thrusts. The Moho surface experiences sharp uprising in the maximum zone, thus corresponding to the mantle diapir with layer velocity of 8.0–8.2 km/s. All of the above testifies to the existence of abyssal non-cool magma focus that roots in the upper mantle. Intrusion breaks the Alpine cover's lower part in the underthrust joint strip of the continental microplates of the Southern and Northern Caucasus, complicated at this section by the Western Caspian abyssal fracture submeridional zone. Incorporation of the Dibrar intrusive focus might have taken place at the Miocene–Pliocene age, when deconsolidation and opening of magma conduits took place in the Western Caspian zone of anti-Caucasus wrench-fault dislocations, with parallel emergence of the submeridional Caspian mega-trough against the background of maximal tangential compression associated with vigorous northward movement of the Arabian Plate.

Therefore, the Shamakhy-Gobustan zone's Cenozoic cover lies on the Mesozoic basement represented by the Vandam-type volcanogenic sedimentary facies. The depth of occurrence of the Cretaceous rock surface is 1–2 km in the west, 5–7 km at the Khilmilli intersection, and 8–10 km or more in the Caspian Sea area. The trough width, with due account for the position of its northern side and concealed under the Govdagh-Sumgayit strip overthrust nappes, is at least 40 km westward and 120 km at the Caspian Sea coast. That is, the zone expands from west to east towards the South Caspian (Janubi Khazar) basin; its land length is over 120 km.

Cenozoic filling of the Shamakhy-Gobustan zone is represented by four lithological and facies rock complexes: the Paleocene–Eocene complex, the Oligocene–Lower Miocene complex, the Middle–Upper Miocene complex, and the Upper Miocene (Pontian)–Pliocene complex. All of them have a predominantly argillaceous composition and subordinate position of psammite and carbonate rock differences. In the east, in syncline troughs, the Pliocene is superstructured by mixed Absheron age molasse; in its turn, this is overlapped erosively by the Pleistocene and Holocene continental and marine sediments (Akhmadov 1957; Khain and Alizadeh 2005).

On the surface, the Paleocene–Miocene formations are gathered into relatively small, highly compressed, isoclinal and comb-shaped anticlines, predominantly southward vergent and complicated by fault-thrusts and separated by brachysynclines. In dramatic unconformity, a considerably thinner cover of Pontian and overlying deposits that level the Paleocene–Miocene structural geometry occur above this complex. Drilling and geophysical research results definitely testify to the development of low-angle thrusts and slice structures in the Cenozoic mass. Frequent repetitions of the same horizons in vertical cross-sections and, as a result, significant increase in the complex thickness are associated with drastic discord and tectonic lamination in folded structures of the Shamakhy-Gobustan trough Cenozoic filling.

In its superficial structure, the zone is characterized by longitudinal and transversal zonation. Based on its stratigraphic section features and tectonic nature, it can be subdivided by its strike into four tectonic segments, subordinate to the transversal zonation: the Shamakhy segment, the Sundu-Maraza segment, the Central Gobustan segment, and the Eastern Gobustan segment. In this process, at least three packages of the thrust slices, corresponding to the general Caucasian strike subzones, participate in the trough formation: the northern Jangichay subzone, the central Keyvandi-Shahgaya, and the southern Langabiz-Alat subzone.

The ***Jangichay subzone*** is represented by para-utochthonous schuppings that are composed of the

Paleocene–Miocene rocks in the deposits (except Pontian age) and stem from the Govdagh-Sumgayit mass overthrust, as they were formed in its frontal part by being torn off their basement autochthone substrate fragments of the southward moving Mesozoic masses (see Figs. 5.15, 5.17 and 5.18). The subzone is traced from the Selibasan River head in the west to the Caspian Sea coast in the east. It looks like an allochthonous plate, compressed between the Garajuzlu and the Jeyrankechmaz thrusts, intensively dislocated and broken into tectonic slices; each of them are represented by a complete fold (comb-shaped anticline without cutoff southern side plus broad syncline), south-confined by the frontal fault-thrust.

The ***Keyvandi-Shakhgaya subzone*** is represented by low-angle thrust slices of the Shamakhy-Gobustan trough central part, which actually result in doubled thickness of the Cenozoic sedimentary cover. According to geophysical data, the root zone is located under the Govdagh-Sumgayit mass overthrust bottom, whereas the thrust slices (with amplitude reaching 20 km) are represented by intraformational strippings in the plastic Paleocene–Miocene mass. Participation of Pontian deposits, and occasionally of productive strata, in the composition of thrust slices and transgressive overlapping of their structural geometry by the Upper Pliocene defines the formation age for these slices as pre-Akchagyl (Rhodanian phase of tectogenesis). The subzone, south-limited by the Sangachal thrust, is nearly totally concealed under the Basgal cover in the west and participates in the Central, partly Southern Gobustan structure in the east. Similarly to the northern subzone, the thrust slices form extensive anticline belts here so that each slice is represented by a complete fold, comprising a comb-shaped anticline but with broader and more low-angled syncline in its northern side.

The ***Langabiz-Alat subzone*** corresponds to the structurally complicated area, generated at the Pirsaat River left bank (Udullu village area) in the Ajichay-Alat thrust hanging wall and tracing along the Kichik Kharami-Alat ridge as a tectonic wedge of thrust sheet, opening towards the Caspian Sea coast and complicating the Lower Kur trough's northern edge. The subzone is of asymmetric structure and is represented by a low-angle and broad, eastward plunging anticline in its northern side. In the south, it consists of echelon-like brachyanticlines, substituting each other and complicated by breaks and transversal thrust-faults. The subzone section is represented by the Pliocene, partly Oligocene–Miocene rocks that along the Ajichay-Alat dislocation are thrust (with block tectonic breccia at the bottom) upon younger formations (up to the Middle Absheron upper rocks) in the dislocation lying wall. Overlapping (with erosion) of tectonic breccia and sedimentary rocks in both thrust sides by the Upper Absheron marine facies displays the pre-Pleistocene formation age in the southern belt thrust slices (Walachian phase). Integrated analysis of geological and geophysical data enabled to define the allochthonous overlapping value for the Lower (Ashaghy) Kur Depression southern edge in various intersections, from 6–8 to 10–15 km and more.

5.3 The Kur Megadepression

The Kur intermountain megadepression corresponds to the central section of the Southern Caucasus continental microplate. Together with the Rioni trough (Georgia), it belongs to the system of internal depressions of the Caucasian segment of the Mediterranean folded belt. In the northwest, the depression is separated from the Rioni trough, plunging towards the Black Sea, by the crystalline basement Dzirula uplift. In the southeast, after opening and deepening, it merges with the Caspian megatrough's Southern Caspian segment. A conventional border between them is drawn by the Shongar-Gyzylagaj abyssal fracture along the Caspian Sea's western coast.

The megatrough was set as an independent structural unit in its present-day outlines only at the final, orogenous Caucasus development stage—that is, during the Oligocene–Quaternary period, when marine conditions with accumulation of a thick series (up to 7–8 km) of molasse deposits were set nearly pervasively within its borders (Aghabekov et al. 1976, 1977; Gamkrelidze 1989; Khain and Alizadeh 2005). These latter lie upon heterochronous structural elements of the trough Meso-Cenozoic base. In its edge sections, they fall outside its limits and overlap the dislocated complexes of the Greater and Lesser Caucasus peripheral structures as well as those of the Mountainous Talysh.

The megatrough's northern border passes along the Ganykh-Ayrichay-Alat thrust. Along this thrust, the structures of the Kakheti-Vandam-Gobustan megazone's southern side overlap the megatrough's dislocated Pliocene–Lower Pleistocene complex along the entire length of its northern edge. The latter (the Chatma-Ajinohur and the Jeyranchol structural zones) is, in its turn, torn off its basement and tectonically shifted southward, overlapping the megatrough's more interior structures. In the south, the Pre-Lesser Caucasus and Pre-Talysh abyssal fracture buried under contemporary molasses limit the megatrough.

The megatrough stretches from west to east at a distance of 450 km, falling into three local depressions (megazones). The extreme western Upper (Yukhary) Kur Depression, composed of the Mesozoic–Upper Miocene deposits, is situated within Georgian borders west of the crystalline basement Dzirula uplift and closes in the east, approximately at the Tbilisi meridian, being separated from the Middle Kur

Depression by the Martkob transversal dam. The Middle and Lower Kur depressions are located west of this latter and are separated by the Kurdamir-Saatly buried high.

Within Azerbaijan's borders, the megatrough's internal structure is two-staged and consists of a pre-Alpine crystalline basement and a relatively poorly dislocated Alpine sedimentary complex. The latter, by its structural and facial characteristics, is subdivided into complexes: the Mesozoic-Eocene lower one and the Oligocene–Quaternary upper one (Mamedov 1973; Bagizadeh et al. 1987; Khain and Alizadeh 2005). In the lower subcomplex composition, two structural levels are distinguished: the lower one is represented by the Lower and Middle Jurassic sediments buried under thick mass of younger formations. The Middle Jurassic deposits stripped by the Saatly superdeep borehole SD-1 and by prospecting boreholes at Jarly, Beylagan, etc. areas are represented by volcanogenic rocks of island-arc type. The Lower–Middle Jurassic formations at the trough edges consist of terrigenic and volcanogenic formations folded as steep folds, complicated by ruptured fractures. The upper structural level includes in terrigenic carbonate and volcanogenic facies of the trough edge parts the Upper Jurassic-Eocene sediments; their overall thickness reaches 3000–3500 m at the trough's northern edge and 2000 m within the trough's southern edge borders. According to the drilling data, the Upper Jurassic deposits have limited occurrence in the trough's internal parts and are represented in the carbonate facies. Similarly to the trough's edges, Cretaceous deposits are represented by volcanogenic-terrigenic-carbonate facies, so that carbonate rocks are essentially confined to the section's upper part. Their thickness in certain trough depressions reaches 1000 m. The Paleocene–Eocene deposits are represented by terrigenic and volcanoclastic differences of rocks.

In turn, the upper structural sub-complex is subdivided into two structural levels. The lower level encompasses the thickness of the Oligocene–Lower Pliocene marine and continental deposits. Therein, three sublevels can be isolated: the lower one of the Oligocene–Lower Miocene argillo-arenaceous deposits, the middle one of the Middle Miocene–Sarmatian terrigenic and carbonate rocks, and the upper one of the Meotian–Lower Pleistocene coarse continental molasses. The upper structural level includes the Upper Pliocene–Quaternary molassic deposits that were formed during the orogenesis final stage in the Caucasus. Two structural sublevels are isolated in its composition; they differ in their rock content and dislocation degree and are separated by momentary interruption and local angular unconformities. The lower sublevel encloses the Akchagyl and Absheron stage marine and continental deposits so that the continental formations have broad development in the trough's western part, while the marine ones are developed in the trough's eastern part. The upper sublevel is represented by the Pleistocene–Holocene continental deposits, consisting of alluvial–proluvial and deluvial formations. The marine origin sediments are predominant only at the Caspian Sea coast.

5.3.1 The Middle Kur Megazone

The Middle (Orta) Kur megazone corresponds to the Kur megadepression's central segment. It is the largest and most structurally complicated element that strikes from west to east (from the Martkob bowling to the Kurdamir-Saatly buried elevation) at a distance of 370 km and maximal width of 130 km. The megazone's northern border passes along the Ganykh-Ayrichay fracture that separates it from the Greater Caucasus Vandam high. In the west, within the Georgian boundaries, the megazone's narrowed end is subdivided by the Trialet high eastern periclinal into the proper Kur branch (in the north) and the Borchaly branch (in the south). The megazone's southeastern boundary is defined by the pre-Lesser Caucasus fault and is delineated by geophysical data examination under the Quaternary continental molasses, a little northward of the plunging strip of the Mesozoic–Cenozoic compositional complexes of the Lesser Caucasus northeastern slope. In the contemporary structure, the megazone is segregated into a number of structural zones that correspond to high-order highs and trough (see Fig. 5.2).

5.3.1.1 The Chatma-Ajinohur Zone

This zone is situated south of the Dashuz-Amirvan elevation zone, related to the southern edge of the Greater Caucasus Vandam uplift. It consists of two subzones: the northern Mirzaani-Arash subzone and the southern Chatma-Goychay one. In its surface structure, the zone forms the Middle Kur trough's northeastern edge, up to the Girdymanchay River; eastward of the river, it plunges under the Ayrichay-Alat thrust bottom. In Azerbaijan's boundaries, the zone length is 265 km, with width of 50 km in the Ganykh flexure area, and of 20 km at the eastern closure.

The ***Mirzaani-Arash subzone*** corresponds to the trough that structurally is the direct southeast extension of the Georgian Kartala trough. It is composed of a thick series of the Miocene–Quaternary molasses. The entire molasses complex is collected here as a series of southward vergent and bent folds that are complicated by such thrusts and faults where the fold's northern sides are generally thrust on their southern ones. The anticlines are represented by small brachyanticlinal folds separated from each other by broad and low-angle synclines. The submeridional flexural zone coincides with the Ganykh River's lower course, where the southern bend's twist takes place for all folds; west of it, the Arash site with a more simplified internal structure is located. Anticline folds are poorly pronounced here and are less asymmetrical, clustering into two belts: the northern Gamigaya belt (composed predominantly of the Pliocene rocks) and the southern Ajinohur-Kurdmashi belt (represented by the Absheron stage), separated by the broad and low-angle Pleistocene–Holocene synclinal strip.

The **Chatma-Goychay subzone** is the most complicated tectonic element of the Middle Kur Depression and is separated from the Arikdar-Bozdagh thrust, accompanying it from the south Jeiranchel zone and surface represented by the Udabno, Arikdar, Palantokan, and Garaja elementary dislocations, which are stepwise substituted by their west-to-east course. Along its entire length, this fractural dislocation zone is accompanied by southward vergent and thrusted folds of the Chatma-Goychay subzone. Similarly to the first case, the submeridional Ganykh flexure subdivides the described folded subzone into two segments: the western Chatma segment and the eastern Goychay one.

The Chatma segment (Fig. 5.19) is composed on the surface by the Upper Oligocene–Miocene deposits. In the Goychay segment boundaries, the Eopleistocene deposits (the Absheron stage) are older surface formations. The Chatma segment folds are larger and more complicated; they originate at the Georgian territory, from the Trialeti high eastern plunge, and stretch in the general Caucasus direction along the Gabyrry (Iori) River right bank. Two large anticlinal belts—the northern Alachyg-Eldaroyughu belt and the southern Udabno-Palantokan belt—participate in the segment's tectonic structure and are separated by a rather structurally complicated synclinal strip. Anticlinal folds are quite large and are surface represented by the northeastern sides, whereas the southern sides are concealed under the thrust. Fold formation took place during two tectogenic phases. The first, the Rhodanian phase, corresponds to the pre-Akchagyl age and is fixed by unconformity between the Upper Pliocene and the Miocene deposits; the second, the Walachian phase, corresponds to the Pleistocene period.

The Goychay segment is also represented by two anticlinal belts. The northern Goychay belt corresponds to the southeastern extension of the Alachyg-Eldaroyughu anticlinal belt. Here, the intensity of rock dislocations decreases; the folds become relatively low-angled, gentle, and diminish in size. In their dome parts, they are composed of younger Upper Pliocene–Holocene deposits. The southern Bozdagh-Garaja-Garamaryam anticlinal belt (which occasionally plunges under modern molasses) is located at the Udabno-Palantokan anticlinal strip extension and is south confined by the Padar thrust, an analog of the Arikdar-Bozdagh dislocation in the east. The Bozdagh, Dayirmandagh, and Garaja anticlinal ridges and the Garamaryam ridge belong to this folded belt and are composed of the Upper Pliocene–Middle Pleistocene deposits.

5.3.1.2 The Jeyranchol Zone

This zone enters Azerbaijan's boundaries in the Jandargol Lake area and stretches along the Kur River up to the Bozdagh ridge at a distance of above 135 km, with 20 km width in the west and 2–3 km width at the southeastern closure. On the surface, the zone is composed of the Upper Miocene–Pliocene deposits folded into a complicated system of sharply asymmetrical and southward vergent folds, grouped into three anticlinal belts: the northern Mammadtapa-Koyrukkeylan, the central Ortagash-Guyrugenchi, and the southern Kur belt, which are severed by relatively broad synclines (Fig. 5.19). On its southern side, along the Kur thrust, the zone is in tectonic contact with Near-Lesser Caucasus zone of the Middle Kur trough's southwestern edge. The zone's structure clearly distinguishes two structural levels, corresponding to independent tectogenic phases: the Lower Miocene–Lower Pleistocene and the upper Pliocene–Holocene, which show sharp angular and azimuthal unconformities. The fold involvement decreases in the southern and southeastern directions. Large-amplitude thrusts and overfolds complicate most anticlines; along them, the low-angle northeastern sides overlap the steep southwestern sides, which results in the presence of only northeastern anticlinal sides. Data from seismic surveys, prospecting, and exploratory drilling indicate the sedimentary complex lamination into a series of southern vergent tectonic slices. In the southeast at the Bozdagh ridge southern bottom, the zone pinches out, plunging under the contemporary deluvial–proluvial sediments; its extension south of the Kur fracture hosts the Yevlakh-Agjabedi zone's buried structures.

5.3.1.3 The Near-Lesser Caucasus Zone

The Near-Lesser Caucasus zone corresponds to the northeastern side of the Lesser Caucasus Artvin-Garabagh high. It forms the Middle (Orta) Kur Depression's southwestern edge from the Georgian border to the Araz River, and it has a complicated geological structure. On the surface, it is composed of the Upper Pliocene–Quaternary complex, which is transgressively lying on the Paleogene–Mesozoic formations. In the contact area of Cretaceous-Paleogene and Quaternary sediments, some steplike substituting fractures can be traced, which are the surface expression of the unified Pre-Lesser Caucasus fault. The latter can be clearly traced by geophysical techniques under the Quaternary complex along the entire length of the Middle Kur's southwestern edge. It serves as a natural border between the Near-Lesser Caucasus and the Lok-Garabagh zones. The northeastern border of the Near-Lesser Caucasus zone is drawn in the west along the Kur thrust and in the southeast along the Barda-Agjabadi-Beylagan line, where geophysical research

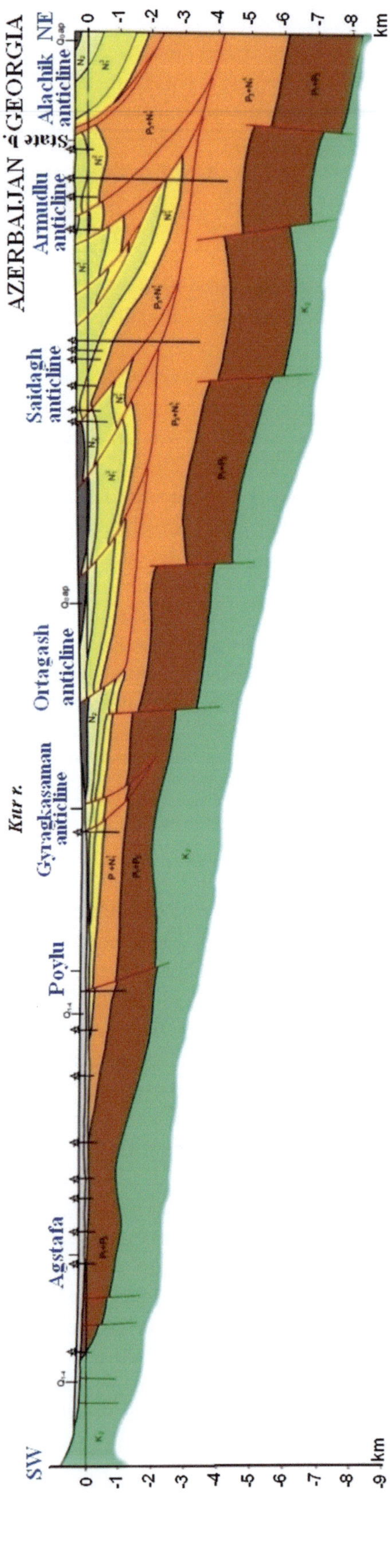

Fig. 5.19 Seismo-geological profile of the Lesser Caucasus–Jeyranchol (Agstafa-Armudlu) regional line (according to SOCAR: State Oil Company of Azerbaijan Republic)

and prospecting drilling found a fracture (the southern Kur fault) of up to 1000 m amplitude. In the southeast, it abuts the transregional Lower Araz's left-side thrust-fault (Palmira-Absheron lineament). Within these borders, the zone length exceeds 300 km, with 8–15 km width.

The Near-Lesser Caucasus zone is characterized by tectonic unconformity of the Pliocene–Quaternary and the Paleogene–Mesozoic complexes. The Pliocene–Quaternary complex is characterized by its monoclinic structure and rather low-angle northeastern slope. Most of anticline folds at the Akchagyl stage bottom are expressed as structural protrusions, slightly complicating the monoclinal structure. At the same time, the buried Paleogene–Mesozoic complex forms hemi- and brachyanticlinal highs at the background of general northeastern and southeastern plunging. The folds are asymmetrical and are characterized by the low-angle and short southwestern sides and by the relatively steep and prolonged northeastern sides, complicated by thrust-type fractures.

5.3.1.4 The Yevlakh-Agjabadi Zone

This zone forms the Middle (Orta) Kur Depression central structure and is located at the Jeiranchel zone's southeastern extension, south of Mingachevir's latitude (Fig. 5.20). It is isolated as an independent tectonic element at the Mesozoic–Paleogene and partially Miocene complexes, as a second-order deep depression complicated by local highs; the Upper Pliocene deposits show only low-angle northeastward bending of layers. In the depression's most bent part, the Alpine cover thickness reaches 16 km. The zone is confined by the South-Kur buried fracture in the southwest, by the Kur and Arikdar-Bozdagh-Padar buried fractures in the north, and by the Goychay-Imishli buries fracture in the northeast. The depression closure takes place in the south, in the Bilasuvar-Garadonlu Mesozoic protrusion region. Within said boundaries, its length is 120 km with 40–50 km width. The occurrence depth of the Mesozoic deposit surface exceeds 7 km at some sites.

From the seismic survey data, structural geometries of the Mesozoic and Paleogene deposits in the zone's southwestern part correspond to one another and are characterized by the presence of numerous local highs—most of them being small brachyanticlinal noses. The Miocene layers are mainly preserved in the fold's northeastern sides. The Mesozoic, Paleogene, Upper Miocene and Pliocene deposits participate in the geological structure of the trough's broader northeastern edge. The Paleogene deposits lean transgressively to the Cretaceous ones and are pinched out in relatively plunged parts of the trough's northeastern side. In the trough's northwestern part, a seismic survey found a chain of brachyanticlines, giving rise to a poorly expressed anticlinal zone; the trough's central part is generally monoclinic. Finally, within the boundaries of the northeastern edge's southwestern part, the structures of the general Caucasian strike are found, which are grouped into indistinct anticlinal belts: the Zardabelob-Muradkhanly and the Shahsunnu-Mil, located to the northeast, both being complicated by transversal and longitudinal fractures. The northeastern side's anticlinal folds become more low-angled up the cross-section. After the Paleogene-Miocene formations filled the trough, the pre-Akchagyl erosion took place and levelled the topographic inequalities. The Upper Pliocene–Quaternary deposits are bent northeastward as a low-angle anticline without any structural complications; in this process, the smallest slope is found in the Muradkhanly high area.

5.3.1.5 The Kurdamir-Saatly Zone

The Kurdamir-Saatly zone separates the Goycha-Imishli fracture from the Yevlakh-Agjabedi trough. It is represented by the buried submeridional Mesozoic elevation of the Middle Kur trough's eastern edge, which in the gravity field corresponds to the known Talysh-Vandam gravity maximum, interpreted as the reflection of horst uplift confined by abyssal fractures.

Data from geophysical research and deep-hole prospecting indicate that the zone is composed of volcanogenic–sedimentary Mesozoic deposits, Sarmatian stage terrigenic-carbonate deposits and Pliocene–Quaternary argillo-arenaceous thickness, which form independent structural levels (Fig. 5.21). Here, the eroded Cretaceous cover is set at 3.0–3.5 km depth, and the assumed disposition depth of the pre-Jurassic basement surface is 9–10 km. The Mesozoic structural level is dislocated with greater intensity, the Miocene one is transitional and less dislocated, and the Pliocene–Quaternary level lies nearly horizontally (Bagirzadeh et al. 1987; Alizadeh et al. 2000; Khain and Alizadeh 2005).

By its Mesozoic level, the Kurdamir-Saatly zone looks like a horst uplift, confined by the West Caspian fracture in the east and northwest and by the Arikdar-Bozdagh-Padar thrust in the northwest. Along these features, it is separated from the Lower Kur and Ajinohur troughs. In the southeast, the uplift can be traced towards the Talysh foothills, then to the Caspian Sea area boundaries. Within said boundaries, the zone length amounts to 200 km, with a width from 50–60 km on the north–northwest up to 15–20 km on the south–southeast. Geophysical data analysis demonstrate that the northwestern extension of the buried elevation is outlined in the Ajinohur foothill's southern edge and in the Kur-Gabyrry interfluve towards the Dzirula massif in Georgia.

The Lower Araz fracture, having a northeastern course, separated the zone into the Kurdamir and the Bilasuvar segments. Within the first one, large positive Bouguer gravity anomalies are prominent; in their limits, Mesozoic brachyhighs are found, which form the Garajaly-Saatly anticlinal zone (Fig. 5.22). The Bilasuvar segment is more plunged.

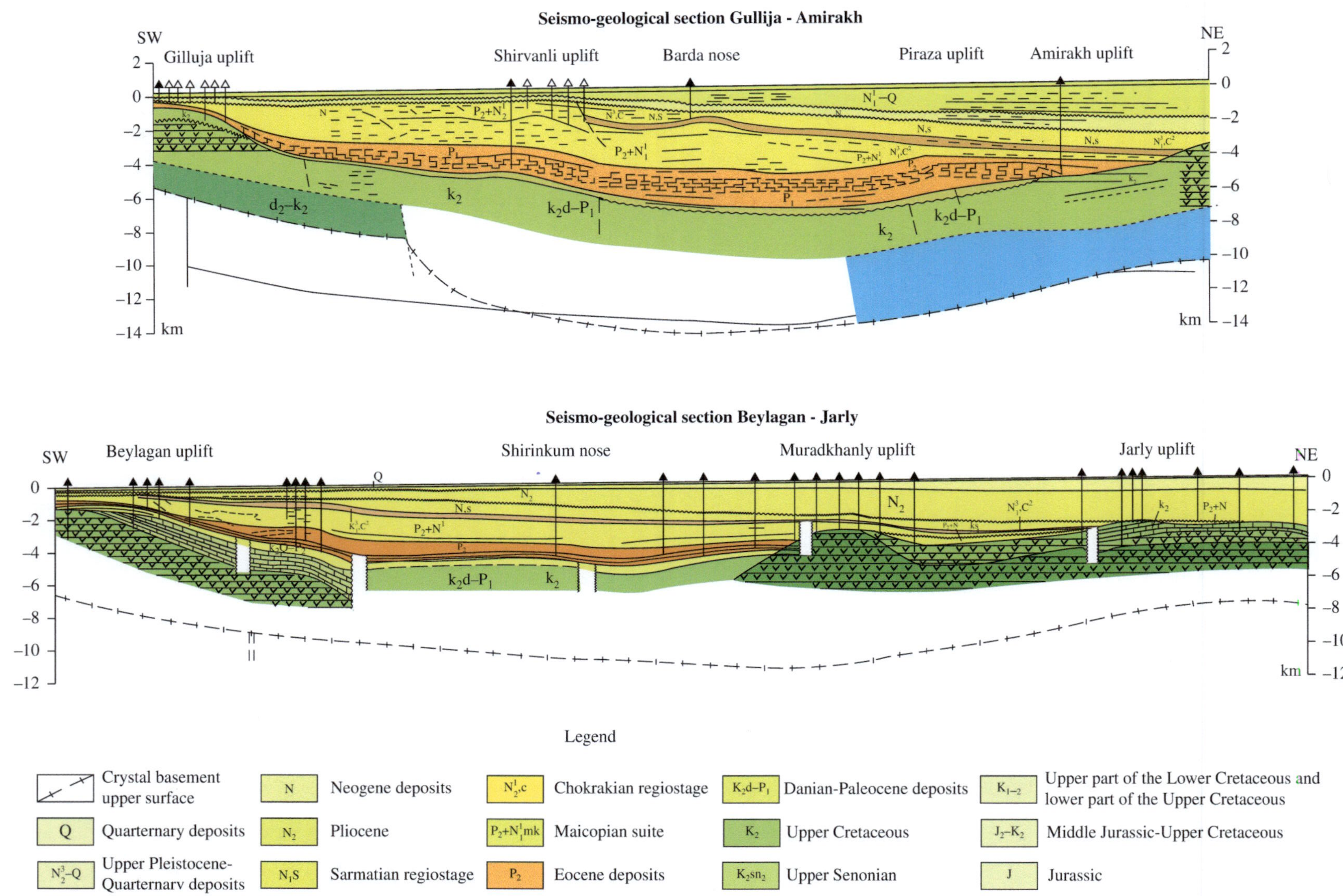

Fig. 5.20 Seismogeological profiles of the Yevlakh-Agjabedi trough (according to SOCAR: State Oil Company of Azerbaijan Republic)

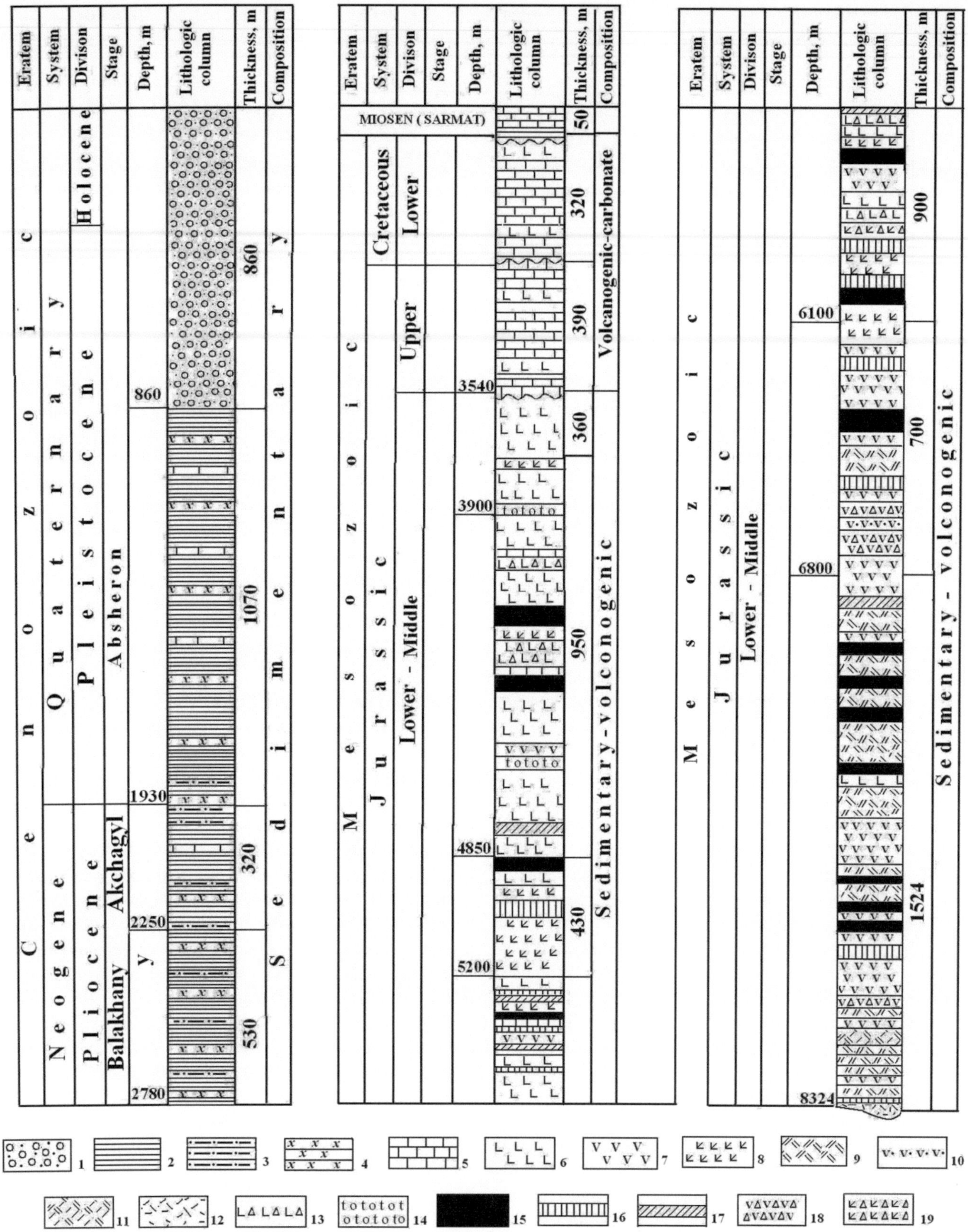

Fig. 5.21 Geological section of the Saatly Super Deep borehole SD-1 (Alizadeh et al. 2000). (*1*) unsorted coarse and fine-clastics; (*2*) clays; (*3*) siltstones; (*4*) sandstone; (*5*) limestone; (*6*) basalts; (*7*) andesites; (*8*) andesite-basalts; (*9*) dacites; (*10*) tuffbreccia of andesites with fauna; (*11*) rhyodacites; (*12*) rhyolite; (*13*) brecciated basalts; (*14*) fine detrital and clastic basaltic lavas; (*15*) dolerites; (*16*) diorites; (*17*) gabbro; (*18*) brecciated andesites; (*19*) brecciated andesite-basalts

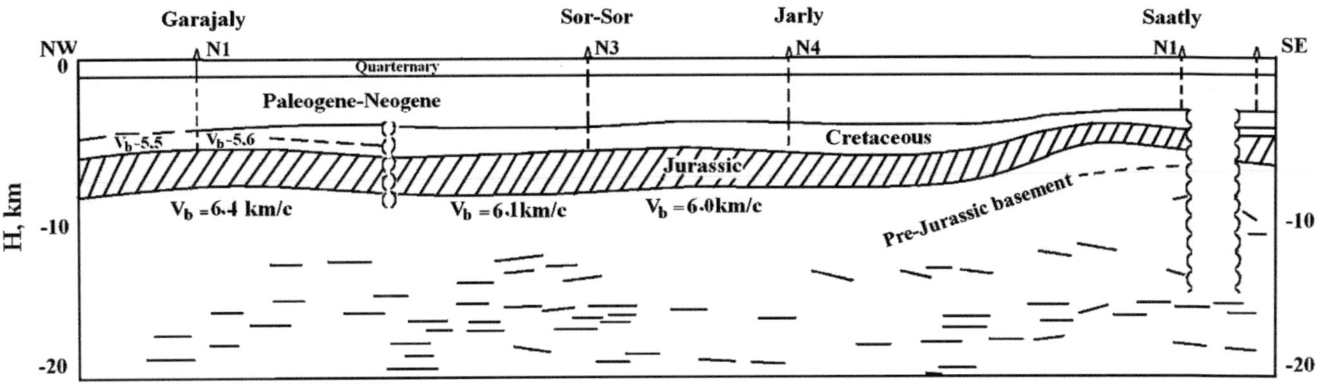

Fig. 5.22 Seismic profile of the Kurdamir-Saatly uplift (Bagirzadeh et al. 1987)

Here, numerous local gravity maxima were found, which reflect the structure of the Mesozoic volcanogenic–sedimentary complex, unconformly thrust by the Sarmatian and Lower Pliocene deposits. The latter fill the Mesozoic deposit's synclinal troughs and topographic inequalities, forming a transitional structural level. A thicker Upper Pliocene–Quaternary complex has a monoclinal structure, gradually passing to the Ashaghy Kur Depression's southwestern edge.

5.3.2 The Lower Kur Megazone

The Lower (Ashagly) Kur megazone corresponds to the structure most deeply plunged in the Kur trough system. Here, the thickness of the Alpine sedimentary cover arrives at 16–20 km; its larger half is attributed to the Pliocene–Quaternary complex (Bagizadeh et al. 1987; Khain and Alizadeh 2005). On the west–southwest, the megazone is confined by the West Caspian fracture; to the northeast, it is confined by the Ajichay-Alat fracture. On the east–southeast, the Shongar-Gyzylaghaj transversal fracture separates it from the Southern Caspian trough. Within said boundaries, the Lower Kur trough is shaped as a triangle of 100 × 50 × 80 km dimensions. At the Pliocene–Pleistocene complex, the trough has a rather complex fold-rupture structure, branching into two submeridionally elongated structural zones: the Shirvan and the Mughan, separated by the Pirsaat-Salyan-Neftchala right-side fault.

5.3.2.1 The Shirvan Zone

The Shirvan zone encompasses the trough's northeastern part and is characterized by sharply pronounced, elongated brachyanticlinal folds that form three elongated anticlinal strips in their echelon arrangement: the northeastern Pirsaat-Hamamdagh strip, the central Kalamaddin-Mishovdagh-Bandovan strip, and the southwestern Kurovdagh-Neftchala strip, which are separated by broad and low-angle synclines. All of these structures find their extension in the structural elements of the Southern Caspian (Janubi Khazar) depression. The anticline domes are complicated by transversal overfault-thrusts, which abut the Ajichay-Alat thrust in the northwest. Along these dislocations, the fold's near-dome parts are thrust on their steeply pitching sides. Transversal and radial dislocations frequently complicate the folds; as a result, they have block-like structures. A typical feature of the Kurovdagh-Neftchala and the Kalamaddin-Mishovdagh-Bandovan anticlinal zones is their co-sedimentational growth in the Upper Pliocene. The fold-generating movements occurred by the early Absheron age (Eopleistocene), and the fold's final formation took place in the Quaternary.

5.3.2.2 The Mughan Zone

The Mughan zone encompasses the Upper Shirvan and the Mughan-Salyan second-order troughs and is limited by the West Caspian fracture to the west and the Pirsaat-Salyan-Neftchala trough from the east. It generally represents the tectonic bench dropped relative to the Kurdamir-Saatly and the Shirvan zones. The Upper Shirvan trough's distinctive feature is a drastic thickness increase (up to 3000 m) in the Upper Pliocene–Quaternary deposits, which in the trough's larger part do not participate in folding. In the trough's southwestern side, a group of echelon-arranged buried brachyanticlines can be traced, which form the anticlinal zone of up to 35 km length. The Mughan-Salyan trough is slightly asymmetrical, with a relatively steep eastern side and a relatively low-angle western side. The latter is known in geological literature as "the Mughan monocline."

5.4 The Lesser Caucasus Mountain-Fold System

The Lesser Caucasus mountain-fold system is composed of the complex of the Paleozoic, Mesozoic, and Cenozoic volcanogenic, volcanogenic–sedimentary and sedimentary rocks intruded by heterochronous intrusives of basic, middle and acid compositions. It is characterized by a heterogenous internal structure caused by the descent of two branches of the Alpine Himalayan mobile belt within its boundaries: the Pyrenean-Lesser Caucasian branch in the north and the Dinar-Zond branch in the south (Khain2001. In the Caucasian region, the Southern Caucasus continental microplate represents the northern branch. Its southern side is related to the major part of the Lesser Caucasus, the Artvin-Garabagh megazone, corresponding to the elevation that continues from the Northern Anatolia Eastern Pontida to the east. In the southeast, the area is limited by the Lower Araz sinistral fault and finds its extension in the Mountainous Talysh folds. The Araz megazone (a large northwestern periphery block of the Central Iranian microcontinent's South Azerbaijan segment) is assigned to its southern branch, while its western extension hosts the Alborz-Binalud fold-cover structure.

In the contemporary structure, the Lesser Caucasus is represented as an area of collisional juncture of the two microplates during the oceanic crust subduction in the Mesotethys Lesser Caucasian branch. For this reason, it is characterized by a complicated fold-cover structure. In the Early Senonian age, the ophiolites that form the Amasiya-Goycha-Hakeri blend zone were obducted onto the northern microplate southern margin. Similarly, the ophiolites that form the Vedi blend zone were obducted onto the Araz block's northern margin. The Lesser Caucasus ophiolites are represented by the northern branch of the Izmir-Ankara-Erzinjan ophiolite zone, which separates the Pontians from the Central Anatolian microcontinent. The carbonate Upper Senonian, the Paleogene flysch and the volcanogenic Eocene seal these structures, forming the neoautochthone cover. The zone of microplate collisional juncture finds its correspondence in the Girratagh fracture that traces the Mesotethys suture. Thus, it can be traced along the Alborz folded system's northern constraint in the east; in the west, it stretches via Northern Anatolia to the Vardar zone on the Balkans.

In the north and northeast, the Artvin-Garabagh high is separated from the Kur megatrough located in the Southern Caucasus continental microplate's central strip by the Pre-Lesser Caucasus fracture. The Late Proterozoic–Middle Paleozoic metamorphic and magmatic formations of the Dzirula, Khram, and Lok massifs and of the Asrikchay nose appear to be a single formation that composes this microplate (Adamia 1984; Gamkrelidze and Shengelia 2005).

The Araz block of the Central Iranian microcontinent, which forms the fold-mountain system's southern part, has a late Precambrian metamorphic basement and an Ordovician-Triassic, Jurassic and Upper Cretaceous shelf-carbonate cover lying upon it. The Iravan-Ordubad trough, which is composed of terrigenic-carbonate and volcanogenic Paleocene–Eocene rocks, is located between the northeastern Miskhana-Zangazur and the southwestern Sharur-Julfa elevated edge structures of the Araz block. The Sharur-Julfa high, in turn, is complicated in its central part by the Nakhchivan superimposed depression, composed of Miocene–Pleistocene molasses.

5.4.1 The Artvin-Garabagh Megazone

The Artvin-Garabagh megazone corresponds to the southern upraised edge of the Southern Caucasus continental microplate, forming the northern and central parts of the Lesser Caucasus. Within Azerbaijan's borders, it consists of three northward archwise-convexed structural zones: the Lok-Garabagh, the Goycha-Hakeri, and the Gafan zones (Khain and Alizadeh 2005). These predominantly Mesozoic structures are characterized by complex interrelations as well as by their specific interior structures (Shikhalibeyli 1956, 1962, 1964, 1966; Hasanov 1971, 1985; Khain and Alizadeh 2005). In the highland's southwestern part, the Gafan and, partially, the Goycha-Hakeri zone structures are overlapped at low angle by the volcanogenic–sedimentary Paleocene–Pleistocene complex that forms the Kalbajar superimposed trough here. In the southeast, they are smoothly plunged under the Lower Araz structural zone's Oligocene–Quaternary deposits. This zone corresponds to the transversal trough that stretches along the Araz River's left bank from the Mehri gorge northeastward up to the merging with the Middle Kur Depression (see Fig. 5.2).

5.4.1.1 The Lok-Garabagh Zone

The Lok-Garabagh zone creates the Artvin-Garabagh high northeastern periphery, stretching within Azerbaijan's borders at a distance of more than 290 km, with maximal width of 45 km in the Zayamchay-Shamkirchay river interfluve. In the west, the zone stretches towards the Somkhit Mts. in Georgia and then continues to Anatolia, to the West Pontian mountains region. In the southeast, under regional plunge conditions, the zone is concealed under the Quaternary continental molasses of the Lower Araz depression. From the northeast, the zone is confined by the Pre-Lesser Caucasian buried fracture. In the southwest, it is separated from the neighboring Goycha-Hakeri zone by large thrusts and overthrusts.

The zone's pre-Jurassic basement is composed of Precambrian-Eopaleozoic crystalline shales. Its Alpine cover includes Jurassic shale, basalt-liparite and limestone (reef)-pyroclastic complexes, Cretaceous terrigenic-marl and volcanogenic complexes, and, partly, the Cenozoic volcanogenic–sedimentary complex. Folded deformations within the zone's boundaries are characterized by their moderate nature. However, in the Upper Jurassic-Cretaceous troughs, the sediments are folded more vigorously than in the Lower and Middle Jurassic ones. The zone's structure becomes more complicated in the area of its juncture with the Goycha-Hakeri zone: here, steep linear folds, sometimes southward and southwestward vergent, can be found in the Jurassic rocks.

In conformity with its structural features and stratigraphic sequence, the Lok-Garabagh zone, in its contemporary northeast-to-southwest structure, is subdivided into high-order longitudinal tectonic elements: the Gazakh-Aghburun, Shamkir, Dashkesan, Murovdagh, Aghdam, Khachynchay-Khojavand and Garabagh subzones with essentially tectonic boundaries between them. In addition, a typical feature of the zone structure is its structural geometry stratification into relatively upraised and downthrown blocks, by means of transversely oriented pass-through faults and transtensional faults, frequently having large (from several hundreds of meters to primary kilometers) horizontal offset—the largest of them being (from northwest to southeast) the Debedchay, Alpout, Tovuz, Asrikchay, Zayamchay, Goshgarchay, Ganjachay, Kurakchay, Buzlug, Injachay, Terterchay, Gargarchay, and Hojavand fractures. In this, the Injachay-Terterchay zone of left-fault dislocations displays a dramatic change in the structural geometry course of the Lok-Garabagh zone, from nearly latitudinal in the west to southeastern in the east, accompanied by the closure of certain structural units and opening of the other ones.

The ***Gazakh-Agburun subzone*** of the Lok-Garabagh northeastern periphery is traced in the surface structure from the Georgian border to the Khachynchay River right bank at a distance of over 210 km. It corresponds to the deep second-order structurally complex trough, composed of the Upper Jurassic and Cretaceous volcanogenic–sedimentary rocks and overbuilt by the volcanogenic–sedimentary Paleogene rocks in the west (the Aghstafachay River right bank) and the terrigenic Paleogene rocks in the east (the Ganjachay-Injachay river interfluve). In the northwest, subvolcanic eruptive bodies of the Upper Cretaceous rhyodacites and dolerites, as well as of the Eocene trachibasalts, participate in the geological cross-section. At the same time, the Upper Cretaceous gabbroid intrusions are exposed in the Goshgarchay, Ganjachay, and Terter river basins. The subzone's northeastern edge is confined by the Pre-Lesser Caucasian fault, and occasionally its central part as well, buried under the Middle Kur Depression's Pleistocene continental molasses. Southeast of the Khachynchay River, the trough is completely plunged under the molasses. However, from the isolated Upper Cretaceous protrusions among alluvial–proluvial formations, it is traced until the town of Fuzuli's latitude. In its northwestern part the trough, along the Yasamal fracture, borders the more southern Shamkir high. In the southeast, it is separated from the Aghdam high by the Aghdara dislocation. Generally, the subzone corresponds to a large asymmetrical syncline (with moderately elevated southeastern side) complicated by secondary folds and fractured dislocations. By its northwestward to southeastward course, the trough is intruded by transversal fractures into some relatively upraised and dropped blocks such as the Dashsalakhly, Gazakh, Zayam, Agjakend (broken into still smaller Ganjachay, Kurakchay, Garachay transversal blocks), Aghdara, and buried Gargarchay and Aghburun blocks.

The ***Shamkir subzone*** is represented by the most elevated and complex structured element of the Lok-Garabagh zone, situated between the Yasamal (in the north) and the Aryghdam (in the south) fractures. Along its strike, the subzone is confined by the Alpout (in the west) and the Goshgarchay (in the east) transversal dislocations, whereas the Tovus fault divides it into two blocks: the western Murguz fault and the eastern Shamkir fault, having a general length of greater than 100 km (above 70 km within Azerbaijan borders) and maximal width of 35–38 km. Being of fold-block structure, within Azerbaijan's borders the subzone corresponds to the sublatitudinally oriented, eastward plunged at low angle, anticlinal elevation with horst-like protruded dome (the Central Shamkir block) and stepwise lowered sides: the Gyadabek (in the south), and the Badakend (in the north) second-order blocks. Within the central high bounds along the Asrikchay River (Boyukgyshlag village) the pre-Jurassic basement is exposed by separate inliers and is represented by the Precambrian-Lower Paleozoic metamorphic shales that are transgressively overlapped by the Lower and Middle (Lower Aalenian) Jurassic basal conglomerates, argillaceous slates and sandstone. According to Hasanov (1985), the latter host the tectonic lamina made of the same metamorphic slates that form the Safarli consedimentary cover (olistoplacks) of 1.2 km^2 area and up to 70 m thickness, which in its northern part is intruded by the Upper Bajocian plagiogranite suberuptive bodies (Fig. 5.23). More to the east, at the extension of this outcrop in the Zayamchay river head, the Lazylar-Arygdam protrusion of the Middle–Upper Carbon tuffaceous-sedimentary rocks is located and is intruded by the Upper Jurassic intrusion of quartz diorites. In the remaining subzone part, the section is represented by the Bajocian volcanogenic-detrital and lava formations (at separate plunged sites, the Bathonian rocks are also present) folded as low-angle folds and interspersed by the Middle Jurassic (Bajocian) and Upper Jurassic–Lower Cretaceous intrusions of acidic (Shamlyg or Asrikchay, Lazylar, Atabek-Slavyanka, Gilanbir, Dashbulag, Goshgarchay, etc.) and median composition. At the plunge of the northern Badakend uplifted side, the Middle Jurassic volcanogenic rocks are transgressively upraised by the Callovian-Kimmeridgian carbonate-pyroclastic rocks and are cut by the borderline Yasamal fracture. Along this fracture, the Upper Jurassic–Lower Cretaceous granitoid intrusives are exposed, such as the Yanygly, Gabagtapa, and Uchtapa intrusives and the smaller ones.

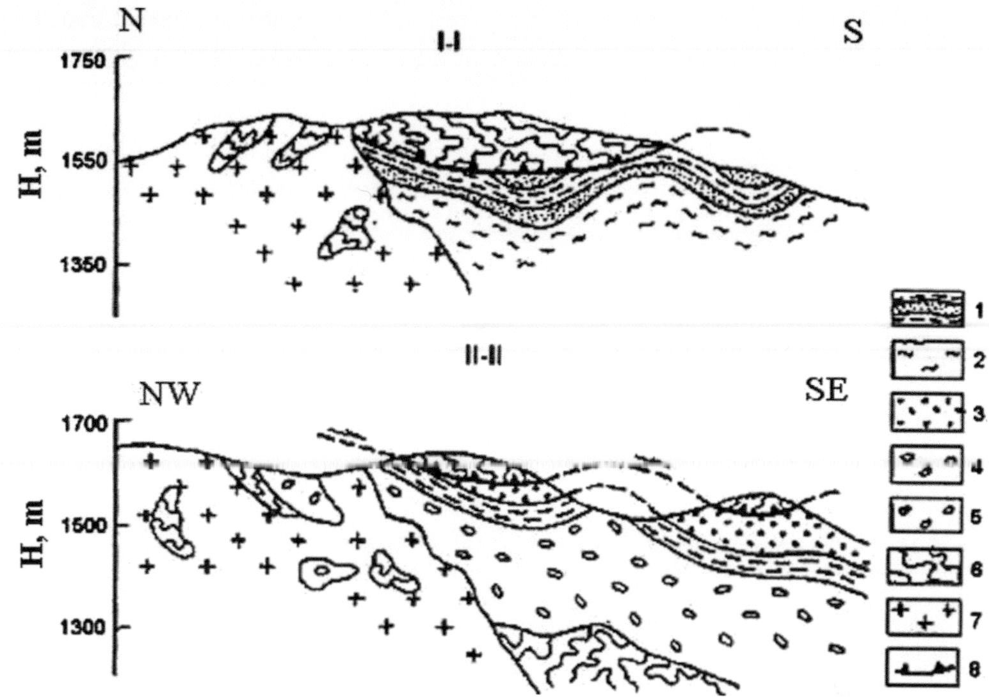

Fig. 5.23 Geological sections (Asrikchay–Akhynjachay rivers) (Hasanov 1985). Sections: *I–I* area of Safarli village; *II–II*—Aghgaya Mt.—Shamlyg village. (*1*) sand-clay schists of the Lower Aalenian; (*2*) shales and mudstones of the Upper Toarcian; (*3*) sandstone and tuff-sandstone of Sinemurian; Hettangian deposits (*4–5*): (*4*) sericite tuffaceous sandstone with impurity, (*5*) basal conglomerates; (*6*) metamorphic schists of Precambrian-Lower Paleozoic; (*7*) subvolcanic rhyolite plagioporphyres of Upper Bajocian-Prebathonian, (*8*) faults in the allochthon basement

The ***Dashkasan subzone*** is generated in the Akhynjachay river head, in the Aryghdam thrust's southern lying side and in the Asrikchay transtensional fault eastern side. It can be traced as a broadening wedge from the Armenian border to the Goshgarchay River, where it sharply increases its northeastern side at the expense of the Shamkir high periclinal closure along the Goshgarchay transtensional fault. On its dropped side, a drastic plunge of the Bajocian-Bathonian formations under the Upper Jurassic and Cretaceous ones takes place. In the east, the subzone border passes along the Injachay dislocation, while in the southwest it is confined by the Arysu-Zivlan-Ballygaya fault. Within these boundaries, the subzone length reaches 105 km, with maximal width of 27–28 km at the Goshgarchay River. The subzone in general corresponds to the complex structured synclinal trough. Its complex outlines take shape along the Bathonian volcanogenic complex bottom, which lies unconformly on the Bajocian rocks and is upraised by the volcanogenic-carbonate Malmian rocks. In the subzone's northeastern side, in the Goshgarchay-Injachay river interfluve, the Cretaceous volcanogenic–sedimentary deposits (the Albian-Senonian period) also participate in the cross-section. The trough bend along its strike undulates twice, forming two transversal highs: the Shamkirchay high in the west and the Goygol high in the east, with interlocated second-order troughs —the western Galakend trough, the central Dashkasan-Zurnabad trough, and the eastern Buzlug-Kapaz trough filled by the Upper Jurassic rocks in the core. The Goygol high dome is the most elevated one, which results in the superficial exposure of the Upper Bajocian formations. The underlying Middle Jurassic volcanogenic rocks demonstrate against the background of the relatively smooth, moderately asymmetrical structure of the Upper Jurassic synclines, a secondary folding, which in the northwest takes the general Caucasus direction. In the southeast, under the impact of anti-Caucasus dislocations (the Goshgarchay, the Ganjachay, the Injachay, etc.), the dislocations often change their course to the northeasternward. The subzone's specific structural feature is the participation of numerous Upper Jurassic-Lower Cretaceous intrusives of granite-diorite and gabbro-diorite composition (the Gedabay, the Shamkirchay, the Dashkesan, the Zurnabad, etc. ones) in the geological sequence. In the subzone's northwestern closure, the Upper Cretaceous Atakhal syenite-diorite intrusive is present.

The ***Murovdagh subzone*** is represented by a southwide vergent anticline composed of Bajocian volcanogenic rocks and compressed between the Arysu-Zivlan-Ballygaya fault to the north and the Murovdagh thrust to the south. It originates in the Zayamchay River head, at the northern side of the like-named thrust, and stretches southeastward as a narrow strip along the northern outskirts of the Shakhdagh and Murovdagh dividing ridges, experiencing a swelling of up to 8–10 km in the Shamkirchay-Kurakchay rivers interfluve and having general length of above 85 km. In the east, by approaching the Injachay transtensional fault, the elevation's narrowed edge strikes an arc, takes the northeastern course, and has tectonic thinning-out in the complicated fracture intersection node of the general Caucasian and anti-Caucasian directions. Along the subzone course, numerous minor outcrops of the Upper Jurassic–Lower Cretaceous quartz diorites and gabbro are observed, as well as subvolcanic bodies of andesidacites and elongated dikes of corresponding composition. In the Boyuk-Gashgachay

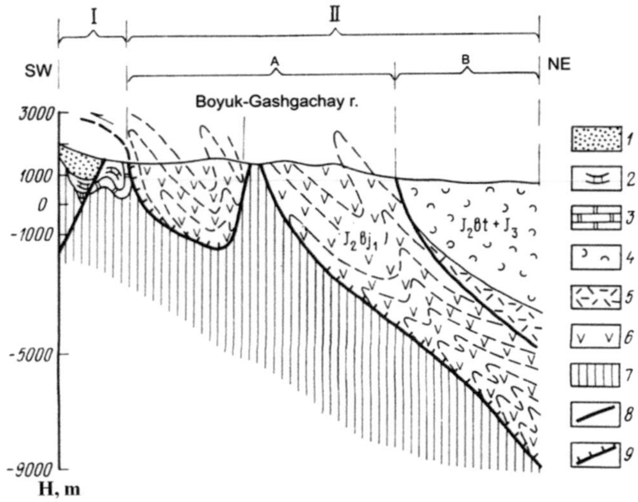

Fig. 5.24 *The* Geological section of Jurassic and Cretaceous deposits on the northwestern margin of Murovdagh subzone (Hasanov 1985). *I* Goycha-Hakeri zone (Toragaychay subzone): *II* Lok-Garabagh zone (*A* Murovdagh subzone, *B* Dashkasan-Galakend subzone). (*1*) sand-siltstone-argillaceous rocks and argillaceous limestone of Albian–Cenomanian; (*2*) sandy-gravel-detrital clayey limestone of Berriasian–Hauterivian; (*3*) slightly sandy, silicified limestone of Tithonian; (*4*) vocanogenic–sedimentary rocks of andesite-basaltic composition and limestone of Bathonian–Upper Jurassic; (*5*) quartz plagioporphyry and their piroclastites of liparit-dacite composition of Upper Bajocian; (*6*) volcanites of basalt-andesite composition of Lower Bajocian; (*7*) brecciated serpentenites and serpentinated hyperbasites; (*8*) faults; (*9*) thrust-overthrust

river head, in the Lower Bajocian volcanite basis (in tectonic contact with them), the ultrabasites (Fig. 5.24) are exposed as separate outcrops, represented by ophiolite fragments from the Goycha-Hakeri zone's northern edge buried under the Murovdagh thrust bottom (Hasanov 1985).

The ***Aghdam subzone*** is generated in the Injachay-Terterchay zone of sinistral-fault dislocations. Being in butt juncture with the Dashkasan and Murovdagh subzones, it is represented by the southward stretching asymmetrical low-angle elevation of the Bathonian volcanogenic complex, complicated by nonsinusoidal secondary folds. The bend twist experiences low-angle undulations controlled by transversal dislocations with segregation of three relatively elevated blocks, replacing each other in a beaded manner along the strike (the Terterchay, the Mekhmana, and the Garajyg ones) and of several downthrown blocks located in between. The subzone in the northeast is confined by the Agdara fault. On the southwestern side, it is superimposed on the Khachynchay-Khojavand subzone structures along the Aghdam thrust. On the subzone's northwestern side, where the folds are less loaded, the Middle Jurassic cross-section is upraised by the Malmian and transgressive Cretaceous rocks and is intruded by the Janyatag and smaller granitoid intrusives of the Upper Jurassic–Lower Cretaceous age. On the Gargarchay River's right bank, on the low side of the like-named transtensional fault, the structures experience flexural plunge. Therefore, on the daylight surface, the Bathonian complex is totally substituted by the Upper Jurassic-Cretaceous one, which is mostly concealed under the Lower Pleistocene molasses cover. The latter completely overlap the subzone's structural geometry at the Garasu River's left bank, and in its outskirts (within the Garajyg block boundaries) the Bathonian formations (intruded by the granitoid intrusives) and transgressively, the Malmian–Neocomian formations, are finally exposed on the surface. Here, they form the periclinal closure of a large anticlinal fold, with the southwestern side cut along the Aghdam thrust. In these boundaries, the zone length exceeds 90 km, with maximal width of up to 18 km at the Terterchay intersection, but it stretches in its buried state up to the Araz River's left bank.

The ***Khachynchay-Khojavand subzone*** can be traced in the Aghdam thrust's southwestern low side, from the Terterchay River mid-channel to the Araz River—first in the general Caucasian direction, and then, east of the Gargarchay River, it makes an arch-like bend and takes the sublatitudional direction. The subzone's southwestern border corresponds to the Meydanchay-Shusha fracture, represented by the zone of elementary faults, with echelons replacing each other along the strike. In the northwest, in the Terterchay transversal dislocation area, the zone is in butt juncture with the Toragaychay and Garabagh subzone structures; in the southeast, near the town of Fuzuli, it plunges under the Lower Araz trough Pleistocene molasses. Within said borders, the subzone length reaches 80 km, with the width growing from 9–10 km in the northwest to 20–25 km in the southeast.

In its northwestern and central parts, the subzone is composed of volcanogenic Bathonian rocks, carbonate-pyroclastic Malmian rocks, and transgressively lying volcanogenic–sedimentary Cretaceous rocks (from the Albian and higher). It is occasionally expanded by terrigenic Eocene rocks in the overthrusts of separate synclines. In the southeast, the Cretaceous cross-section of the bottom is expanded by the terrigenic-carbonate Neocomian rocks, preserving the overlying interval transgressive nature. The subzone's bend undulated at low angle, separating the upthrown and downthrown blocks controlled by transversal dislocations. However, in general, the strike's structural geometry experiences a step-like plunge, except for the extreme southeastern upthrown block, where the Jurassic and Lower Cretaceous formations again protrude on the surface from under the Upper Cretaceous shell and are intruded by granitoid intrusives.

In general, the subzone corresponds to the southwestward sloping, structurally complex anticline with narrow southwestern and broad northeastern synclinal flanks complicated by secondary folds and fractured dislocations. The median anticline composed by the volcanogenic Bathonian rocks in its core, and its southern synclinal flank composed mostly of the Upper Jurassic and Cretaceous rocks, appear in the Meydanchay-Shusha trough's downthrown side in the Khachynchay-Gargarchay interfluve and can be traced southwestward along the subzone's southern edge. Only the northern synclinal strip dislocated by additional folds shapes its outlook among its entire length.

The ***Garabagh subzone*** corresponds to the anticlinal high that stretches from northwest to southeast for over 100 km, having width of 10–15 km. It is composed of the Bajocian volcanogenic formations and the Malmian volcanogenic–sedimentary complex, with transgressive overlapping of the latter by the Cretaceous terrigenic-carbonate and volcanogenic–sedimentary rocks (the Albian–Senonian age) in the southeast. In its northwestern part (at the Terterchay River left bank), the subzone has sublatitudinal course, replacing it by the southeastern one in the Injachay-Terterchay area of transversal dislocations. In continuation, the subzone, similarly to more northern Lok-Garabagh zone structures, experiences southward vergent arched flexure, changing its course in the Ishkhanchay River head from southeastern to sublatitudinal and plunging under the Lower Pleisocene continental molasses at the Lower Araz troughs outskirts. The subzone's northeastern border is determined by the Meydanchay-Shusha fault; on its southwestern side, along the Garabagh dislocation, it is overthrust onto the Sarybaba subzone's northern edge.

On the Terterchay River's left bank, the subzone is represented by a large southward vergent anticline composed of Bathonian volcanogenic rocks built up by the Malmian formations in its northern edge. Eastward, the anticline closes periclinally and has tectonic thinning-out as a result of convergence of the subzone's confining dislocations. Southeastward of the Terterchay River valley, it opens and breaks into a series of broad and low-angled second-order folds. Along the subzone course, a number of small Upper Jurassic quartz-diorite intrusives are exposed among the Bathonian volcanogenic rocks. Along its northeastern edge, the Upper Bathonian subvolcanic intrusions of liparite-dacite porphyries are located; the largest of them, the Khachynchay intrusion, stretches at a distance of up to 15 km. The transversal block divisibility is displayed rather poorly, expressing itself in the general structural geometry rise in the Terterchay-Gargarchay river interfluve and in the smooth structural plunging bilaterally of the upraised block.

5.4.1.2 The Goycha-Hakeri Zone

The Goycha-Hakeri ophiolite zone is the northern branch of the Izmir-Ankara-Erzinjan ophiolite belt within the Lesser Caucasus boundaries. It is situated south- and southwestward of the Lok-Garabagh zone, in the Murovdagh and Garabagh thrust lying side. It has an extremely complicated tectonic structure, caused by the broad-scale development of ophiolite association rocks in the Upper Cretaceous cross-section. They compose the extensive allochthone plates of longitudinal and transversal asymmetry (Hasanov 1985). The zone lies in the Murovdagh and Garabagh thrust's southern lying side and occupies the Shakhdagh Ridge, near-dividing parts and southwestern slopes of the Murovdagh and Garabagh ridges. It stretches from the northwest to the southeast (within Azerbaijan's boundaries) at a distance of more than 190 km (Fig. 5.25).

In the contemporary structure, the zone looks like a deep Cretaceous trough, preserved at the Lok-Garabagh zone's southern side. It was established in the mid-Cretaceous (the Aptian-Albian age) at a dislocated base that formerly experiences its uprise and erosion (Austrian phase of tectogenesis). It is isolated as an independent structural unit due to the fact that a thick ophiolite complex—absent in the adjacent Artvin-Garabagh megazone structures—takes part in its structure.

The Lachyn-Bashlybel fracture is of faulty character. It can be traced from the town of Jabrail to Bashlybel village; westward, it plunges under the levelling Middle Eocene sedimentary–volcanogenic mass of the Kalbajar trough, serving as the zone's southwestern boundary. The most upraised part of its contemporary structure is the Goycha-Hakeri zone's southwestern part, where the Albian-Senomanian deposits are centroclinally closed in the Allychay River head (the Alyjan village area), giving on the surface place to the Middle–Upper Jurassic formations of the Zamzur high, which protrude from under the Cretaceous rocks. This high is complicated by folded and ruptured dislocations and strikes towards the Araz River, being overlapped on its left bank by the Lower Pleistocene's continental thickness of the Lower Araz's superimposed depression.

The Jurassic, Cretaceous, and occasionally Paleogene and Neogene formations take part in the zone structure (Fig. 5.26). The cross-section's upper part, characterized by the largest areal development, is predominantly represented of the Aptian-Albian-Cenomanian terrigenic flyschoids, by the Upper Cenomanian–Early Santonian ophioliteclastic olistrostrome thickness, by serpentinite mélange and by allochthonous plates of gabbro-ultrabasites and effusive-radiolarite series rocks, and by the Later Senonian terrigenic-carbonate-tuffaceous flyschoids.

In the western part, the Paleocene terrigenic flyschoids and the Eocene volcanogenic rocks are also present. As compared to the adjacent Lesser Caucasus zones, this zone has a more complicated structure and is characterized by intensive dislocations of heterochronous compositional complexes into predominantly linear compressed folds,

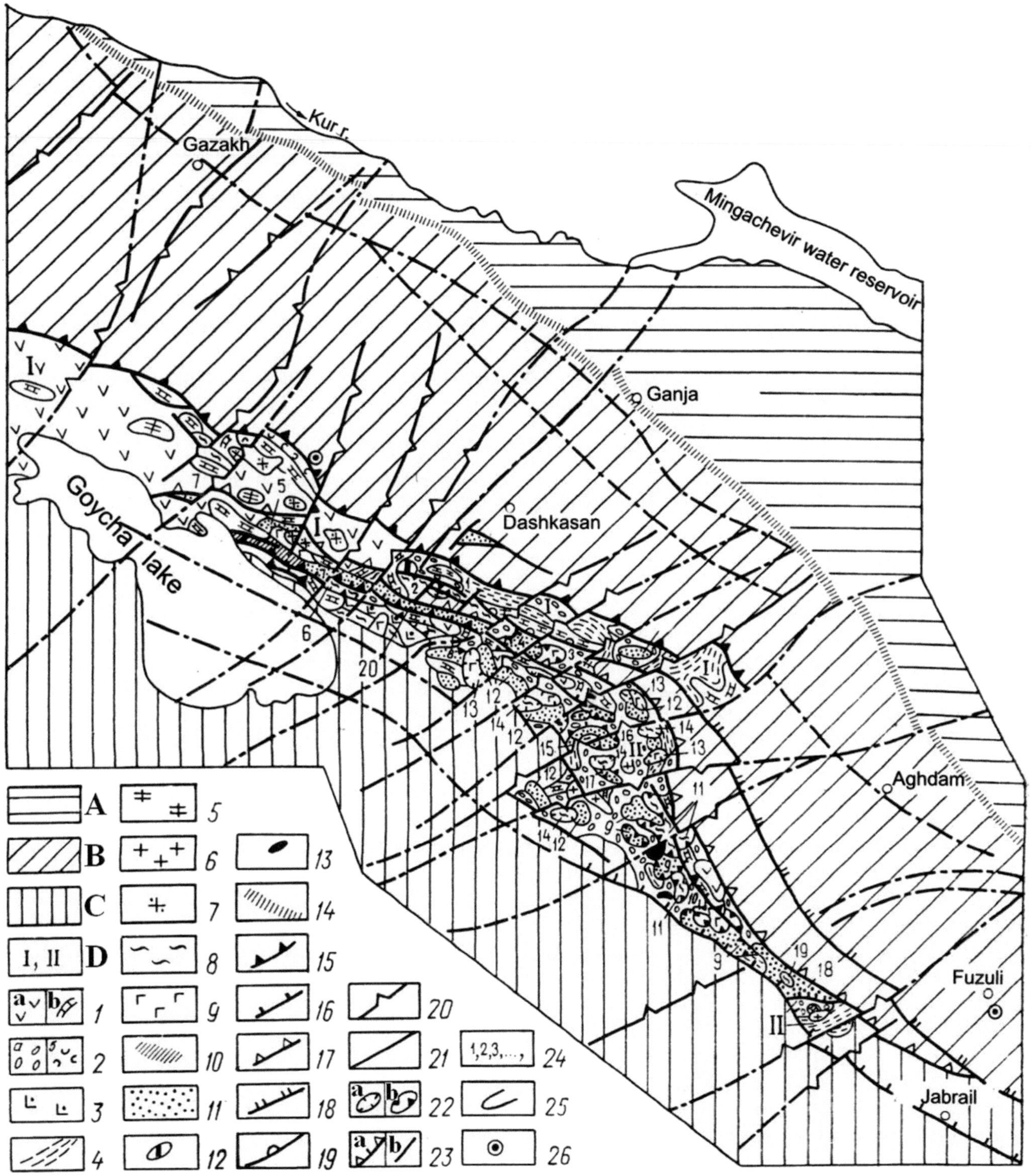

confined by extended thrusts and faults. The ophiolite and radiolarite overthrust sheets are contorted together with underlying deposits, thus generating antiforms and synforms.

Similarly to the region's more northward structures, the zone is broken into upthrown and downthrown blocks by means of transversal dislocations, looking predominantly like low-amplitude faults and transtensional faults of the northeastern strike. In this process, the downthrown blocks are more typical for the zone's northwestern and southeastern parts, with upthrown more common for its central

◀ **Fig. 5.25** Tectonic scheme of the Goycha-Hakeri zone of the Lesser Caucasus (Hasanov 1985). ***A*** Near-Lesser Caucasus through. Zones: ***B*** Lok-Garabagh, ***C*** Gafan, ***D*** Goycha-Hakeri (*I–II*). (*1*) neoautochthone: *a* volcanogenic, clastic sedimentary and carbonate rocks of Paleogene, *b* limestone of Upper Santonian-Maastrichtian; (*2*) autochthone: *a* olistostrome stratum of Upper Cenomanian-Lower Santonian, *b* Coniacian clastic sediments; (*3*) Upper Coniacian conglomerates; (*4*) conglomerates and sandy-argillaceous schists of Apt-Lower Cenomanian; (*5*) quartz diorites, quartz syenite-diorites; (*6*) pegmatoid gabbroids and granitoids; (*7*) gabbroids; (*8–10*) allochthone (ophiolithic association): (*8*) effusive-radiolaritic series of Jurassic-Lower Cretaceous, (*9*) gabbro and gabbro-amphibolites, (*10*) hyperbasites; (*11*) serpentinite melange; (*12–13*) inclusions in melange: (*12*) diabases and andesites of Middle and Upper Paleozoic, (*13*) metamorphic schists of Precambrian-Lower Paleozoic; (*14–23*) faults: (*14*) Pre-Lesser Caucasus deep fault, (*15*) Murovdagh thrust, (*16*) Lachyn-Bashlybel thrust-overthrust, (*17*) Garabagh thrust, (*18*) Meydanchay overthrust, (*19*) Susuzlug trust-overthrust, (*20*) faults and overthrusts dividing transverse blocks, (*21*) faults in basement by gravity data, (*22*) boundaries of tectonic nappes (*a* post Late Eocene, *b* Precampanian), (*23*) faults (*a*), overthrusts (*b*); (*24*) tectonic nappes and plates (*nappes of the Toragaychay subzone*: *1* Bogdadagh, *2* Sarysu, *3* Chaply, *4* Lev; *nappes of the Sarybaba subzone*: *8* Inakdagh, *9* Ipak (Precampanian age), *10* Ipak (Postlate Eocenian age), *11* Naghdaly, *12* Goydara, *13* Chaykend, *14* Gendara, *18* Kechaldagh (Lysogorskiy), *19* Kohnakend; *tectonic plates*: *5* Chatindara, *7* Jil, *15* Zulfugarly, *16* Almalyg, *17* Kechiligaya, *20* Gongur; (*25*) stratigraphic boundary; (*26*) boreholes discovered ultrabasic rocks

part. The upthrown sites are mainly composed of the ophiolite association rocks; the downthrown blocks host, in addition to them, the Senonian and even Paleogene deposits.

In the Gamyshly village area, east of the Levchay River, the Goycha-Hakeri zone is dissected into two branches or subzones: the northern Toragaychay branch and the southern Sarybaby branch, which are separated by the Middle Jurassic complex of the Garabagh high western closure.

West of Gamyshly village, the border between the Toragaychay and the Sarybaby subzones passes along the Garabagh thrust extension. It can be traced in the Shakhdagh ridge's near-dividing strip toward the Mt. Alagollar branch, going westward outside the Republic borders.

The establishment of large overthrust nappes within the Goycha-Hakeri zone that are composed of ophiolite association rocks allows consideration of its contemporary structure as a result of the generation and interaction of three tectonic complexes: the autochthone, the allochthone, and the neoautochthone (Hasanov 1971, 1985). The structural and compositional rock complex of the Goycha-Hakeri zone's visible section, located below the ophiolite complex rocks, are taken as the autochthonous complex. The neoautochtone cross-sections began from the Upper Senonian sedimentary deposits, overlapping (together or in separate) the rock allochthonous and autochthonous complexes.

In the autochthonous complex, the upper part is formed by the olistrostrome complex that was generated during the Late Cenomanian–Early Santonian period. Olistostrome masses are composed of erosion and destruction products of the ophiolite complex rocks; that is, they are ophiolitoclastic. In their composition, one can isolate proper landslide horizons (olistostrome and olistolites) and normal sedimentary stratified strata and packs composed of conglomerates, gravelites, sandstone, aleurolites, argillites, marls and limestone. Three types can be isolated among the landslide horizons: breccia, boulder-gravel argillites (tillite-like conglomerates) and argillaceous slates "littered" by debris. Inter-olistostrome deposits alternate with olistostrome horizons or are grouped in a certain cross-section part. They also substitute for each other without any definite regularity along the course. Their interrelation also differs—from the predominant stratified rocks to their absolute disappearance and replacement by landslide horizons.

The Early–Late Santonian boundary corresponds to the generation period of the allochthonous complex, which forms two independent nappe plates: the mélange lower one and the effusive-radiolarite upper one (Knipper 1975; Sokolov 1977; Hasanov 1985). *The lower plate*, named the Ipak nappe, is composed of serpentinite and gabbroid massifs or of serpentinite mélange; it overlaps the olistrome mass in all places. Serpentinites are grouped by dunites, peridotites, harzburgites, lherzolites and pyroxenites. In the nappe composition, serpentinites and serpentinized ultrabasites occupy the section's lower part, whereas the gabbroids and plagiogranites occupy its upper part.

Serpentinite mélange is a giant-thickness tectonite composed of a chaotic mixture of heterochronous and differently originating rocks, plunged into serpentinite matrix (groundmass). Melange inclusions consist of peridotites, dunites, serpentinites, gabbro-amphibolites, gabbro-diabases and diabases, spilites and their pyroclastolites, aleurotuffites, silicious-argillaceous slates argillites and radiolarites, pillow lava of andesite-basalt composition, poorly and highly marmorized limestone and marble, laminated limestone, metamorphic slates, and, occasionally, small and large blocks of quartz diorites and quartz syenite-diorite with ultrabasite xenolites. All of these rocks are chaotically compressed, crushed, flared and consolidated by serpentinites. Melange is represented by monomictic and polymictic differences, with the latter having greater occurrence.

Inside the serpentinite mélange as well as inside the olistostrome complex, numerous inclusions are found; some of them are classified as detached because these rocks have not been detected either in the Goycha-Hakeri zone or in its edges. Among these, particularly, are chlorite and amphibolite metamorphic slates with radiological age dating back to the pre-Devonian epoch, as well as detached blocks and

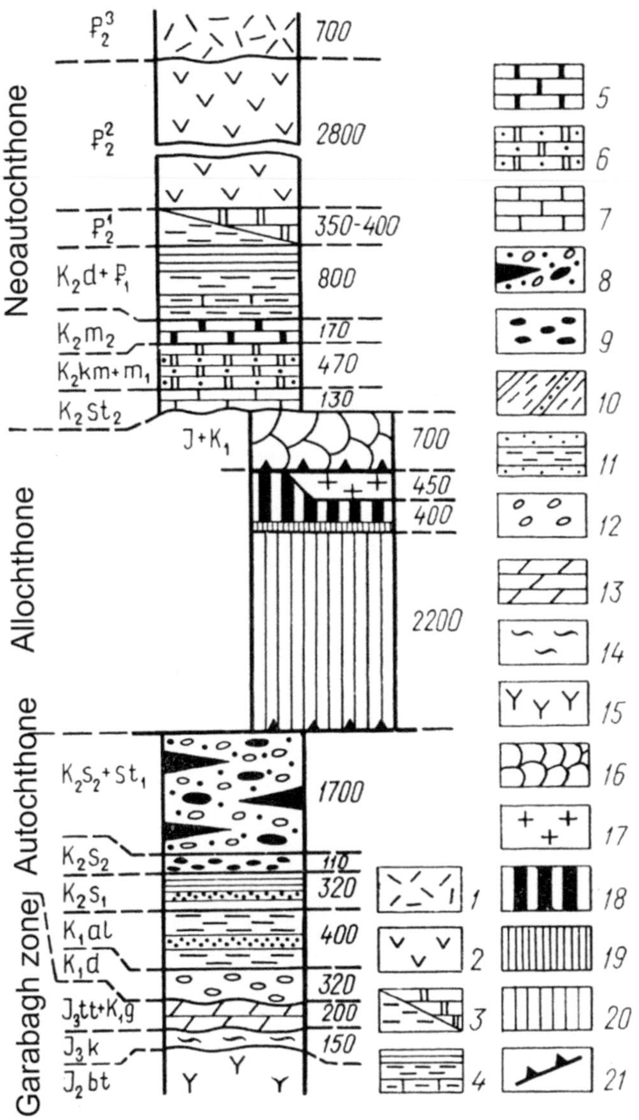

Fig. 5.26 Generalized geological scale of the Goycha-Hakeri zone (Hasanov 1985). (*1*) dacites, liparite-dacites and their volcanic tuffs of Upper Eocene; (*2*) andesites and sedimentary–pyroclastic rocks of Middle Eocene; (*3*) limestone and sandy-argillaceous schists of Lower Eocene; (*4*) sandy-argillaceous schists, mudstone and argillaceous limestone of Danian-Paleocene: (*5*) gravel and sandy limestone of Upper Maastrichtian; (*6*) limestone and marls of Campanian-Lower Maastrichtian; (*7*) limestone of different composition of Upper Santonian; (*8*) olistostrome stratum of Upper Cenomanian-Lower Santonian; (*9*) tilloid facies of Upper Cenomanian; (*10*) sandy-argillaceous schists, limestone and conglomerates of Lower Senomanian; (*11*) sandy-argillaceous schists, argillites and argillaceous limestone of Albian; (*12*) conglomerates and clastic sedimentary facies of Aptian; (*13*) limestone of Tithonian-Hauterivian; (*14*) clastic sedimentary facies of Callovian; (*15*) volcanic tuffs of andesite composition of Bathonian; ophiolithic complex (including allochthon) (*16–20*): (*16*) effusive-radiolaritic series of Jurassic-Lower Cretaceous, (*17*) quartz diorites and gabbro-diorites, (*18*) gabbro and gabbro-amphibolites, (*19*) troctolites, (*20*) hyperbasites and serpentenite melange; (*21*) faults in the allochthon basement

masses of two epochs: the Upper Jurassic-Neocomian and the Devonian (Knipper 1971a, b; Sokolov 1977; Hasanov 1985).

Serpentinite mélange also contains lenses and blocks of pillow lava of basalt-andesite-basalt composition, blocks of Cenomanian sandy shales, limestone debris and blocks of olistostrome complex (Upper Cenomanian–Lower Santonian rocks), debris and masses of the Upper Coniacian–Lower Santonian sandstone and conglomerates, limestone with the Upper Maastrichtian fauna, and the Lower Eocene nummulite limestone. In addition to the previously mentioned inclusions, the Sarybaby subzone serpentinite mélange contains numerous debris and blocks of the Middle Eocene andesites, andesite-dacites and dacites, as well as quartz diorites and quartz syenite-diorites with ultrabasite xenoliths of the Upper Senonian and Middle Senonian age. It should be specifically emphasized that large serpentinite mélange inclusions participate as olistoliths in the Upper Cretaceous olistostrome complex composition. The given facts indicate that the generation of described tectonic formation began during the Late Cretaceous at least, while its final establishment stage coincides with the Pyrenean (late Upper Eocene–Oligocene) tectogenesis phase. All of the above demonstrates the polycyclicity of serpentine mélange formation (Knipper 1971a).

The *allochthonous gabbroid complex* is essentially represented by gabbro-amphybolites and hornblende gabbro, which give facies transitions to norites and gabbro-diorites. Of lesser occurrence are troctolites, quartz diorites, as well as anorthosites and pegmatitic varieties of gabbro and hornblende. The largest massifs composed of gabbroids and of effusive radiolarite series rocks host the bodies of plagiogranites, gabbro-diabases and andesite porphyrites, which are not genetically associated with the Mesozoic ophiolite association rocks. The results of radiometric age determination show that combining series of gabbro-amphybolite and gabbro-diabase rocks are a complicated metamorphic complex; the former are the transformation results of the Early–Middle Paleozoic ultrabasic rocks, probably Precambrian, whereas the latter are the result of the Carbon age diabases. Their observed interrelation in the exposed cross-sections can be explained by the capture of diabase formations by older ultrabasites, when the latter are cold squeezed to the Earth crust's higher horizons out of the depths.

The upper plate, named the Naghdaly nappe, is composed of the effusive-radiolarite series rocks; the most developed of them are tholeiitic basalts, diabases, aphyric basaltoids, argillaceous-silicious, carbonate-silicious and silicious rocks, radiolarites, jaspers, micritic limestone and calcarelites (calcareous sandstone, debris limestone). The nappe composition displays two volcanogenic–sedimentary formations that constitute the independent, frequently spatially discrete tectonic lamina: lower effusive-radiolarite and upper effusive-carbonate. The effusive-radiolarite formation is characterized by alternations of lava beds, tuff breccias, lava breccias, hyaloclastites and laminated siliceous rocks with the horizons and lenses of gravelites, sandstones and microfragmental breccias in the succession's lower part. In breccias, the fragments consist of diversified rocks of the effusive-radiolarite series; they can be seen higher and lower in the succession, as well as of serpentinites, plagiogranites, and gabbroids. The fragments of bioclastic and algal limestone are frequently met. The matrix (groundmass) is, as a rule, ferruginized, with a large number of radiolarian shells. The effusive-radiolarite formation is represented in its lower part by the horizon alternations of silicious, carbonate-silicious, silicious-carbonate rocks, micritic limestones, calcarenites, lava flows and tuff; in its upper part, micritic limestones are predominant.

The ***Toragaychay subzone*** within Azerbaijan's boundaries stretches as a broad strip (7–10 km) in the general Caucasus direction at a distance of about 110 km, from the Chambarak pass in the Akhynjachay River head to the Terterchay River's left bank, where it closes centroclinically —or to be more precise, is cut by the northeastern fault of the same name. Generally, the subzone is a structurally complicated megasyncline filled with the Jurassic, Cretaceous and Paleogene deposits. The subzone's northern border passes along the Murovdagh thrust, while in the south it is confined by the Sarsang fault in its eastern part and by the Garabagh thrust (after the dislocation thinning-out in the Levchay River head) in its western part. The post-Neocomian interval of the Cretaceous compositional complex lies transgressively on the Jurassic–Neocomian formations, which is typical for the Lok-Garabagh high Garabagh subzone. Because the subzone is an underthrust structure (with regard to the Murovdagh thrust), it experienced the influence of north overthrust masses; for this reason, it has an asymmetrical structure with a steep, northern side, folded as isoclinal, compressed and vergent folds, and with more smooth southern side. The folds are usually complicated by thrusts and overfaults in their domes and southern sides. In the Garabagh and Murovdagh thrust zones, a highly involved series of overfaults and thrusts frequently dissects the rocks into tectonic lamina. The subzone internal structure, in addition to longitudinal dislocations, is also complicated by transversal (predominantly of northeastern strike) flexures, faults and overfaults. In the west, the subzone's Upper Cretaceous rocks are raised by the Paleocene–Eocene volcanogenic–sedimentary complex of the Shakhdagh synclinal strip, intruded by the small intrusives of quartz-diorite, granodiorite, syenite-diorite, and gabbroid composition and complicated by additional folds, generally southward vergent and broken by transpressional faults.

Within the subzone boundaries, the *autochthonous complex* includes the Middle Jurassic–Upper Cretaceous formations that separate two stratigraphic subcomplexes: the Middle Jurassic-Neocomian and the Aptian-Upper Cretaceous, which differ greatly from one another by their formation conditions. The autochthone's lower part is represented by the Middle and Upper Jurassic volcanogenic and volcanogenic–sedimentary masses, where effusives are represented by andesites, rhyolites, albitophyres, quartz porphyries, etc. Higher in the succession, the Tithonian-Valanginian pelitomorphic, flagstone, petrified limestones and the Hauterivian gravelite, sandy, and clay limestones are bedded. The upper part section is represented in the basement by the Aptian-Lower Senomanian terrigenic flyschoids, which are later replaced by the Late Cenomanian tilloid facies. The latter gradually passes in the top into a thick olistostrome detrital complex, represented by debris, blocks and lenses of ophiolite association rocks (ultrabasites, gabbroids, effusive-radiolarite series), as well as marbles, marbleized limestones, diabases, and metamorphic shales buried in the argillite-limestone-sandstone matrix (Knipper 1971b; Hasanov 1985). The olistostrome complex is established in two age ranges: the Late Cenomanian–Early Coniacian and the Late Cenomanian–Early Santonian. By the course from southeast to northwest, the olistrostrome complex experiences a sharp increase in ultrabasic composition inclusions and decrease in the fragments of radiolarites, jaspers and spilites. Olistostrome mass has in many places tectonic contact with serpentinite mélange, the Upper Jurassic–Lower Cretaceous effusive-radiolarite series and the Tithonian-Neocomian limestones. In the Kukurdchay River head, it is transgressively overlapped by the Upper Santonian limestones.

The *allochthonous complex* is represented by northern fragments of the Ipak and Naghdaly nappes as well as by the independent Boghdadagh overthrust sheet, which is composed of Tithonian-Valanginian limestone. It tectonically overlaps the olistostrome-containing Albian-Lower Santonian complex and is in turn "sealed" by the Upper Santonian-Maastrichtian limestones that define the cover formation age (Fig. 5.27). The Boghdadagh nappe is the northwestern fragment of the allochthone Susuzlug limestone massif, isolated as a result of the latest disintegration processes (Hasanov 1985). The Ipak allochthone analog is the Chaply nappe, represented by the Levchay and Meydanchay shuppens, composed of serpentinite mélange and superimposed on the autochthonous olistostrome complex; it determines the cover formation stage (sub-Hercynian phase of tectogenesis). At the same time, on separate areas, serpentinite is thrust at a low angle upon the faunistically characterized Upper Maastrichtian and Middle Eocene rocks (Fig. 5.28), which evidences the preservation of horizontal compression efforts after the cover formation principal stage,

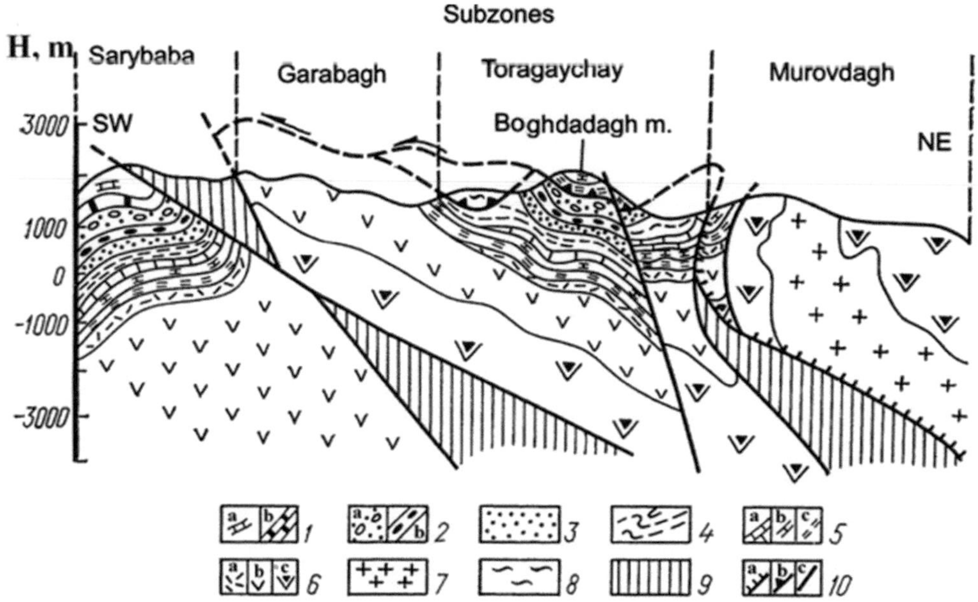

Fig. 5.27 Geological section of the Boghdadagh nappe (Hasanov 1985). Rocks: (*1*) limestone (*a* Campanian–Maastrichtian, *b* Upper Santonian); (*2*) olistostromes (*a* Upper Senomanian–Lower Santonian, *б* tilloid facies of Upper Senomanian); (*3*) sandstone, mudstones, conglomerates of Lower Senomanian; (*4*) sandy-argillaceous schists of Albian; (*5*) limestone (*a* Hauterivian, *b* Berriasian–Valanginian and *c* Tithonian); (*6*) volcanoclastic sediments (*a* mudstones of Callovian–Kimmeridge, *b* Bathonian andesites, *c* andesite-basalts and liparite-dacites of Bajocian); (*7*) gabbroids and gabbro-diorites of Lower Bajocian. Ophiolithic association: (*8*) effusive-siliceous series of Upper Jurassic-Lower Cretaceous; (*9*) serpentenites and serpentenitic melange; (*10*) faults (*a* Murovdagh thrust, *b* faults in the allochthone basement, *c* overthrusts)

with repeated stripping of allochthonous masses and their overlapping of the neoautochthonous complex.

The Naghdaly allochthone analog is situated in the Meydanchay and Toragaychay rivers interfluve, where it forms the Sarysu nappe that tectonically overlaps the Levchay and the Meydanchay parts of the Chaply nappe. Occasionally, the nappe is located on the autochthonous olistostrome complex or directly on argillo-arenaceous formations of its basement. The nappe effusive-radiolarite formation is represented by interbedding of pillow basalts, spilites and their pyroclastolites, radiolarite jaspers and flints with the Late Jurassic–Early Cretaceous fauna with nipped blocks of more ancient metamorphic shales, marbles and marblized limestones.

The *neoautochthonous complex* is situated sharply transgressively both on the autochthonous deposits and on the allochthone series, as it is composed by sedimentary and volcanogenic formations from the Upper Santonian to the Upper Eocene inclusively. In the subzone's eastern and central parts, the neoautochthone is preserved as separate spots on the autochthone and allochthone surface and is represented solely by the Upper Senonian rocks. In the west, in the Shamkirchay river head, the neoautochthone Upper Cretaceous completely overlaps the subzone structural geometry and, being upraised by the volcanogenic–sedimentary Paleocene–Eocene rocks, defines its shape in the Shakhdagh synclinal strip.

The ***Sarybaba subzone*** forms the Goycha-Hakeri zone southern branch and can be traced from the northeast (Jil village) through the Shakhdagh ridge, the Zod pass, the Tutgunchay River, the Hakeri River head, the Kechaldagh pass and the northeastern outskirts of the plunging Garabagh ridge up to Shishdagh mountain meridian. Eastward, the zone conceals itself under the Lower Araz superimposed trough's Lower Pleistocene rocks. The subzone's length within Azerbaijan's boundaries is up to 160 km, so that its largest width (up to 24 km) is at the Tutgunchay intersection and the smallest one (1.5 km) is at the Kechaldagh pass. Generally, in the contemporary folded structure, the subzone has a complicated syncline-like shape. Its late Neocomian Cretaceous complex has no principal differences compared to that of the Toragaychay subzone, and it has subjacent Middle–Upper Jurassic formations, similar to those of the Garabagh subzone.

The Sarybaba subzone's internal structure is characterized by great complexity: the presence of linear folds, confined by extended thrusts and overfaults, extensive development of large overthrust nappes, composed of ophiolite association rocks and dislocated concertedly with underlying autochthonous deposits. Neuautochthonous

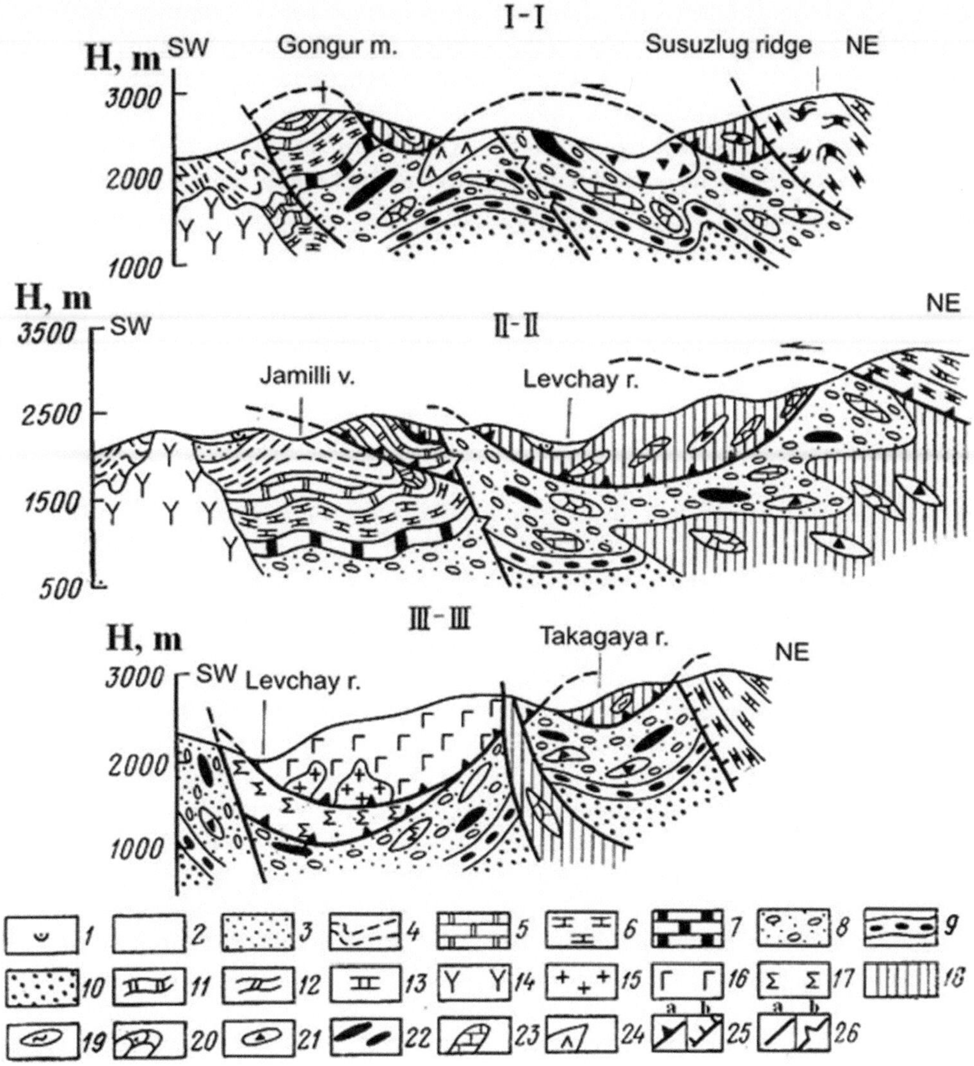

Fig. 5.28 Geological map of the interfluves of the Soyutluchay and Levchay rivers (Hasanov 1985). (*1*) landslide deposits; (*2*) undivided Quaternary deposits; (*3*) terrigeneous-carbonate flysh of Middle Eocene; (*4*) flysh of Paleocene-Lower Eocene; (*5*) gravel and sandy limestone of Upper Maastrichtian; (*6*) limestone of Campanian-Lower Maastrichtian; (*7*) gravel and sandy limestone of Upper Santonian; (*8*) olistostrome complex of Upper Senomanian–Lower Santonian; (*9*) tilloid facies of Upper Senomanian; (*10*) sandy-shales and conglomerates of Lower Senomanian; (*11*) gravel limestone of Hauterivian, (*12*) sandy limestone of Valanginian, (*13*) silicified and gravel limestone of Tithonian-Berriasian; (*14*) subvolcanic bodies of andesite-dacites of Middle Eocene; ophilothic associations (*15–18*): (*15*) quartz diorites, (*16*) brecciated and schistose gabbro and gabbro-amphibolites, (*17*) serpentenites, (*18*) serpentinitic melange; blocks, lumps and lens (*19–24*): (*19*) radiolarites and jaspers, (*20*) spilites, diabases and their tuff breccias, (*21*) gabbro-amphibolites, (*22*) serpentinites, serpentinitic breccia and serpentinitic melange, (*23*) marble and marbleized limestone, (*24*) listvenites; (*25*) overthrusts and thrusts: *a* thrusts in the allochthonous basement, *b* thrusts-overthrusts; (*26*) faults and upthrusts: *a* faults, *b* faults and upthrusts which are boundaries of transverse blocks

deposits are also compressed into complicated folds so that its structural geometry does not coincide with those of autochthone and allochthone.

The subzone, as well as more northern and more southern structures of the Lesser Caucasus mountainous-fold system, are complicated by transversal flexural raptures that dissect its structural geometry into transversely upthrust and downthrust blocks. West of the Tutgunchay River meridian, the subzone's southern side undergoes plunge under the Kalbajar superimposed trough's Paleogene–Neogene formations. In the southeast, the post-Neocomian complex is cut by the Hojavand fault. It closes centroclinally in the Alyjanly syncline, thus promoting the outcropping of the Jurassic upraised block (the Zamzur high), stretching towards the Araz River. The subzone's peculiar feature is the presence of the Upper Eocene granitoid intrusions—the largest of them being the Gazykhanly intrusive, intruding the Albian-Cenomanian and the Middle Eocene deposits in the Tutgunchay River head.

In the *autochthonous complex* composition, two lithological age groups are isolated, which reflect two tectonic stages of the subzone development. The autochthonous lower part is exposed at the subzone southeastern end, within the Zamzur anticlinal high, and is represented by the Bathonian sedimentary pyroclastic thickness, by the Callovian-Lower Oxfordian tuffaceous-carbonate formations and by the Kimmeridgian-Tithonian limestones; these zones have unconformal contacts. The high has dramatically asymmetrical structure with a low-angle northeastern side and steep, nearly vergent southwestern side; it is complicated by second-order folds and ruptured dislocations.

The autochthonous upper part is similar to that of the Toragaychay subzone and begins with the terrigenic Aptian-Lower Cenomanian rocks (with basal conglomerates in the basement), replaced by the Late Cenomanian tilloid facies. Higher in the succession, the latter are gradually substituted by the coarse olistostrome complex, subdivided into two subcomplexes that differ from each other only by their formation termination period (Sokolov 1974, 1977; Hasanov 1985). The Late Cenomanian–Early Coniacian olistrostomes are distributed very focally and are prominent in the subzone's central part, the Tutgunchay river lower course (Kilisali and Alychaly villages). Here, olistrostrome horizon (whose matrix contains the Cenomanian, the Upper Turonian and the Lower Coniacian fauna) occurs in the anticlinal fold side on tilloids and is overlapped by the Upper Coniacian–Lower Santonian limestones with basal conglomerates (Fig. 5.29). The Late Cenomanian–Early Santonian olistostromes have broad areal distribution. Their typical cross-section can be found at the right bank of the Okuzuchanchay River (the Tutgunchay River tributary), in the Otagly village area (Fig. 5.30); here, tilloid facies gradually transform into the olistostrome, whose matrix contains numerous Upper Cenomanian, Upper Turonian-Lower Coniacian and Lower Santonian foraminifera. A specific feature of the Upper Olistostrome subcomplex is the participation of large (up to 1.0–2.5 km^2) serpentinite mélange olistoplaques in its succession. In addition, separate olistostrome horizons directly participate in the Upper Coniacian–Lower Santonian conglomerate–sandy series cross-section, where they are of the Early Santonian age.

The allochthonous complex is represented by large nappes, such as the lower serpentinite–gabbro Ipak nappe and the upper effusive-radiolarite Naghdaly, which determine the Sarybaba subzone tectonics along its entire length (Figs. 5.31 and 5.32). The Ipak nappe generally tectonically overlaps the olistostrome mass in all places and is represented by isolated plates (or plate packs)—the proper Ipak, Goydara, Chaykend, Charaktar, Shorbulag, Zulfugarly, Almalyg, Kechili, Kechaldagh, Chatindara, Ipakdagh, Jarchygabri and Jil plates (the latter three are outside the Republic borders). In the nappe cross-section, its lower part is occupied by serpentinites and serpentinized ultrabasites; its upper part is characterized by gabbroids and plagiogranites; in some places, a gradual transition can be seen, expressed by the feldspar rock zone. The nappe basement in serpentinites displays both rolled blocks made of the underlying olistostrome mass rocks and gabbroid, gabbro-amphybolite, and rodingite bodies. The serpentinite mélange and conglomerated serpentinite zones are widely distributed. In gabbroid massifs of the nappe upper parts, inclusions of different-scaled blocks of plagiogranites, gabbro-diabases and their breccia can be found, as well as basic-composition diabases and volcanogenic rocks. The Naghdaly nappe, whose structure involves the effusive-radiolarite series rocks such as spilities, diabases, keratophyres, aphyric basaltoids, argillaceous-silicious, carbonate-silicious and silicious rocks, jaspers, mucritic limestones, is represented within the subzone by strongly deformed, brecciated and broken into different-scaled blocks fragments—the largest of them being the proper Naghdaly, Seyidlar, Kandyeri, Kohnakend and Goygol plates. The nappe bottom generally tectonically overlaps various horizons of the gabbro-serpentinite Ipak nappe, and the upper nappe frequently occurs on olistostrome mass only at the sides.

In separate plots, in addition to ophiolites, the allochthone composition includes the terrigenic Albian-Lower Coniacian age mass that transgressively lies on the lower and upper nappe rocks; in turn, this is transgressively overlapped by serpentinite mélange with overlying neoautochthonous mantle (Sokolov 1977). Therefore, the ophiolite complex age is determined as pre-Albian. Moreover, the results of radiological studies indicate that ultrabasic rocks and gabbroids are of pre-Carboniferous age, while the effusive-radiolarite mass is related to the Jurassic-Lower Cretaceous age.

The neoautochthonous complex is represented by two different facies thicknesses—the terrigenic-carbonate Upper Senonian–Lower Paleocene and the volcanogenic–sedimentary Middle Eocene ages, looking like synclinal bowls complicated by additional folds and ruptured dislocations. It occurs transgressively on the autochthone and allochthone structures. In the north, in the Levchay River head, a tectonic sheet composed of the neuautochthone Cretaceous-Eocene formations outcrops at the Gonur latitudinal ridge and extends west- and eastward (Fig. 5.33). It is located in the suture zone between the Toragaychay and the Sarybaba zubzones (the Garabagh thrust lying side), thus witnessing the post-Eocene movements that took place in the strip (Pyrenean phase of tectogenesis). Synclinal overthrusts of

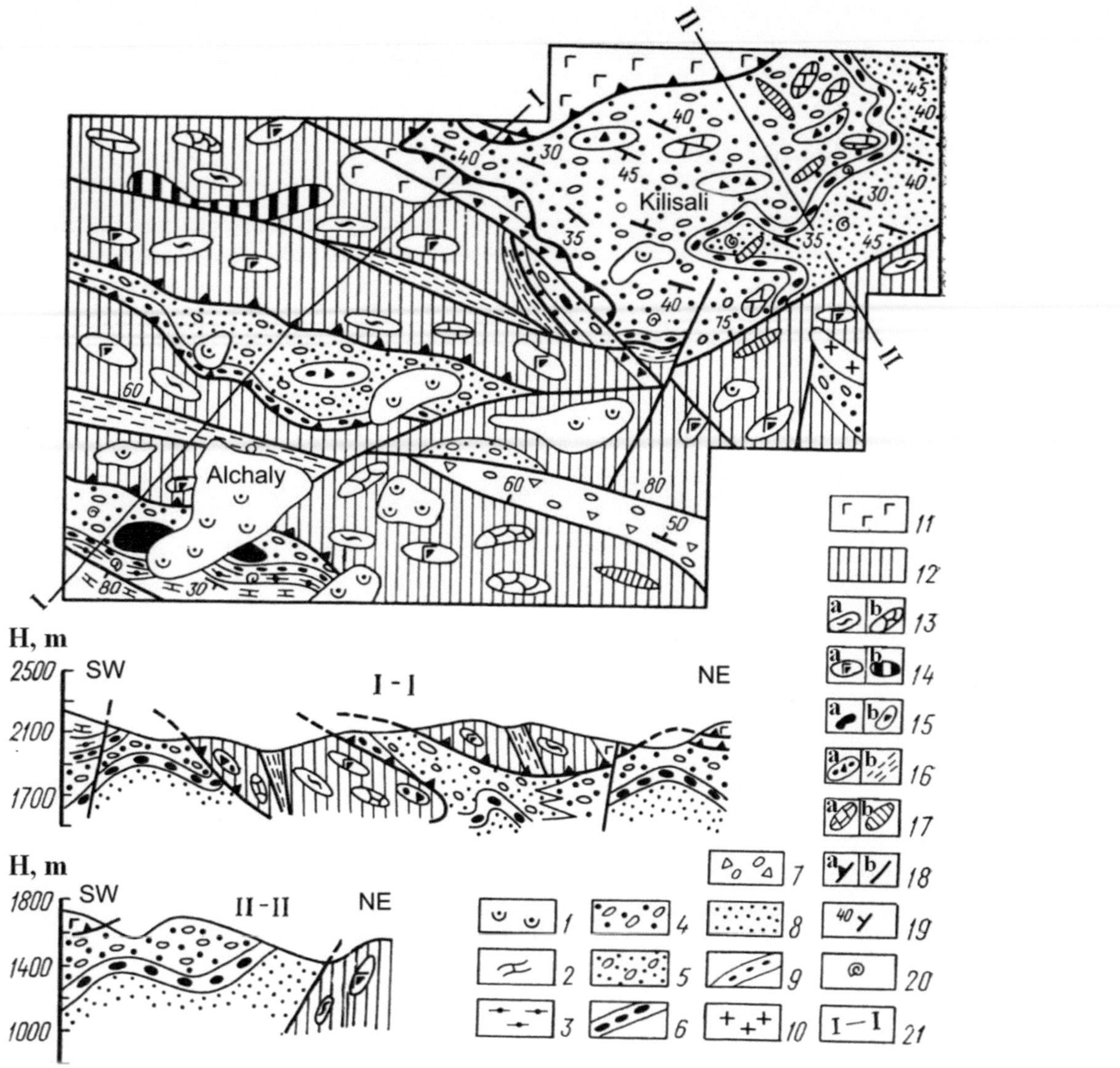

Fig. 5.29 Geological map of the Kilisali area (Hasanov 1985). (*1*) landslide deposits; (*2*) limestone of Campanian-Maastrichtian; (*3*) sandstone, mudstone and gritstone of Upper Coniacian; (*4*) olistostrome stratum of Upper Cenomanian–Lower Coniacian with argillaceous cement; (*5*) serpentinitic-sandy-gravel material; (*6*) tilloid facies of Upper Cenomanian; (*7*) ophiolite clastic conglomerate breccias of Cenomanian; (*8*) sandy-argillaceous schists of Albian-Lower Cenomanian; (*9*) basal conglomerates; (*10*) subvolcanic bodies of andesite-dacitic composition (middle Eocene); ophiolithic association (*11–12*): (*11*) gabbro-amphibolites, (*12*) serpentinitic melange; lumps, fragments and lenses (*13–17*): (*13*) serpentinitic melange; lumps, fragments and lenses of Upper Jurassic–Lower Cretaceous (*a* radiolarites, *b* spilites and their tuff breccias), (*14*) the same (*a* gabbro-amphibolites, *b* serpentinized dunites); (*15*) the same (*a* serpentenites, *b* serpentinitic breccia); (*16*) the same (*a* serpentinitic breccia, *b* schistose serpentinites), (*17*) the same (*a* marble and marblized limestone, *b* listvenites); (*18*) faults (*a* allochthon basement, *b* overthrusts); (*19*) bedding items; (*20*) fauna; (*21*) lines of geological sections

the subzone southern side still host the Miocene–Pliocene volcanogenic formations that unconformly lie on the Eocene and, extending beyond its borders, on the Senonian rocks. Westward plunging autochthone and allochthone structures of this subzone side completely subside at the Tutgunchay River's left bank under the Upper Senonian neoautochtone mantle, which further westward is upraised by the volcanogenic–sedimentary Eocene rocks of the transversely oriented southeastern edge of the Kalbajar superimposed depression.

5.4.1.3 The Gafan Zone

The Gafan zone forms the Artvin-Garabagh high southwestern side; its geological succession peculiarities are comparable with the Lok-Garabagh zone of the high northeastern side. The zone's northern boundary is the Lachyn-Bashlybel fault; in the southeast, it is confined by the Girratagh fracture and the Mesotethys suture in the Lesser Caucasus. Within Azerbaijan's boundaries, the zone can be traced from the Tutgunchay River head (where it outcrops from the Kalbajar zone transversely overlapping Paleocene–Eocene volcanogenic–

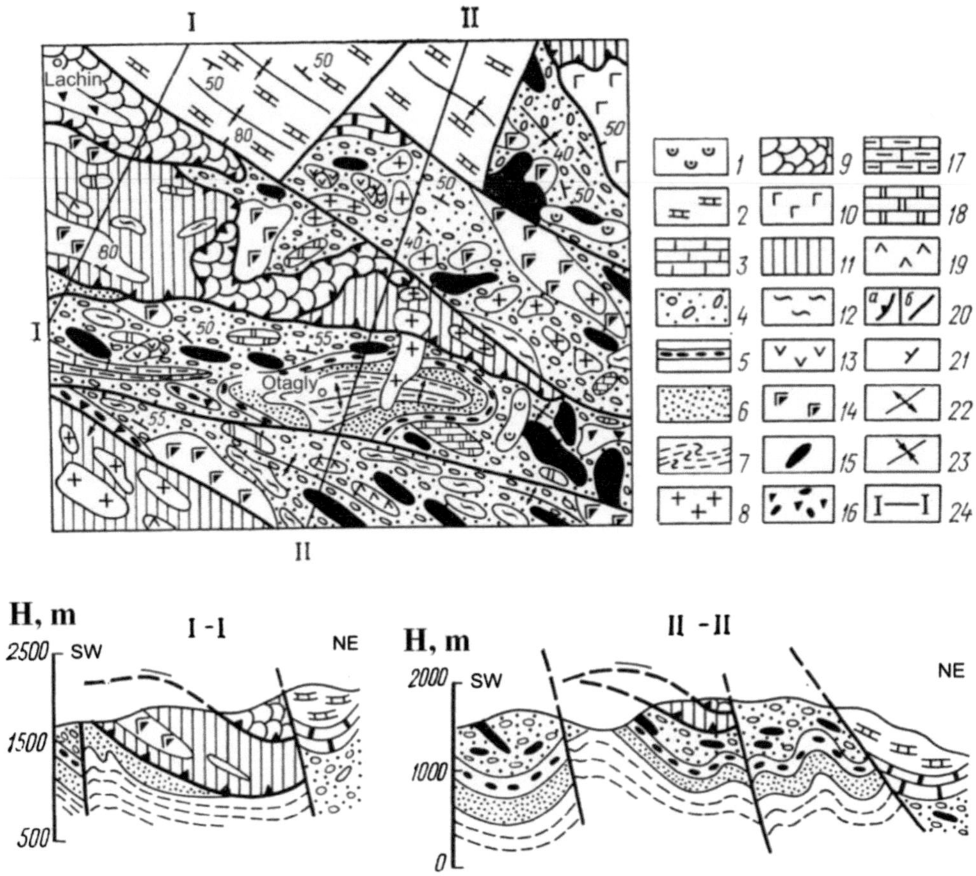

Fig. 5.30 Geological map of the Otagly area (Hasanov 1985). (*1*) modern landslide deposits; (*2*) Campanian-Maastrichtian limestone; (*3*) sandy, fragmental, gravel light-pink limestone of Upper Santonian; (*4*) olistostrome complex of Upper Cenomanian–Lower Santonian; (*5*) tilloid facies of Upper Cenomanian: mudstones, grits and fine detrital conglomerates with inclusion of hyperbasite pebbles and radiolarites; (*6*) sand-shale and siltstone of Lower Cenomanian (?); (*7*) shales, mudstones; argillaceous limestone and sandstone of Upper Albian; (*8*) Lower Santonian quartz diorites; diorite porhyry, gabbro-diorites with serpentinite xenoliths; (*9*) rocks of effusive-radiolaritic series; (*10*) gabbro-amphibolites, brecciated and schistose (including allochthon); (*11*) serpentinitic melange (including allochthon); inclusions in olistostrome and serpentinite melange of Jurassic-Lower Cretaceous (*12–14*): (*12*) radiolarites, (*13*) basalts, (*14*) gabbro-amphibolites; (*15*) hyperbasites; (*16*) serpentinite breccias; (*17*) argillaceous limestone; (*18*) marblized limestone; (*19*) listvenites; (*20*) faults; *a* in allochthon basement, *b* overthrusts; (*21*) bedding items; fold axes (*22–23*): (*22*) anticlines, (*23*) synclines; (*24*) line of section

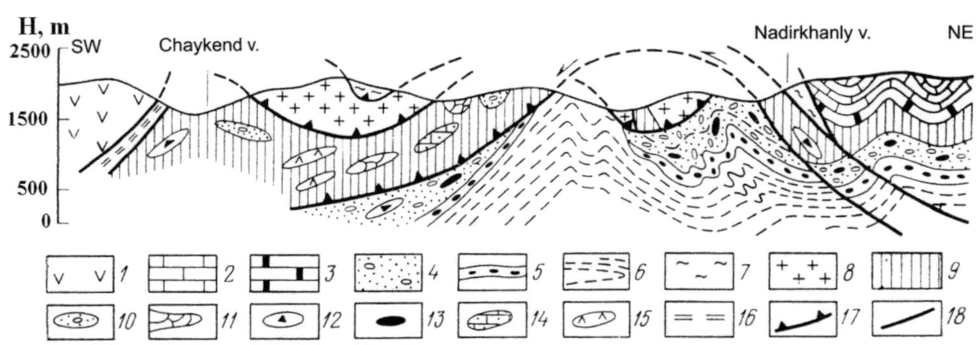

Fig. 5.31 Geological section of left river bank of Terterchay (Hasanov 1985). (*1*) sedimentary–pyroclastic rocks and subvolcanites of andesite and andesite-dacitic composition of Middle Eocene; limestone (*2–3*): (*2*) clastic of Campanian-Lower Maastrichtian, (*3*) sandy-gritstone associations of Upper Santonian; (*4*) olistrostrome complex of Upper Cenomanian–Lower Santonian; (*5*) tilloid facies—mudstones, grits and fine-clastic conglomerates with inclusion of pebbles, hyperbasites and radiolarites of Upper Cenomanian; (*6*) sandy-shales and siltsones of Albian-Lower Cenomanian: ophiolitic complex (*7–13*): (*7*) Jurassic-Lower Cretaceous effusive-radiolaritic series, (*8*) gabbro-amphibolites, (*9*) serpentinitic melange; blocks, lumps and lenses (*10–15*): (*10*) olistostromes, (*11*) jasper radiolarites and rarely spilits, (*12*) gabbro-amphibolites, (*13*) serpentinized dunites and peridodites, serpentinized sandstone and breccia, (*14*) marbles and marblized limestone of Precambrian-Lower Paleozoic (?) and Upper Jurassic–Lower Cretaceous, (*15*) listvenites; (*16*) limonitized and strongly schistose hyperbasites and serpentinites; (*17*) faults in allochthone basement; (*18*) thrusts-overthrusts

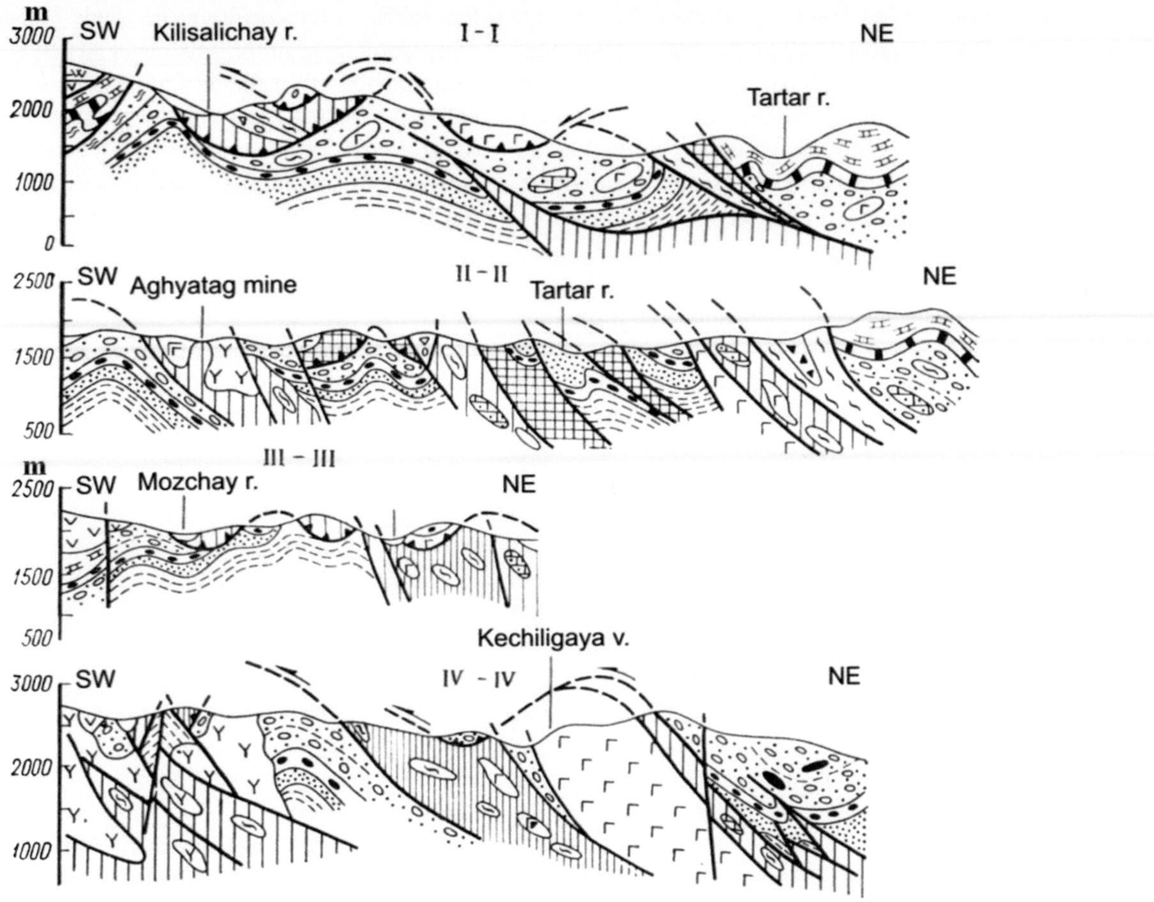

Fig. 5.32 Sections of a geological map of the Tutgunchay River's basin (Hasanov 1985). *I–I* and *II–II* lower, *III–III* and *IV–IV* middle river's stream. Captions and locations are shown in Fig. 5.27

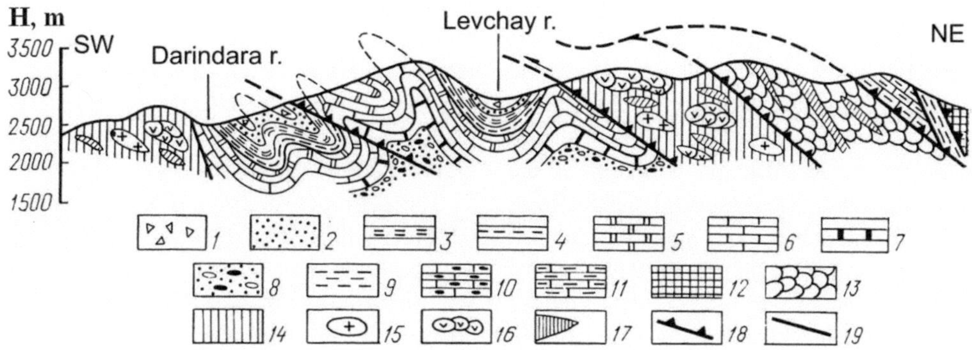

Fig. 5.33 Schematic section of Jurassic, Cretaceous and Paleogene deposits in the upper reaches of the Levchay and Gonurchay rivers (Hasanov 1985). (*1*) siltstones, sandstone, limestone and conglomerates of Middle Eocene; (*2*) terrigeneous and terrigeneous-carbonate flysh of Lower Eocene; (*3*) terrigeneous flysh of Upper Paleocene; (*4*) sandy-shales and argillaceous limestone of Lower Paleocene; limestone (*5–6*): (*5*) sandy-grit of Upper Maastrichtian, (*6*) weakly sandy of Campanian-Lower Maastrichtian; (*7*) girt limestone of Upper Santonian; (*8*) olistrostrome complex of Upper Cenomanian–Lower Santonian; (*9*) sandy-shales and sandstone of Albian-Lower Cenomanian; (*10*) sandy and grit limestone of Berriasian-Hauterivian; (*11*) limestone of weakly sandy Tithonian; ophiolithic association (*12–16*): (*12*) hyperbasites and serpentinites, (*13*) effusive-radiolarite series of Jurassic-Lower Cretaceous, (*14*) serpentenite melange, (*15*) fragments and lumps of gabbro and gabbro-amphibolites, (*16*) the same, volcanogenic-siliceous series; (*17*) the same, marbles and marblized limestone; (*18*) thrusts in allochthon foundation; (*19*) overthrusts

sedimentary complex) as an expanding southeastward wedge up to the Araz River, having 120 km length and maximal width of 75 km at the river's left bank. The zone's structure involves the Middle Jurassic age basalt-liparite formation and the Upper Jurassic age reefogenic-pyroclastic formation, the Cretaceous carbonate-pyroclastic and carbonate-volcanogenic formations, the Paleogene submarine-pyroclastic (of andesite composition), tuffaceous-terrigenic and terrigenic formations, the Neogene and the Quaternary terrestrial-volcanogenic and continental-subaerial formations (Shikhalibeyli 1964).

Within the Gafan zone, the Lachyn and the Gafan-Basitchay anticlinal subzones can be isolated, as well as the interlocated Gochaz synclinal subzone.

The Lachyn subzone looks like a narrow Jurassic rock high with an occasionally preserved Albian-Cenomanian transgressive mantle, having a length of over 75 km and width of 2–8 km. It can be traced from the Pichanis and Shalva rivers interfluve in the northeast to the Araz River left bank in the southeast. There, at the northwestern margins of the town of Jabrail, it plunges under the Lower Pleistocene continental gravels. The subzone represents a linear structure (the Lachyn-Kohnataghlar anticline), northwest confined by the Lachyn-Bashlybel fault fracture and southwest confined by the Deykhan fault fracture; they converge in the northwest and thus result in its tectonic thinning-out. In the southeast subzone, the Bathonian sedimentary-pyroclastic thickness is overlapped by the Callovian-Lower Oxfordian tuffaceous-carbonate formations; on the southern side, it is upraised by the terrigenic Albian-Cenomanian rocks and by the Senonian volcanogenic-sedimentary rocks, thus creating the Khuzabirt synclinal strip. The Middle Jurassic volcanogenic formations in the area of Lachyn-Shusha highway are intruded by the Bilulduz plagiogranite intrusive of the same age. West of Khyrmanlar village, plagiogranites and the volcanites enclosing them are intruded by the Shalva Upper Jurassic-Neocomian age granitoid intrusive, which, in its turn, is intruded by the Keshishtapa intrusion of the Upper Eocene quartz diorites. Along several dislocations, smaller bodies of diorite and gabbroid composition are also exposed. The anticline located southeast of the town of Boyuk Kirs experiences vergency and forms scaly structures that thrust over one another and have southward vergency. The Khuzabirt synclinal strip is also complicated by the second-order echelon-arranged structures. Within its borders, the Upper Eocene age granitoids are exposed.

The ***Hochaz subzone*** is the largest Gafan zone structural element. It stretches from the Araz River to the Tutgunchay River head and is confined by the Deykhan fracture from the northeast and by the Ishygly fracture from the southwest. The subzone length is above 115 km; its width increases from northwest to southeast, achieving 40 km at the Araz River left bank. Compositional complexes of the Cretaceous, Paleogene, Miocene–Pliocene and Pleistocene–Holocene rocks participate in the subzone's structure. The Cretaceous rocks composed of the terrigenic Albian-Cenomanian rocks and of the transgressive volcanogenic–sedimentary Senonian rocks plunge in the southeast under the Lower Pleistocene continental formations of the near-Araz sloping plain and outcrop only in the subzone's western side, along the Bargushad River. Here, the succession's upper part is predominantly created by Eocene volcanogenic–sedimentary rocks as well as by the terrigenic Oligocene rocks that again are exposed from under the Absheron age (Eopleistocene) dislocated conglomerates, sandstone and clays on the Araz River left bank scarps. In the northwest, the Cretaceous complex plunges under the Paleocene–Eocene tuffaceous-flysch and pyroclastic-lava deposits of the Kalbajar superimposed depression, which compose the laterally oriented Mykhtokan ridge in the Tutgunchay-Hakeri river interfluve. They are again exposed in the Tutgunchay River head erosion window as a result of the Cenozoic complex deep-seated washout. At the same time, the subzone's southwestern side is concealed under the Miocene–Holocene lavas, pyroclasts and subaerial formations of the Garabagh-Alagollar volcanic upland, which corresponds to the southeastern part of the vast Goycha trough of recent subsidences.

In general, the subzone looks like a vast flat-bottomed asymmetrical trough filled by heterochronous sediments and various facies in various cross-sections. The trough's southwestern side is low-angled and broken by faults, while the northeastern one is steep and complicated by linear folds, split faults and thrusts. In this process, the echelon dropped southern Basharat synclinal strip and the Gubadly northern synclinal strip are plotted in the subzone's transversal sections, which are dissected by the Hakerichay anticline. In the Hochazsu River head, the anticline periclinal plunge is intruded by the polyphase Dalidagh Late Eocene–Oligocene granite-syenite intrusion and covered by the Miocene–Pleisocene lava flows.

The ***Gafan-Basitchay*** subzone generally corresponds to the brachy-high, whose considerable part is located beyond the country's bounds. It defines the geological structure of the Bargushad, Okhchuchay and Basitchay rivers' lower courses. Being of submeridional strike, the subzone is confined by the eastern Hakeri fracture and by the western Girratagh fracture, plunging along its strike both northwestward and southeastward and being overlapped by the Bargushad-Hakeri interfluve Pleistocene rocks in its eastern side. The subzone's width does not exceed 25 km, with a length of up to 55 km within Azerbaijan's boundaries. The subzone's section is represented by basalt-rhyolite, pyroclast-reef and carbonate-terrigenic complexes of the Lower Bajocian, Upper Jurassic-Hauterivian and Barremian-Aptian, respectively (beyond the western borders of the area—also by the thin terrigenous-marly-pyroclastic complex of the Senonian); in the northwest, they are overlapped by the Miocene–Lower Pliocene lavas and pyroclasts of the Garabagh-Alagollar volcanic plateau.

Within the subzone's boundaries, low-angle, eastward and south-eastward, widely opening swell-like anticlines are developed, alternating with flat-bottomed trough-like synclines filled in troughs by the Barremian-Aptian formations. At the same time, in the subzone's southern side, the rocks along the Girratagh fracture are folded into a number of narrow and steep folds. At the Araz River's right bank, the subzone's Lower Cretaceous rocks (as well as the boundary Girratagh fracture) experiences a steep northeastern flexural monoclinal fold, thus reflecting the left-side shift in the Lower Araz fracture, having up to 15 km amplitude. Further on, changing its strike to the sublatitudional, the subzone tectonically thins out along the Girratagh fracture in the Sutagh River head.

5.4.1.4 The Kalbajar Zone

Within Azerbaijan's boundaries, the zone occupies the heads of the Hakeri and Terterchay rivers and their tributaries, looking like laterally oriented low-angle structures, levelling the structural geometry of the Hochaz, and partially, the Sarybaba subzones in the Artvin-Garabagh high southwestern side. The zone is represented as a fragment of the Goycha-Ordubad graben eastern side—that is, one of echelon located segments of the Neotethys backarch Paleogene rifts, which is obliquely superimposed on the convergence zone of the Southern Caucasus continental microplate and Central Iranian microcontinent in the Lesser Caucasus (Rustamov 1995). The zone is composed by the Upper Paleocene–Eocene tuffogenous-flysch and pyroclastic-lava deposits, as well as of the Miocene–Holocene land rhyolite-dacite subformation, and (in its southern part) by the Pleistocene–Holocene andesibasalt subformation. The zone stretches at a distance of 50–55 km between the Goycha and Mykhtokan ridges. Northeastward of the Garabagh-Alagollar highland, it is unfolded for 30–35 km. Smoothly plunging northeastward, the structure is step-like dropped northeastward. The anti-Caucasian (northeastern) strike folds as well as the Caucasian strike folds can be seen in the sediments that fill this trough, and they are inherited. Of transversal structures, the Kalbajar trough composition includes the anticlinal (the Chovdar and the Terterchay) and the synclinal (the Mykhtokan, Sarybulag, and Katidagh) strips. At the same time, along the Kalbajar trough southeastern margin in the eastern spurs of the Mykhtokan and Sarybulagdagh ridge, respectively, the inherited Paleogene synclines of general Caucasian strike are distinguished (the Chalbayir and the Uzunyal). The zone's Eocene formations are sometimes intruded by the Upper Eocene granitoid intrusions, by subvolcanic bodies of diorite porphyrites (Upper Eocene) and andesite (Lower Pliocene), as well as by the submeridional belt of andesite and andesidacite dykes that are controlled by a system of faults complicating the Terterchay high dome.

5.4.1.5 The Lower Araz Zone

The Lower Araz zone is a transversal superimposed depression controlled by the transregional sinistral fault of abyssal nature of the same name (the Palmira-Absheron lineament) plunged under the Araz River valley's alluvial formations. The zone stretches northeastward along the Araz River valley until it merges with the Middle Kur Depression. Mesozoic structures of the Artvin-Garabagh high tectonic areas smoothly plunge under the Quaternary continental molasses of the Ashaghy Araz superficial filling that also conceal the structure of its more ancient compositional complexes. However, they can be judged both from the behavior of the trough surrounding Mesozoic structures and from the separate Paleogene and Neogene outcrops in the river valley.

It is characteristic that by approaching the Araz River, the strike of the Lesser Caucasus structures changes from the southeastern to the sublatitudional one, finding its extension at the river right bank with northeastward displacement along the system of elementary sinistral faults, serving as the superficial expression of the Lower Araz fracture and defining its location in the Mesozoic structure plunging band. The zone's Pleistocene mantle appears in the Basitchay River (the Araz River's left tributary) debouchment, and as an expanding wedge spreads northeastward, penetrating into the Lesser Caucasus outskirts and overlapping older formations, including the Bathonian ones. The Eopleistocene dislocated molasses underlying the Pleistocene rocks are outcropped in the Araz River's left bank slopes, lying uncorformly on the Oligocene (the Diridagh ridge) and the Miocene's (Mashanly village) steeply lying formations. The latter are cut by a transcurrent fault on the river right bank. The fault also displays itself in the left bank scarps. The Paleogene deposits stand out, deeply protruding into the Hochaz trough boundaries along the Hakeri and Bargushad Rivers' lower courses (see above).

Recent continental molasses intrude the Araz River's right bank, north of the Iranian Injachay River inflow; in their eastern strike, they fringe here as a broad arch of the latitudionally oriented Oligocene-Lower Miocene complex of the Lerik-Yardymly and the Buravar zone western extension, completely concealing the Jalilabad zone's Miocene rocks of Talysh. By the Araz River, the Paleogene rocks plunge under the Pleistocene molasses but partially display themselves (as indicated above) in separate Eocene–Oligocene outcrops at the Hochaz synclinal subzone's southwestern edge. Based on the results of a seismic survey (Khain and Alizadeh 2005), the northward structures of the Artvin-Garabagh megazone also find their extension under the superimposed depression Quaternary nappe; they can be traced through the Bilasuvar-Garadonlu buried Mesozoic inlier towards the Mountainous Talysh outskirts, where they find their expression in the Jalilabad and the southern Mughan structural zone's buried structures. Thus, the described depression corresponds to a recent structure superimposed on the transit zone controlled by the Lower Araz's sinistral fault and located between periclinally plunging structures of the Artvin-Garabagh high southern edge and western centroclinal closure of the Talysh folded structures.

5.4.2 The Araz Megazone

The Araz megazone constitutes the Dinar-Zond branch of the Mediterranean (Alpine-Himalayan) folded belt and corresponds to the northwestern block of the Southern Azerbaijan segment of the Central Iranian microcontinent. From the north and northeast, it is confined by the Girratag abyssal fracture; from the southeast, by the Lower Araz abyssal fracture; and from the southwest, by the Tabriz abyssal fracture. In the northwest, the border passes along the abyssal fracture zone (one of the Transcaucasian high dislocations at the Arabian Plate northwestern protrusion front) concealed under the Miocene–Pleistocene lavas and molasses. It is traced from the Chaldyran Mts. (Turkey) through the western Arpachay River valley in the border between Turkey and Armenia, to Georgian borders (Gadjiev et al. 1989). The fracture (or one of its branches) is exposed in the Araz river gorge, below the Garbi Arpachay River debouchment, where the northwestern course fault is plotted, with the southwestern side dropped for over 1000 m (Azizbekov et al. 1972). The megazone's central part determines the structure of Nakhchivan AR of the Azerbaijan Republic; its northwestern side stretches along the Armenian territory, and its southwestern edge is located within the Iranian southern Azerbaijan, and, partially, in the Turkish border district.

The megazone's Precambrian basement is exposed in fragments along the megazone's northern confinement—that is, the Miskhana (Arzagan) massif in the northwest and the Nuvadi massif in the southeast. It is represented by two compositional complexes: (1) the lower volcanogenic–sedimentary one composed of gneisses, metamorphic shales, amphibolites, and marbles; and (2) the upper one composed of poorly metamorphized argillaceous-silicious and arcosic shales that contain acidic pyroclastic material. The complexes are penetrated by large granitoid intrusives and are overlapped by rhyolite flows in the roof. The Precambrian rocks are outward projecting in the megazone southwest as well, in the Iranian and Turkish border districts, where they form the basement Aghgol protrusion. The megazone's sedimentary cover is composed of the Paleozoic, Mesozoic, and Cenozoic sedimentary and magmatic complexes, with their cumulative thickness exceeding 14 km. The most ancient Ordovician-Silurian volcanogenic formations border the Aghgol protrusion, plunging southward and eastward under the Devonian carbonates. The sedimentary cover structure is inhomogeneous and is characterized by the complex interaction of the megazone's constituent structural zones: the Sharur-Julfa high, the Iravan-Ordubad trough, the Miskhana-Zangazur high and the Nakhchivan-Aghrydagh suporimposed depression, which differ in their geological history, lithologic-stratigraphic succession peculiarities, and folded-ruptured deformations, and, in their turn, have fractional disaggregation into lesser structural units (subzones and blocks). Within the Nakhchivan AR boundaries, these structures are represented only partially (see Fig. 5.2), separated as the Sharur-Julfa, Ordubad, Zangazur, and Nakhchivan structural zones (Azizbekov 1961; Azizbekov et al. 1972; Khain and Alizadeh 2005).

5.4.2.1 The Sharur-Julfa Zone

The Sharur-Julfa zone corresponds to the northeastern relatively dropped side of a large high that can be traced in Iranian borders along the Araz megazone's southwestern edge by the Maku-Marand traverse. In the northeast, the zone Paleozoic deposits plunge under the Agrydagh (Ararat) trough's Upper Miocene rocks. Within Azerbaijan's boundaries, the zone geological structure involves the vigorously dislocated Devonian-Triassic terrigenic-carbonate complex of the Gondwana northern margin platform mantle, as well as the occasionally unconformly lying, thin Jurassic, Upper Cretaceous, Eocene, and Lower Pliocene rocks. Recent formations of the Nakhchivan superimposed depression that southward plunge under the zone borders subdivide it into the northwestern Sharur and the southwestern Julfa segments. Their interrelation is confirmed by the deep borehole materials, as the borehole has outcropped the Upper Triassic limestones in the Oligocene–Miocene deposit bottom. Thus, in its longitudinal section, the zone resembles a double-humped high with median saddle filled by the Miocene rocks and the Pleistocene continental molasses.

The Sharur segment corresponds to the transversal structural protrusion in the zone's northwestern part and is represented by the longitudinally oriented folded blocks that are step-like dropped and uplifted along transgressional faults and faults. These blocks are north arched and additionally complicated by faults and transtensional faults, transversally oriented against the general course of the structures. The structure of this area of the Sharur-Julfa zone involves the Devonian–Lower Carbonaceous terrigenic-carbonate complex and the Upper Permian-Triassic carbonate complex transgressively lying on it. At the northeastern periphery segment, ancient deposits are unconformly covered by the Upper Cretaceous carbonate-terrigenic rocks, by the Eocene sedimentary rocks, and by the Lower Pliocene volcanogenic mass. In some spots, the Liasic effusive-pyroclastic formations and the Middle Jurassic terrigenic complex outcrop as fragments. The Lower–Middle Devonian deposits protrude as a low-mountain ridge among the near-Araz sloping plain Pleistocene gravels and form the Sharur segment's most elevated Validagh block. It is dislocated into folds of fan-shaped orientation with a relatively stringent northeastern strike and typical arch-like camber of northwestward convexity. In its southwestern extension, northeastward of the Validagh block, in the Arpachay River basin, geological structures of the Daralayaz range are determined by the Arpachay block, consisting of the blocks of the second order upthrusted on each other from south to north.

Each of them generally represents an antisymmetric anticline complicated by secondary folds, with a synclinal southwestern side and frequently vergent and tectonically cut northeastern side; for example, the Sharur segment's northeastern side is characterized by an imbricated structure (Fig. 5.34). Folded structures are of sublatitudinal strike in the west (within the Armenian boundaries). However, in the Arpachay River basin, they sharply change their azimuth for the southeastern one, up to the southern, at the Daralayaz range water divide that meridianally drops to the near-Araz sloping plain. Further on, they flexurally plunge under the Nakhchivan superimposed depression's northern side Miocene rocks along the abyssal fracture of the same name. In its northeastern side, along the Havush-Nehram (Anabadgadik) deep fracture, it borders the Ordubad trough's Shahbuz segment. The length of the Sharur segment reaches 55 km from the northwestern border of Karki enclave of Nakhchivan AR up to the tectonic thinning-out of structures in eastern areas of Chalkhangala village.

The Julfa segment within Azerbaijan's borders stretches southeastward as a narrow strip (2–5 km) along the river left bank, from the Araz water storage dam to the Alinjachay River, its left tributary junction, at a distance of up to 20 km. Here, the segment is represented by the northeastern side of anticlinal high bearing the same name (its dome is located in Iranian territory), composed of the Upper Permian, Triassic, Lower, Middle Jurassic and Upper Cretaceous rocks that steeply plunge under the low-angle Lower–Middle Eocene sedimentary complex in the Havush-Nehram (Anabadgadik) deep fracture area. Additional folds, broken by longitudinal transpressional faults and transversal fault-thrusts with high horizontal offset, complicate the uplift's wing. At the Araz River's right bank, the uplift's surface is traced by the Devonian-Triassic deposits at a distance of over 30 km, plunging under Middle Eocene in the southeast, and sharply changing its course to the northern one in the northwest. There, it plunges under the transgressive Miocene rocks of the Nakhchivan depression.

5.4.2.2 The Ordubad Zone

The Ordubad zone corresponds to the southeastern part of the Iravan-Ordubad trough, which has been inheritedly developed in the northwest Central Iranian microcontinent since the Early Jurassic age and is complicated by the obliquely bedded Goycha-Ordubad graben, a fragment of the Neotethys backarch Paleogene rift. The zone is northeast confined by the Eastern Ordubad abyssal fracture, southeast confined by the Lower Araz abyssal fracture and southwest confined by the Havush-Nehram (Anabadgadik) abyssal fracture. In the northwest, it plunges under the Lower Pliocene volcanogenic mass that shapes the Daralayaz and the Zangazur ranges water partings. The preferentially sedimentary Jurassic and Cretaceous deposits are exposed along the zone's southwestern (on the plunge of the Julfa high northeastern side and in the Jahrichay River) and northeastern (the city of Ordubad area) raised edges. They are also plotted at a significant depth based on the results of a seismic survey of the Araz River's left bank (the Nakhchivanchay and the Garadara rivers interfluve). In its remaining part, the zone is represented on its surface structure by the Paleocene terrigenic flyschoids and by the Lower and Middle Eocene thick volcanogenic–sedimentary mass (with participating volcanites of the basalt-andesite-dacite-rhiolite and trachybasalt-latite-trachyandesite formations), which in its most sagged structures is upraised by the terrigenic Upper Eocene and volcanogenic–sedimentary Oligocene rocks. In the lower course of the Nakhchivanchay and Alinjachay Rivers interfluve, the zone's southwestern edge and the Havush-Nehram (Anabadgadik) deep fracture that confines it are overlapped by the eastern closure Miocene complex of the Nakhchivan superimposed depression. The folds within the zone borders are moderately folded, persistent, and strike predominantly in a sublatitudinal direction in the north and a southeastern direction in the south.

The sublatitudinal Goygol dislocation merges with the Nakhchivan abyssal fracture of northern constraint of the same-named trough at the Alinjachay River. It dissects the Ordubad zone into two folded fragments, the northern Shahbuz fragment and the southern Shurud fragment; these, in turn, are subdivided by transversal and longitudinal dislocations into higher order blocks.

The Shahbuz segment is delineated in the Havush-Nehram (Anabadgadik) deep fracture's northeastern lying side, thus corresponding to the Ordubad zone's northwestern part. It generally looks like an eastward plunging sublatitudionally oriented second-order trough broken by transversal and longitudinal dislocations into

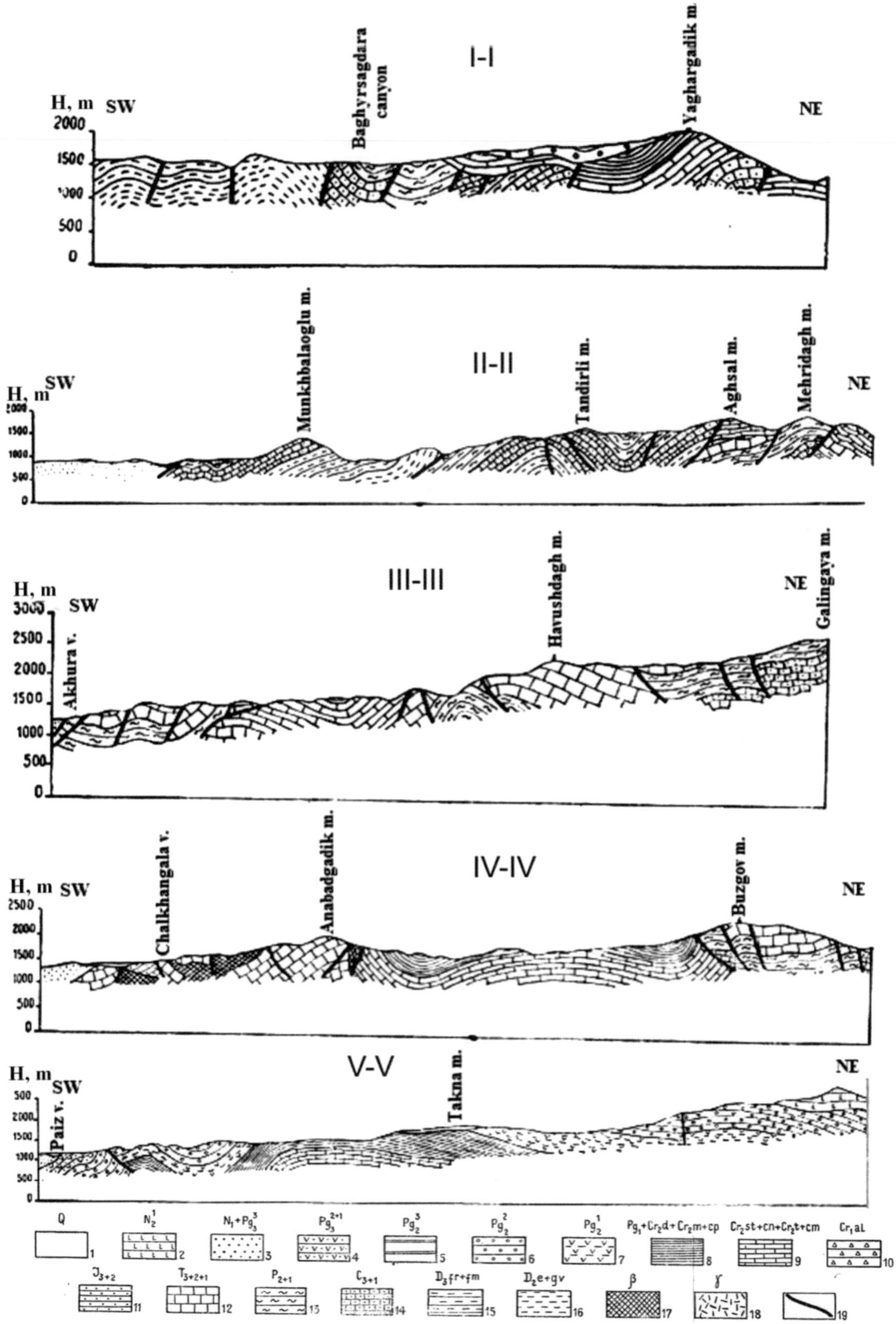
I-I
H, m
SW
NE
2000
1500
1000
500
0
Baghyrsagdara canyon
Yaghargadik m.
II-II
Munkhbalaoglu m.
Tandirli m.
Aghsal m.
Mehridagh m.
III-III
3000
2500
Akhura v.
Havushdagh m.
Galingaya m.
IV-IV
Chalkhangala v.
Anabadgadik m.
Buzgov m.
V-V
Paiz v.
Takna m.
Q
1
2
3
4
5
6
7
8
9
10
11
12
13
14
15
16
17
18
19
β
γ

◀ **Fig. 5.34** Geological profiles of the northern part of the Nakhchivan Autonomous Republic (Azizbekov 1961). Profiles: *I–I* Baghyrsagdara-Yaghargadik; *II–II* Munkhbalaoglu-Mehridagh; *III–III* Akhura-Galingaya; *IV–IV* Chalkhangala-Buzgov; *V–V* Badamly-Kuku (*1*) Quaternary deposits; (*2*) effusive stratum of Lower Pliocene; (*3*) saliferous limestone-clay-sandy stratum of Miocene-Upper Oligocene; (*4*) volcanogenic stratum of Lower and Middle Oligocene; (*5*) sedimentary deposits of Upper Eocene; (*6*) tuff conglomerate and volcanogenic–sedimentary sequence; (*7*) volcanogenic stratum of Lower Eocene; (*8*) clastic-carbonate deposits of Paleocene, Danian stage and Upper Senonian; (*9*) tuffogenic carbonate deposits of Lower Senonian, Turonian and Senomanian; (*10*) quartz porphyry, tuffs and tuff breccia of Albian; (*11*) sandy-argillaceous deposits of Middle and Upper Jurassic; (*12*) dolomites and limestone of Lower, Middle and Upper Triassic: (*13*) bituminous and massive limestone of Lower and Upper Permian; (*14*) limestone shales and quartz sandstone of Lower and Upper Carboniferous; (*15*) limestone, shales, quartzites and quartzite-sandstone of Upper Devonian; (*16*) limestone, sandstone, quartzites and shales of Middle Devonian; (*17*) Lower Pliocene andesite-dacite intrusives; (*18*) Lower Jurassic basaltic intrusives; (*19*) faults

mixed-mode blocks. The folds within the segment boundaries are essentially of the general Caucasian, and partly, sublatitudinal strike. They are in large part controlled by longitudinal overfaults and faults and are complicated by oblique slip-faults of northwestern, meridional and northeastern (Fig. 5.35). In the transversal dislocation zone, the folds sometimes experience contrasting flexural bends, interpreted as nonsinusoidal brachyfolds transversal to the general structural geometry. In Nakhchivan AR, the length of the zone reaches to 80 km at the maximal width of 32–35 km within the Nakhchivanchay basin.

The visible cross-section's lower part is exposed within the northwestern part of the segment and is represented by the Middle–Upper Jurassic and Upper Cretaceous terrigenic and carbonate rocks, having both tectonic and stratigraphic contacts with the Permian-Triassic carbonates and with transgressively upraised terrigenic Paleocene rocks. In the remaining segment area, the Eocene volcanogenic–sedimentary complex participates in surface structure, and the complex cross-section demonstrates unconformities (transgressive contacts) on the Paleocene–Lower Eocene, Lower–Middle Eocene, Middle–Upper Eocene borders as well as in the Middle Eocene composition. The Cretaceous-Middle Eocene formations are intruded by small Middle Eocene intrusives with andesite and diorite composition.

In the segment's eastern part, in the Alinjachay River head, the Lakatagh polyphase granitoid intrusive is exposed. Large and small intrusives of the Lower Miocene diorites and quartz diorite-porphyrites are confined to the Nakhchivan and Goygol fractures and to the complicated joint of their intersection with the Alinjachay fault. The zone's structural geometry is levelled in the Nakhchivanchay basin's northern part by the effusive, vent, and subvolcanic facies of the Lower Pliocene subaerial andesite volcanism.

The Shurud fragment is differentiated in the Goygol sublatitudinal fracture's southern downcast side, corresponding to the Ordubad zone's southeastern part. In the contemporary structure, the segment's southwestern edge and the Havush-Nehram (Anabadgadik) fracture's southwestern fragment that limits it are buried under the Nakhchivan superimposed depression's Miocene and Pleistocene molasses. For this reason, the segment's western border in the surface structure is defined by the Alinjachay strike-slip fault, stretching along the same-name river valley from the northeast to the southwest. At the same time, in its northeastern side, the segment, along the Eastern Ordubad fracture, borders the Zangazur high; it is confined by the Lower Araz disjunctive dislocation in the southeast. In these boundaries, the segment length reaches 45 km, with a width of 20–25 km.

The Duylyun-Saltag pivotal fault dissects the fragment into two longitudinal blocks, experiencing common uplift from north to south with outcropping of the Paleocene and Upper Senonian sedimentary deposits on the Darydagh ridge and in the Araz River valley. In its remaining part, the segment is represented by the Eocene (volcanogenic–sedimentary in the lower–middle area and terrigenic in the upper area) rocks unconformly overlapped by the volcanogenic Oligocene rocks in its southwestern side. On the same side, the Nakhchivan superimposed depression Miocene complex, by passing to the Alinjachay river left bank, penetrates deeply into the segment boundaries, and routing to the Araz River right bank, separates it from the Sharur-Julfa zone Julfa horst uplift. The folds within the segment boundaries have pronounced northwestern strike. In the southwestern part of the Shurud segment, the folds strike to the northwest and submerge under the gentle rocks of the Lower Miocene of Nakhchivan trough in the right bank of Alinjachay River. Here, they are poorly expressed in the mentioned superimposed complex. Similar to the north Ordubad zone, large anticlinal folds within the segment boundaries are complicated by axial-longitudinal overfaults and faults. They are also broken by transversal dislocations, essentially of northeastern strike. Injections of dacites, andesite-dacites, andesite-basalts (Lower Eocene), andesites and andesite porphyrites (Middle Eocene) take part in the Eocene complex succession in the segment eastern part; diorites and diorite-porphyrites (Lower Miocene), teschenites and trachyandesite-basalts (Middle Miocene) participate in its western part (see Fig. 5.35).

5.4.2.3 The Zangazur Zone

The Zangazur zone is the Miskhana-Zangazur high southern block for the Araz megazone's northeastern edge and corresponds to the double-humped horst elevation intruded in its central part by the Middle Eocene–Early Miocene

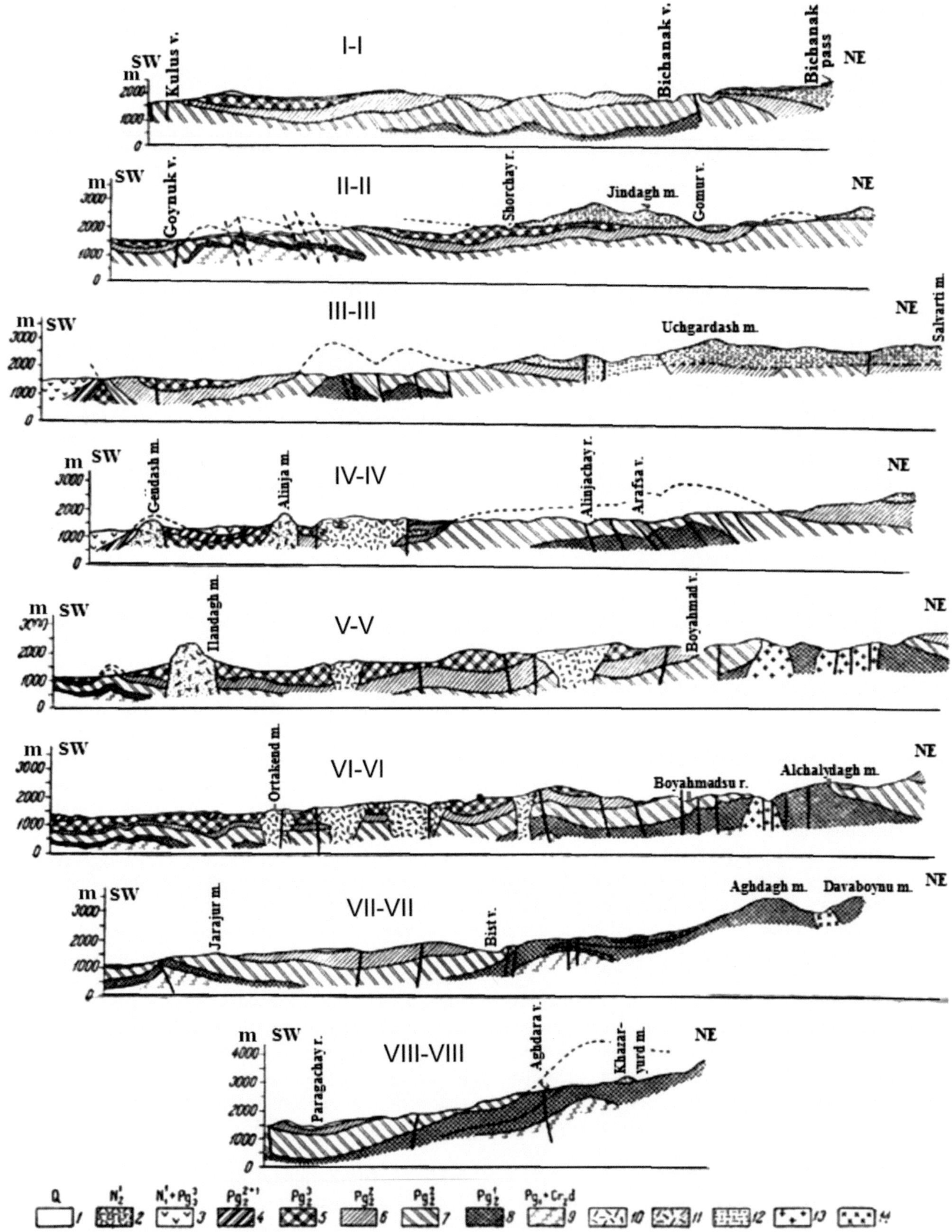
I-I
SW
NE
Kulus v.
Bichanak v.
Bichanak pass
II-II
Goynuk v.
Shorchay r.
Jindagh m.
Gomur v.
III-III
Uchgardash m.
Salvarti m.
IV-IV
Gendash m.
Alinja m.
Alinjachay r.
Arafsa v.
V-V
Ilandagh m.
Boyahmad v.
VI-VI
Ortakend m.
Boyahmadsu r.
Alchalydagh m.
VII-VII
Jarajur m.
Bist v.
Aghdagh m.
Davaboynu m.
VIII-VIII
Paragachay r.
Aghdara v.
Khazar-yurd m.

◀ **Fig. 5.35** Geological profiles of the eastern part of the Nakhchivan Republic (Azizbekov 1961). Profiles: *I–I* Kulus-Bichanak; *II–II* Goynuk-Gomur; *III–III* Gyrkhlardagh-Salvarty; *IV–IV* Gendagh-Arafsa; *V–V* Ilandagh-Boyahmad; *VI–VI* Paradash-Alchalydagh; *VII–VII* Jarajur-Davaboynu; *VIII–VIII* Paraghachay-Khazaryurd. (*1*) Quaternary deposits; (*2*) Lower Pliocene, effusive stratum; (*3*) Lower Miocene–Upper Oligocene, Kashirdagh volcanogenic stratum (tuff conglomerates, tuff breccias and tuffs); (*4*) volcanogenic stratum (tuff-sandstone, tuff conglomerates, tuff breccia and andesites) of Middle and Lower Oligocene; (*5*) sedimentary deposits of Upper Eocene; (*6*) Middle Eocene, tuff-conglomerate stratum; (*7*) volcanogenic-sedimentary deposits of Middle Eocene; (*8*) volcanogenic stratum of Lower Eocene; (*9*) clays and sundstone of Paleocene and Danian stage; (*10*) Lower Pliocene andesite, andesite-dacite and dacitic extrusions; (*12*) Lower Pliocene doleritic intrusive Aradzhi; (*13*) post Oligocene–Pre-Miocene monzonite-dioritic Lyakatag intrusive; (*14*) post Oligocene–Pre-Miocene adamellite Sarrapsu

polyphase granitoid batholith and its satellites (the Saggarsu and the other ones). The zone boundaries are confined by the Girratagh deep-laid fracture in the northeast, by the Lower Araz fracture in the southeast, by the Eastern Ordubad fracture in the southwest, and by the Goygol deep-laid fracture in the northwest. Within the Nakhchivan AR, the zone is represented by the southwestern side composed of the Senonian terrigenic-carbonate complex and of the Paleocene-Lower Eocene volcanogenic (and partly terrigenic) deposits striking for 45 km from the town of Ordubad area to the Gilanchay River head (with maximal width of 10 km on the Araz River's left bank), and embracing the Zangazur range water parting and slopes. The zone's southwestern side forms a broad, rather steep monocline complicated by the Kilit-Kotam anticlinal high in the Eastern Ordubad uplifted side. The fractural zone controls the Lower Eocene small subvolcanic bodies of dacites, andesidacites, rhiolites, and rhiodacites in the north, and large Agrydagh intrusion of dioritic composition in the south. The southeastern extension of the latter reveals an extensive swarm of diorite-porphyrite and diabase dikes along tension cracks among the Upper Cretaceous formations. A borehole drilled on the Araz River's left bank, in the Kilit village area, exposed the metamorphic Precambrian formations in the Upper Cretaceous bottom at 600 m depth. The Precambrian formations bloom out downstream, generating the Nuvadi massif on the Armenian-Iranian border. They are also represented by the magma-borne differently sized xenolites (the largest of them being the gigantic Pazmara block) that are exposed in the Megri-Ordubad batholith body.

5.4.2.4 The Nakhchivan Zone

The Nakhchivan zone corresponds to the Agrydagh (Ararat)-Nakhchivan superimposed depression's major southeastern part. It is separated from the Agrydagh segment by the protrusion of the Bagyrsag ridge Paleozoic rocks and by partially levelling the Sharur-Julfa and the Ordubad zone's structural geometries. On the Araz River's left bank, its length from the Nakhchivan AR western border to the Alinjachay River exceeds 92 km, with the width varying from 15 to 25 km at various intersections. In the north and in the east, it is confined by uplift areas; in the south, southwest and west, it extends beyond the Nakhchivan AR boundaries. The depression owes its origin to the movements along the Nakhchivan abyssal fracture, passing along its northern edge. It is distinguished both by extremely high power gradients of the Miocene molassic deposits that fill the trough and by the nature of its junction with the Sharur-Julfa and Ordubad zone structures. Compositional complexes, from Paleozoic to Oligocene, participate in the geometry of superimposed depression basement. In the east, in the area of the Nakhchivan trough juncture with the Ordubad one, intrusive bodies of the Early Miocene quartz-diorite formation are outcropped, as was shown above. The superimposed depression compositional complex is represented in its basement by terrigenic (at the depression extreme west, by partially volcanogenic–sedimentary) deposits. Effusive and intrusive assemblages of trachyandesite-teschenite formation stemming from magmatism in graben-like plunge and tension are related to the Middle Miocene. The Miocene cross-section upper part is composed of the so-called "salt-bearing section" (Azizbekov 1961) that was deposited preferentially in shallow-lagoon conditions. The Quaternary complex is represented by the rocks of exclusively continental origin: alluvial-proluvial, deluvial–proluvial, eluvial, fluvioglacial, partly eolic and travertine deposits. According to the available geological and geophysical data (Khain and Alizadeh 2005), the thickness of the Miocene–Pleistocene deposit complex is variable in the trough northwestern and central parts (from 200 to 1500 m) and is relatively stable in the southwest (500–800 m) with reduction to complete thinning-out towards the northeastern and southwestern edges. According to the same data, the complexes underlying the trough basement are characterized by the complicated folded-block structure with the participation of buried bodies of intrusive nature. The trough's surface structure is characterized by a peculiar folded geometry: here, arch-like curved in plan folds have developed, which, however, at the trough's periphery have a course conforming to its general course.

In the trough's central part, the Lower Miocene deposits with deep Paleogene erosion lie directly on the Triassic eroded surface, which indicates the existence of a significant depositional break in the Sharur-Julfa high area at the boundary between the Oligocene and the Miocene periods.

5.5 The Talysh Folded Region

Being situated in the region between the Lower Araz (in the northwest) and the Shongar-Gyzylaghaj (in the southeast) fractures, the Talysh folded system corresponds to a large structural unit (megazone) of Azerbaijan. It is northeast adjacent, along the Girratagh fracture (beyond the state border), to the Alborz-Binalud folded-cover structure, and thus constitutes the Pyrenean-Lesser Caucasian branch of the Alpine-Himalayan folded belt, corresponding to its extreme eastern onland block. Unlike the Lesser Caucasus folded structures, the north vergent dislocations predominate here, complicating the cross-section of the outcropping compositional complexes: the Upper Cretaceous carbonate complex, the Paleogene-Lower Eocene tuffaceous-sedimentary one, the Middle–Upper Eocene volcanogenic (subalkaline and alkaline volcanites of andesite-basalt composition) complex, and intrusive (mafite-ultramafite alkaline intrusions), the Oligocene–Miocene terrigenic, and the Upper Miocene–Pliocene molassic complexes (Azizbekov et al. 1972, 1979; Khain and Alizadeh 2005). The Eocene complex has the largest areal development and cross-section completeness; it occurs with angular unconformity on the tuffaceous-sedimentary Paleogene rocks and is overlapped by the Oligocene–Miocene marine molasses. The zone represents the region of alkaline-basalt volcanism vigorous manifestation and probably corresponds to the eastern segment of the Neotethys backarch Paleogene rifts within the Azerbaijan boundaries, joining the Goycha-Ordubad graben via Iranian Garadagh.

These complexes are collected in folds constituting the region structural zones: Astara, Lerik-Yardymly, Buravar, Jalilabad, and Southern Mughan zones, which having general Caucasian strike, are dissected by a system of penetrating transverse faults into a series of different-scaled uplifted and downthrown blocks. These blocks have a significant effect on the nature and composition of sedimentary and volcanogenic formations: the largest of them are (from southeast to northwest) the Hamarat uplifted block, the Gosmalion downthrown block, the Masally uplifted block, and the Deman downthrown block (Fig. 5.36).

The oblique cut of northern structural zones by the Pre-Talysh fracture is also observed in the folded region contemporary structure. Along this fracture, the region's eastern sinking took place in the Late Miocene-Pliocene time. This sinking promoted the separation of two morphologically defined units instead of one folded system: the Mountainous Talysh and the Lenkaran plain.

5.5.1 The Astara Zone

The Astara zone is a horst-like elevation of the southeastern strike with a clearly defined block structure. The zone's northwestern boundary passes along the like-named fracture of transpressional fault nature. In its southwestern side (beyond the state border), it is step-like dropped along the Girratagh deep fracture, forming in general an arch-wise northeastward curved folded structure at the Alborz-Binalud folded system front. Within the country's boundaries, the zone stretches at 80 km, with maximal width of over 12 km at the Astarachay River. At the river's right bank, the zone accepts submeridional course and is concealed under the Caspian Sea silts.

Within the Hamarat uplifted block, broken into second-order blocks, the zone's cross-section is represented by the highly dislocated tuffaceous-sedimentary Paleocene rocks and by the occasionally outcropped Campanian carbonates. At the block's southwestern edge, the leucite-sanidine tuffs of trachyandesites are noted, which lie with angular unconformity on the Paleocene deposits. Here, a number of subvolcanic bodies that correspond to the linearly located Eocene paleovolcanic centers can be met (Shandan-Galasy, Laji, Divashi, Kalaputu, etc.), and they are represented by trachybasalts, trachyandesites and andesibasalts. In addition, the intrusions of the Late Eocene ultrabasic subalkaline magma represented by gabbro-pyroxenites and gabbro-syenites of the Hamarat-Aliabad intrusive group participate in the Hamarat block's northwestern edge section. Steep, frequently isoclinal, folds complicated by transversal and longitudinal fractures represent a folded structure. In the zone's southern part, the folds of the northeastern strike are also noted. They form vergent folds, often with fault lines—the fact being related to the vigorous movements that took place here starting from the Late Eocene. More westward, the Gosmalion downthrown block is filled by the Lower and Middle Eocene volcanogenic masses and is represented by two synclinal folds separated by brachyanticlines. Typical for the Early and Late Eocene volcanic centers, plotted here, is their confinedness to the intersection nodes of northwestern and northeastern fractures. The consequent Masally uplifted block reveals itself in sharp bends of rectilinear outlines of the masses and in decrease of their thicknesses northwestward. The Gyz-Galasy subvolcanic intrusive corresponding to the Late Eocene volcanism center is also associated with the elevation. Finally, the most westward Deman block is characterized by the reduced Middle Eocene thicknesses when the Upper Eocene volcanites achieve the zone maximal thickness (450 m).

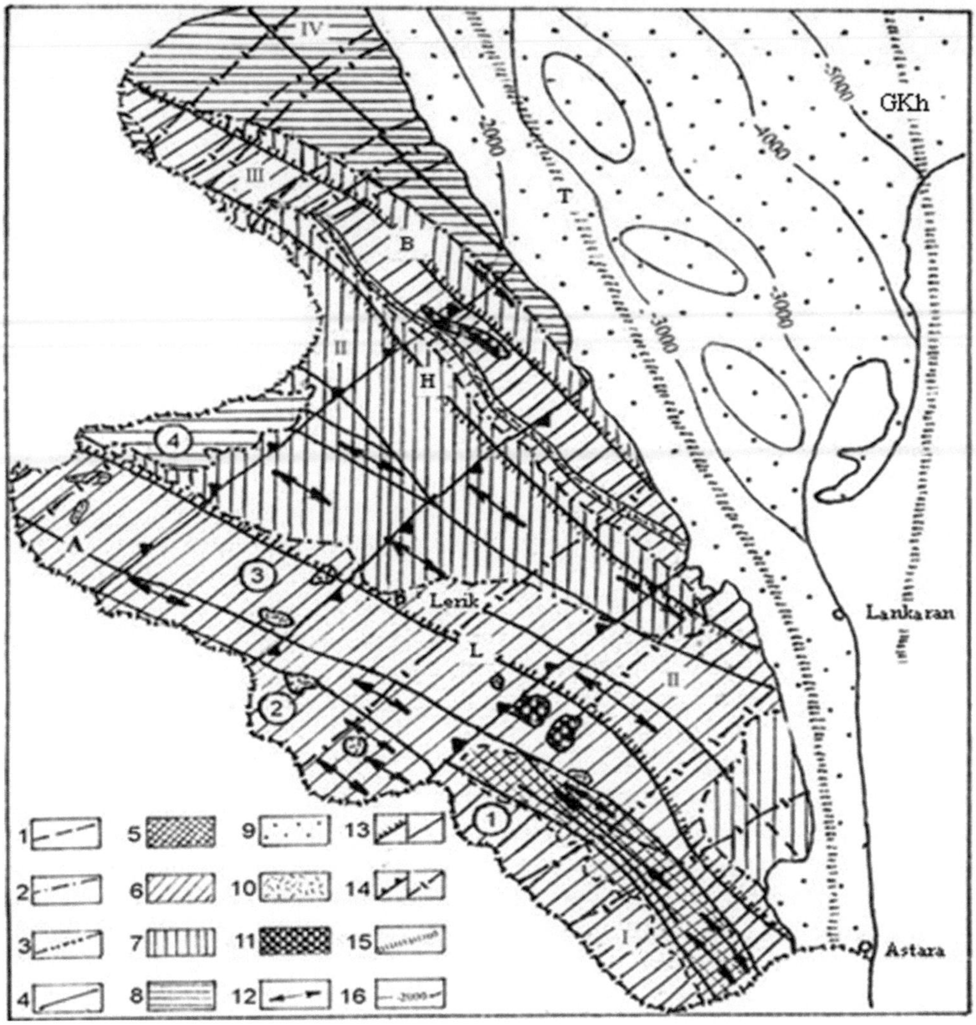

Fig. 5.36 Tectonic scheme of the Mountainous Talysh (Azizbekov et al. 1979). (*1–4*) boundaries of structural levels: (*1*) by Eocene's bottom, (*2*) by Early Oligocene's bottom, (*3*) by Middle Miocene's bottom, (*4*) by Quaternary's bottom; (*5–9*) formations and subformations: (*5*) Upper Cretaceous–Paleocene sedimentary-tuffaceous, (*6*) Eocene volcanic (alkali basaltoids), (*7*) Oligocene–Early Miocene marine molasses, (*8*) Middle Miocene lagoon-marine molasses, (*9*) Quaternary continental-marine molasse; (*10*) subvulcanic masses; (*11*) alkali ultrabasal intrusions; (*12*) anticlinal folds; (*13*) longitudinal faults of the first (*a*) and the second (*b*) ranks (overthrusts, upthrusts): *A* Astara; *L* Lerik; *H* Haftoni; *B* Buravar; (*15*) geophysically detected deep faults: *T* Pre-Talysh (Talyshonu); *GKh*Western Caspian (GarbiK-hazar); (*16*) isolines along the top of Upper Cretaceous deposits; Structural zones: *I* Astara; *II* Lerik-Yardymly; *III* Buravar; *IV* Jalilabad; *V* Mughan. Latitudinal blocks (*numbers in circles*): *1* Hamarat (horst); *2* Gosmalian (graben); *3* Masalli (horst); *4* Deman (graben)

5.5.2 The Lerik-Yardymly Zone

The The Lerik-Yardymly zone corresponds to the fault trough on the Astara high northern side filled with the volcanogenic and volcanogenic–sedimentary Eocene rocks and with terrigenic Oligocene–Miocene rocks. Contemporary outlines of the zone's steep southwestern edge are visible quite clearly against the Middle Eocene sedimentary-tuffaceous thickness and are traced by the Astara overfault. The Haftoni fracture on the plunge of the Buravar high southwestern side serves as the zone's northeastern boundary. In this process, the zone experiences broadening from 15–20 km in the southeast to 38–40 km in the northwest; its length within the country's borders is up to 75 km. At close range to the Caspian Sea coast, the zone plunges under the coastal plain Pleistocene–Holocene deposits, and in the northwest it expands and can be traced towards the Araz River. There, it adjoins the Lesser Caucasus southeastern plunge structures along the Lower Araz fracture and actually represents the Hochaz trough's eastern extension. The zone's Eocene complex structure is asymmetrical with the development of echelon-like wide synclines and northeast biased narrow synclines situated in between. Generally, the folds are oriented from northwest to southeast, parallel to the trough edges, and are complicated by the northwestern and northeastern fractures. In the zone's southeastern part, structurally corresponding to the Hamarat transversely uplifted block, the Late Eocene–Early Oligocene intrusives of the Hamarat group gabbro-pyroxenites and gabbro-syenites are observed, as well as subvolcanic bodies of trachydolerite basalts. In the zone's northwestern and southeastern dropped parts, these folds periclinally plunge under the transgressively lying Oligocene-Lower Miocene argillo-arenaceous molasses, which, in turn, are upraised in the Yardymly trough by the transgressive conglomerate-arenaceous Upper–Lower Miocene molassic complex. Folding in lower molasses is expressed by broad and low-angle synclines, separated by relatively narrow anticlines, which by their northwestward strike experience periclinal bend and plunge under the upper molasse formations of northeastern direction.

5.5.3 The Buravar Zone

The Buravar zone corresponds to a large anticlinal high created in the Late Eocene–Early Oligocene period and exposed to folding caused by neotectonic movements. Within the zone's boundaries, the high is represented by its southeastern side composed of the Upper Eocene tuffaceous-sedimentary rocks and traced as a broad anticline (with low-angle southwestward drop) from the state border in southeastern direction at a distance of over 75 km, with varying width of 8–10 km. In so doing, the high northeastern side is cut by the Buravar fracture; along this fracture, the southeastern side is thrust upon the Jalilabad zone's Oligocene-Miocene complex. Starting from the Vilashchay River's right bank, the zone's northeastern side is obliquely cut by the coastal plain's Pleistocene continental molasses and completely plunges under them on the Veravulchay River. At the plot structurally corresponding to the Masally transverse uplift, a subalkaline-ultrabasic intrusive (Alashar-Yelagaj) is located, similar to the Hamarat group of intrusions. In addition, a differentiated sille of trachydolerites of variable thickness (60–180 m) and length of over 45 km is confined to the Upper Eocene median tuff-arenaceous-argillite horizon. On the Bolgarchay River's right bank, the zone experiences plunge along its northwestward strike. For this reason, the Upper Eocene rocks are upraised by the transgressive Oligocene, Lower and Middle Eocene rocks with basal conglomerates in the basement.

5.5.4 The Jalilabad Zone

The Jalilabad zone represents a trough located in the Talysh northwestern part and filled with the thick argillo-arenaceous Middle and Upper Miocene mass, which is mostly overlapped by the Middle Pliocene deluvial–proluvial piedmont plain sediments. According to the outcropped Miocene deposits, the zone length from the state border to the Vilashchay River is 50 km, with maximal width of 15 km in the northwest (the Bolgarchay River). By its northwestward strike, on the Bolgarchay River's left bank, the Miocene complex completely plunges under the Lower Araz's superimposed trough continental molasses. The trough's southwestern edge is clearly outlined by its tectonic contact (the Buravar thrust) of the Middle–Upper Miocene deposits with the Buravar high Upper Eocene aleurite-arenaceous rocks. In the Bulgarchay and Goytapa rivers interfluve, the fracture lying side displays tectonic slices composed of the Oligocene argillaceous-arenaceous molasse. The trough's northeastern boundary is the buried southern Mughan high edge, confined by the Pre-Talysh fracture. The folding geometry within the zone borders is inherited, with northeastward flattening of all structures towards the buried high. In general, the zone is represented by two broad synclinal folds, dissociated anticlines, and complicated ruptures (thrusts, faults, dislocations) that show structural shifts and northward vergences.

5.5.5 The Southern Mughan Zone

The southern Mughan zone corresponds to a buried high situated in the pre-Talysh fracture northeastern side. The zone was delineated on the basis of indirect geological and geophysical data. They include, specifically, the gravimetric data (Gadjiev 1965; Tzimelzon 1965) that support the existence of a pre-Talysh maximum in the Talysh lowland (between Uzuntapa (Novogolovka) village and the Astara town) interpreted as the submeridional Meso-Cenozoic high with upraised pre-Jurassic basement. These facts are supported by the drilling data in Uzuntapa village area: there, the Upper Cretaceous sedimentary rocks of up to 1000 m thickness were found, overlapped transgressively (basal conglomerates in drill-hole cores) by the Upper Eocene sedimentary formations. The latter contain bedding intrusives of trachydolerites, similar to those of the Buravar high. All of these findings indicate that this part of Talysh for a long time interval was a dryland that limited the Lerik trough from the northeast and experienced another plunge in the second half of the Late Eocene period. At the northwestern extension, the buried high can be traced towards the buried Bilasuvar-Garadonlu Mesozoic protrusion at the border between the Middle and Lower Kur depressions. This protrusion serves as a connecting link between the southern Mughan zone and the northeastern edge structures of the Lesser Caucasus southeastern plunge.

5.6 The Caspian Megadepression

In the Earth crust's contemporary structure, the Caspian Sea represents the intracontinental megatrough laid on a heterogeneous basement during the Lower Miocene age and confined to the submeridional depression area that is superimposed on a series of structural elements belonging to the southeastern suburb of the Precambrian Eastern European platform, to the Scythian-Turanian epi-Hercynian continental plate and to the Alpine-Himalayan folded belt. The megatrough is dissected into three segments by the interbasin high sublatitudinal zones and is represented in the Azerbaijan sector offshore by the southwestern part of the Middle Caspian (Orta Khazar) trough and by the western part of the Southern Caspian (Janubi Khazar) trough. The boundary between them in the submarine relief passes along the Absheron sill. The latter corresponds in the contemporary structure to the Absheron-Balkhanyany (Absheron-Nearbalkhan) elevation zone—a connecting link between the Greater Caucasus folded systems and the Kopetdagh (Azizbekov et al. 1972; Khain and Bogdanov 2003; Khain and Alizadeh 2005; Babayev and Hajiyev 2006). These segments superimposed transversally on differently featured, eastward plunging structural elements of the Greater Caucasus, the Kur trough, and Talysh correspond in the region contemporary geological framework to the tectonic megazones consisting of the higher-order structural units (zones and subzones) and of separate local structures. In this process, the Absheron-Nearbalkhan elevation at the border between the Middle and Southern Caspian troughs is the natural extension of accretion prism and of the Greater Caucasus structural zones that confine it the from south to north. The fact is demonstrated by the recent results of marine seismic surveys (Figs. 5.37, 5.38, 5.39 and 5.40).

Geological interpretation of geophysical data (Khain and Bogdanov 2003; Khain et al. 2007; Mamedov 2008, 2010; Green et al. 2009; Kangarli 2012) demonstrate the action of pseudo-subduction (S-subduction) mechanism in this strip, where the Southern Caucasian microplate (within the offshore area, it corresponds to the Southern Caspian depression basement) is underthrust under the Middle Caspian a segment of the Scythian-Turanian epi-Hercynian platform. The allochthonous accretion prism clamped between the Absheron-Nearbalkhan abyssal dislocation in the north (an extension of the Main Caucasus thrust under the seafloor) and the Kbaad-Zangi abyssal dislocation, separates the Middle Caspian block, with relatively high pre-Jurassic basement position, from the deeply plunged (9–10 km amplitude) and moved far northward Southern Caspian block. In this process, the marine geophysical data show that the substrate of the Greater Caucasus Southern slope zone is

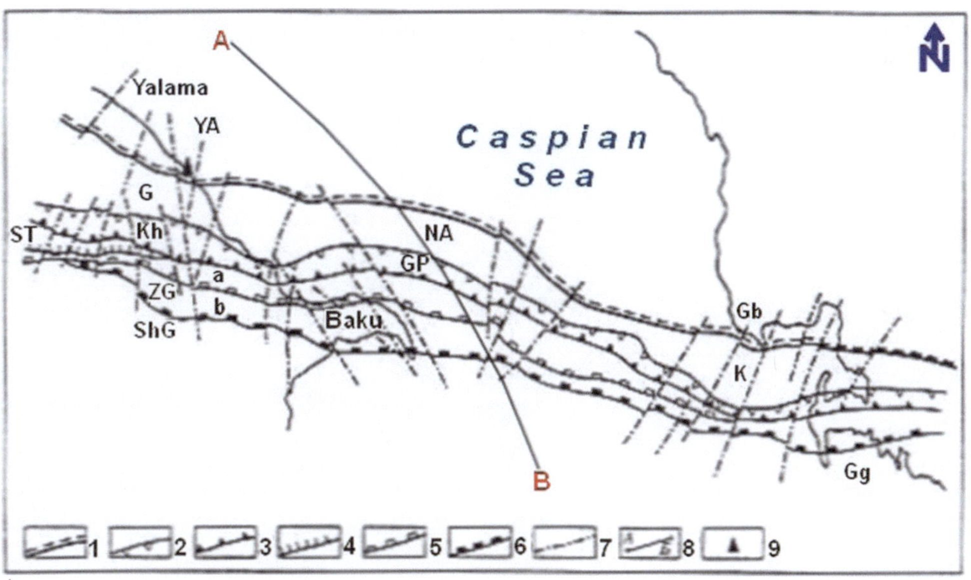

Fig. 5.37 Tectonic scheme of the Absheron sill (Khain et al. 2007; T. N. Kangarli) **Structures:** *YA* Yalama-Aghzybirchala uplift; *Gb* Gubadagh-Boyuk Balkhan (Gubadagh-Large Balkhan) uplift; *G* Guba depression; *NA* Northern Absheron depression; *K* Kelkor depression; *Kh* Khizi zone; *GP* Gilazi-Neargubadagh zone; *ST* Speroza-Tufan zone; *ZG* Zagatala-Govdagh zone (*a* Zagatala-Dibrar subzone; *b* Govdagh-Sumgayit subzone); *ShG* Shamakhy-Gobustan zone; *Gg* Gyzylgum zone. Fractures at the boundaries of structural zones and subzones (*1–6*): (*1*) Imamgulukend-Khachmaz-Yanybibalkhan, (*2*) Siyazan, (*3*) Main Caucasus (Germian, Absheron-Balkhanyany), (*4*) Gamarvan, (*5*) Ilisu-Aladash, (*6*) Kbaad-Zangi (Zangi-Garajuzlu), (*7*) latitudinal fractures (determined and expected); (*8*) lines of the geological-geophysical profiles (see Figs. 5.38, 5.39 and 5.40); (*9*) Agzybirchala well

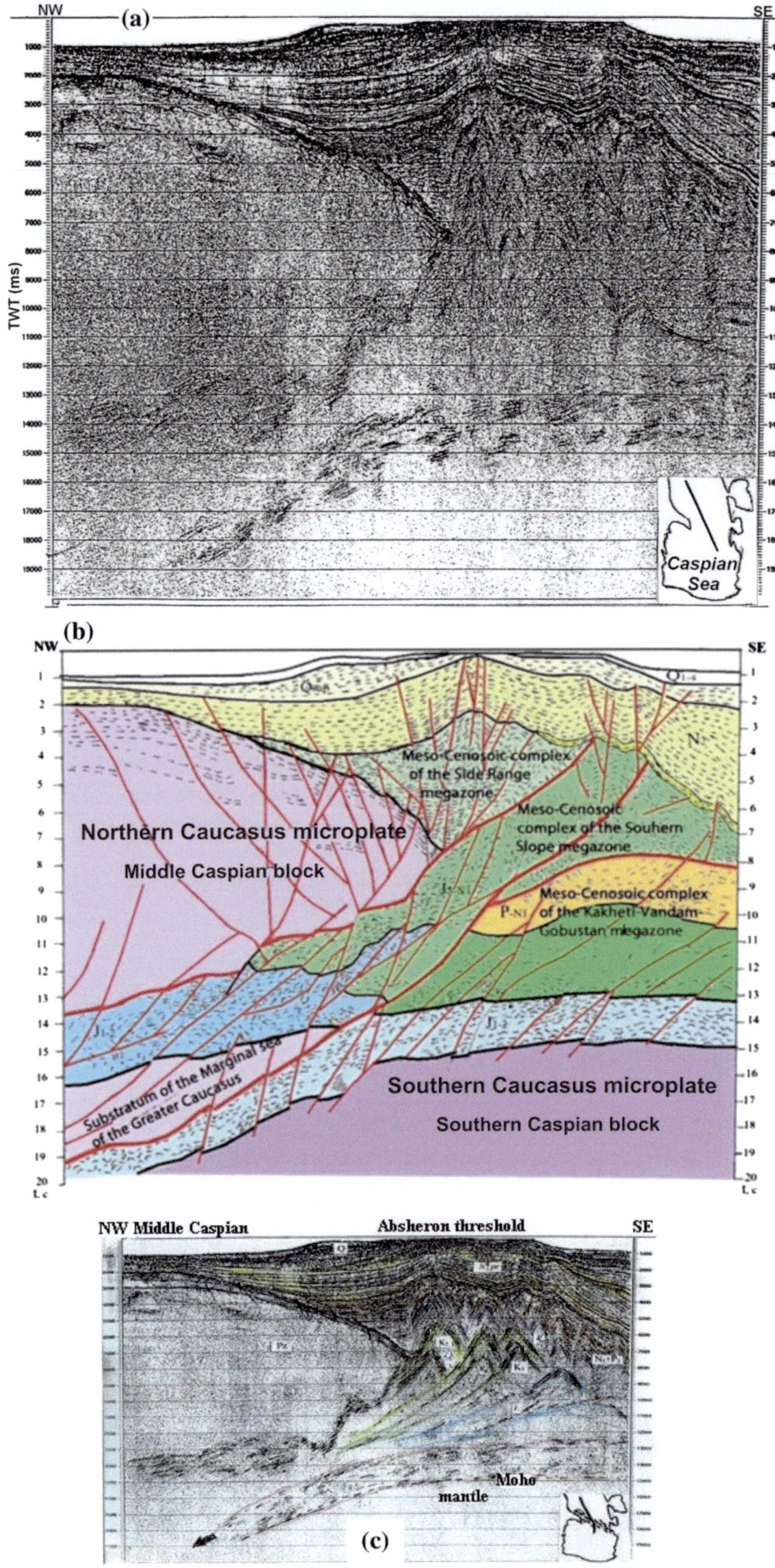

Fig. 5.38 Submeridional seismic section (profile *A–B* in Fig. 5.37) across the Absheron threshold (**a**) and its interpretations according to T.N. Kangarli (**b**) and P.Z. Mamedov (**c**)

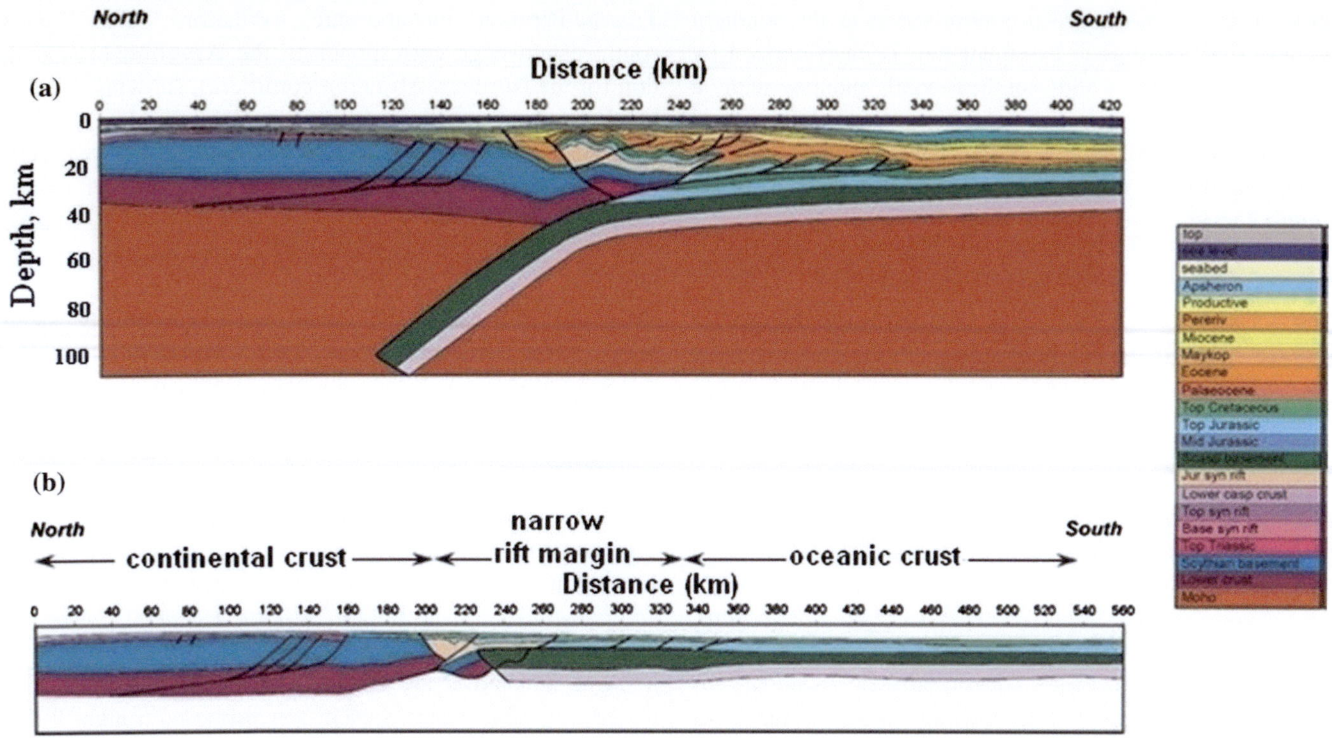

Fig. 5.39 **a** Tectonic balanced cross-section and **b** restoration through the Middle and Southern Caspian areas (profile *AB* in Fig. 5.37) (after Green et al. 2009)

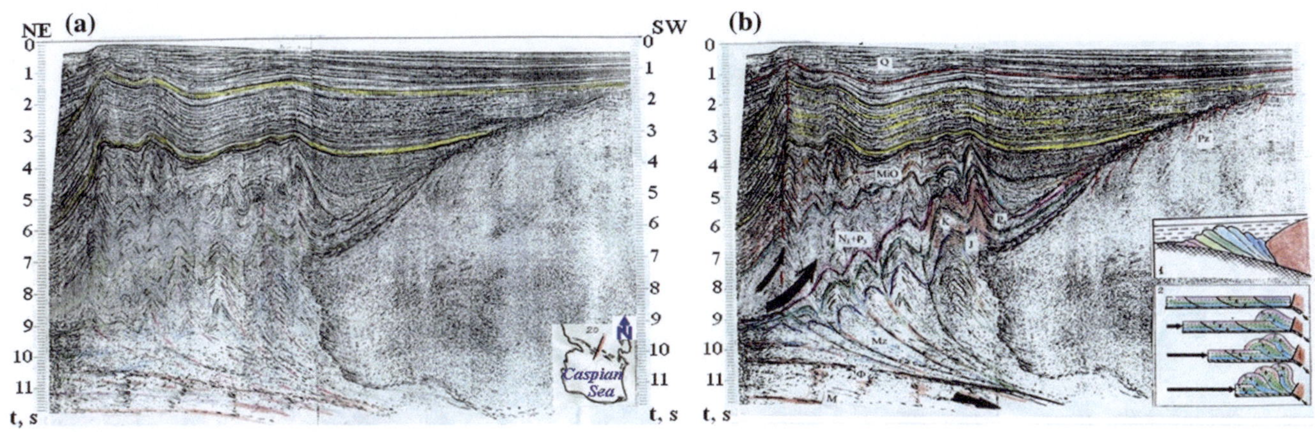

Fig. 5.40 Submeridional seismic (**a**) and seismo-geological (**b**) cross-section across the Absheron threshold and their interpretation(according to Mamedov 2010)

squeezed and absorbed in pseudo-subduction area, which is in complete conformity with geophysical materials for the Caucasian dryland.

5.6.1 The Middle Caspian Megazone

The Middle Caspian megazone corresponds to the marine extension of the Greater Caucasus megastructures (the Gusar-Devechi, the Side Range, and the Southern Slope megastructures) within the Azerbaijan sector of the Caspian offshore area and is represented (from north to south) by the Yalama-Agzybirchala, Northern Absheron, Gilazi-Neargubadagh and Absheron-Nearbalkhan structural zones (see Fig. 5.37). The megazone's southwestern boundary passes along the Yalama-Khyrdalan transtensional fault that continues the Darband fracture within the Azerbaijan boundaries, while marine extension of the Kbaad-Zangi thrust in the accretion prism bottom serves as the southern border. The pre-Jurassic basement roof is characterized by regional southward and southeastward plunge with a simultaneous increase in sedimentary mantle thickness, from 3–4 km northward (in the Samur-Kumdi Muis elevation zone central part) to 14–15 km in the Northern Absheron trough's most dipped area.

5.6.1.1 The Yalama-Agzybirchala Zone

The Yalama-Agzybirchala zone corresponds to the southwestern plunged part of the Samur-Kumdi Muis elevation that southward confines the Greater Caucasus Terek-Caspian foredeep. A number of bending folds (Yalama-deniz, Khachmaz-deniz, Guba-deniz, etc.) stands out within the zone boundaries at the pre-Jurassic basement and Jurassic complex surface. Sedimentary mantle is characterized as a medium stratified into structural compositional complexes of varying content, such as the Pliocene–Quaternary, Oligocene–Miocene, Paleocene–Eocene, Upper Cretaceous, Lower Cretaceous, Upper Jurassic, and Lower–Middle Jurassic (Fig. 5.41). The Triassic–Jurassic boundary is characterized by regional unconformity caused by the Early Cimmerian tectonic activization and depositional break. According to deep drilling data onshore (Agzybirchala area) and offshore (Yalama-deniz area), the Triassic rocks are characterized by volcanogenic and volcanogenic–sedimentary composition with the underlying Lower Triassic-Permian metamorphic formations. The zone's southwestern edge corresponds to the Agzybirchala elevation (or the Northern Absheron cordillera), showing itself in the Bouguer gravitaty anomalies, which stretches 200 km southeastward. Interpretation of seismic survey materials indicates the possibility of the Upper Jurassic reef massif seating within the elevation boundaries (Fig. 5.42). Marine extension of the Imamgulukend-Khachmaz fracture, featuring high gravity step and moderate magnetic peak, can be traced by the elevation's southwestern, relatively steep slope. In the Turkmenian part of the Caspian Sea area, the dislocation is named the Southern Balkhan fracture; by tracing inland, it south confines the Gubadag-Large Balkhan elevation.

5.6.1.2 The Northern Absheron Zone

The northern (Shimali) Absheron zone is located in the Imamgulukend-Khachmaz fracture's lowered southern side and constitutes a deep trough, striking a general Caucasian direction up the Turkmenistan shores at a distance of over 380 km, situated at the Guba lowered zone's southeastern extension in the Gusar-Devechi superimposed trough. It has a similar asymmetrical structure with a relatively low-angle northeastern edge and steep southwestern edge. The zone's Cretaceous complex is represented by five anticlines with surface Cretaceous depth of 7900–8300 m. The Paleocene–Miocene and the Pliocene complex structural geometries in

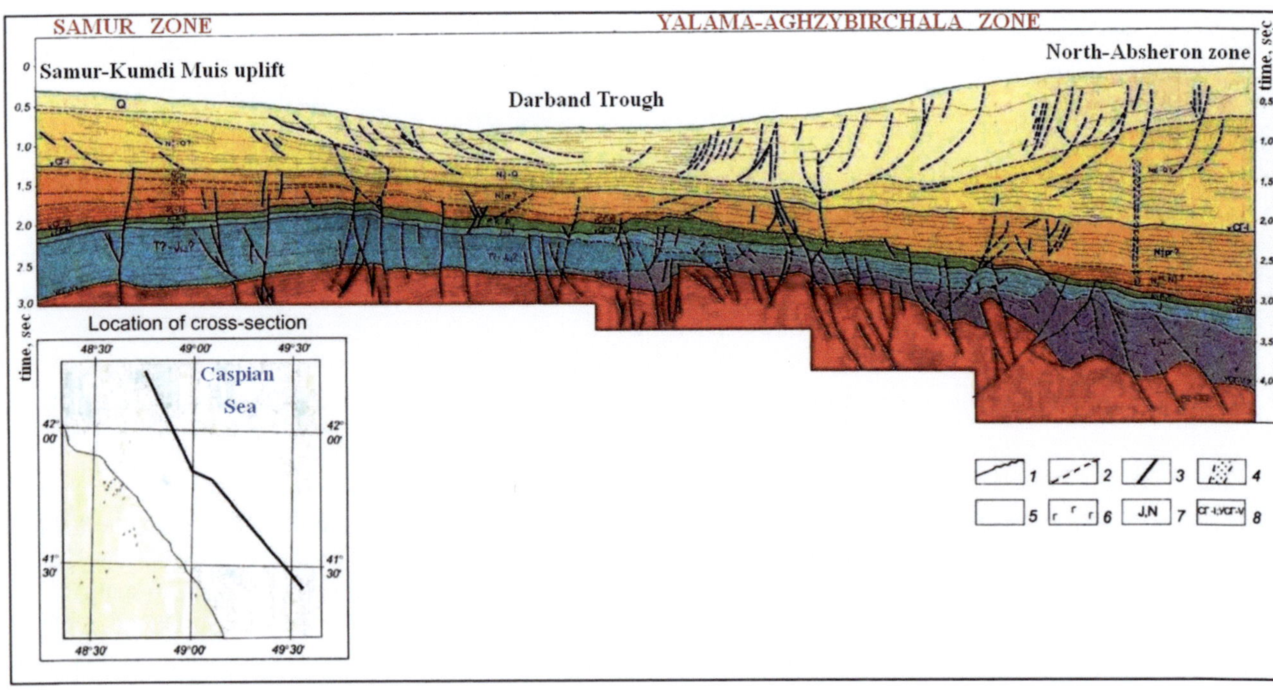

Fig. 5.41 Seismogeological interpretation of composite seismic section across the Samur-Kumdi Muis zone of elevations according to Trust "Khazardanizneftgeofizikikeshfiyyat" (Glumov et al. 2004). (*1*) seismogeological boundaries; (*2*) seismic horizons of undefined geological origin; (*3*) faults; (*4*) zones of limit velocity inversion; (*5*) zones of possible bioherm development; (*6*) zones of possible development of volcanogenic–sedimentary complexes; (*7*) geochronological indexes; (*8*) indexes of regional seismic horizons

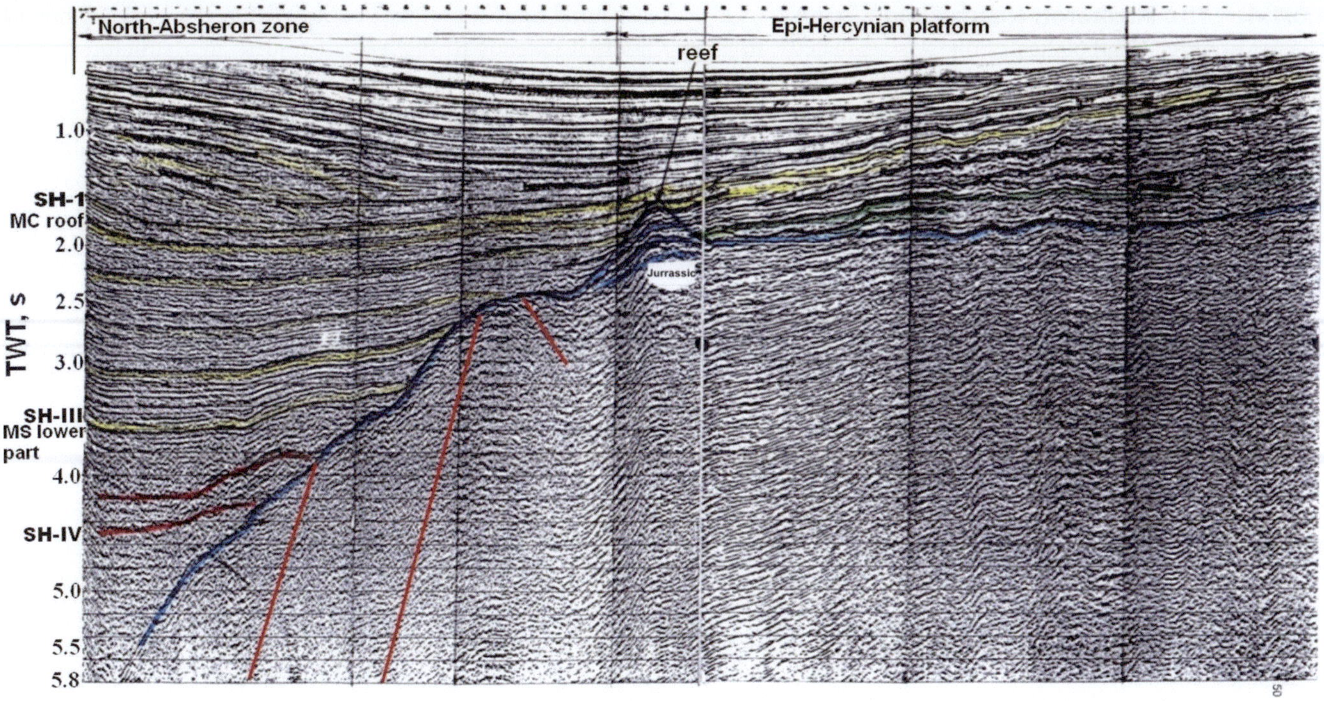

Fig. 5.42 Geological interpretation of seismic section #8605216 developed by Association "Khazardanizneftgeofizikikeshfiyyat" (Babayev and Hajiyev 2006)

their general outlines preserve a similarity to those of the Upper Cretaceous deposits. For the Pliocene–Quaternary complex, a chain of isometric depressions separated by saddles can be traced. By comparing structure maps for seismic horizons confined to various structural levels (the Mesozoic surface, the Maykop series, the Paleocene lower rocks, etc., down to the Upper Pliocene roof), the 74-km northwestward migration of the trough central part is observed. Along the troughs south–southwestern edge, the Siyazan fractural zone's Mesozoic deposits are uplifted and folded as anticlinal folds (Zarat-deniz, Gyzylburun-deniz, etc.). At the zone's extension in Turkmenian water areas and adjacent inland areas, the Kelkor trough and the east-joining Inter-Balkhan and Kopetdagh troughs are located. The South Turkmenian fracture, south confining them, probably corresponds to the Syiazan dislocation eastern extension.

5.6.1.3 The Gilazi-Gubadaghyany Zone

The Gilazi-Neargubadagh zone looks like an upraised tectonic step at the Shakhdagh-Khizi zone's marine extension of the Greater Caucasus Side Range megazone, which stretches towards the Turkmenian borders at a distance of over 330 km. According to the seismic survey and deep drilling data, the zone emerges in the Lower Pliocene tops as a vast (up to 20 km) and poorly dislocated strip of elevations, plunging by its eastward strike and dissected by a narrow structural saddle into two anticlinal lines. The northern Shuraabad-deniz-Ufug strip is controlled by the Siyazan fracture marine extension, while the southern Yashmadeniz-Vurgun strip is controlled directly by the Absheron-Balkhanyan (in the land, the Main Caucasus thrust). The zone is characterized by decreased thickness of the Pliocene–Miocene–Paleogene deposits and by high hypsometric and stratigraphic level of folding plane in Cretaceous deposits. Local structures are distinguished by their small dimensions, low-angle side dipping, and an echelon joining. Structural domes are wide and are not broken by ruptures, but some structural complications are noted with depth. The Pliocene eastward folding plunges and gradually fades at the Gubadagh elevation remote periphery, where the Pliocene strata obtain low-angle monoclonal bedding. Due to diversified unconformities and breaks found at the Triassic–Jurassic, Jurassic–Cretaceous, Lower–Upper Cretaceous, Upper Cretaceous–Paleocene, Oligocene, Miocene–Pliocene interfaces, the anticlinal folds here have multiple-deck structure. Within the Turkmenian offshore area, the zone finds its extension in the Livanovo-Shanlik-Nearcheleken anticlinal line and in the onshore Cheleken-Garatapa anticlinal zone, which are located in the Absheron-Nearbalkhan dislocation upthrown side.

5.6.1.4 The Absheron-Nearbalkhan Zone

The Absheron-Nearbalkhan zone is situated in the same-named thrust lying side. On its northern and southern sides, it represents the marine extension of the Zagatala-Dibrar and the Govdagh-Sumgayit nappe complexes, respectively, which generate the accretion prism structure (the southern slope megazone) in the Southeastern Caucasus (see Fig. 5.37). Within the offshore, the zone stretches at a distance of over 250 km and is represented by two chains of local elevations of general Caucasian strike, corresponding to structural subzones. In the contemporary structure, the zone is represented as an intense high-amplitude uplift of the Miocene-Quaternary complex. The zone's typical features are asymmetrical, linearly elongated ridge-like anticlines complicated by transversal and longitudinal ruptures and by diapirism. The northern Kurkachidagh-deniz-Kapaz anticlinal strip, distinguished as the Yashma-Chyrag subzone (see Fig. 5.2), is positioned on the Zagatala-Dibrar allochthone buried marine extension. In its western section, the zone is the most uplifted; as a result, the anticlinal structure crests (Khali, Chilov and Neft Dashlary) are eroded, and the Eocene–Miocene deposits outcrop at the seafloor. Structural analysis leads to the conclusion that the subzone thins out in the Absheron-Nearbalkhan thrust's southern side at the meridian of Turkmen dryland situated north of the Gubadagh-Large Balkhan uplift western closure (Khain et al. 2007). The southern Absheron-Gurgan subzone (see Fig. 5.37) is expressed by the Hovsan-deniz-Goygol line of anticlinal elevations and crosses the offshore, joining the Kopetdagh elevation western pericline. According to the existing concepts (Kleschev et al. 1992), the Western Turkmenian Gubadagh and the Kopetdagh uplifts are of allochthone nature, and the Absheron-Balkhanyany zone allochthonous masses participate in abyssal structure of the Gograndagh-Okarem elevation zone that is isolated here.

5.6.2 The Southern Caspian Megazone

In the contemporary Earth crust structure, the Southern Caspian megazone represents an area of recent intense plunging, constituting the Southern Caucasus continental microplate trough system and corresponding to the marine extension of the inland Shamakhy-Gobustan and Lower Kur troughs and to the structures of the Talysh folded area. It is divided from them by the Shongar-Gyzylagaj abyssal fracture (the Western Azerbaijan), according to Shikhalibeyli et al. (1984), of transtensional fault nature. The megazone's northern side, as indicated above, is buried in the Greater Caucasus accretion prism bottom, while its southern side hosts the structures of the Iranian pre-Alborz-Gorgan structural zone's northern edge. Within the offshore boundaries, it extends to the Talysh folded area structures at the Alborz-Binalud fold-nappe system front.

In accordance with the Iranian territory data (Priestley et al. 1994; Allen et al. 2000, 2001, 2002; Axen et al. 2001; Jackson et al. 2002; Khain and Bogdanov 2003), the Southern Caspian block experiences subduction under the Central Iranian microcontinent along the pre-Alborz dislocation. The process commencement was assumed to take place during the Oligocene (Pyrenean phase of tectogenesis), and it continues recently as well. Thus, it can be concluded that the Southern Caspian Depression exists in the conditions of bilateral convergence with the adjacent Earth's crust bocks, such as the Scythian-Turanian northern plate and the Central Iranian southern microcontinent. This is confirmed by the variations in the Earth crust thickness, where sublatitudional zonality can be traced: peak values (45–50 km), recorded at the northern and southern edges, are reduced to 30–40 km in its central part that connects the Kur and the Transcaspian (in Turkmenistan) troughs (Fig. 5.43). The Alpine cover is composed of the Mesozoic–Eocene, Oligocene–Miocene, and Pliocene–Holocene compositional complexes of maximal 20–25 km thickness, over 7 km of them being attributed to the latter one. Many researchers are of opinion that the oceanic-type Earth crust of 15–20 km thickness is located under the sedimentary sheath and is characterized by increased seismic wave velocities. The Southern Caspian block should be considered as an eroded rigid massif of continental character (Southern Caucasus) continental microplate constituent) that has experienced underthrust (S-subduction) under the continental plates that confine it south- and northward and was exposed to rifting accompanied by the Earth crust reduction and basification and transfer from the continental to oceanic type.

The dryland structural elevations east of the Shongar-Gyzylagaj fracture find their natural marine extension but change their course from eastern and southeastern to south–southeastern, and further on, recover their former course. This can be explained by the effect of submeridionally oriented rupturing dislocations that control the Caspian Basin's western edge on the folds (and accompanying longitudinal dislocations).

Structural facial and tectonic peculiarities of heterochronous compositional complexes and fold morphogenetic properties assist in isolating the external near-edge margins and internal depression structures as the megazone's basic elements. Among the former ones are the marine extension structures of the Shamakhy-Gobustan structural zone in the north and the Talysh structural zone in the south. The internal depression structure is the Lower Kur trough marine part. The megazones contemporary structure is determined by a make-up of the dislocated Pliocene–Pleistocene complex that

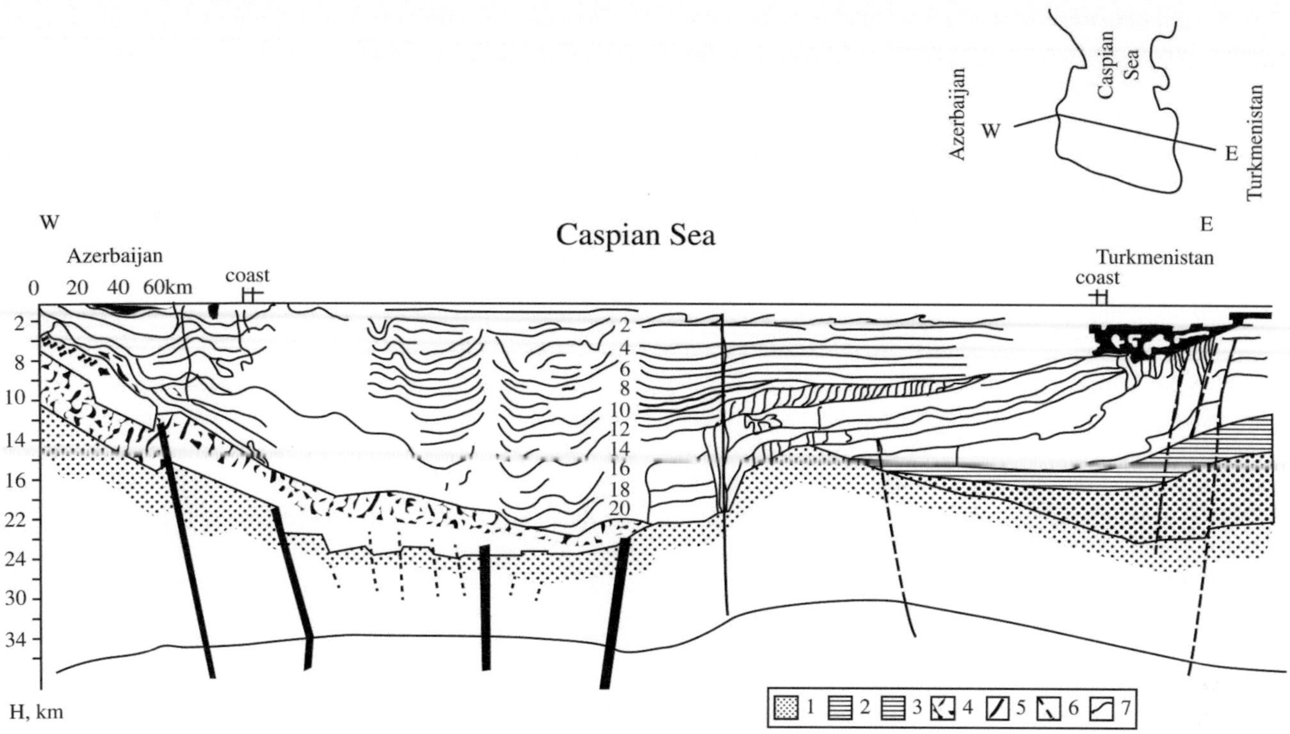

Fig. 5.43 Geological-geophysical section across the Southern Caspian Depression (Khain2001; I.S. Guliyev). (*1*) Paleozoic basement; (*2*) "granitic" layer; (*3*) "basaltic" layer; (*4*) Mesozoic magmatites; (*5*) deep faults; (*6*) other faults; (*7*) discordance unconformities

unconformly covers the Mesozoic–Miocene deposits and is complicated by disjunctive dislocations, diapirism and mud volcanism. Local buried highs determined by geological-geophysical techniques usually form the anticlinal belts that are archwise south cambered; some of them have submeridional strike, lining up along the same-course fractures. In separate anticlinal structures, their crest sections eject the Pliocene–Eopleistocene formations onto the seafloor. In the strip of structure plunging towards the basin central part, the anticlinal belts are vergent, changing their strike from southeastward to eastward and northeastward. The depression's largest and longest (over 150 km) structural element is the Sefidrud-Ataturk-Garaboghaz fractural zone that controls a number of buried anticlinal highs of northeastern strike.

The megazone's Azerbaijan sector of the Caspian Sea offshore is subdivided by the Pliocene–Quaternary complex into the following structural zones (from north to south): (1) southern Absheron, (2) Bandovan, (3) Gyzylagay, and (4) Lenkaran-Shahagaj. The first one encompasses the Absheron-Jeirankechmez depression marine part at the Shamakhy-Gobustan trough extension and the Baku archipelago elevation area's northeastern side. The Bandovan and the Gyzylagaj zones correspond to the Baku archipelago elevation area's central and southwestern parts and the Lower Kur trough marine part. Finally, the Lenkaran-Shahagaj zone encompasses the southeastern segments of the northeastern side structural zones in the Talysh folded region.

5.6.2.1 The Southern Absheron Zone

The southern Absheron zone includes the Absheron-Jeirankechmez depression marine part at the Shamakhy-Gobustan trough extension as well as the Baku archipelago elevation area's northeastern side. It is represented by wide synclinal troughs composed of the Eopleistocene–Quaternary deposits and separated by brachyanticline elevations with eventually outcropping Pliocene rocks in the seafloor crests. Structurally, the zone is subdivided into three subzones: the Puta, the Sangachal, and the Baku Archipelago. The subzones correspond generally to the eastward plunging undulating monocline made of the Akchagyl-Absheron age rocks. In its bottom, elder formations are complicated by intense folding.

The Puta subzone continues the Shamakhy-Gobustan trough Jangichay subzone within the offshore boundaries and is represented in its southwestern side by the Lokbatandeniz-Amirov belt of buried anticline elevations. In the subzone's western part, an independent Bakhar brachy-elevation is found, controlled by a submeridional dislocation.

The Sangachal subzone is a natural marine extension of the Shamakhy-Gobustan trough Keyvandi-Shahgaya subzone and can be traced along the extensive Sangachal-deniz-Ataturk belt of buried elevations, controlled by the Sangachal fracture zone and composed of the Lower Pliocene rocks in the crest. In the subzone's central part, an independent Mushfig brachy-elevation is situated in the anticline belt's northern side.

The Baku archipelago subzone corresponds to the marine extension of the Shamakhy-Gobustan zone's extreme southern element and is represented in the contemporary structure as the Hamamdagh-deniz-Alov anticline belt at the Ajichay-Alat thrust upthrown side. The subzone's extreme southeastern part hosts independent brachy-elevations (Zafar, A. Nazim, and Sharg) that are controlled by the anti-Caucasian direction ruptures.

5.6.2.2 The Bandovan Zone

The Bandovan zone is situated at the Shirvan zone's southeastern plunge at the Lower Kur's northeastern edge and corresponds to the central and southeastern regions of the Baku archipelago elevation area. The zone displays itself as a strip of anticline highs at the Ajichay-Alat thrust's southern lying side, southwestward confined by the Pirsaat-Salyan-Neftchala fracture. The inland mappable anticline belts, being traced to the offshore boundaries, are separated from one another by relatively narrow (as compared to the on-shore observed and to more northern marine structures) synclinal troughs and, having southeastward plunging, are vergent with the emergence of additional buried branches. In doing this, the Pirsaat-Hamamdagh, the Kalamaddin-Mishovdagh-Bandovan and the Kurovdagh-Neftchala anticline elevation belts find their extension, in the Bandovan-deniz-Araz, Yanantava-Gulustan and Kurdashy-Chanaggala anticline zones, respectively. In addition, a number of independent brachyfolds are mapped between these major anticlinal fold belts (Sh. Azizbekov, Gantemir, Nizami Northern, etc.).

5.6.2.3 The Gyzylagaj Zone

The Gyzylagaj zone corresponds to the Mughan zone marine extension at the Lower Kur trough's southwestern edge and is a recent trough filled by the Pleistocene–Holocene molasses at the seafloor, southwest confined by the Bilasuvar dislocation and northeast confined by the Pirsaat-Salyan-Neftchala dislocation. According to the marine seismic survey data, two buried anticlinal belts participate in the zone formation: the northern Gyzylagaj-deniz-Vidadi belt and the southern Saradeniz-Hasrat belt, situated in the Bilasuvar fracture northeastern side. The latter belt continues the Padar folding group in the offshore area and, being of sublatitudinal strike, comes in conformity with the southern Chikishlar zone of Turkmenistan seacoast.

5.6.2.4 The Lenkaran-Shahagaj Zone

The Lenkaran-Shahagaj zone is situated at the southeastern extension of the Talysh folded region and is separated by the Western Caspian (Garbi Khazar) and Shongar-Gyzylagaj fractures of transtensional fault character (see Fig. 5.2). In the east, it is traced along the Iranian Caspian seacoast and the adjacent offshore, participating in the structure of the near-Alborz-Gorgan megazone's northern side. The zone's northern boundary is defined by the Bilasuvar fracture and by the near-Alborz fracture that prolongs it eastward. Due to the buried northeastward striking Sefirud elevation, the zone is separated into the western near-Alborz and the eastern Gorgan segment. The latter stretches towards the Hasangulu-deniz anticlinal zone at the Turkmenistan seacoast, which can be taken as the extension of Talysh structures in the Transcaspian area.

References

Adamia, S. A. (1984). Pre-alpine basement of the Caucasus—composition, structure and formation. In *Tectonics and metallogeny of the caucasus*. Metsniereba, Tbilisi, pp. 3–104 (in Russian).

Adamia, S., Chuhotua, T., Kekelia, M., et al. (1981). Tectonics of the Caucasus and adjoining region: Implications for the evolution of the Tethys ocean. *Journal of Structural Geology 3*(4), 437–447.

Aghabekov, M. G., Allakhverdiyev, R. A., & Moshashvili, A. B. (1977). *Structural interralations of Mesozoic and Cenozoic complexes in the central part of Kur Depression and Shamakhy-Gobystan region*. Baku: Elm (in Russian).

Aghabekov, M. G., Karimov, K. M., Moshashvili, A. B., & Khain, V. Y. (1976). New data on the structure of the central part of Kur depression. *Geotektonika, 5*, 75–81 (in Russian).

Akhmadov, G. A. (1957). *Geology and oil and gas bearing of Gobystan*. Baku: Azgosizdat (in Russian).

Akhmedbeyli, F. S., Ismailzadeh, T. A., & Kangarli, T. N. (2002). Geodynamics of the Eastern Caucasus in Alpine Tectonic-Magmatic Cycle (Azerbaijan). *Transactions of the Institute of Geology, Azerbaijan National Academy of Sciences, 30*, 36–48 (in Russian).

Alizadeh, A. A., Khain, V. E., & Ismailzadeh, A. D. (2000). *Saatly superdeep. Analysis of deep structure of the Kur intermontane depression by the data of Saatly superdeep Borehole SG-1 drilling*. Baku: Nafta-Press (in Russian).

Allen, M., Flecker, R., Bartholomev, I., Ismailzadeh, A., et al. (2000). Tectonostratigrafic evolution of the South Caspian basin. In *Transactions of the AAPG Annual Meeting*, April 2000, New Orlean.

Allen, M., Jackson, J., Hovius, N., Ghassemi, M., & Ismailzadeh, A. (2001). Neotectonics of the South-Caspian region: Incipiend subduction of a trapped ocean basin. In *Transactions of the EUG XI abstracts* (p. 399). Oxford: Oxford University Press.

Allen, M., Jones, S., Ismailzadeh, A., Simmons, M., & Anderson, L. (2002). Onset of subduction as the cause of rapid Pliocene-Quaternary subsidence in the South Caspian basin. *Geology, 30*(9), 775–778.

Axen, G., Lam, P., Grove, M., Stockli, D., & Hassanzadeh, J. (2001). Exhumation of the West-Central Alborz mountains, Iran, Caspian subsidence and collision-related tectonics. *Geology, 29*(6), 559–562.

Azizbekov, S. A. (1961). *Geology of Nakhchivan Republic*. Moscow: Gosgoltehizdat (in Russian).

Azizbekov, S. A., Alizadeh, K. A., Shikhalibeyli, E. S., & Gadjiev, T. G. (Eds.). (1972). In *Geology of the USSR* (Vol. 47). Azerbaijan Republic. Part 1: Geological description. Moscow: Nedra (in Russian).

Azizbekov, S. A., Bagirov, A. E., Valiyev, M. M., Ismailzadeh, A. D., Nijeradze, N. S., Yemelyanova, E. N., & Mamedov, M. N. (1979). *Geology and volcanism of Talysh*. Baku: Elm (in Russian).

Babayev, D. G., & Hajiyev, A. N. (2006). *Deep structure and perspectives of oil and gas bearing of the Caspian Sea basin*. Baku: Nafta-Press (in Russian).

Babayev, R. Y., Taghiyev, R. E., & Metaxa, X. P. (1978). Main peculiarities of the deep structure of the Alazan-Ayrichay depression by a set of geological-geophysical data. *Azerbaijan Oil Industry, 11*, 20–25 (in Russian).

Bagizadeh, F. M., Karimov, K. M., & Salayev, S. G. (1987). *Deep structure and oil and gas bearing of South Caspian megadepression*. Baku: Azgosizdat (in Russian).

Baranov, G. I., Belov, A. A., & Dotduyev, S. I. (1990). The Greater Caucasus. In: *Tectonic schistosity of lithosphere and regional geological investigations* (pp. 196–215). Moscow: Nauka (in Russian).

Dotduyev, S. I. (1986). On the nappes structure of the Greater Caucasus. *Geotektonika, 5*, 94–106 (in Russian).

Dotduyev, S. I. (1989). Mesozoic-Cenozoic geodynamics of the Greater Caucasus. In: *Geodynamics of the Caucasus* (pp. 82–91). Moscow: Nauka (in Russian).

Gadjiev, R. M. (1965). *Deep geological structure of Azerbaijan*. Baku: Azerneshr (in Russian).

Gamkrelidze, I. P. (1977a). Tectonic evolution of the Anatolian-Caucasian-Iranian segment of the Mediterranean belt. *Geotektonika, 3*, 25–37 (in Russian).

Gamkrelidze, I. P. (1977b). *Tectonic Nappes of the Southern slope of the Greater Caucasus (within Georgia Territory)*. Tbilisi: Metsniereba (in Russian).

Gamkrelidze, I. P. (1989). Geodynamic evolution and mechanism of forming structure of the Mediterranean folded belt. In: *Geodynamics of the Caucasus* (pp. 28–36). Moscow: Nauka (in Russian).

Gamkrelidze, I. P., & Shengelia, D. M. (2005). *Precambrian-Paleozoic regional metamorphism, granitoid magmatism and geodynamics of the Caucasus*. Moscow: Nauchnyi Mir (in Russian).

Glumov, I. G., Malovitskiy, Y. P., Novikov, A. A., & Senin, B. V. (2004). *Regional geology and oil and gas bearing of the Caspian sea*. Moscow: Nedra (in Russian).

Green, T., Abdullayev, N., Hossack, J., Riley, G., & Roberts, A. (2009). Sedimentation and subsidence in the South Caspian basin, Azerbaijan. South Caspian to Central Iran basins. *Geological Society Special Publication, 312*, 241–260.

Guschin, A. I., & Panov, D. I. (1992). Structure of the Side Range zone in the Eastern Caucasus. *Bulletin of the Moscow State University, Series Geology, 2*, 25–37 (in Russian).

Gadjiev, T.G., Nechayev, Yu.V., Potapova, E.I., and Sattarova, V.M., Gadjiev, T.G., Nechayev, Yu.V., Potapova, E.I., and Sattarova, V. M., 1989. The Map of Deep Structure of the Caucasus According to Cosmic Data. Map Printing Factory of the Ministry of Geology of the USSR, Moscow.. The Map of Deep Structure of the Caucasus According to Cosmic Data. Map Printing Factory of the Ministry of Geology of the USSR, Moscow.

Hasanov, T. A. (1971). Tectonic Nappes in the Shakhdagh ridge (Lesser Caucasus). *Doklady Academy of Science in USSR, Series Geology, 201*(4), 920–922 (in Russian).

Hasanov, T. A. (1985). *Ophiolites of the Lesser Caucasus*. Moscow: Nedra (in Russian).

Isayev, B. M., Gadjiev, T. G., Alizadeh, S. A., & Kangarli, T. N. (1981). Tectonic Nappes and Olistostrome complexes of the South-Eastern Caucasus. *Geotektonika, 1*, 70–84 (in Russian).

Ismailzadeh, A. D. (1999). Geodynamic situations of Alpine magmatism of the Caucasus. In: *Thesis of the International Conference* on *Geodynamics of the Black-Caspian Sea Segment of the Alpine Folded Belt and Perspectives of Searching Economic Deposits* (pp. 145–146). Baku: Nafta-Press.

Jackson, J., Priestley, K., Allen, M., & Berberian, M. (2002). Active tectonics of the South Caspian basin. *Geophysical Journal International, 148*, 214–245.

Kangarli, T. N. (1999). Nappe tectonics of the oil and gas bearing regions of the South-Eastern Caucasus. In *Transactions of the International Conference on Modern Tectonics and Its Influence to Forming and Location of Oil and Gas Deposits* (pp. 86–90). Baku: Nafta-Press (in Russian).

Kangarli, T. N. (2005). Stages of formation of nappes of tectonically Schistous Alpine cover of the Greater Caucasus within Azerbaijan. *Izvestiya Academy of Sciences* Azerbaijan, *Series*: *Earth* Sciences, *4*, 37–44 (in Russian).

Kangarli, T. N. (2006). Differentiated-mixed tectonic slabs in the infrastructure of Govdagh-Sumgayit Nappe Complex of the South-Eastern Caucasus. *Izvestiya Academy of Sciences* Azerbaijan, *Series*: *Earth* Sciences, *1*, 28–35 (in Russian).

Kangarli, T. N. (2007). Jurassic allochthone complexes of the Greater Caucasus (Azerbaijan). *Izvestiya Academy of Sciences* Azerbaijan, *Series*: *Earth* Sciences, 3–11 (in Russian).

Kangarli, T. N. (2009). *Tectonic stratification of the Alpine cover of the Greater Caucasus within the territory of Azerbaijan*. Baku: Doctor of Science Thesis, Institute of Geology (in Russian).

Kangarli, T. N. (2011). Nappes in structure of the South-Eastern Caucasus as indicator of accretion interaction of Northern and Southern Caucasian microplates. In *Transactions of the International Conference on "Modern conditions of Earth Sciences" Suggested to the Memory of V.E. Khain*. Moscow State University, pp. 849–854 (in Russian).

Kangarli, T. N. (2012). Mass overthrust within the structure of the Greater Caucasus (Azerbaijan). *Modern problems of geology and geophysics of the Caucasus* (pp. 163–202). Baku: Nafta-Press.

Kangarli, T. N., & Akhundov, A. B. (1988). Surface and deep structures of the southern slope of the Greater Caucasus. *Soviet Geology, 10*, 42–52 (in Russian).

Kangarli, T. N., & Aliyev, A. M. (2005). Subsurface structural architecture of the Absheron Peninsula (based on recently conducted geologic/geophysical studies). *News of Azerbaijan Geophysics, 4*, 43–46 (in Azerbaijanian).

Kangarli, T. N., Akhundov, A. B., Aliyev, A. M., & Abdullayev, V. D. (1994a). The deep geological mapping of the oil and gas bearing regions of the South-Eastern Caucasus. *Türkiye 10 Petrol Konqresi ve Sergisi Bildirileri* (pp. 204–212). Istanbul, Türkiye (in Turkish).

Kangarli, T. N., Shekinsky, E. M., & Zamanov, Y. C. (1994b). Tectonic stratification of the earth's crust of the Greater Caucasus and the problems of prospecting of oil and ore deposits. *Türkiye 10 Petrol Konqresi ve Sergisi Bildirileri* (pp. 92–103). Istanbul, Türkiye (in Turkish).

Khain, V. E. (1937). *Geological investigations and searching oil in Lagyj Mts*. Baku: Aznefteizdat (in Russian).

Khain, V. E. (1950). *Geotectonic evolution of the South-Eastern Caucasus*. Baku: Aznefteizdat (in Russian).

Khain, V. E. (1952). Tectonic structure of Azerbaijan. In *Proceedings of the Conference on Geological Problems of the Transcaucasian. Azerbaijan* Academy of Science, Baku, pp. 162–175 (in Russian).

Khain, V. E. (1975). Main stages of tectono-magmatic evolution of the Caucasus: Experience of geodynamic interpretation. *Geotektonika*, *1*, 13–27 (in Russian).

Khain, V. E. (1984). *Regional geotectonics. Alpine Mediterranean Belt*. Moscow: Nedra (in Russian).

Khain, V. E. (2001). *Tectonics of continents and oceans (Year 2000)*. Moscow: Nauchnyi Mir (in Russian).

Khain, V. E., & Alizadeh, A. A. (Eds.), 2005. (authors: Aliyev, G. A., Akhmedbeyli, F. S., Ismailzade, A. D., Kangarli, T. H., & Rustamov, M. I.). Geology of Azerbaijan (Vol. IV: Tectonics). Baku: Nafta-Press (in Russian).

Khain, V. E., & Bogdanov, N. A. (Chief Eds.). (2003). *Explanation notes for the international tectonic Map of the Caspian sea and its framing. Scale 1:2,500,000*. Moscow: Nauchnyi Mir (in Russian).

Khain, V. E., Hajiyev, A. N., & Kangarli, T. N. (2007). On the tectonic nature of the Absheron threshold of the Caspian Sea. *Doklady Russian Academy of Science, 414*(4), 523–527 (in Russian).

Khalifazadeh, C. M. (1967). On the facial-paleogeographic conditions of accumulation of Bathonian deposits of East and South-Eastern Caucasus. *Scientific Notes of Azerbaijan State University, Series Geology-Geography*, *3*, 14–19 (in Russian).

Kleschev, K. A., Shein, V. S., & Slavkin, V. S. (1992). Novel conception of geological structure and oil & gas bearing of the Western Turkmenistan. *Geology of Oil and Gas*, *5*, 1–8 (in Russian).

Knipper, A. L. (1971a). Serpentinite melange of the Lesser Caucasus (internal structure and age). *Geotektonika*, *5* (in Russian).

Knipper, A. L. (1971b). History of evolution of serpentinite melange of the Lesser Caucasus. *Geotektonika*, *6*, 87–100 (in Russian).

Knipper, A. L. (1975). Oceanic crust in the structure of Alpine folded region (Southern Europe, Western Asia and Cuba). In *Proceedings of the Geology Institute, Acadamy of Science USSR* (Vol. 267). Moscow: Nauka (in Russian).

Kopp, M. L., & Kurdin, N. N. (1980). Regional displacements in the south-eastern Caucasus. *Izv. VUZov, Series Geology and Prospecting, 11*, 30–37 (in Russian).

Kurdin, N. N. (1987). Pirsaat-Salyan fault zone of the south-eastern Caucasus (morphokinematic analysis). *Bulletin Moscow Association of Naturalists Investment, Series Geology*, *62*(2), 31–39 (in Russian).

Makarovskiy, O. V. (1982). Structure of the main thrust of the Greater Caucasus within Abkhazia. *Izvestiya of USSR Academy of Sciences, Series: Earth Sciences, 5*, 111–116 (in Russian).

Mamedov, A. V. (1973). *Geological structure of the Middle Kur depression*. Baku: Elm (in Russian).

Mamedov, P. Z. (2008). Deep structure and tectonic evolution of the South Caspian Depression. In A. V. Mamedov, A. A. Alizadeh et al. (Ed.), *Geology of Azerbaijan, oil and gas* (Vol. VII, pp. 6–158). Baku: Nafta-Press (in Russian).

Mamedov, P. Z. (2010). Modern architecture of the South-Caspian Megabasin as result of multistage lithosphere evolution in the central segment of the Alpine-Himalayan mobile belt. *Izvestiya of USSR Academy of Sciences, Series: Earth Sciences, 4*, 46–72 (in Russian).

National Atlas of Azerbaijan Republic. (2014). Baku: Baku Cartography Factory.

Panov, D. I. (2002). Tectonic structure of the Jurassic terrigenous complex of the Greater Caucasus. Mechanism and time of its forming. In *Collection of papers of the North Caucasian State Technical University, Series: Tectonics and Geodynamics* (Vol. 1, pp. 60–70). Stavropol (in Russian).

Priestley, K., Baker, C., & Jackson, J. (1994). Implication of earthquake focal mechanism data for the active tectonics of the South Caspian basin and surrounding regions. *Geophysical Journal International, 118*, 111–141.

Romanov, N. T. (1983). New data on the Main Caucasus fault in the Eastern Caucasus. *Doklady Academy of Sciience USSR, 272*(5), 1203–1205 (in Russian).

Rustamov, M. I. (1995). Collision-Riftogenic magmatism of palaeogene of the central segment of the mediterranean belt. In *Proceedings of the Russian Petrography Meeting on Magmatism and Geodynamics*, Ufa, pp. 179–181 (in Russian).

Rustamov, M. I. (2005). *South Caspian Basin—geodynamic events and processes*. Baku: Nafta-Press (in Russian).

Shardanov, A. N. (1953). On the problem of effects of nappe tectonics in the South-Eastern Caucasus. *Doklady Academy of Science Azerbaijan, 9*, 439–444 (in Russian).

Shardanov, A. N., & Molchanov, M. F. (1954). Astrakhan tectonic Nappe in the South-Eastern Caucasus. *Izvestiya, Academy of Science Azerbaijan*, *11*, 39–49 (in Russian).

Shekinsky, E. M., Isayev, B. M., Alizadeh, S. A., & Kangarli, T. N. (1985). Chaotic complexes of the Northern slope of the South-Eastern Caucasus. *Geotektonika*, *4*, 69–77 (in Russian).

Shempelev, A. G. (1978). On a deep stating of the Main Caucasian thrust. *Geotektonika, 6*, 57–65 (in Russian).

Shikhalibeyli, E. S. (1956). Tectonics of Sevan-Akera Synclinorium within Azerbaijan territory. In *Transactions of the Tectonic Symposium on Alpine Geosynclinal Zone of the South of the USSR. Azerbaijan Academy of Science*, Baku (in Russian).

Shikhalibeyli, E. S. (1962). Geology of Sevan-Akera Zone (basin of the middle flow of Hakeri River). In *Proceedings of the Geology Institute, Azerbaijan Academy of Science* (Vol. XXII). Baku (in Russian).

Shikhalibeyli, E. S. (1964). *Geological structure and history of tectonic evolution of the Eastern part of the Lesser Caucasus* (Vol. 1). *Stratigraphy of Meso-Cenozoic deposits. Azerbaijan Academy of Science*. Baku (in Russian).

Shikhalibeyli, E. S. (1966). *Geological structure and history of tectonic evolution of the Eastern part of the Lesser Caucasus* (Vol. 2), *Tectonics and Magmatism. Azerbaijan Academy of Science*. Baku (in Russian).

Shikhalibeyli, E. S., Hasanov, A. G., Taghiyev, R. E., & Metaxa, X. P. (1984). On the problem of structure and forming Southern Caspian Sea (by modern data). In *Sketch on geology of Azerbaijan* (pp. 61–84). Baku: Azerneshr (in Russian).

Sokolov, S. D. (1974). Upper Jurassic Olisostromic stratum of the South-Eastern part of the Sevan-Akera ophiolithic zone. *Bullentin Moscow Association of Naturalist Investment, Series Geology, 49* (3), 33–48 (in Russian).

Sokolov, S. D. (1977). Olisostromic layers and ophiolithic nappes of the Lesser Caucasus. In *Proceedings of the Geology Institute, Academy of Sci. USSR* (Vol. 296). Moscow: Nauka (in Russian).

Solovkin, A. N. (1945). On the Basgal nappe and facial-tectonic regioning. *Izvestiya, Academy of Science USSR, 7*, 84–94 (in Russian).

Tzimelzon, I. O. (1965). Deep structure of Earth's crust and tectonics of Azerbaijan by data of geophysical investigations. *Soviet Geology*, *4*, 103–111 (in Russian).

Vassoyevich, N. B. (1938). New data on stratigraphy of upper Jurassic and Neocomian of the North-Eastern Azerbaijan. *Doklady Academy of Science USSR, New Series, 21*(3), 135–140 (in Russian).

Vassoyevich, N. B., & Khain, V. E. (1940). Phenomena of nappe tectonics in Lahyj Mts. *Izvestiya, Academy of Science USSR, Series Geology, 1*, 76–80 (in Russian).

Voskresenskiy, I. A. (1958). On the Basgal Nappe in the South-Eastern Caucasus. *Soviet Geology, 7*, 62–84.

Voskresenskiy, I. A., Khain, V. E., & Shurygin, A. M. (1963). Tectonic Nappes of the South-Eastern Caucasus and conditions of their forming. *Herald of the Moscow State University, 4*, 15–33 (in Russian).

6 Neotectonics

Research in the field of neotectonics of Azerbaijan covers a period of almost 70 years, since Obruchev (1948) first proposed the concept of "neotectonics" as reflecting "structures of the Earth's crust formed by the most recent movements". He related to "neotectonics" all movements of the Neogene–Quaternary periods causing the formation of the various present-day tectonic structures. Subsequent studies, particularly in the Alpine fold belt, showed that the neotectonic processes played a key role in the formation of both the present geological structures and the terrain relief.

Research in the field of neotectonics in Azerbaijan is carried out on the basis of integrated analysis of geotectonic, geomorphological, geophysical (first of all, seismological) and geodesic data, and later aerospace materials as well. This analysis has revealed regional neotectonic patterns in the Greater and Lesser Caucasus, Kur Depression, Talysh and Nakhchivan, and has allowed scientists to trace the development of organic processes in the neotectonic (late orogenic) Alpine stage. In addition, a number of correlations have been established between the recent geological structural and topographical features, and numerous local folds and rupture dislocations and abrupt changes in the folded structures orientation.

The initial border of the neotectonic stage, as it is known, has different time definitions for various regions within the Alpine-Himalayan mobile belt (Caucasus, Central Asia, Himalayas, etc.). Such "sliding" of the time boundaries is likely associated with the concrete geological evolution of each region (area) under study (Aliyev et al. 2005).

After the first 15 years of research, the initial border for the territory of Azerbaijan was accepted as the Middle and Upper Pliocene (i.e. the time boundaries of the neotectonic stage in Azerbaijan correspond to Middle (Late) Pliocene–Quaternary). The following combined criteria were taken into account: sharp change of paleogeographical conditions, increased plunge intensity in depression zones and uplifts within the Greater and Lesser Caucasus; strengthening of denudation processes and accumulations of thick strata of sedimentary rocks (mainly of coarse-grained material); change of orientation of the Upper Pliocene–Quaternary folding series (mainly to the east of the West Caspian deep fault); activation of mud volcanoes and magmatic volcanism (Lesser Caucasus); and presence of sharp angular unconformities between the Upper Pliocene and underlying deposits. On this basis, the first neotectonic map of Azerbaijan (scale 1:600,000) was composed (Azizbekov et al. 1972). This map included known for that time the main peculiarities of the newest and modern geological structures.

Recent researches enabled to revise a time of the newest and modern tectonics of Azerbaijan. Due to novel data obtained, the initial border of the neotectonic stage was reconsidered and redefined as the end of the Late Miocene; correspondingly, this stage covers the entire Pliocene–Quaternary interval. This definition was based on well-defined borders of mountainous and depression zones, sharp activation of tectonic movements and other geological phenomena since the Miocene. The major peculiarities of the neotectonic processes, including vertical movements, the newest folded-ruptured deformation and recent horizontal movements, are presented in the corresponding sections of Chap. 5 "Tectonics" of this Volume.

6.1 Vertical Movements of the Neotectonic Stage

A considerable volume of new geological-geophysical, geomorphological and satellite data have enabled to differentiate both recent and early-formed first-order tectonic elements of the vertical movement development at the newest and recent stages. All these data were generalized on a new neotectonic map of Azerbaijan (scale 1:500,000; Akhmedbeyli et al. 1991).

More than 40 large first-order tectonic structures were delineated in the Azerbaijan territory, with the most

A.A. Alizadeh et al., *Geosciences of Azerbaijan*, Regional Geology Reviews,
DOI: 10.1007/978-3-319-27395-2_6

concentration in mountainous zones of the Greater and Lesser Caucasus, Talysh and Nakhchivan, and partially in the depression zones. The internal structures of these constructions and their conjugation zones are sufficiently complext. They differ in the composition of mountain rocks forming the series, the thickness of sedimentary deposits, by lithological characteristics of the volcanogenic and volcanogenic-sedimentary rocks, as well as other geological features.

6.1.1 The Greater Caucasus

Existing tectonic zonation indicates the occurrence of Tufan (Tfan), Tengi-Beshbarmag and Vandam structural uplifts; Shakhdagh-Khizi, Zagatala-Govdagh and Shamakhy-Gobustan troughs; and Gusar-Shabran (Devechi) superimposed depression in the eastern part of the Greater Caucasus (Azerbaijan). The Devechi depression is located in the interface of the Alpine fold area with the Epihercynian platform. According to the manifestation regime of the newest and modern tectonic movements, the mentioned structures are developed by different ways (Akhmedbeyli 1966).

The *Tufan uplift is* represented by a neotectonic zone of stable intensive roof block uplifts. Research carried out by Shikhalibeyli (1996) substantiated inherited development of this structure—the pre-Alpine basement occurring below it is represented possibly by the Hercynian complex which is significantly lifted and located at depths of 4–5 km. The block nature of the pre-modern structural plan of the pre-Alpine basement according to recently geophysical data for the Azerbaijanian part of the Greater Caucasus is connected with deep faults of the common Caucasian and transverse direction (Shakhdagh-Hermian, Siyazan, Malkamud Hamarvan, Altyagach, Zangin, etc.). Numerous earthquake epicenters are connected with these faults and this connectivity is supported by their activity at a modern stage.

Amplitudes of the Tufan structure uplift at the neotectonic stage (along the axial plane) is 4500 m and more, taking into account the subsequent erosion processes, and decreases on slopes up to 3000–3200 m. This structure is complicated by regional fractures dividing numerous local structures and disjunctives. At the modern stage, the tectonic movement rate increase. According to the instrumental observations, the rates of uplifts consist now more than 8 mm/year in the arch-like area and about 6–8 mm/year on slopes.

The *Tengi-Beshbarmag structure* corresponds to the neotectonic zone in the eastern area of the Greater Caucasus which differs from other structures by a particular regime. At the neotectonic stage, this structure was subjected to moderate differentiated raising with further transformation of the movement direction—vertical and horizontal. On the basis of paleotectonic reconstructions, it was defined that the rise existed as a narrow and extended ridge for a long period of time, hypothetically in the Late Cretaceous or Early Cenozoic. The studied structure is bounded from the north to south by deep faults—Siyazan and Garanour. Its northwestern ending is fixed on the junction of the above-mentioned faults, continues to the Mt. Shakhdagh area, and here is referred to as the Gryz overthrust.

According to Kangarli (1978, 1999), the Tengi-Beshbarmag structure cannot be related to anticlinal type. The Tengi-Beshbarmag, Budug, Gyzyl-gaya and Shakhdagh calciferous massifs are spatially connected and form a common structural zone along the strike. Fault planes (Siyazan and Garanour) bounding the Tengi-Beshbarmag zone, fall beneath it and form structural complications in the overlapping zone of the northern flanks of the Shakhdagh-Khizi tectonic bench by the southern slope of the Sudur bench.

From this point of view, Kangarli (1999) emphasized an important role of the Siyazan deep fault as a zone with the greatest tensions. High activity of this fault remains and at a neotectonic stage. On the basis of instrumental and macroseismic data examination, a number of earthquake epicenters (with different focal depths) are registered along the fault, including such strong earthquakes as the Gyzyl-Burun (Siyazan) (6.0 M, July 25, 1931) and Siyazan (7.0 M, March 26, 1962).

The newest uplifts within the Tengi-Beshbarmag neotectonic zone have amplitudes of 800–1200 m. Modern tectonic movements (vertically oriented) are continuing at a rate of 1–3 mm/year.

The *Vandam structure*, by its regime of neotectonic movements, is characterized by manifestation of stable, moderate and differentiated uplifts. On the basis of regional and tectonic research (Kangarli and Akhundov 1988) and geological mapping, it was established that the structure joint undulates on all its extension.

The more raised parts of the structure occur along the axis between the Gabala town and the Girdymanchay River (eastern part) and between the Sheki town and the Mazymchay River (western part). In the western area, Bajocian volcanogenic formations are overlapped by Valanginian and Tithonian formations, but in the eastern part, they occur lower and overlapped by Cretaceous formations. This is why the Vandam structure is thought to represent relatively early tectonic elements of the eastern area of the Greater Caucasus (Akhmedbeyli 1966). According to the available data on the area of this structure, a considerably erected uplift existed in pre-modern times. In the intermediate zone, from Sheki to Gabala, where the structure joint subsides, Mesozoic deposits (Bajocian and Cretaceous volcanic associations) are overlapped by coarse Upper Pliocene–Quaternary deposits. As it is seen, neotectonic movements occurred unsteadily along the Vandam structure. Maximal values of uplift amplitudes are 1600–2000 m in east and west, and in the zone between them, 600–800 m, and sometimes, 1200 m. Recent tectonic movements (uplifts) occur relatively steadily along this structure, reaching 2–5 mm/year.

The *Shakhdagh–Khizi and Shamakhy-Gobustan* troughs, according to the regime of neotectonic movements, belong to the inverted type of structures which are delineated in the neotectonic map of Azerbaijan (Akhmedbeyli et al. 1991). The following investigations determined that these structural elements differ from one another both by the modern structure and on the regime of their development. The Shakhdagh-Khizi trough is narrow structure (10–15 km in width), extends from the NW to SE (140–150 km) and is bounded by deep faults: in the north, by the Garabulag-Gryz (Gazma-Gryz), and in the south, by the Shakhdagh-Hermian. According to the available data (Geomorphology of Azerbaijan 1959; Azizbekov et al. 1972), this structure is divided into Shakhdagh and Khizi segments. However, even taking into account the different morphologies of some tectonic elements, such division is an unconvincing one. At the neotectonic stage, the Shakhdagh-Khizi structure developed in a regime of differentiated uplifts and their amplitudes reduced from the NW to SE.

Another situation is observed for the Shamakhy-Gobustan trough, where thick Cenozoic deposits and mud volcanoes are widespread. Numerous local folds of different sizes expressed in differently aged layers of these deposits are developed in its eastern part; they sharply change their direction from sublatitudinal to near-meridional. The structure has the form of a semi-oval with transverse sizes in the WNW of 6–8 km and in the ESE of nearly 30 km.

Its northern border coincides with the Gozluchay fault, and the southwestern border—with the Ajichay-Alyat fault, and the conjugation zone—with the Absheron periclinal trough passes on the Yashmin flexure. The essential peculiarity of the Shamakhy-Gobustan trough is the Pirekishkul-Basgal thrust block overlapping most of its northern part (Govdag-Sumgayit allochthonous complex). The peculiarities of the block are exhaustively covered in literature (e.g., Vassoevich and Khain 1940; Alizadeh 1945; Voskresensky et al. 1963). In central and southern areas, Cenozoic beds have a complex dislocation due to a wide development of regional and local thrusts, overthrusts and tectonic scales. The newest sedimentary complex (Pliocene–Quaternary) forms a number of particular thrusts (synclines) —Zogalovaichay, Maraza (Gobustan), and Jeirankechmez—divided by transverse uplifts which are geomorphologically represented by ridges, plateaus and highs.

From the above-mentioned follows that neotectonic movements occur in time and space differentially and this notion can be supported by numerous fractures, activities of mud volcanoes, sharp changes of modern deposit thicknesses and directions of folded structures. Taking into account peculiarities of tectonic deformations of recent (Pliocene–Quaternary) deposits, there is no doubt that compression stress plays a principal role in the neotectonic processes. Amplitudes of the recent uplifts in the western area of the Shamakhy-Gobustan structure do not exceed 1200 m (along the frontal zone of the Pirekishkul-Basgal cover). The rest of the eastern area was subjected to very differentiate dsubsidence: from 400 to 2400 m.

So, the highland area of the Greater Caucasus, in a new tectonic aspect, possesses all the peculiarities typical for the mountain belt (especially of young age): linear form, sedimentation and structural zonation, complex inner structure accompanied by large thrusts, well-defined structural-formational complexes, crystal basement of continental origin and active seismicity in marginal parts of the mountain zone.

6.1.2 The Lesser Caucasus

Compared with the Greater Caucasus, within the Lesser Caucasus, neotectonic movements occurred more differentially and, that is seen well, in its surface structure and terrain relief. Ridges with different orientation correspond to large (first order) positive tectonic elements, but mountainous plateaus and volcanic uplands of different forms correspond to troughs.

Another distinctive feature is the fact that amplitudes of total uplifts in the Lesser Caucasus at the newest and modern stages are lesser than in the Greater Caucasus. In a historical-geological aspect, intensive volcanism of Jurassic, Cretaceous and Paleogene periods covers a considerable area with displacement of activity from the central zone of this megastructure to the southwest. Volcanic activity have continued and at a late orogenic stage, i.e. at the neotectonic stage (Early Pliocene–Quaternary).

One of the typical properties of the Lesser Caucasus megastructure is a complex mosaic morphostructural chart. According to the pattern and morphology of tectonic structures, rock composition, rate direction and amplitudes of newest and modern movement and erosion processes occurred irregularly whereby the delineated morphostructures have mainly small sizes, various forms and orientations (Geomorphology of Azerbaijan 1959; Azizbekov et al. 1972). Similar elements of the Greater Caucasus are much larger and extended conformably to main structures including the Shamakhy-Gobustan trough.

According to the tectonic zonation of the Azerbaijan territory, a number of separate structures are distinguished within the megastructure of the Lesser Caucasus: the Murovdagh, Garabagh, Agdam, Shamkir, Zamzur, Sharur-Julfa, Zangazur structural uplifts, the Dashkesan, Agjakend, Tourgaichay, Agdere, Khojavand and Gazakh troughs, and the Kalbajar superimposed trough. According to the regime of the newest and modern tectonic movement manifestation, all of the above-mentioned structures correspond to definite neotectonic zones. The Murovdagh, Garabagh and Zangazur structures belong to zones subjected to stable, intensive roof block inherited uplifts at the newest and modern stages.

The *Murovdagh structure* has a complex architecture. For the segments which occur to the east and west of the top of the Gyamysh mountain (fold-natured, conjugation of different age series, echelon shifting of thrusts, etc.), it is typically diverse. That is why the process of uprising occurred unevenly at the newest and modern stages, more intensive along the Murovdagh thrust, which corresponds to the suture line of deep occurrence. Maximum amplitudes of the uplifts reach 2800 m (without the denudation section).

The *Garabagh structure* can be characterized by a complex internal structure, and its compound structural elements are complicated by thrusts and numerous overthrusts, frequently of transverse orientation. The Dalidagh-Mekhman transverse uplift and Gargarchay transverse trough are also identified along the structure. These and other peculiarities of this structure played a significant role in differentiated manifestation of tectonic movements at the newest and modern stages. Amplitudes of uplifts are from 2400 to 800 m at different areas, decreasing from NW to SE.

The *Zangazur uplift* has a complex internal structure which developed at a neotectonic stage in the same regime as the two previous zones. This structure occurs between two deep faults (Girratag and Nyuvedi-Debaklin) and its rocks complex is broken by the Megri-Ordubad batholith with a granitoid composition. Based on the available data (Azizbekov et al. 1972), the batholith age is defined as post-Upper Eocene–pre-Miocene. This period is considered as the beginning of the early orogenic processes for all of the Azerbaijan territory. This information suggests that the Zangazur structure has risen intensively at the neotectonic stage, as evidenced by uplift amplitudes reaching 3600 m.

The *Agdam uplift* was formed in another regime of the newest tectonic motions which relates to the neotectonic zone of the moderate, differentiated uplifts changed to the end of the appropriate stage by partial subsidence of the internal troughs (an analogous regime has been determined for the Touragaychay, Agdara and Khojavand troughs). The amplitudes of the newest and modern uplifts, not only of the Agdam uplift, but also other uplifts located along a boundary with the Kur Depression (Pre-Lesser Caucasus marginal trough), do not exceed 400–800 m.

The *Shamkir uplift* represents the most ancient large structure of the Lesser Caucasus with a folded-blocky internal structure. There are widely developed small folding, transversely orientated anticlines and intermediate synclines, and regional rupture dislocations. Important peculiarities of this structure are the protrusions of granitoid intrusive rock. The newest tectonic movements sufficiently manifested differentially and, in contrast to the Murovdagh structure, inherited moderately from the ancient folded basement. The fragments of the peneplain of the Early Quaternary age preserved at different heights indicate various amplitudes of uplifts from 400 to 1200 m, and rarely up to 1400–1500 m.

The neotectonic processes within the *Dashkasan* and *Agjakend* troughs are identical. Both structures are zones of moderate differentiated uplifts arising at the place of pre-newest ones. Akhmedbeyli et al. (1991) concluded that a partial subsidence was occurred at the end of the neotectonic stage. However, critical analysis of geological and geomorphological data leads to a somewhat different conclusion. The subsidence has likely not occurred at this stage, but a delay of uplift rates within the structures is evidenced by analysis of different age rock outcrops. Taking into consideration the pre-newest structural peculiarities, one can assume that the separation of the Dashkasan and Agjakend troughs as independent structures is somewhat artificial. According to Shikhalibeyli (1956, 1996), Cretaceous deposits of the Agjakend structure overlap Middle–Upper Jurassic formations of Dashkasan having a superimposed nature. The uplift amplitudes of the newest stage increase from the foothills to the internal parts of the Lesser Caucasus, from 200–400 m to 1200–1600 m.

The *Gochaz trough*, according to its regime of neotectonic movement manifestation, relates to a zone of moderately differentiated uplifts accompanied by volcanism. The uplift amplitudes are about 400–800 m. As for the *Kalbajar superimposed trough*, it significantly differs from many structures of the Lesser Caucasus. The beginning of the neotectonic stage is characterized by the manifestation of intensive sagging sufficiently differentiated and inherited from the Post-Cretaceous intensive movements. Towards the end of the stage, sagging has changed by raising, and at the same time, volcanism (surface) has actively manifested, forming rhyolite-dacite and andesite-basalt subformations. An important role in the formation of the Kalbajar trough played the transversely oriented faults and deep faults which were sufficiently active and at the neotectonic stage. In the Pliocene–Quaternary, the uplift amplitudes were 2000–2800 m. The maximum velocity values of the recent vertical movements, 8–10 mm/year, were fixed within the Murovdagh uplift and Kalbajar trough but in the northeastern and eastern subsidence, they decreased up to 2 mm/year.

These data attest to the fact that the observed geological structure and relief of the mountain structures of the Greater and Lesser Caucasus, as large tectonic and geomorphological elements, were formed in conditions of absolute (stable, intensive high amplitudes) and relative (stable, less intensive and small amplitudes) uplifts.

6.1.3 Talysh Region

According to the tectonic zoning of Azerbaijan (Azizbekov et al. 1972; Alizadeh et al. 1980), the Astara and Burovar uplifts and the Lerik and Yardimli troughs are distinguished in the Talysh zone, and the Pre-Talysh marginal trough is identified along the northeastern and eastern foothills. Paleogene volcanogenic and volcanogenic-sedimentary rocks participated in the formation of these structures. This entire zone has been divided by transverse faults; therefore, all of these structures have been divided into blocks displaced relative to one another.

According to the regime of the newest tectonic movement manifestation, almost all of the structures are similar: Astara, Burovar and Lerik have had moderate differentiated uplifts. The Yardimli structure is characterized by similar uplift occurring in the site of moderate downwarping. The exception is the Pre-Talysh trough, which sagged steadily during the neotectonic stage, with weak uplifts occurred only at the end of this stage. The amplitudes of the newest uplifts increase appropriately from the Pre-Talysh fault (300–400 m) in a direction towards the mountainous part (1600–2000 m). The velocities of recent movement vary in the same direction from 1 to 6 mm/year.

6.1.4 Depression Zones

The *Gusar-Shabran (Devechi)* is the northernmost superimposed depression, the specificity of which is in its tectonic position, in the area of Alpine folding and epi-Hercynian platform conjugation. According to earlier publications (Akhmedbeyli 1954, 1955), it was considered to have had stable subsidence accompanied by folding at the neotectonic stage. Moreover, partial structures of this depression—the Gusar buried uplift, Zeykhur-Shabran (Devechi) syncline and Guba marginal trough—were defined as zones of moderate subsidence changed by uplifts at the end of the latest stage. The Yalama-Khudat buried uplift is slightly different from them.

More recent geological and geophysical data allow to reconsider of the cardinal manifestations of the newest tectonic movements. The Gusar-Shabran (Devechi) trough at the neotectonic stage was developed in a regime of stable, irregular sagging. The Zeykhur-Shabran (Devechi) and Guba marginal troughs subsided continuously due to an accumulation of thick strata of Pliocene–Quaternary deposits, about 4600 and 4300 m, respectively. Within the Zeykhur-Shabran trough, there was no change in the direction of tectonic movement at the end of the newest stage, and some relief uplift arose due to sedimentation overcompensation. A similar situation is revealed for the zone of the Gusar buried uplift, but with lower subsidence values, not exceeding 1700–1800 m. The zone of the Yalama-Khudat buried uplift in Pliocene–Quaternary time subsides steadily without changing the tectonic movement sign. Such differentiated manifestation of the neotectonic movements in the Gusar-Shabran (Devechi) trough is associated with the periodic activation of the northern Azerbaijan, southern Gusar and Siyazan deep faults.

In the Gusar-Shabran (Devechi) trough, is suggested that folding is not expressed in Pliocene–Quaternary strata, but the layers structurally form a sufficiently gentle monocline with northeast dipping. Geophysical data obtained during recent years indicate that the thickness of all the newest deposits is significantly complicated by the fold-disjunctive deformations. Moreover, a number of regional ruptures—not only longitudinal but also transverse (sub-meridional, NE–SW directions)—have been determined, some of which are continuations of deep faults crossing the sections of Miocene–Paleogene and Mesozoic strata. The Pliocene and Lower Quaternary layers were divided into numerous tectonic blocks of different sizes (mostly small) and forms; there were defined local anticlinal folds as well. Such a structure is typical for the southern and central parts of the Gusar-Shabran (Devechi) trough; at the same time, the intensity of deformations is considerably weakened in the coastal area.

The Kur Depression, or intermountain depression, is the largest structure of this type in the region under study. Its internal structure, geotectonic development, conjugation with the Greater and Lesser Caucasus, deep structure and other problems have been investigated by many geologists and geophysicists (Khain and Shardanov 1952; Shikhalibeyli et al. 1984; Khain and Shikhalibeyli 2000; Aliyev et al. 2005; Alizadeh 2012; Eppelbaum and Khesin 2012). Existing information has been considerably enriched by geophysical studies and deep drilling, especially the Saatly superdeep borehole SD-1.

The Kur Depression was developed in a regime of stable subsidence at the neotectonic stage. These movements occurred unevenly, as evidenced by significant changes in Pliocene–Quaternary sediment thickness. Analysis of these thicknesses has shown subsidence amplitudes of 400 m in the western part of the Kur Depression. In its central part, the newest sediment complex to a small degree is affected by folding and fractures. In the Lower Kur Depression, the folded structures of sub-meridional direction, grouped into linear anticline zones and complicated by local fractures, were formed towards the end of the neotectonic stage.

The Lower Kur Depression is characterized by the development of mud volcanism that is almost absent in the central and western parts of this depression. There are certain differences in morphostructural zonation. The Central Kur segment of the depression is flat, low-inclined alluvial plain reflecting the structure of Quaternary and Pliocene strata. The combination of various morphostructures is typical for the eastern and western parts of the depression. Each of the mentioned Kur Depression segments includes individual structures, the largest of which—the Yevlakh-Agjabedi trough and Talysh-Kurdamir buried protrusion—are located in the central part of the depression.

The folded zones of the interfluve area of the Kur-Gabyrry, Ajinour and Lyangyabiz-Alat ridges are located along the northern flank of the Kur Depression. They are similar according to the regime of the newest tectonic movement manifestations but differ slightly in intensity of the movements. The Pre-Lesser Caucasus trough is similar to the above-mentioned structures, but with moderate movement manifestation stretches on the southern flank along the Lesser Caucasus foothills.

The boundaries between the Greater Caucasus and the Kur Depression and between the last one and the Lesser Caucasus are marked out definitely at the neotectonic stage. Concerning their position and origin time, the existing views are at variance. The viewpoint of some authors is that these borders go along the foot of the mountain ranges. According to other scientists, the borders coincide with the southern margin of the folded zones of the interfluve area of the Kur-Gabyrry, Ajinour and Lyangyabiz ridge in the north and with the external margin of the Pre-Lesser Caucasus trough (monocline) in the south.

In the territory of Nakhchivan [Nakhchivan Autonomic Republic (AR)], neotectonic processes were complex and differentiated ones. They were preceded by active Late Eocene intrusive, subvolcanic and volcanic activity accompanied by the formation of the granitoid andesites, andesite-dacites, rhyolites and other intermediate effusives in the Lesser Caucasus.

The Nakhchivan territory at the neotectonic stage was developed in the regime of stable, differentiated subsidence with accumulation of normal sedimentary formations occurring against the background of roof-block inherited uplifts. Among the morphostructures reflecting the character of neotectonic process manifestations are Zangazur, a long-continued, intensive roof-block; Sharur-Julfa, a moderately differentiated uplift; and the Nakhchivan superimposed trough, weak subsidence with periodic relative raising. At the recent stage, the Nakhchivan depression is characterized as an area of weak (0–2 mm/year) subsidence, but the mountainous part located to the northeast is characterized as an area of moderate rise (2–8 mm/year).

It is important to note that many of the latest tectonic elements of the Azerbaijan territory are developed inheriting the early (pre-neotectonic) structural chart. At the same time, a significant distribution of young structures expressed in the thickness of the Pliocene–Quaternary deposits by small folding and a large quantity of local rupture dislocations have been determined. In the eastern areas, the young folding relative to the earlier structural chart changes the strike sharply to the NW–SE and even the sub-meridional one.

6.2 The Newest Fold-Thrust Deformations

It should be emphasized that neotectonic investigations were carried out for many years from the perspective of the decisive role of vertically directed tectonic movements. The role of horizontal tectonic stresses has scarcely been taken into account. However, many large and small deformations (sheets, tectonic overlappings, shifts and scaly structures) are associated with horizontal movements. This knowledge gap has been filled to some degree by the recent investigations, the results of which are presented below.

According to previous data (Azizbekov et al. 1972; Aliyev et al. 2005) from the western borders of Azerbaijan to the West Caspian deep fault at the Greater Caucasus, the steeply dipping rupture dislocations, faults, overfaults (and to the east, low-angle overthrusts) in the young geological formations, have been developed. This may be caused by active movements along this fault not only at the pre-newest stage but also at the neotectonic stages. The subsequent detailed geological analysis showed that numerous faults and overfaults of the southern slope and watershed area of the Greater Caucasus present the observed featherings of large regional thrusts on the surface that are traced along the borders of the main structural elements. The thrusts planes have north (northeast) dipping, and with depth (near the "granite" layer occurring), their steepness increases almost up to vertical in the sedimentary stratum.

Active motions along thrusts have caused a significant reduction of the lateral dimensions of Zagatala-Govdagh and Vandam structures that is clearly fixed from the meridian of Gabala town to the west–northwest direction. Along the frontal parts of the powerful thrusts it has been determined a presence of numerous isoclinal-scaly folds overturned to the south in the Middle Jurassic, Lower Cretaceous, and in the eastern areas, in the Upper Cretaceous deposits.

The cover (nappe) elements are evidence of intensive horizontal stresses. The known Basgal nappe (Vassoevich and Khain 1940; Voskresensky 1958; Voskresensky et al. 1963) has been identified as a small fragment of such a structure. The subsequent investigations have shown that overthrusting or folding-overlapping of structures is widespread in the eastern part of the Greater Caucasus (for example, the Basgal-Pirekishkul structure). This structure occurs in the form of individual scales (preserved from the subsequent washout) extending almost to the valley meridian of the Goychay River (Kangarli 1978, 1999). Tectonic overlap has considerable amplitudes to the west of the Zagatala city.

It should be noted that the formation of the Basgal-Pirekishkul overthrust began in the early period of the neotectonic stage. According to Kangarli (1978, 1999), in the eastern (Azerbaijan) part of the Greater Caucasus, the processes of folded–sheet complexes generation were gradually manifested and earlier—in the Middle Cretaceous, Late Cretaceous–Paleocene and Miocene.

The intensity of the horizontally oriented tectonic movements has clearly manifested in the piedmont region of the southern slope of the Greater Caucasus including the Ganykh-Ayrichay young trough and the folded zones of Adjinour and the Kur and Gabyrry riversinterfluve. This zone differs from the high mountain part by presence of scaly-thrust structures expressed in the Pliocene–Quaternary series of the deposits.

Two pairs of anticlinal lines (Akhmedbeyli 1962) have been determined in the Adjinour zone: in the north, the Dashyuz-Amirvan and Gemigaya and in the south, the Surkhaykhan and Godjashen-Goychay. They are quite narrow and elongated structures with sublatitudinal orientation. Their internal structure is complicated by numerous anticlinal folds often located shingly and by sheared (along the axis) local fractures–overthrusts. The small folding manifested in clay strata of Upper Pliocene–Lower Quaternary is widely developed.

Over a large longitudinal fault extending along the Dashyuz-Amirvan anticlinal line (Ganykh-Alnjanchay rivers interfluve), its north wing (Akchagyl strata) has overthrusted onto the southern and even overlaps Absheron deposits of the north wing of the adjacent Gemigya anticlinal line from the south. The thrust amplitudes are about 1.5–2 km, and rarely up to 3–5.5 km. In turn, the Gemigaya structure is also complicated by longitudinal thrust on which the northern wing overlaps completely the south one. It is clear that the northern pair of the anticlinal lines of Ajinohur–Dashyuz-Amirvan and Gemigaya in the modern geological structure represent a large tectonic scale formed at the neotectonic stage.

A similar situation has been determined for the Surkhaykhan and Gojashen-Goychay anticlinal lines that formed the southern pair of structures. The structures contact on thrust fragments which can be traced to the south of the Surkhaykhan mountain peak and also on the valleys of the Goychay and Turianchay rivers. The amplitude of thrust along which the northern slope of the Surkhaykhan structure overlaps its southern slope and part of the northern slope of the Goychay anticline reaches 2.0–2.5 km.

The folded zone of the interfluve of the Kur and Gabyrry rivers, including the Jeyranchel-Palantekyan trough and Chatma uplift, has a structure similar to that of the Ajinohur folded zone. Here, three anticlinal lines are distinguished (Akhmedbeyli 1962), composed of anticlinal folds expressed in Pliocene–Quaternary formations. The northern part of the zone (Chatmin structure) is the exception, where the Maykop and Middle Pliocene layers on the surface compose the cores of a large anticline. The vast majority of anticlinal folds and anticlinal lines are complicated by overthrusts-thrust faults, which are southwardly directed and have amplitudes from 100–150 up to 1000–1500 m.

It is necessary to note that the folded zone of the Kur-Gabyrry interfluve relative to the same Adjinour zone is significantly displaced southward (along the meridional segment of the Ganykh River valley). Due to this, the most northern—Chatmin structure of the first zone and the Gojashen-Goycha of the second zone—are situated along the same narrow area and are similar both tectonically and in their relief features. Such displacement (shift) is distinctly fixed on the geological and tectonic maps of Azerbaijan and coincides with a large Goygel-(Khanlar)-Khachbulag-Ganykh fault of a NE–SW direction (near-meridional). This fault was delineated by integrated analysis of geological-geophysical data (Shikhalibeyli et al. 1984). According to the author of this work, "within the Kur Depression this fault zone can contribute to the formation of the linear ledge of west margin of the Mingechavir water reservoir where laterally oriented Ganykh River bed (Alazani), knee-shaped curving in the lower course takes the meridional direction".

By the structural pattern and development degree of the folded-disjunctive deformations in the Pliocene–Quaternary series of deposits, the studied folded zones (extension about 240 km) markedly differ from other parts of the Kur Depression (west and central parts). It is well known that within the depression, the tectonics of young (newest) molasse formations is quiet, i.e. plicated and disjunctive dislocations are practically absent.

According to the above-mentioned, we can conclude that the intense folding and overthrusts in the Adjinour and Kur-Gabyrry areas associated with horizontal compression were formed at the neotectonic stage. The role of large regional overthrusts, the Vandam, Ganykh-Ayrichay and North Kur, is significant in the development of the geological structure of this region. The seismicity data of the south slope of the Greater Caucasus and piedmont area (seismostatistic and macroseismic data) show their activity at the neotectonic stage. It should be noted that the formation and the following development of sheet-overthrust structures and tectonic scales have been occurred at different stages of the Alpine tectogenesis cycle. From the nature of conjugation of large tectonic structures is evident that each north structure overlaps on a considerable extension of adjacent south structure that indicates the displacement of horizontal compression activity in the direction from north to south.

In the Lesser Caucasus, the overthrusts and scaly structures are not as well defined as in the Greater Caucasus. According to Shikhalibeyli (1996), a large Murovdagh overthrust is distinguished incoming to the system of Goycha-Garabagh fault zones. The overthrust extends along the border of the Murovdagh uplift and Toragaychay trough. The Murovdagh structure has a complex, asymmetrical structure and is overturned to the south; its amplitude consists of 4.5–5.0 km. At the area where the overthrust cuts the axial zone of structure (Shamkirchay River—the top of the Gyamysh mountain), the southwest wing of the last is sunken and covered by Cretaceous and Eocene sediments. A large scaly-thrust structure is established by Shikhalibeyli (1996) within the Toragaychay trough which complicated the structure of the Benevshelin anticline which is isoclinal compressed and southward thrusted. Tithonian-Valanginian deposits of the fold core overlie the Lower Cenonian layers of the southern wing of the adjacent syncline from the south. The overthrust extension is 20 km.

The Garabagh uplift on the southwest borders the Sarybaba trough along the deep fault expressed by overthrust where the Bathonian sediments of the uplift are thrust over the Upper Cretaceous and over the Eocene of the trough. In addition, ophiolite cover determines the internal structure of the entire Goycha-Hakeri zone. Thrust structures are also revealed within the Shakhdagh synclinal area, along the northeast wing of the Gochaz trough, and tectonic scales occur on the southeastern periclinal subsidence of the Lachin uplift in Upper Jurassic deposits. The thrust system of longitudinal and cross orientation was distinguished by Azizbekov and Shikhalibeyli (1966) in the zone between the Sharur-Julfa uplift and the Ordubad trough.

In the Lesser Caucasus, the fold-thrust scaly shapes of deformations are irregularly distributed, complicating the internal structure of the separate parts of large structural elements or zones of their conjugation. Orientations of these deformations do not often coincide; they are elongated in sublatitudinal northwest or northeast directions, i.e. in contrast to the structures of the Greater Caucasus, definite lineation of the regional thrusts (along which large tectonic plates [corresponding to the first-order structure] overlap each other southward) is not clearly shown.

In the territory of Azerbaijan, deformations of horizontal compression that formed at the Alpine stage, particularly at the orogenic stage, have a definite zonation. A considerable extension, lineation, frequently large amplitudes of overthrusts, sheet-overthrust structures, and tectonic scales which are typical of the Greater Caucasus, are markedly weakened southward, although these elements are available on the Lesser Caucasus. The intensity of the horizontal compression moves from north to south with time, reaching its maximum in the fold zones of the foothills of the south slope of the Greater Caucasus. Apparently, the compression front zone (south) coincides with the North-Kur deep fault. According to the geophysical data and deep drilling results, the studied kinds of deformation have not been found either in Cenozoic deposits, or on the surface of the Mesozoic series to the south, within the Kur Depression. The same situation is recognized for the northeastern and eastern foothills of the Lesser Caucasus.

The deep faults played an important role in the neotectonic processes and during the formation of the modern geological structure of Azerbaijan. Here, a significant quantity of similar deformations has been established, part of which is considered as ancient (pre-Alpine) formation and another part relates to the Mesozoic–Cenozoic age. Shikhalibeyli (1996) indicated the presence of separate "subjacent" faults, i.e. active during almost the entire Phanerozoic (Arpa-Samur fault zone). The author noted that the faults of longitudinal direction (common Caucasian) relate to Alpine formations, and the transverse (northeast orientation) relates to Precambrian (West Caspian deep fault) formations. It is known that the Caspian Basin as a major tectonic structure was formed in its modern shape (inland sea) at the neotectonic stage when rapid and large-scale tectonic movements occurred due to a sharp softening of the continental lithosphere (Artyushkov 2003).

Seismic data indicate activity of the deep faults at the modern time. The analysis of seismostatistic and macroseismic field data (1960–1998) gives reason to note that a large amount of earthquake focuses ($K \geq 9$) with different depths of seismic sources is associated with many deep-seated faults. Many epicenters are confined to the lines dividing adjacent areas of vertical movements of the opposite sign (uplift, subsidence) which represent the areas of plastic bending of the earth's crust. According to Belousov (1958), the formation hereafter of new deep faults is connected with similar areas. However, it can be considered that these areas also mark a transition from the quiescent state to fault activation which was formed at the pre-neotectonic stage. The above-mentioned facts are clearly reflected in the neotectonic map of Azerbaijan (Akhmedbeyli et al. 1991), where the correspondence of the uplift and subsidence margins of the main structural elements with deep faults (characterized by significant displacement amplitudes) is traced.

In the uplift areas experiencing horizontal compression, the interrelation between earthquakes and crustal deformations is expressed in the formation of faults or fault systems in the marginal parts of the uplift, where the tensions (compression) are high (Akhmedbeyli 2001).

Such faults are classified as high-active seismic structures. Due to this fact, the following deep faults are considered as the most active ones in Azerbaijan at the neotectonic stage: Siyazan, Dashagil-Myudresin, Vandamian, Ganykh-Ayrichay, North-Kur, Pre-Lesser Caucasus and Murovdagh.

Except for the last, which is situated in the interior of the Lesser Caucasus, all other faults are extended along the marginal parts of the Greater and Lesser Caucasus uplifts. The faults formed at the pre-modern time are significantly activated.

The late orogenic processes are characterized by the activity of many deep faults, by contrast and differentiation of vertical movements and conditions of horizontal compression. The combination of their manifestations reflects the geodynamic setting of the earth's crust, for which a quantitative assessment of neotectonic movements is an important indicator. To solve this problem, maps of neotectonic deformations for two sub-stages, the Late Miocene–Holocene and Late Pliocene–Holocene were utlized. For each sub-stage, rates and rate gradients of tectonic movements were determined, and a correlation between indicators of two sub-stages was determined. Moreover, data of active deep faults at the neotectonic stage (due to amplitudes of block displacement), seismicity, density of folded deformation distribution and tectonic knots at the intersection of differently oriented faults are analyzed.

Lastly, it is noteworthy that the recent relief and tectonic structure were formed on Azerbaijan territory in the post-Sarmatian period, i.e. during the last 9–10 Ma (Geomorphology of Azerbaijan 1959; Azizbekov et al. 1972). This period accepted as the neotectonic stage and is characterized by a sharp increase of vertical movement rate, activation of volcanism and folded and disjunctive deformations. Within Azerbaijan, the total range of vertical neotectonic movements can be estimated as 13–14 km.

The neotectonic stage comprises three sub-stages, which differ by amplitudes and rates of vertical movement: (1) Mio-Pliocene (Late Miocene–Early Pliocene), (2) Late Pliocene–Eopleistocene, and (3) Pleistocene–Holocene. The amplitudes of the Greater and Lesser Caucasus uplift at the first sub-stage is estimated as 1.6–1.7 km and 0.8–1.9 km, respectively; at second −1.0–1.2 km and 0.5–0.8 km, respectively; and at third −1.0–1.4 km and 0.3–0.7 km, respectively. In both the Caucasian Mountains, the rate of uplift increases from the early to late sub-stages. In the Mio-Pliocene, the rate of the Greater and Lesser Caucasus uplift is estimated as 20–30 cm/t.y. and 10–12 cm/t.y., respectively, and in the Pleistocene–Holocene, as 100–170 cm/t.y. and 40–100 cm/t.y., respectively. Undoubtedly, this is a result of strong horizontal compression providing "clustering" of lithospheric mass in the convergence zone between the Eurasian and Arabian plates and also formation of modern mountainous sheet-folded constructions in Azerbaijan.

6.3 Recent Movements on GPS Data

Recent horizontal movements are well traced on the GPS results and measurements and interpretations of gravity anomalies of different orders.

In 1998, the Institute of Geology (now the Institute of Geology and Geophysics) of Azerbaijan National Academy of Sciences (ANAS) (F.A. Kadirov) and Massachusetts Institute of Technology (MIT) (R.E. Reilinger) set up together Azerbaijan test sites of GPS measurements on the basis of available gravity-geodetic sites among the Azerbaijan geodynamic field polygons. Obtained measurements were processed jointly by the MIT and ANAS by using the software package GAMIT/GLOBK (Reilinger et al. 2006; Kadirov et al. 2012, 2014, 2015).

Figure 6.1 demonstrates a map with rates of GPS sites for Azerbaijan and adjacent areas for the period of 1998–2014. Arrows in the figure show the direction of vector velocity and values of these velocities that indicated by the length of arrows (according to the scale in the right lower corner of the map). Vectors of velocities of the GPS sites are shown in the Eurasian coordinate system, the start of which corresponds to the Onsala site (Sweden). Obtained velocity vectors of horizontal movements enable to represent a relatively full image of modern horizontal displacements of numerous structural elements in Azerbaijan on the basis of instrumental data.

Very large velocity vectors (9–12 mm/year), oriented in a northeasterly direction, are monitored in the southeast area of the Lesser Caucasus. The same velocity vectors are also observed at the GOSM (Gosmalyan-Lerik) and YARD (Yardimli) sites in the Talysh. These values are also similar to the observed in the BILE (Bilyasuvar) site.

The region in which the KUDI (territory of Georgia), KATE (Katekh) and SHEK (Sheki) sites are located is the most mobile one in the Greater Caucasus. At the KATE site, a large velocity (12 mm/year) is observed but the vector is turned more towards the northeast. At the SHEK site, the velocity is about 8 mm/year and the direction is also oriented parallel to the vectors of sites located in the south-eastern part of the Lesser Caucasus. There are almost no velocities vectors observed at the KEBE (Gabala), SAMU (Samur), SIYE (Siyazan), MEDR (Medrasa) or SHIK (Shikhlar) sites. According to McKenzie (1974), the protrusion of the Arabian Plate as a "plunger" forced the splitting of some fragments from the Eurasian Plate and pushed them away. This is evidence of the strong compression at the plates' boundaries in the Caucasus, and accumulation of the considerable deformation energy in the northern sides of the KEBE, SAMU, SIYE, MEDR and SHIK sites.

A decrease in velocity and large accumulation of elastic energy are observed in the south of the Absheron peninsula. A strong earthquake occurred in the Caspian Sea on November 25, 2000 (M = 7.0), and its aftershocks may be estimated as a response to the continuous deformation processes in recent years and associated stress accumulations in the foothills of the Greater Caucasus, Absheron peninsula and the Middle Caspian. The tendency for horizontal movements on the Azerbaijan territory is a predetermining factor for activation of seismic processes in areas of elastic stress accumulation and in the adjacent areas. As a result of horizontal movements of the lithosphere, the plastic deformation of the crust is occurring. At the same time, the Kur Depression subsides, and the Greater and Lesser Caucasus rise (Fig. 6.2).

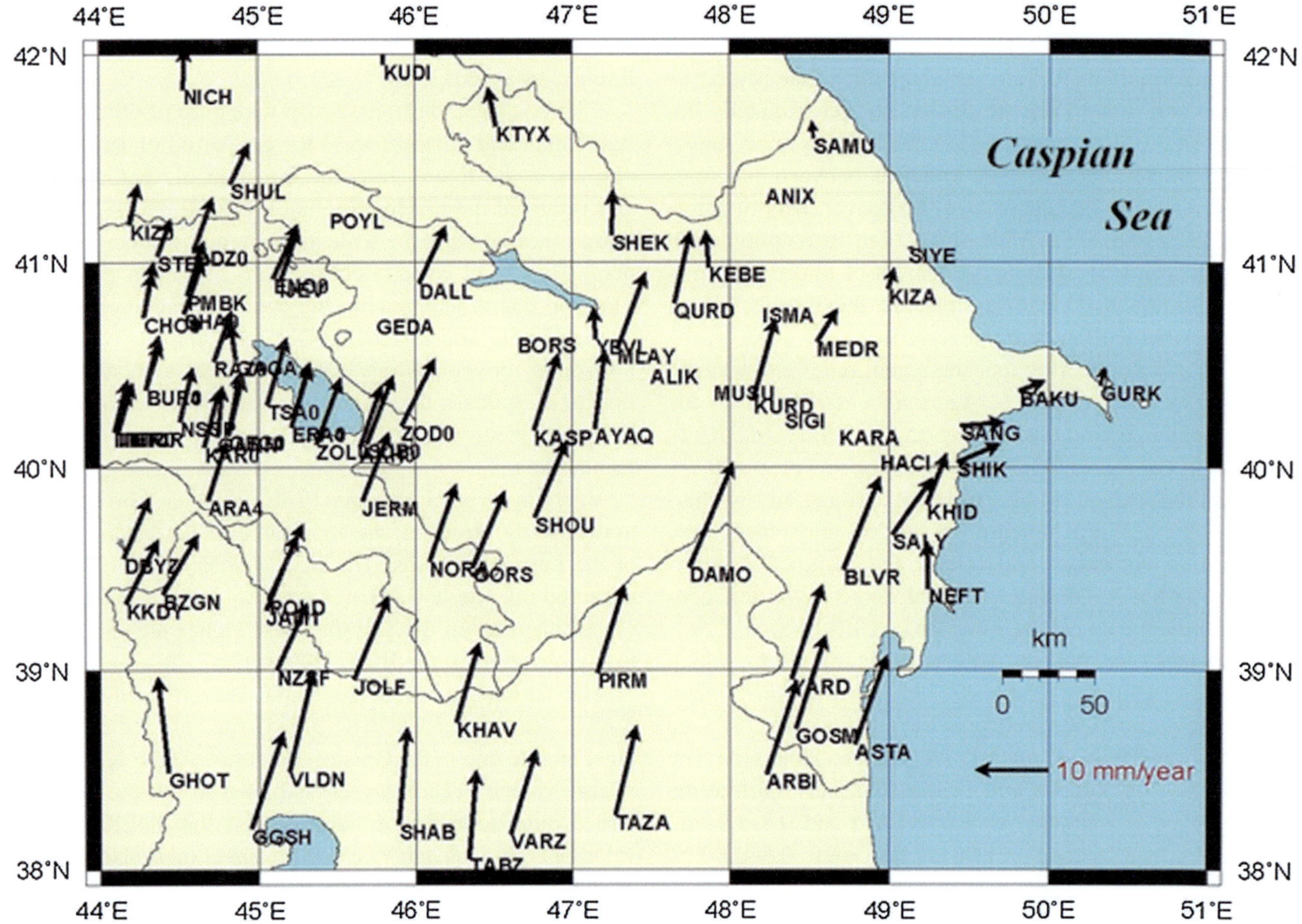

Fig. 6.1 Map of the GPS velocity of the region under study. Data about the site velocities disposed outside the territory of Azerbaijan were taken from the works of Reilinger et al. (2006), Masson et al. (2006) and Nilforoushan et al. (2003)

The boundaries between depressions and surrounding mountain blocks are marked by clear faults. The maximum velocity of uplifts reaches 10 and 6 mm/year, respectively, in the Greater and Lesser Caucasus and also in the Talysh Mountains, whereas intensive sagging of the Kur Depression is observed with a velocity of 5 mm/year (Lilienberg and Matskova 1980; Philip et al. 1989; Akhmedbeyli et al. 1991)

It is interesting to note that the velocity of modern vertical movements in the zone with a 30-km width, covering the population places of Gabala-Balaken (Katekh) on the southern slope of the Greater Caucasus, lags behind the velocity of the total uplift. The modern movements along the line stretching from the Lesser Caucasus up to the Greater Caucasus (from south to north) have a wave nature which is the result of interference of different tectonic waves, i.e. the result of complex combination of horizontal and vertical movements of the earth's crust (possibly, the movement's asymmetry is caused by simultaneous wave manifestations with different lengths and amplitudes). Therefore, the wave-like deformations are not a straight line that determine all the main peculiarities of the region's neotectonics.

The obtained data comparison for the GPS measurements with regional gravity anomalies indicates that the sites located in the zone of intense negative gravity anomaly in the south of the Lesser Caucasus (about −160 mGal) and in the zone of the Talysh Mountains (about +95 mGal) almost equally are moving in a northeast direction (Kadirov and Safarov 2013). On the other hand, the closed gravity anomalies from deep buried sources in the Lesser and Greater Caucasus, Talysh Mountains and Kur Depression are characterized by a form of deformed circles compressed in the northeast direction. This is an evidence that the Lesser Caucasus and Talysh Mts. combined together to a single block in horizontal movement. This block has been moving in a northeast direction and embraces all anomalous sources of the gravity field. Therefore, it can be concluded that the southeastern parts of the Lesser Caucasus and Talysh are moving on the viscous asthenosphere towards the northeast under the influence of the northern drift of the Arabian Plate.

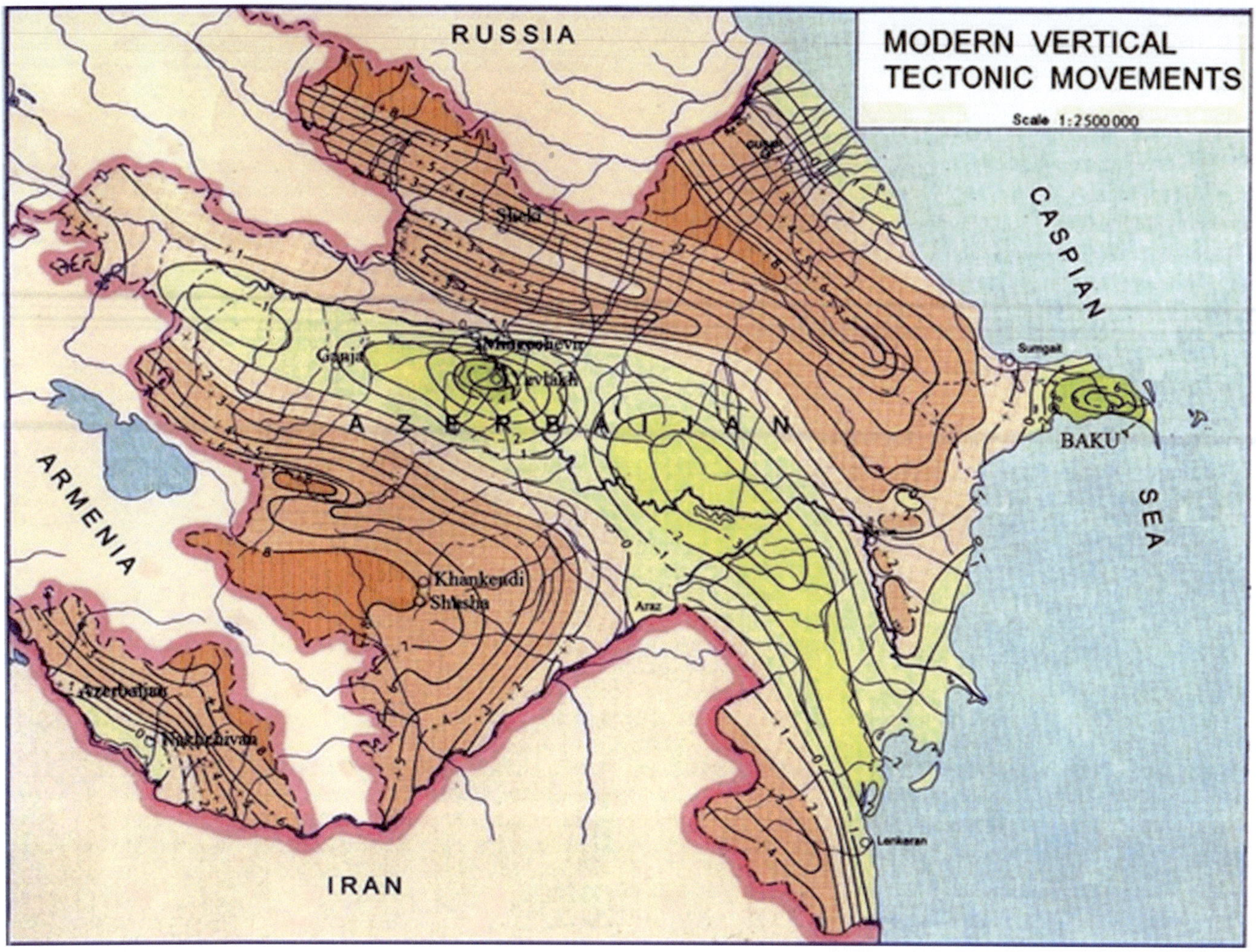

Fig. 6.2 Vertical motion map of Azerbaijan determined from re-measurements of geodetic polygons (Authors: N.Sh. Shirinov, D. A. Lilienberg, V.D. Haciyev and V.R. Yaschenko). Measurements were made during periods 1949–1953 and 1970–1973 by Lilienberg and Matskova (1980). Contour values are in mm per year. Negative and positive values indicate subsidence and uplift, respectively

References

Akhmedbeyli, F. S. (1954). Geological history of Kusar-Devetchi synclinorium in Cenozoic. *Izvestiya* of *Azerbaijan Academy* of *Sciences, 5* (in Russian).

Akhmedbeyli, F. S. (1955). Tectonics of strip of Cretaceous deposits of the Tagerdjalchay and Karachay interfluve (north-eastern Azerbaijan). *Izvestiya* of *Azerbaijan Academy* of *Sciences, 6*, 3–15 (in Russian).

Akhmedbeyli, F. S. (1962). *Neotectonic movements and their reflection in structure of the foothills of the Greater Caucasus (within territory of Azerbaijan)*. Baku: Azgosizdat. (in Russian).

Akhmedbeyli, F. S. (1966). *Neotectonics of the eastern part of the Greater Caucasus (within territory of Azerbaijan)*. Baku: Azerbaijan National Academy of Sciences. (in Russian).

Akhmedbeyli, F. S. (2001). Regional seismicity of Azerbaijan territory due to tectonic activity of the central segment of the Alpine-Himalayan belt. *Proceeding of Geology Institute, Azerbaijan Academy* of *Sciences, 29*, 40–46 (in Russian).

Akhmedbeyli, F. S., Mamedov, A. V., Shirinov, N. Sh., & Shikhalibeyli, E. Sh. (Eds.). (1991). *Neotectonic map of Azerbaijan*. Scale 1:500,000. Baku Map Factory.

Aliyev, G. A., Akhmedbeyli, F. S., Ismailzade, A. D., Kangarli, T. H., & Rustamov, M. I. (2005). In V. E. Khain & Ak.A. Alizadeh (Eds.), *Geology of Azerbaijan*, Vol. IV: Tectonics. Baku: Nafta-Press (in Russian).

Alizadeh, A. A. (1945). The oil fields of the Pre-Caspian region. In: *Proceedings of the Geology Conference of AzNeft*. Aznefteizdat, Baku, 3–18 (in Russian).

Alizadeh, A. A., Agabekov, M. G., Shekinsky, E. M., & Shikhalibeyli, E.Sh. (Eds.). (1980). *Tectonic map of Azerbaijan*. Scale 1:500,000. VSEGEI, Leningrad Map Factory.

Alizadeh, Ak. A. (Ed.). (2012). *The modern problems of geology and geophysics of the Eastern Caucasus and South Caspian Depression*. Baku: Nafta-Press.

Artyushkov, E.V. (2003). The sharp softening of the continental lithosphere as a condition of existence of rapid and large-scale movements. *Geotektonika* (*Geotectonics*), No. 2, 39–56 (in Russian).

Azizbekov, Sh. A., Alizadeh, A. A., Shikalibeyli, E. Sh., & Gadjiev, T. G. (Eds.). (1972). *Geology of the USSR, Azerbaijan*, Vol. XLVII. Moscow: Nedra (in Russian).

Azizbekov, Sh. A., & Shikalibeyli, E. S. (1966). Tectonic evolution of the Lesser Caucasian segment of the Alpine geosynclinal belt. *Geotectonics, 6* (in Russian).

Belousov, V. V. (1958). Types and origin of folding. *Soviet Geology, 1*, 40–64 (in Russian).

Eppelbaum, L. V. & Khesin, B. E. (2012). *Geophysical Studies in the Caucasus*. Springer, Heidelberg-N.Y.

Geomorphology of Azerbaijan. (1959). Baku: Azerbaijan National Academy of Science (in Russian).

Kadirov, F., Floyd, M., Alizadeh, A., Guliyev, I., Reilinger, R., Kuleli, S., et al. (2012). Kinematics of the Eastern Caucasus near Baku, Azerbaijan. *Natural Hazards, 63*(2), 997–1006.

Kadirov, F. A., Floyd, M., Reilinger, R., Alizadeh, Ak. A., Guliyev, I. S., Mammadov, S. G., et al. (2015). Active geodynamics of the Caucasus Region: Implications for earthquake Hazards in Azerbaijan. *Izvestiya, Azerbaijan Academy of Sciences, Series: Earth Sciences, 3*, 3–17.

Kadirov, F. A., Guliyev, I. S., Feyzullayev, A. A., Safarov, R. T., Mammadov, S. K., Babayev, G. R., et al. (2014). GPS-based crustal deformations in Azerbaijan and their influence on seismicity and mud volcanism. *Izvestiya (Proceedings), Russian Academy of Sciences Series: Physics of the Solid Earth, 50*(6), 814–823.

Kadirov, F. A., & Safarov, R. T. (2013). Deformation of the Earth's crust in Azerbaijan and adjacent areas inferred from the GPS measurements. *Izvestiya (Proceedings), Azerbaijan Academy of Sciences, Series: Earth Sciences, 1*, 47–55 (in Russian).

Kangarli, T. N. (1978). Cover and oilisostrome formations of the south-eastern Caucasus. In *Transgender first Conference of Young Scientists of Georgia*. Tbilisi (in Russian).

Kangarli, T. N. (1999). Alpine geodynamics of the Azerbaijan Earth's crust. In *Proceedings of the 5th International Congress of International Ecoenergetic Academy*. Baku, 199–205.

Kangarli, T. N., & Akhundov, A. B. (1988). Surface and deep structures of the southern slope of the Greater Caucasus. *Soviet Geology, 10*, 42–52 (in Russian).

Khain, V. E., & Shardanov, A. N. (1952). *Geological history and structure of the Kur Depression*. Baku: Azerbaijan National Academy of Science. (in Russian).

Khain, V. E., & Shikalibeyli, E. Sh. (2000). Position of Saatly SuperDeep borehole SD-1 in structure of the Kur Intermontane Depression. In Ak. A. Alizadeh, V. E. Khain, & A. D. Ismailzadeh (Eds.), *Saatly Superdeep. investigation of deep structure of the Kur Intermontane Depression by drilling Data of Saatly SuperDeep Borehole SG-1*. Nafta-Press, Baku, 6–19 (in Russian).

Lilienberg, D. A., & Matskova, V. A. (1980). *Detailed map of vertical movements in Azerbaijan for periods from 1949–1953 to 1970–1973 years*. Nauka, Moscow: Present-Day Motions of the Earth's crust. (in Russian).

Masson, F., Van Gorp, S., Chery, J., Djamour, Y., Tatar, M., Tavakoli, F., et al. (2006). Extension in NW Iran driven by the motion of the South Caspian Basin. *Earth and Planetary Science Letters, 252*, 180–188.

McKenzie, D. P. (1974). Active tectonics of the Mediterranean region. *Geophysical Journal of the Royal Astronomical Society, 30*(2), 109–185.

Nilforoushan, F., Masson, F., Vernant, P., Vigny, C., Martinod, J., Abbassi, M., et al. (2003). GPS network monitors the Arabia-Eurasia collision deformation in Iran. *Journal of Geodesy, 77*, 411–422.

Obruchev, V. A. (1948). The main peculiarities of kinetics and plastics of neotectonics. *Izvestiya, Academy of Sciences USSR, Series: Geology, 5*, 13–24 (in Russian).

Philip, H., Cisternas, A., Gvishiani, F., & Gorshkov, A. (1989). The Caucasus an actual example of the initial stages of continental collision. *Tectonophysics, 161*, 1–21.

Reilinger, R., McClusky, S., Vernant, P., Lawrence, S., Ergintav, S., Cakmak, R., et al. (2006). GPS constraints on continental deformation in the Africa-Arabia-Eurasia continental collision zone and implications for the dynamics of plate interactions. *Journal of Geophysical Research, 111*(B05411), 1–26.

Shikhalibeyli, E Sh. (1956). *Geological structure and evolution of Azerbaijanian part of the Southern slope of the Greater Caucasus*. Baku: Azerbaijan National Academy of Sciences. (in Russian).

Shikhalibeyli, E. Sh. (1996). *Some problems of geological structure and tectonics of Azerbaijan*. Baku: Elm (in Russian).

Shikhalibeyli, E. Sh., Mamedov, A. V., Aliyev, A. A. a.o. (1984). *Geotectonic evolution of the Kur Depression*. Baku: Elm (in Russian).

Vassoevich, N. B., & Khain, V. E. (1940). Phenomena of cover tectonics in the Lagich Mts. *Izvestiya, Academy of Sciences USSR, Series: Geology, 1*, 76–80 (in Russian).

Voskresensky, I. A. (1958). On the Basgal nappe in the South-Eastern Caucasus. *Soviet Geology, 7*, 62–84 (in Russian).

Voskresensky, I. A., Khain, V. E., & Shurygin, A. M. (1963). Tectonic nappes of the south-eastern Caucasus and conditions of their generation. *Bulletin of the Moscow State University, 4*, 15–33 (in Russian).

7 Mud Volcanism

Mud volcanism presents one of the most interesting and unusual phenomena, with periodical eruptions of large amounts of gases (mainly hydrocarbons) and breccia with inclusions of rock fragments from small sizes up to blocks of several meters. Eruption products accumulate near the orifice and crater volcano cone. The size of the volcano cone depends on the eruption frequency and character of the erupted material (Guliyev 2005).

Mud volcanoes are widespread in the world both on land and in marine basins, in collision and transtensional settings (e.g. Kovalevsky 1928; Gubkin and Fedorov 1938; Yakubov 1971; Gorin and Biniat-Zade 1971; Dadashov and Mekhtiev 1974; Guliyev and Feizullayev 1997; Kholodov 2002; Limonov 2004; Guliyev 2005; Mamedov et al. 2008; Aliyev et al. 2015). Their presence is often an indicator of deep-seated hydrocarbon accumulations (Gorin and Buniyat-zade 1971; Yakubov 1971; Rakhmanov 1987).

At the same time, mud volcanism represents a great environmental hazard that must be taken into account in the design of oil and gas pipelines and other constructions (Eppelbaum and Khesin 2012).

The issue of mud volcanism is one of the essential investigation trends in geological sciences, and mud volcanoes are quite an interesting and mysterious natural phenomena because they carry enormous insightful information about the earth's interior. Study of mud volcanoes allows one to clarify the structure of deeps, of occurring geochemical processes, in order to solve practical tasks such as assessment of deep hydrocarbon potential. As compared to magmatic volcanoes, mud volcanoes are globally distributed on a limited scale: they are located within the Alpine-Himalayan, Pacific and Central Asian mobile belts in 42 world countries.

The Azerbaijan territory is a unique and classical region of mud volcanism development. About 400 of the 800 mud volcanoes known on the earth are found within the South Caspian oil and gas-bearing basin, and more than 300 are found in Eastern Azerbaijan onshore and in the adjacent Caspian water area (Fig. 7.1). Thus, it is not by chance that Azerbaijan is considered the birthplace of mud volcanoes. Furthermore, all types of mud volcano manifestations occur here (active, dormant, immersed, submarine, island, oil-bearing). There is no area in the world similar to Azerbaijan in the amount of volcanoes, their diversity and active state.

Guliyev and Panahi (2004) developed a tectonic scheme of mud volcanoes location in Azerbaijan and adjacent regions (Fig. 7.2).

Edifices over 200 m high exist in Azerbaijan, with large eruptions in some cases producing flames of a similar scale. The eruptions are driven from a deep mud reservoir which is connected to the surface even during dormant periods, when seeping water still comes from the depths.

The study of Azerbaijan mud volcanoes is closely connected to the name of G.V. Abich. The cause for his investigations of this natural phenomenon was the colossal eruption of "Chigil-deniz" Bank in the southern Caspian Sea in 1861. More than a quarter of all known mud volcanoes are concentrated within the Caucasus (e.g. Kadirov et al. 2005; Mamedov et al. 2008; Aliyev et al. 2015), and most (more than 220; Kholodov 2002) are located within the "Abich triangle" (Abich 1863). In the following century, as interest in mud volcanoes rose, numerous papers and a number of fundamental publications appeared by such authors as M.V. Abramovich, M.G. Agabekov, D.V. Golubyatnikov, V.A. Gorin, I.M. Gubkin, I.S. Gulyev, S.A. Kovalevsky and A.A. Yakubov. Yakubov (1971) not only described mud volcano eruptions and their activity products but also examined the issues of volcano genesis and their connection to oil and gas fields. This analysis was successfully continued by Guliyev and Feizullayev (1997) and Aliyev et al. (2009).

The Institute of Geology and Geophysics of the Azerbaijan National Academy of Sciences (ANAS) has conducted detailed and dedicated geological and geochemical research of mud volcanoes since 1966, when the Mud Volcanism Laboratory was established; the lab was later

A.A. Alizadeh et al., *Geosciences of Azerbaijan*, Regional Geology Reviews,
DOI: 10.1007/978-3-319-27395-2_7

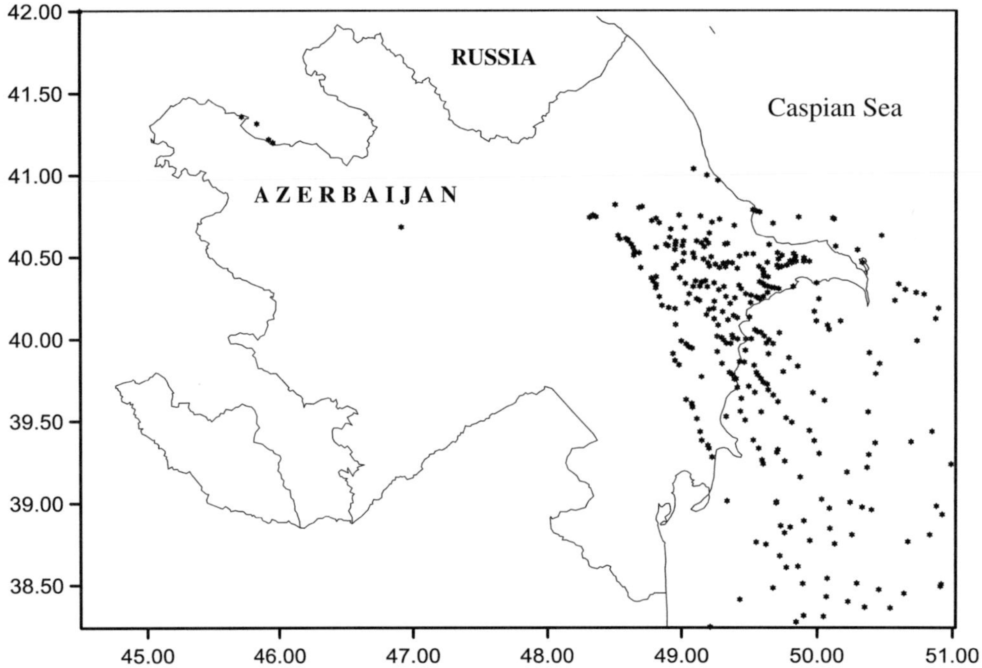

Fig. 7.1 Distribution of mud volcanoes in Azerbaijan

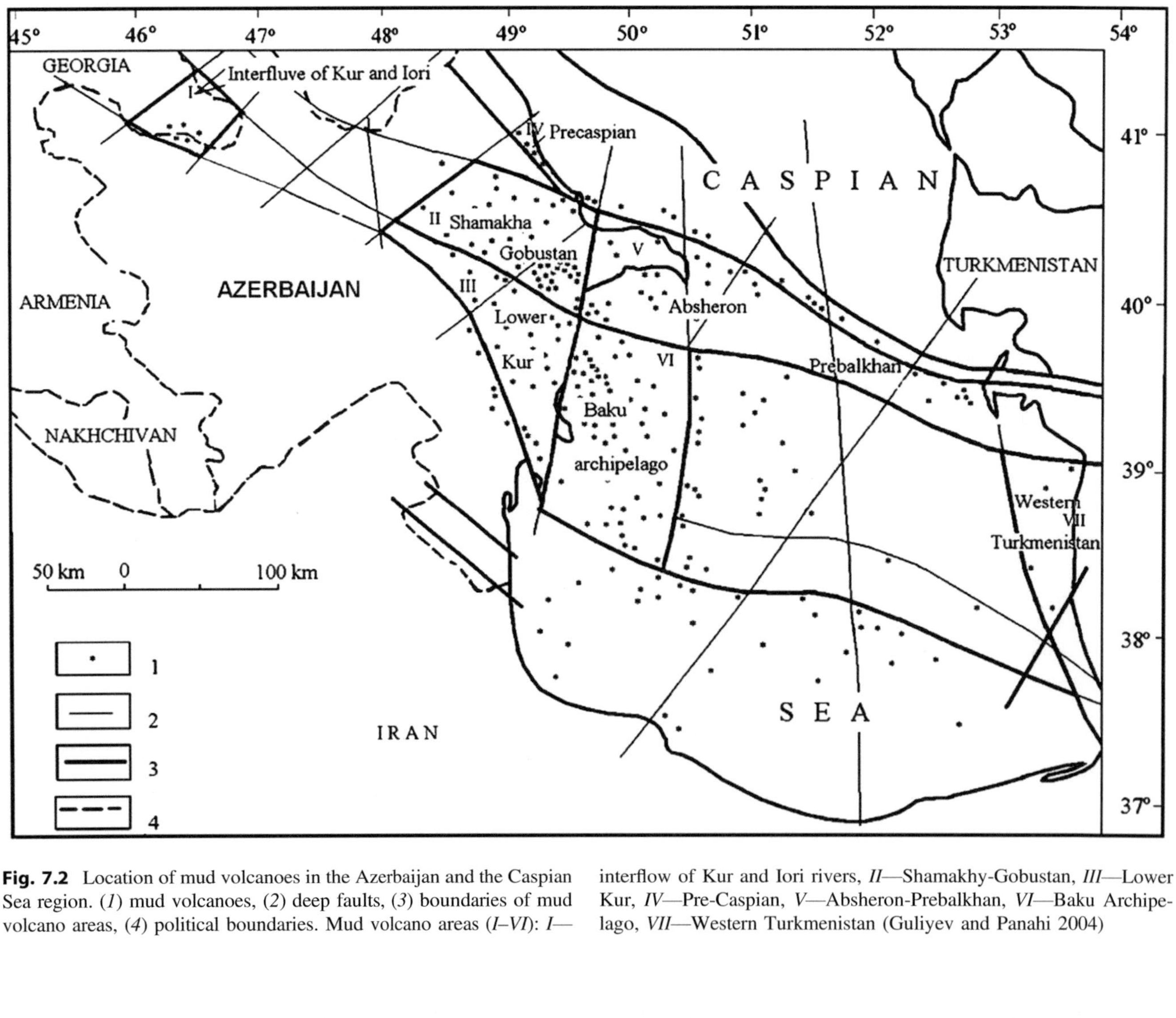

Fig. 7.2 Location of mud volcanoes in the Azerbaijan and the Caspian Sea region. (*1*) mud volcanoes, (*2*) deep faults, (*3*) boundaries of mud volcano areas, (*4*) political boundaries. Mud volcano areas (*I–VI*): *I*—interflow of Kur and Iori rivers, *II*—Shamakhy-Gobustan, *III*—Lower Kur, *IV*—Pre-Caspian, *V*—Absheron-Prebalkhan, *VI*—Baku Archipelago, *VII*—Western Turkmenistan (Guliyev and Panahi 2004)

expanded and transformed into the Mud Volcanism Sector and Standing Research Expedition.

Throughout these years, alongside the development of theoretical and practical mud volcanism issues, extensive attention has been paid to volcanic mud eruptions, their recording, the characterization of natural processes and the study of volcanic ejecta products (breccia, fluids). Because of the fugacity of the mud volcano eruption process, it is impossible to observe and track the entire mechanism from start to finish.

Spontaneously emerging eruptions of mud volcanoes are quite picturesque, sometimes scaring persons in close vicinity. The eruptions comprise brontides, or thunderous noise, then explosion, ejection of volcanic mud breccia and self-ignition of hydrocarbon gases, generating pillars of fire of up to 200–500 m high. Gas combustion temperature at that moment reaches 1000–1200°. Surrounding small rock particles turn into slag. Periodically, as though in separate "portions", a huge amount of breccia is ejected together with fire high to the skies through the vent that connects the crater with the volcano focus.

Mud volcano cones existing today are the result of their multiple eruptions that started millions of years ago. Activity of mud volcanoes in the southeast Caucasus first started 30–35 Ma, as evidenced by findings of a stratal mud volcanic breccia in the Upper Maykop (i.e. the Lower Miocene) deposits at some Gobustan areas. In addition to Gobustan, stratal breccia is traced in cross-sections of the Miocene–Pliocene and Quaternary deposits of other oil- and gas-bearing Azerbaijan regions: at the Absheron Peninsula, in the Lower Kur Depression and at the Baku archipelago.

Below is a brief characterization of some typical mud volcanoes, along with an indication of their morphology, structural fixing, eruption climaxes, and volume of volcanic-drifted breccia and its distribution areas.

Many interesting ideas about the genesis of mud volcanoes have been given in Pilchin (1979), Khalilov and Kerimov (1981), Brown (1990), Guliyev and Feizullayev (1997), Dimitrov (2002), Kopf (2002), Planke et al. (2003), Yusifov and Rabinowitz (2004), Mellors et al. (2007), Manga et al. (2009), Roberts et al. (2010), Bonini (2012), Feyzullayev (2012) and others.

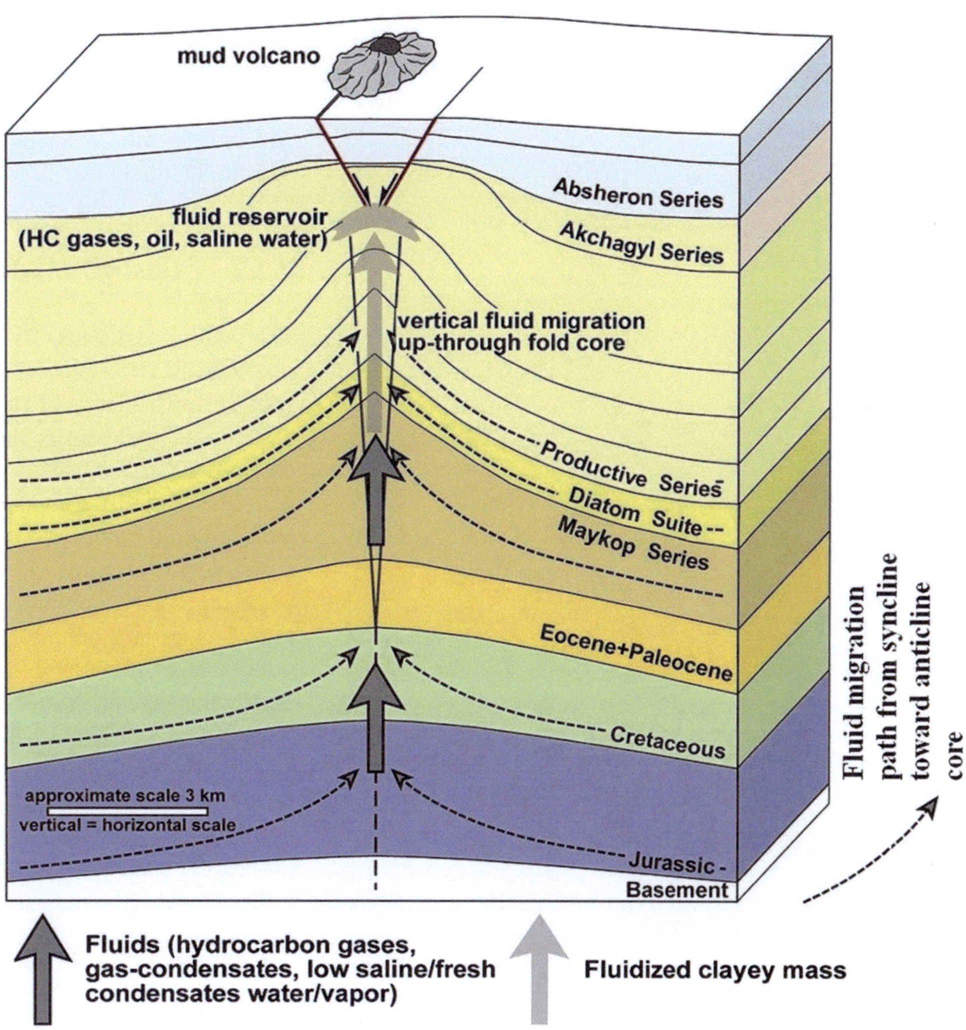

Fig. 7.3 Conceptual setting of Azerbaijan mud volcanism, which typically localizes at tight buckle folds detached from the overpressured Maykop shales (slightly modified from Bonini et al. 2013 and Antonielli et al. 2014)

Based on the analysis of numerous publications, Antonielli et al. (2014) constructed a conceptual model of Azerbaijan mud volcanism, key elements of which are the overpressured Maykop shales (Fig. 7.3).

And finally, with regard to the chemical composition of fluids ejected by mud volcanoes, waters and gases from marine and dry land volcanoes are nearly identical in composition. Gases differ in their isotope content. In general, the identity of genesis and manifestation of marine and dry land volcanoes can be emphasized (Lavrushin et al. 2009, 2015).

New data on chemical and isotopic properties of waters from 35 mud volcanoes of Azerbaijan show that they are represented by two contrasting types, characterized by different mineralization: Cl–Na (M = ~30–80 g/L) and HCO_3–Cl–Na (M = ~8–15 g/L). Waters of the last type are usually enriched with alkaline metals, B, and Br. Waters of the Cl–Na type are similar to seawater in their Cl/Br coefficient. According to Mg–Li and Na–Li geothermometers, waters were formed at temperatures varying from ~20 to ~140 °C (Fig. 7.4). Taking into consideration these estimates, "roots" of the volcanoes under consideration should be located at depths of 6.0–7.5 km. The fluid generation temperatures exhibit a tendency to increase towards the Greater Caucasus orogen. The δD, $\delta^{18}O$ and $\delta^{13}C$ (TDIC) values in dissolved inorganic carbon vary from −32 to −12 and −0.6 to +10.4 ‰ (V_SMOW) or −12.9 to +37.3 ‰ (V_PDB), respectively.

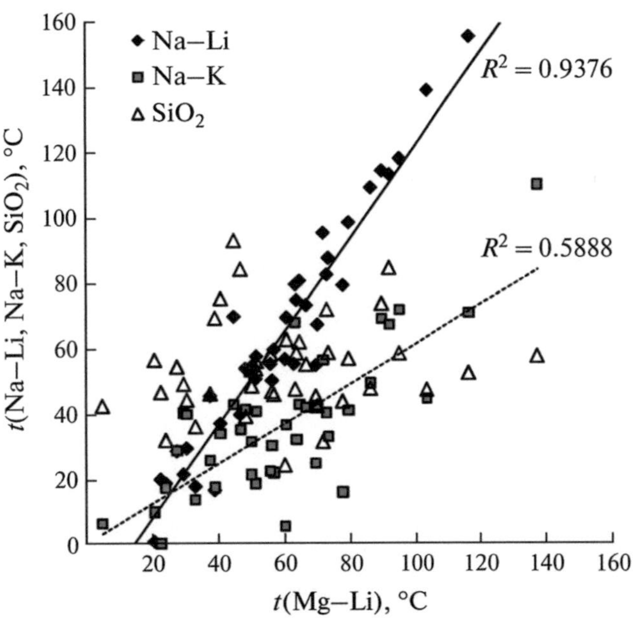

Fig. 7.4 Correlation of formation temperatures of fluid generation derived from Mg–Li, Na–Li and SiO_2 geothermometer data. *Lines* show trends of correlation with Mg–Li (*solid*) and Na–K (*dashed*) temperatures (Lavrushin et al. 2015)

The highest δD, $\delta^{18}O$ and $\delta^{13}C$ values are typical of HCO_3–Na waters. The rise in formation (Mg–Li) temperatures is accompanied by a significant increase in $\delta^{18}O(H_2O)$ and $\delta^{13}C(HCO_3^-)$ values, along with HCO_3^- concentrations. The degree of water enrichment with Br, B, and alkaline metals also increases. Thus the formation of soda waters from mud volcanoes with elevated ^{18}O and ^{13}C content can be attributed to relatively high temperature transformations of mineral and organic matter from the alteration of sedimentary rocks during the process of catagenesis.

Let us consider some of the most typical mud volcanoes of Azerbaijan on land and in the Caspian Sea.

7.1 Land Mud Volcanoes

7.1.1 Lokbatan Mud Volcano

Lokbatan is the most active mud volcano of Azerbaijan, and the "world champion" in number of eruptions (Fig. 7.5a, b). Since 1829 (the first geological record), it has erupted 24 times, the latest in September 2012. All volcanic eruptions were accompanied by outbursts of large volumes of volcanic breccia, gas ignition and combustion, pillars of fire, occasionally reaching 300–400 m, and outpouring of breccia along the slopes, forming flow "tongues" greater than 200–250 m in length. Volcanic breccia cover occupies an area of 425 ha; its thickness, according to drilling data, is 150 m.

7.1.2 Dashgil Mud Volcano

The volcano is located 2–2.5 km north of Alyat town. Morphologically, it is a rather flat and latitudinally elongated upland, with a relative altitude of 100 m above the surrounding area.

Its crater field hosts numerous active and a number of extinct bald peaks, salsas and gryphons that vigorously liberate gas and small amounts of mud with thick oil films. Volcano bald peaks and salsas have many cones, with relative height varying within 0.5–2 m. Salsas bubble over, excreting a large amount of gas, and every 3–5 min eject turbid water at a height of 1–1.5 m. According to calculations, one gryphon evolves about 600–700 m^2 gas per day, and this gas creates whitish steam around the peak, which is well seen at a distance. The volcano crater rampart diameter is 200–220 m. Eruption products ejected by the volcano form a wide and long flow (Fig. 7.6). Prominent among the volcano breccia are gray fine-grained, occasionally oil-bearing sandstones, gray shale clays and solid marls.

Two large salsas are found in the crater field central and eastern parts (Fig. 7.7). One of them, more than 15 m in

(a)

(b)

Fig. 7.5 **a** Lokbatan mud volcano eruption. **b** Lokbatan mud volcano view after the eruption

Fig. 7.6 Fanciful tongues of volcanic mud eruption (Dashgil)

diameter, vigorously excretes gas and muddy water. Southward of it, a range of burnt mud volcano breccia of the 1926 eruption can be traced. The overall volume of volcano-drifted breccia is 253 mln. m^3, its average thickness is 55 m, and the breccia cover area is 460 ha. In all, six eruptions have been recorded for the volcano, in 1882, 1886, 1908, 1926, 1958 and 2001.

Sand collectors of the productive strata are potential oil and gas targets. The Miocene deposits are also of certain interest.

7.1.3 Toragay Mud Volcano

It is a large volcano, with an absolute altitude of 400 m, with numerous radiating ravines and gullies on its slopes (Fig. 7.8). The volcano's southern and eastern slopes are steep. It is located 15 km westward of Sangachal town. The volcano crater, 500–550 m in diameter, is fringed by the circular crater rampart of up to 8 m high, and from its southwestern side, it is broken by volcanic mud breccia outflows of up to 1 km long and 250–300 m wide. The volcanic mud cover area is 753 ha and the overall volume is 753 mln. m^3, with an average breccia thickness of 100 m.

Tectonically, the volcano is confined to the near-axial part of the Toragay fold's northeastern margin composed of the Pliocene deposits at the intersection point of longitudinal and transversal ruptured dislocations.

Ten volcano eruptions have been recorded (1841, 1901, 1924, 1932, 1947, 1950, 1984, 1985, 1987 and 1988). Nearly all eruptions were characterized by gas combustion and by pillars of fire (Guliyev and Feizullayev 1997).

Fig. 7.7 Crater field of Dashgil volcano

Fig. 7.8 Toragay volcano: a general view

7.1.4 Boyuk Kyanizdagh Mud Volcano

This mud volcano is located 55 km southwestward of Baku, between the Sangachal and Duvanny rail stations, and looks like a large orographic unit looming as a truncated cone above the nearly plain area (Fig. 7.9). Its slopes are highly dissected by deep gullies with sheer and scarp walls, radiating from the top and descending to the near-Touragain Plain and towards the Caspian Sea. These gullies widen towards the volcano bottom, and are sometimes inaccessible even for pedestrians. The volcano's contemporary crater resembles a plowed ground, with dimensions of 200 × 300 m. The volcano bottom diameter is 2 km. The crater rampart has one active and two extinct gryphons. The active gryphon, pulsing periodically, discharges small amounts of muddy water.

Relatively fresh mountain breccia was ejected during the latest violent eruption in the spring of 1950, after the sesquicentennial calm. During this eruption, two long and narrow eruption tongues were generated, 150–170 m wide and about 600 m long. Cracks of differing widths that appeared during the eruption can be seen at the crater rampart edge. Tectonically, the volcano is located in the brachyanticline fold stretching from the northwest to the southeast, whose central part is complicated by transversal and longitudinal dislocations with a displacement amplitude of 125–150 m. This mud volcano is exactly associated with these dislocations (Fig. 7.9).

Mountain breccia of the latest eruption, when intense gas discharges were observed, has accumulated debris of the Middle Pliocene oil-bearing sandstones, shale clays and clayey sandstones typical of the Middle Miocene. The distribution area of Boyuk Kyanizdagh mountain breccia is 980 ha. Cover thickness is about 100–150 m. Its average thickness is 75 m, and overall breccia volume is 735 mln. m^3.

The area's potential oil and gas-bearing capacity are associated with delineation of the revealed oil field in the VII horizon suite and with prospecting of oil and gas deposits in the productive strata VIII horizon suite, in more subsided sections of the fold's eastern plunge. The Oligocene and Miocene deposits are of certain interest concerning their oil and gas-bearing capacity.

Fig. 7.9 Boyuk Kyanizdagh volcano

7.1.5 Cheildagh Mud Volcano

The volcano is situated 50 km westward of Baku. In appearance, it is a ridge-like upland of latitudinal strike, looming 110 m above the surrounding terrain. Tectonically, the volcano is confined to the Cheildagh fold dome part, composed of the Oligocene–Miocene deposits.

Four independent groups are prominent on the volcano, one of them being closer to the western periclinal and being characterized by oil-releasing gryphons that form small kir fields on the surface. Here, the flows of volcanic mud breccia are observed, occasionally more than 1 km long. In its northern part, the crater is fringed by a 2-m-high rampart. More southward, a crushing zone of 180 × 250 m is prominent. More westward, a large circular fissure is tracked. The overall distribution area of the volcanic breccia is 212 ha with an average thickness of 80 m and volume of 170 mln. m^3.

Cheildagh, one of the largest and activity-attractive mud volcanoes of Azerbaijan, is characterized by the eruption looking like gradual, long-standing, dome-shaped protrusion of volcanic mud breccia (Fig. 7.10). There, phenomena occurred in 1870 and 1970. On June 4, 1970, a 60 × 50 m-sized and 4-m-high dome consisting of semidry breccia advanced on the old breccia dome. During the early days, mass movement speed exceeded 1 m per day. In late July, the protrusion stopped. Cracks 700 m long, 0.5 m wide and 2.5 m deep have formed.

It is peculiar that burning of hydrocarbon gases rising through cracks from the depths and originating from multiple (about 20) foci 3–4 m high continued for 3 months.

In all, six volcano eruptions were recorded, all with gas burning and pillars of fire. The latest eruption with breccia protrusion took place in June 2004. This time, a 35 × 50-m dome 3–35 m high was created that is clearly looming in the area terrain. In its central part, it is vigorously broken by cracks and is westwardly elongated. Protrusion of the Maykop-age clayey rocks can be traced along the cracks.

Fig. 7.10 Cheildagh volcano. Protrusion of mud volcanic breccia

7.1.6 Goturdagh Mud Volcano

It is located 70 km southwestward of Baku, 3 km from the Airanteken Volcano. Its relative height is 150 m, and morphologically, it is a truncated cone with a subsidence caldera over 150 m in diameter and as deep as 15–20 m. Tectonically, the volcano is confined to the Goturdagh fold near-dome part, composed of the productive strata deposits.

Goturdagh is one of the interesting mud volcanoes of Azerbaijan, differing from the others by its manifestations. For more than 100 years, permanent mud squeezing (protrusion) has taken place on the volcano, which outcrops from the crescent-shaped crater crack 50 m long and 15 m wide. Semidry breccia cracks and descends along the slope of the volcano like toothpaste from a tube. This phenomenon was first observed in 1926, when the breccia flow rate was 42 m per year. This process continued until 1966, and was followed by the common eruptions of October 1966 and spring 1970. In parallel, the breccia protrusion process continued, as an ellipsoid shape up to 25 wide and more than 10 m high. The squeezing rate was 2.5 m/month. This mud-squeezing process continues today. Outflow tongues are observed on the volcano, with a northward course at a distance of 1 km and greater. The breccia cover area is 408 ha; the overall volume of volcano-erupted breccia exceeds 530 mln. m^3, and its average thickness is 130 m.

7.1.7 Galendarakhtarma Mud Volcano

It is located 60 km southwest of Baku, near the Umbaki oil field, and is confined to the Umbaki anticlinal eastern periclinal, composed of Miocene deposits. Morphologically, the volcano looks like an upland of a longitudinal strike with a steep eastern slope and a height of 160 m above the earth's surface. It is one of the largest and rather beautiful Azerbaijan volcanoes with a huge crater field and numerous active bald hills, gryphons and salsas of various sizes and diameters deployed at the area of above 600 m. Certain bald hills are as high as 15 m. The overall area of volcanic mud breccia is more than 562 ha, with the thickness of breccia cover averaging 30 m (drilling data), and volume of about 169 mln. m^3.

7.1.8 Akhtarma-Pashali Mud Volcano

One of the largest Azerbaijan volcanoes, with an absolute altitude of 300 m, is the Akhtarma-Pashali mud volcano, located 35 km north-northwest of the city of Shirvan. Tectonically, the mud volcano is confined to the abutment joint of Kalamadyn, Large and Small Harami folds, to the tectonic convergence node of their longitudinal ruptures. The volcanic crater field is over 10 km^2. The area occupied by volcanic breccia is about 400 ha, the volume of ejected breccia is 16 billion m^3 and breccia cover thickness varies from 100–300 m (at the periphery) to 600 m (in the center). The width of the volcanic orifice reaches 700–750 m. The cover surface is shaped like a plano-convex shield, 150–200 m in diameter.

The volcano is cone-shaped and is characterized by profuse oil discharge. It is classified as periodically eruptive, and five volcanic eruptions have been recorded, in 1948, 1962, 1969, 1982 and 1986. During the September 1948 eruption, thick smoke rose from the crater, and a black pillar of oil and gas rose to a height of 20–30 m. During the July 1969 eruption, breccia average volume amounted to 30 thou. m^3.

7.1.9 Shikhzarli Mud Volcano

The volcano is situated in the Maraza District, near the village of the same name, 38 km southeast of Shamakhy, and is confined to the near-dome part of Shikhzarly fold, composed of Paleogene deposits. It is the largest and the most active Gobustan mud volcano. Its absolute height is 640 m, and its height relative to the surrounding area is 150 m. Since 1848, 21 volcano eruptions have been recorded, the latest in March 2004. Many eruptions were accompanied by gas combustion, pillars of fire reaching heights of 100–300 m, and a comparatively small volume of ejected volcanic breccia, on the order of 7.5–12 thou. m^3. The strongest eruption, which continued for 2 days and ejected a large volume of breccia, was on February 12, 1902, 15 min after the famous catastrophic Shamakhy earthquake.

Over the past two decades, the Shikhzarli mud volcano has erupted rather frequently (1986, 1987, 1991, 1992, 1997), basically without flames, with a small volume of ejected breccia (1.0–2.0 thou. m^3). The volcano crater field after the 1997 eruption is shown in Fig. 7.11. Volcanic mud breccia, composed of the Eocene–Miocene terrigenous-carbonate rocks, is characterized by an abundance of diversified shapes and sizes of pyrite and chalcopyrite crystals. During the latest eruption, also without gas combustion, the volcano ejected 2.7 thou. m^3 of breccia, which, having average thickness of 0.9 m, covered an area of 3.0 ha.

Fig. 7.11 Shikhzarli Volcano. Crater field after eruption (June of 1997)

7.1.10 Otmanbozdagh Mud Volcano

It is situated at the Absheron peninsula, 35 km southwestward of Baku. It is one of the largest Azerbaijan volcanoes with an altitude mark of 403 m (Fig. 7.12). It is confined to local anticline of the Ramazan zone eastern part and is composed of the Pliocene deposits. The first recorded and rather strong volcanic eruption took place in February 1854, with subsequent eruptions in 1904, 1922, 1951, 1965, 1985 and 1994.

The volcano's latest eruption occurred on June 21, 2004 at 18:00, and continued for 1.5 h. As usual, the eruption was accompanied by brontides, explosions but without gas combustion; however, volcanic breccia was ejected to heights of 20–25 m. The breccia ejected by the volcano, 52 thou. m^3 in volume, has covered an area of 4.3 ha at a thickness of 0.4–1.8 m. Mud volcanic breccia of a nearly round pancake shape and a 240 × 195-m diameter only covered the volcano crater field. This volcano eruption was weaker than the previous one of 1994. The breccia ejected by the volcano did not cover the old nappe. Two narrow circular cracks and a series of radial ones were generated on the latter. And after the eruption of 1985, and particularly of 1994, three large tracks have been traced, more than 400 m long, 3 m deep and 2 m wide. The average breccia cover thickness is 120 m, and its total volume is 1.2 billion m^3.

It should be noted that each subsequent volcano eruption is usually weaker than the previous one. With rare exceptions, when a mud volcano is dormant for a long period (for instance, more than 40–50 years), it accumulates energy. During the 1854 eruption, over just a 3-hour period, the volcano ejected 20 mln. m^3 of breccia, covering an area of 1000 ha. In the most recent three or four eruptions, the volumes of ejected breccia have not exceeded 50–60 thou. m^3.

Of interest are some peculiar features of previous volcanic eruptions. All occurred in autumn or winter, with the exception the most recent, which occurred in early summer. The 1922 eruption was gaseous, with a pillar of fire over 200 m high, without breccia ejection. Crackle of old breccia cover took place and cover overthrust from the crater center to its periphery. Nearly all Otmanbozdagh eruptions have been characterized by explosions, pillars of fire, generation of differently oriented cracks on the old breccia cover and, in 1985, by crushing zone formation.

Fig. 7.12 Otmanbozdagh volcano. Crater field in summer 2003

7.1.11 Suleimanakhtarma Mud Volcano

The volcano is located in the eastern part of the Nardaranakhtarma Ridge, 13–14 km north of the Umbaki oil field. In plan, its sizes are within 2.0–5.0 km. Its relative height is 74 m.

The volcano is confined to the Suleiman anticlinal fold dome area and is associated with longitudinal dislocation of a thrust manner, through which the Maykop deposits contact with the Upper Miocene rocks. Two volcanic fields are isolated (the Western and the Eastern Suleimanakhtarma). Some gryphons within the eastern field excrete oil profusely, forming kir covers around them. The volcanic breccia distribution area is 35 ha, with an average thickness of 25 m. Breccia volume is 8.7 mln. m^3. Eocene oil shales are present in the mud volcano breccia as well.

In summer 2004, the first recorded eruption of the Eastern Suleimanakhtarma volcano was noted. The volcano ejected 91 thou. m^3 of breccia, with an average cover thickness of 1.3 m over an area of 7 ha.

7.2 Marine Mud Volcanoes

The activity of mud volcanoes has resulted in the generation of islands, banks, shoals and submarine ridges in the sea. In the Baku Archipelago, eight islands (e.g., Zenbil, Gil, Khara-Zire, Garasu, Sangi-Mughan) and many underwater banks originated from mud volcanoes. In addition, underwater ejections of volcanic breccia form vast covers on the sea bottom. Submarine volcanoes occur at depths ranging from 2 to 900 m; their cone heights also vary. There are volcanoes in a "buried" state. Similar to those on dry land (Toragay, Otmanbozdagh, L. Kyanizdagh), hypothetical volcanoes can achieve absolute depth marks of 400 m or greater. In the abyssal subzone found at the Southern Caspian western continental slope, submarine volcanoes with heights as great as 500 m (Dadashov and Mekhtiyev 1974) have been revealed. The largest number (above 300) and most massive of volcanoes are concentrated at the northwestern margin of the South Caspian abyss (Yakubov and Aliyev 1978; Huseynov and Guliyev 2004; Aliyev 2004, 2006).

New marine volcanoes (Neft Dashlary-2, Gyuneshli, etc.) have been found within the Absheron rock bar (Gyuneshli, Chirag, Azeri, Kyapaz, etc.), and in the adjacent Southern Caspian Sea area. They are characterized by isolated distribution zones of mud volcanic breccia on the sea bottom.

In light of their location (under water), information about marine mud volcanoes was obtained mainly from analysis of seismic images.

7.2.1 Khara-Zire Island Mud Volcano

It is located 13 km east of Alyat Cape, and 5 km southeast of Zenbil Island. It is oval-shaped, its length together with a sandbar is 3.3 km and its area is 3.5 km^2. The island comprises a crater and a mud volcano crater rampart. The crater field is raised 30 m above the sea level.

Khara-Zire Island is an asymmetrical Pliocene fold of a southeastern strike, located somewhat an echelon with regard to the Duvanny-deniz fold. The elevation is complicated by two longitudinal dislocations, stretching nearly parallel to the fold axis.

Khara-Zure is a frequently erupting volcano. Ten eruptions have been recorded from 1810 till 1997. The 1810, 1857, 1859, 1940 and 1997 eruptions were accompanied by a pillar of fire that was particularly high in 1857 and in 1940. The 1940 eruption was violent; several billion m^3 of volcanic breccia was erupted that covered an area of 156 ha. Black volcanic ash covered the entire island. A comparatively high volume of breccia was observed during the 1960 eruption (270 thou. m^3, 18 ha). The recent 1997 eruption was also quite powerful. The volcano ejects amounted to more than 100 thou. m^3 of breccia, with simultaneous ejectment of "lapilli" (melted, vitreous, baked tiny particles of mud).

7.2.2 Garasu (Los) Island Mud Volcano

The volcano is located 70 km southeast of Baku, within the Hamamdag-Dashly Island (Ignaty's Stone) anticlinal zone and is confined to the NW–SE orientation fault. It is one of the largest and most active island mud volcanoes of the Baku Archipelago, characterized by an abundance and diversity of gryphons and salsas that discharge gas and water and ooze mud with an oily film. The first recorded eruption of the volcano dates back to the second half of the nineteenth century, when in February and March 1876, two eruptions with gas ignition and high pillars of fire took place. The volcano eruption of 1923 ejected a 60–70 thou. m^3 of breccia. The ejection of April 1977 was the most intense and prolonged (2 days; Fig. 7.13). And, finally, the latest volcano manifestation, without fire and breccia ejecta, was recorded in 1993, when a crushing zone originated on the old breccia cover (200 × 25–50 m) in the island's central

Fig. 7.13 Garasu Island volcano. Eruption of April 28, 1977

and eastern parts, as well as many cracks down to 1-m depths, and occasionally, of up to 700 m long.

Below is presented a description of the 1977 eruption (Yakubov and Aliyev 1978).

The volcano was "dormant" for more than half a century, and upon awaking, thrust a pillar of fire skyward to a height of more than 500 m, its blaze seen by many Baku residents.

According to witnesses, the eruption began in the evening as powerful brontides. Black smoke and a mushroom-like cloud of steam appeared above the island. Oil workers on drilling rigs 800 m from the islands felt the heat of the strong. Even Bulla (Khara-Zire) Island, situated at a distance of 15 m from the volcano, was blanketed with small particles of clay rocks and volcanic dust. The Baku Seismic Station devices recorded local soil tremors.

In the island's northwestern part, a series of deep concentric and radial cracks formed around the crater, making the island surface difficult to travel. Numerous holes more than 1 m in diameter and depth are visible on the fresh breccia upper layer, the as result of large rock debris spewed from the depths of the volcano. The volume of ejected mud volcano breccia exceeded 1.5 mln. m^3, covering nearly one-fourth of the island area. Among the solid ejecta were found many oil-permeated Paleogene–Miocene sandy rocks.

Within a couple of days, the volcanic crater looked like a pancake-shaped upland generated by semi-liquid mud

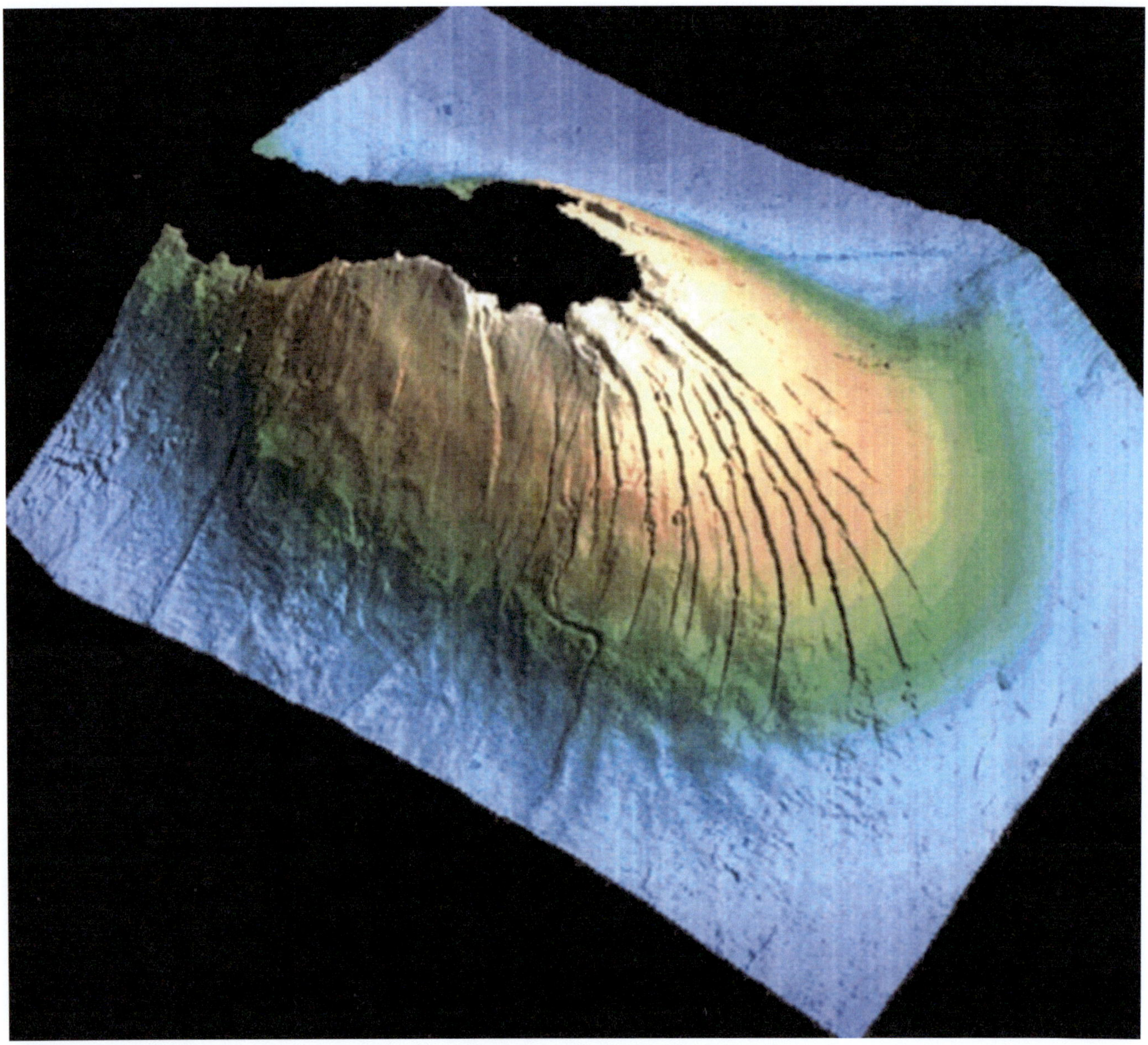

Fig. 7.14 Underwater mud volcano in the Nakhchivan block (Corthay and Aliyev 2000)

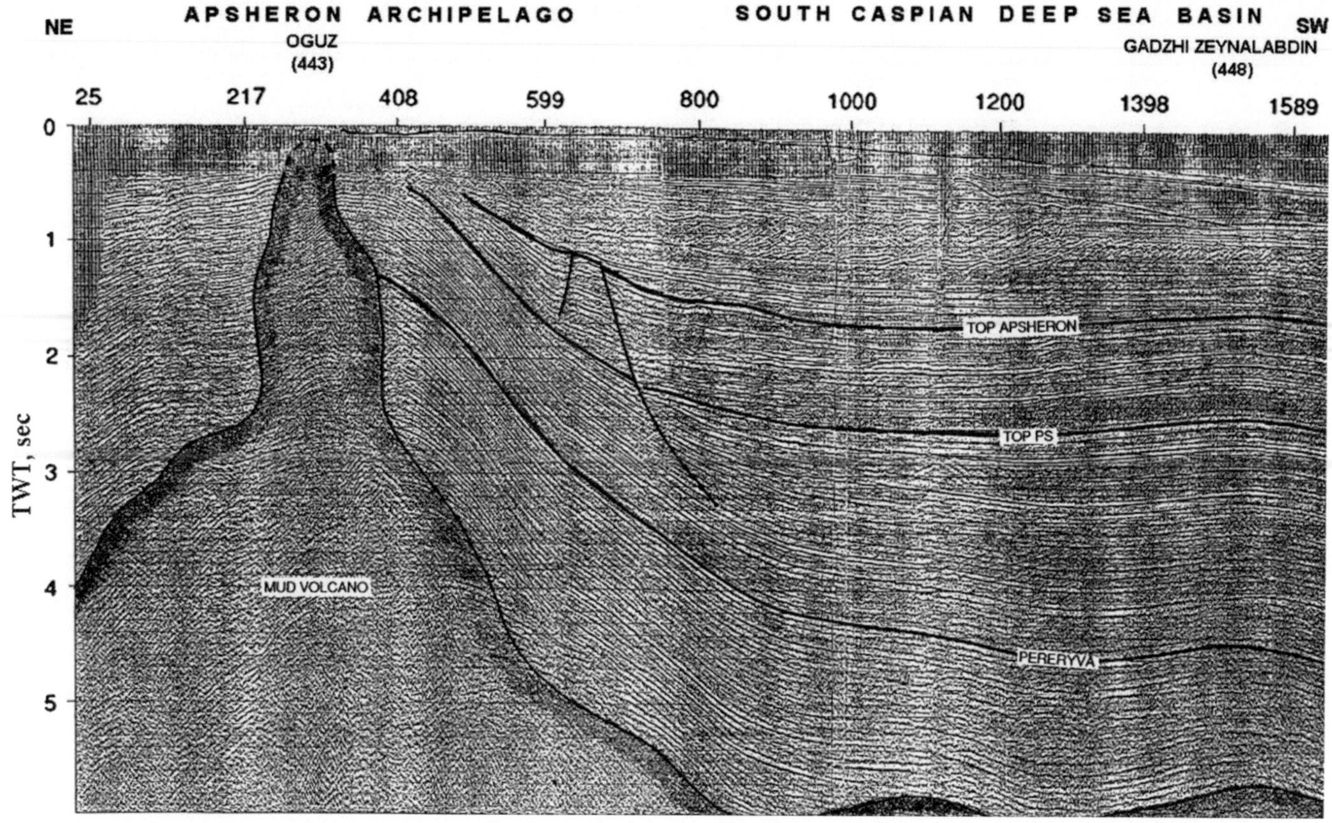

Fig. 7.15 Structure "Oguz": seismic images of marine mud volcanoes of the South Caspian Basin (from the data of "Kaspmorneftegeofizika" Assoc.)

volcanic breccia. Gas discharge was vigorous, and at certain intervals, breccia jetted with characteristic effervescence at heights of 7–10 m. A strong smell of oil was reported on the island, particularly near the crater.

Later, individual bursts of burning gas jets up to 200 m high were also observed. Local areas of caked breccia were prominent near the crater. Heat originating at the depths was still felt here, and it was associated with underground combustion of hydrocarbon gases. "Interbedding burning" after eruption is a rather common phenomenon for mud volcanoes, continuing sometimes for many months until the influx of gas from below ceases.

7.2.3 Chigil-Deniz (Kumani) Mud Volcano

On May 30, 2001, the underwater mud volcano named "Kumani Bank" (Chigil-deniz) of the Baku archipelago, situated 38 km northeast of the Neftchala town, awoke from a 40-year "sleep". The volcano is classified as active and periodically erupting. In this process, each volcano eruption results in generation of varisized islands that with time, in several months, are washed out by the Caspian waves and the volcano body subsides under water, turning into an underwater bank.

The first recorded Kumani Bank volcano eruption dates back to 1861; the last but one eruption was observed in 1959. The eruption of May 2001 was the eighth. During the eruption of May 19, 1861, the volcano ejected a huge amount of breccia (1 mln. m^3). On May 26, the island that was derived was 87 × 66 m in size, 3.5 m high, and occupied a 0.57-ha area. In early 1862, the island was eroded by water. Volcano eruptions of the previous century (1927, 1950 and 1959) were accompanied by pillars of fire 100–300 m high. The eruption that occurred on November 5, 1928, was of long duration (nearly 2 days). According to witnesses, clouds of smoke escaped the depths, then a white cloud appeared. On November 7, an island 144 m in diameter and 3.5 m high was formed (6.5-ha area). The strongest eruption of the Kumani Bank volcano took place in 1950, when a huge island 700 × 500 m in area and 6 m high was created, and numerous radial cracks appeared. By 1959, eruption of a large amount of "lapilli" (tiny vitreous particles of volcanic mud) was observed.

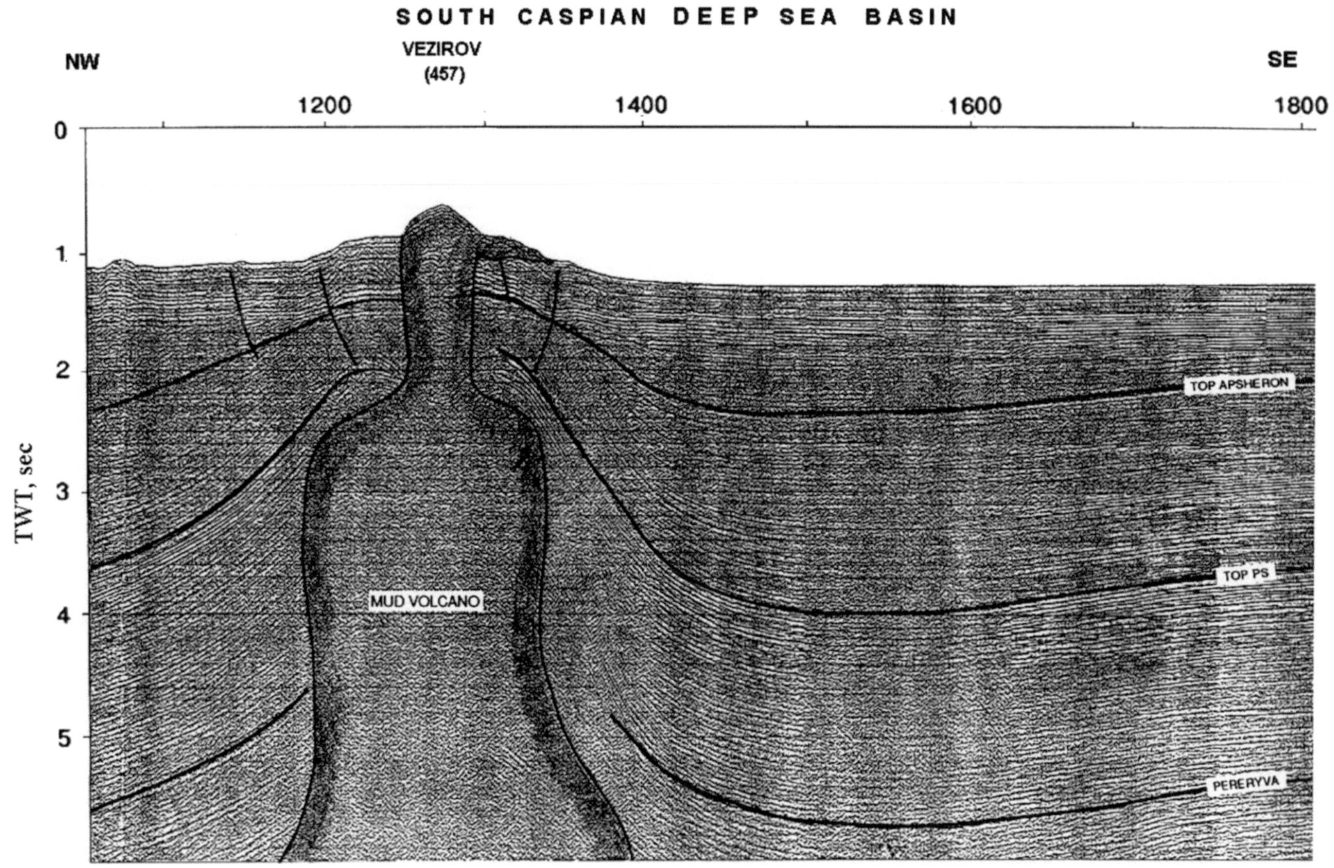

Fig. 7.16 Structure "Vezirov": Seismic image of a dry land mud volcano related to the South Caspian Basin (from the data of "Azneftegeofizika" Assoc.)

7.2.4 Mud Volcano in the Marine "Nakhchivan Block"

A large new marine volcano has been found in the south of the Baku archipelago, in the Nakhchivan block (Corthay and Aliyev 2000). Figure 7.14 shows a 3D image of the underwater mud volcano in the Nakhchivan block.

Marine mud volcanoes were of great importance in geological history of the Caspian Sea development and in its contemporary history as well. Volcanic activity in marine conditions results in the generation of terrain-positive elements. Their ejecta participate in formation of micro relief for the surrounding sea bottom areas and influence sea dynamics and the composition of bottom sediments.

Chaotic orientation of debris in a volcano crater field is recorded by seismic profiles as a decrease in the number of reflecting boundaries and their disorderly distribution (Figs. 7.15 and 7.16). Mass specific weight and induced magnetization decrease in the mud volcano crater field and vent filled with volcanic breccia, due to the debris deconsolidation. The cumulative negative gravitational and magnetic effect of a mud volcano is higher than that of a crushing zone, ruptured dislocations and alternative factors. In this process, the area of tectonic dislocation, fractures has an elongated anomaly shape, while a mud volcano has an isometric anomaly shape. Seismic profiles that pass through a mud volcano display a sharp drop of reflecting boundaries, and their total absence at the volcano orifice (Aliyev et al. 2009).

7.2.5 Mud Volcano Oguz

In the seismic time section (Fig. 7.15), the location of the Oguz mud volcano (occurring within the Absheron and PS suites) is clearly detected.

7.2.6 Mud Volcano Vezirov

Similar seismic time section presented in Fig. 7.16 undoubtedly indicates the presence of the mud volcano Vezirov.

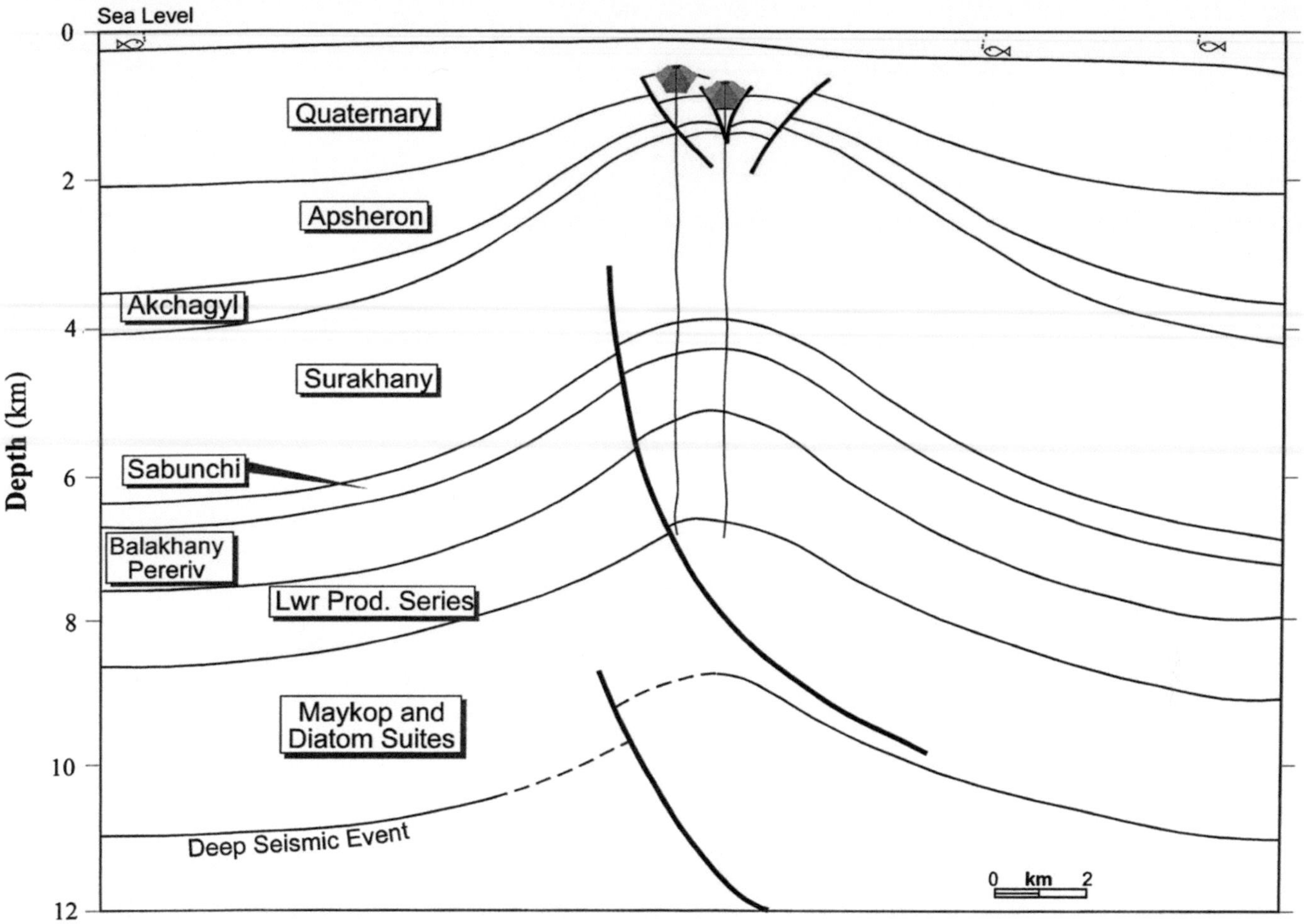

Fig. 7.17 Schematic cross-section across Shah Deniz showing the relationship of the deep structure with the shallower structure and mud volcanoes (after Fowler et al. 2000)

7.2.7 Mud Volcano Shah Deniz

Shah Deniz is a large dip, closed structure in the South Caspian located approximately 35 km to the southeast of Bakhar and 70 km southwest of the Guneshli–Chirag–Azeri structures. A large 3D survey—800 km^2—was acquired in this area (Fowler et al. 2000). Figure 7.17 clearly indicates the relationship of the deep structure with the subsurface structure and mud volcanoes.

7.2.8 Mud Volcano Azeri

An area of the seabed at the volcano's summit (Azeri mud volcano, near the Absheron peninsula) is expressed as a negative polarity reflection in multiple seismic surveys (Fig. 7.18a–d) (Evans et al. 2006). An amplitude map of the seabed reflection shows that the area of negative polarity is irregularly shaped in plain view and extends about 2.0 km^2 (Fig. 7.18b). Two circular amplitude anomalies (1.1 and 0.6 km in diameter) can be easily detected; they marked V in Fig. 7.18b. Both these anomalies are characterized by an outer zone of relatively high negative reflection amplitude surrounding a zone of relatively low reflection amplitude (Evans et al. 2006). The largest zone is located on a flat area of the seabed at the summit of the Shah Deniz volcano, slightly exceeding the surface area of the circular anomaly. A high-resolution 2D seismic line through the smallest circular area shows a high amplitude, positive polarity and convex-down reflection underlying the area of phase-reversed seabed reflection (marked X in Fig. 7.18d). The maximum distance between this reflection and the overlying seabed is 11 millisec TWT. Evans et al. (2006) chose an acoustic velocity for gas-bearing seabed mud of 984 m s^{-1}, and depth-converted this distance as approximately 5 m.

Integrated geophysical studies of mud volcanoes in Azerbaijan are briefly presented in subsection Chapter 9 of Volume II.

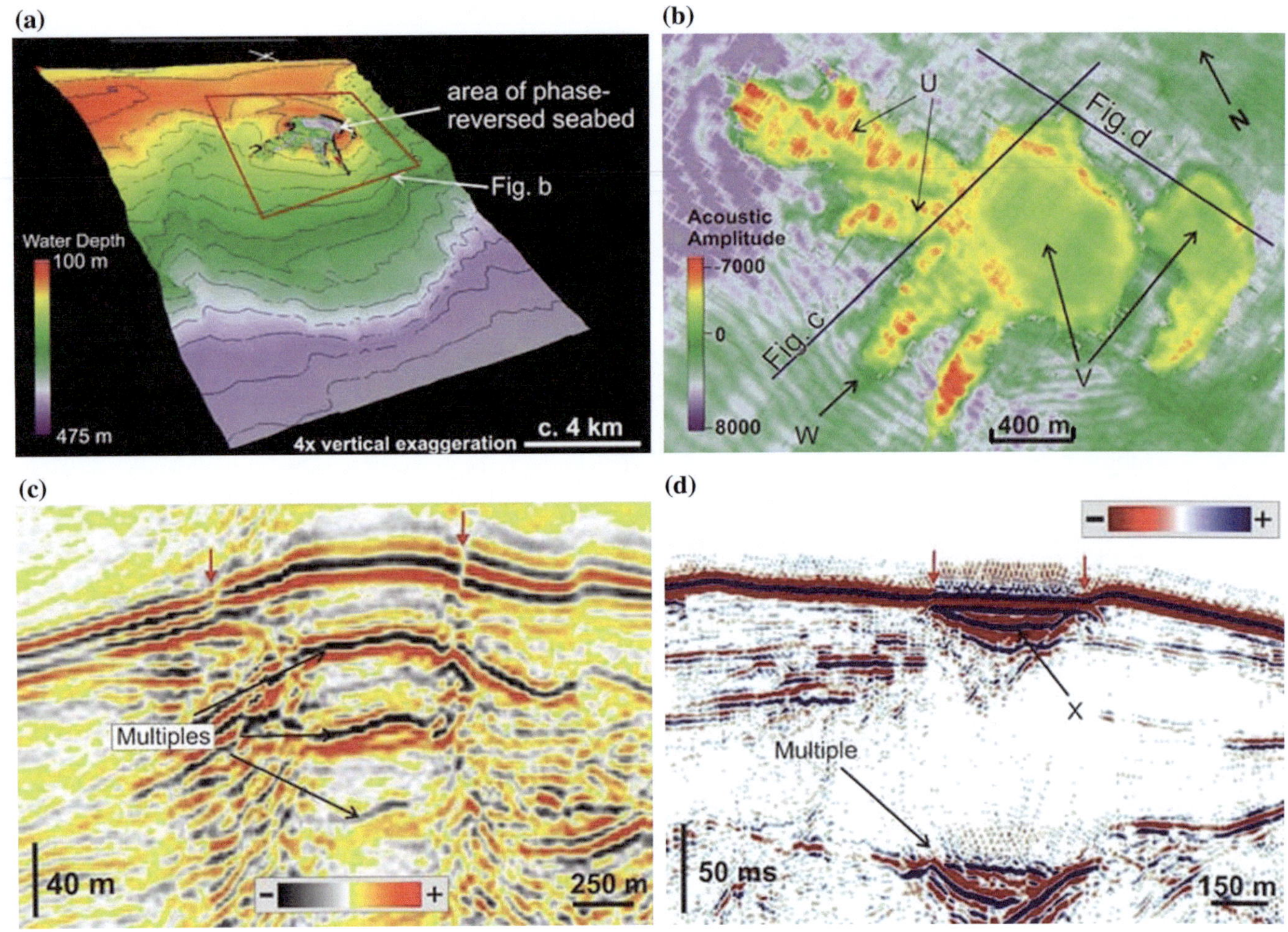

Fig. 7.18 The Azeri mud volcano. **a** Three-dimensional bathymetric map of the seabed at the site of the Azeri mud volcano. Note the conical shape of the volcano's summit and the area of phase-reversed seabed reflection at the volcano summit. The phase-reversed section is colored according to its acoustic amplitude (see **b** for further details). **b** Acoustic amplitude map of the phase-reversed section of the seafloor at the Azeri mud volcano summit. *U*—linear negative amplitude anomalies interpreted as seafloor mudflows, *V*—circular amplitude anomalies of relatively low reflection amplitude interpreted as seafloor sales, *W*—"halo" of reduced positive reflection amplitude surrounding the phase-reversed section of seafloor reflection; this is interpreted to be the result of the progressive degassing of gas-bearing sediment within the phase-reversed area (map location shown on **a**). **c** Representative seismic section through the Azeri volcano, showing the character of the phase reversal. The seabed phase reversal occurs in the area between the vertical arrows (see **b** for section location). **d** "High-resolution" seismic profile through the Azeri volcano showing the character of the phase reversal, which occurs between the vertical arrows. *X*—convex-down positive polarity reflection interpreted as the base of a seafloor salse (see **b** for section location) (after Evans et al. 2006, with small modifications)

References

Abich, G. V. (1863). New Islands on the Caspian Sea and the cognition of Mud Volcanoes of the Caspian Region. *Memories Academy Science, Petersburg, Series, VIII, 6*(5).

Aliyev, Ad. A. (2004). Mud volcanism of the South Caspian oil and gas Basin. In *Proceedings of the Geology Institution, Azerbaijan National Academy of Sciences*, (No. 31, pp. 21–47). Baku: Nafta-Press (in Russian).

Aliyev, Ad. A. (2006). Mud volcanism of the South Caspian oil-gas depression. *Geology and Economic Minerals of the World Ocean, 3*, 35–51 (in Russian).

Aliyev, Ad. A., Guliyev, I. S., Dadashov, F. H., & Rakhmanov, R. R. (2015). In Ak. A. Alizadeh & A. D. Ismailzadeh (Eds.), *Atlas of the world mud volcanoes*. Baku: Sandro Teti Editore and Nafta-Press.

Aliyev, Ad. A., Guliyev, I. S., & Rakhmanov, R. R. (2009). *Catalogue of Mud Volcano Eruptions of Azerbaijan* (1810–2007). Nafta-Press (in Russian).

Antonielli, B., Monserrat, O., Bonini, M., Righini, G., Sani, F., Luzi, G., et al. (2014). Pre-eruptive ground deformation of Azerbaijan mud volcanoes detected through satellite radar interferometry (DInSAR). *Tectonophysics, 637*, 163–177.

Bonini, M. (2012). Mud volcanoes: Indicators of stress orientation and tectonic controls. *Earth-Science Reviews, 115*, 121–152.

Bonini, M., Tassi, F., Feyzullayev, A. A., Aliyev, C. S., Capecchiacci, F., & Minissale, A. (2013). Deep gases discharged from mud volcanoes of Azerbaijan: new geochemical evidence. *Marine and Petroleum Geology, 43*, 450–463.

Brown, K. M. (1990). The nature and hydrogeologic significance of mud diapirs and diatremes for accretionary systems. *Journal of Geophysical Research, 95*, 8969–8982.

Corthay, J. E., & Aliyev, A. A. (2000). Delineation of mud volcano complex, surfical mud flows, slump blocks, and shallow gas reservoirs offshore Azerbaijan. In *Transaction of the offshore technology conference* (pp. 1–4). Houston, Texas, U.S.A.

Dadashov, F. G., & Mekhtiyev, A. A. (1974). Mud volcanoes of the Caspian Sea. *Izvestiya, Azerbaijan National Academy of Sciences, Series: Earth Science*, *5*, 26–32 (in Russian).

Dimitrov, L. I. (2002). Mud volcanoes—the most important pathway for degassing deeply buried sediments. *Earth Science Reviews, 59*, 49–76.

Eppelbaum, L. V., & Khesin, B. E. (2012). *Geophysical studies in the Caucasus*. N.Y., Dordrecht, London: Springer.

Evans, R. J., Stewart, S. A., & Davies, R. J. (2006). Phase-reversed seabed reflections in seismic data: Examples related to mud volcanoes from the South Caspian Sea. *Geo-Marine Letters, 27*, 203–212.

Feyzullayev, A. A. (2012). Mud volcanoes in the South Caspian basin: Nature and estimated depth of its products. *Natural Science, 4*(7), 445–453.

Fowler, S. R., Mildenhall, J., Zlova, S., Riley, G., Elsley, G., Desplanques, A., & Guliyev, F. (2000). Mud volcanoes and structural development on Shah Deniz. *Journal of Petroleum Science and Engineering., 28*, 189–206.

Gorin, V. A., & Buniatzadeh, Z. A. (1971). *Deep faults, oil-gas Volcanism and oil and gas deposits of the Western Margin of the South Caspian Basin*. Baku: Azerneshr(in Russian).

Gubkin, I. M., & Fedorov, S. F. (1938). Mud Volcanoes of the Soviet Union and their relationship with genesis of oil deposits of the Krum-Caucasian geological province. In *Proceedings of the Academy of Science*, Moscow. (in Russian).

Guliyev, I. S. (2005). Mud volcanism in Azerbaijan. In: (A.T. Ismailzadeh, Ed.) Recent geodynamics, georisk and sustanaiable development in the Black Sea to Caspian Sea Region. In *Proceedings of the International Workshop, AIP Conference Proceed* (Vol. 825, pp. 11–18), Baku.

Guliyev, I. S., & Feizullayev, A. A. (1997). *All about Mud Volcanoes*. Baku: Nafta-Press. (in Russian).

Guliyev, I. S., & Panahi, P. (2004). Geodynamics of the deep sedimentary basin of the Caspian Sea region: Paragenetic correlation of seismicity and mud volcanism. *Geo-Marine Letters, 24*, 169–176.

Huseynov, D. A., & Guliyev, I. S. (2004). Mud volcanoes natural phenomena in the South Caspian basin: Geology, fluid dynamics and environmental impact. *Journal of Environmental Geology, 46*, 1012–1023.

Kadirov, F. A., Lerche, I., Guliyev, I. S, Kadyrov, A. G., Fezullayev, A. A., & Mukhtarov, A. Sh. (2005). Deep structure model and dynamics of mud volcanoes, southwest Absheron Peninsula (Azerbaijan). *Energy Exploration & Exploitation, 23*(5), 307–332.

Khalilov, N. Yu., & Kerimov, A. A. (1981). On the genesis of mud volcanism and diapirism. *Izvesiya Vuzov*, Ser.: Geology and Prospecting, No. 11, 23–28 (in Russian).

Kholodov, V. N. (2002). Mud volcanoes, their distribution regularities and genesis: Communication 1. Mud volcanic provinces and morphology of mud volcanoes. *Lithology and Mineral Resources, 37*(3), 197–209.

Kopf, A. J. (2002). Significance of mud volcanism. *Reviews of Geophysics*, **40**, No. 2, 1–51, doi: 10.1029/2000RG000093.

Kovalevsky, S.A., 1928. On the genesis of mud volcanoes of the Eastern Transcaucasian. *Azerbaijan Oil Industry*, No. 1, 27–34, No. 2, 31-39 (in Russian).

Lavrushin, V.Yu., Guliyev, I.S., Kikvadze, O.E., Aliyev, Ad.A., Pokrovsky, B.G. and Polyak, B.G. (2015). Waters from Mud Volcanoes of Azerbaijan: Isotopic–Geochemical Properties and Generation Environments. *Lithology and Mineral Resources, 50*(1), 1–25.

Lavrushin, V. Yu., Polyak, B. G., Pokrovsky, B. G., Kopp, M. L., Buachidze, G. I. & Kamenskii, I. L. (2009). Isotopic–geochemical peculiarities of gases in Mud Volcanoes of Eastern Georgia. *Lithology and Mineral Resources*, 44(2), 183–197.

Limonov, A. F. (2004). Mud volcanoes. *Soros' Education Journal, 8* (1), 63–69 (in Russian, abstract in English).

Mamedov, P. Z., Guliyev, Q. Q., Kadirov, F. A., Shikhalyev, Y. A., Guliyev, I. S., Aliyeva, E. H., et al. (2008). In Ak. A. Alizadeh (Ed.), 2008. *Geology of Azerbaijan*, Vol. VII: *Oil and Gas*. Baku: Nafta-Press, (in Russian).

Manga, M., Brumm, M., & Rudolph, M. L. (2009). Earthquake triggering of mud volcanoes. *Marine and Petroleum Geology, 26*, 1785–1798.

Mellors, R., Kilb, D., Aliyev, A., Gasanov, G., & Yetirmishli, G. (2007). Correlations between earthquakes and large mud volcano eruptions. *Journal of Geophysical Research, 112*, B04304. doi:10. 1029/2006JB00448.

Pilchin, A. N. (1979). On the genesis of mud volcanoes. *Soviet Geology*, 10, 78–81 (in Russian).

Planke, S., Svensen, H., Hovland, M., & Banks, D. A. (2003). Mud and fluid migration in active mud volcanoes in Azerbaijan. *GeoMarine Letters, 23*, 258–268.

Rakhmanov, R. R. (1987). *Mud volcanoes and their importance in oil and gas bearing prognosis*. Moscow (in Russian): Nauka.

Roberts, K. S., Davies, R. J., & Stewart, S. A. (2010). Structure of exhumed mud volcano feeder complexes, Azerbaijan. *Basin Research, 22*, 439–451.

Yakubov, A. A. (1971). *Mud volcanoes of Azerbaijan*. Atlas (in Russian): Baku.

Yakubov, A. A., & Aliyev, A. A. (1978). *Mud Volcanoes*. News in Science and Technics, Series: Earth Sciences, No. 8, Znanie, Moscow (in Russian).

Yusifov, M., & Rabinowitz, P. D. (2004). Classification of mud volcanoes in the South Caspian basin, offshore Azerbaijan. *Journal of Marine and Petroleum Geology, 21*, 965–975.

Index

A.A. Alizadeh et al., *Geosciences of Azerbaijan*, Regional Geology Reviews,
DOI: 10.1007/978-3-319-27395-2

MIX
Papier aus verantwortungsvollen Quellen
Paper from responsible sources
FSC® C105338

If you have any concerns about our products,
you can contact us on
ProductSafety@springernature.com

In case Publisher is established outside the EU,
the EU authorized representative is:
Springer Nature Customer Service Center GmbH
Europaplatz 3, 69115 Heidelberg, Germany

Printed by Libri Plureos GmbH
in Hamburg, Germany